AQA GCSE 9-1

Combined Science Trilogy

Nick Dixon, Nick England, Richard Grime, Nora Henry, Ali Hodgson and Steve Witney. Edited by James Napier

Approval message from AQA

This textbook has been approved by AQA for use with our qualification. This means that we have checked that it broadly covers the specification and we are satisfied with the overall quality. Full details of our approval process can be found on our website.

We approve textbooks because we know how important it is for teachers and students to have the right resources to support their teaching and learning. However, the publisher is ultimately responsible for the editorial control and quality of this book.

Please note that when teaching the *AQA GCSE Combined Science Trilogy* course, you must refer to AQA's specification as your definitive source of information. While this book has been written to match the specification, it cannot provide complete coverage of every aspect of the course.

A wide range of other useful resources can be found on the relevant subject pages of our website: www.aqa.org.uk.

Although every effort has been made to ensure that website addresses are correct at time of going to press, Hodder Education cannot be held responsible for the content of any website mentioned in this book. It is sometimes possible to find a relocated web page by typing in the address of the home page for a website in the URL window of your browser.

Hachette UK's policy is to use papers that are natural, renewable and recyclable products and made from wood grown in well-managed forests and other controlled sources. The logging and manufacturing processes are expected to conform to the environmental regulations of the country of origin.

Orders: please contact Hachette UK Distribution, Hely Hutchinson Centre, Milton Road, Didcot, Oxfordshire, OX11 7HH. Telephone: +44 (0)1235 827827. Email education@hachette.co.uk. Lines are open from 9 a.m. to 5 p.m., Monday to Friday. You can also order through our website: www.hoddereducation.co.uk

First published in 2016 by
Hodder Education,
An Hachette UK Company
Carmelite House, 50 Victoria Embankment
London EC4Y 0DZ

Impression number   10 9 8

Year   2024

Cover photo © Getty Images/Flickr RF
Typeset in ITC Officina Sans Book 11.5/13 by Aptara, Inc.
Printed by CPI Group (UK) Ltd, Croydon CR0 4YY

A catalogue record for this title is available from the British Library

ISBN 978 1 4718 8328 6

Contents

Contents

Get the most from this book

Welcome to the AQA GCSE Combined Science Trilogy Student Book.

This book covers the Foundation and Higher-tier content for the 2016 AQA GCSE Combined Science Trilogy specification.

The following features have been included to help you get the most from this book.

Prior knowledge

This is a short list of topics you should be familiar with before starting a chapter. The questions will help to test your understanding. Extra help and practice questions can be found online in our AQA GCSE Science Teaching & Learning Resources.

KEY TERMS

Important words and concepts are highlighted in the text and clearly explained for you in the margin.

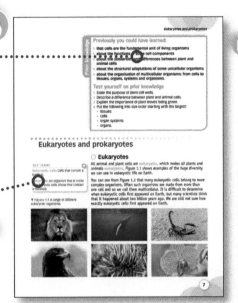

Higher-tier only

Some material in this book is only required for students taking the Higher-tier examination. This content is clearly marked with the blue symbol seen here.

Practical

These practical-based activities will help consolidate your learning and test your practical skills.

Required practical

AQA's required practicals are clearly highlighted.

TIPS

These highlight important facts, common misconceptions and signpost you towards other relevant chapters. They also offer useful ideas for remembering difficult topics.

Examples

Examples of questions and calculations that feature full workings and sample answers.

Show you can...

Complete the Show you can tasks to prove that you are confident in your understanding of each topic.

Test yourself questions

These short questions, found throughout each chapter, allow you to check your understanding as you progress through a topic.

Chapter review questions

These questions will test your understanding of the whole chapter. They are colour coded to show the level of difficulty and also include questions to test your maths and practical skills.

Simple questions that everyone should be able to answer without difficulty.

These are questions that all competent students should be able to handle.

More demanding questions for the most able students.

Answers

Answers for all questions and activities in this book can be found online at:

www.hoddereducation.co.uk/aqagcsecombinedscience

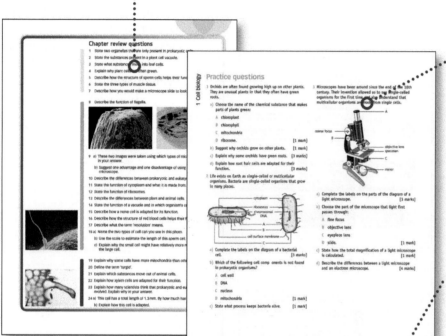

Practice questions

You will find Practice questions at the end of every chapter. These follow the style of some of the different types of questions you might see in your examination and have marks allocated to each question part.

Working scientifically

In this book, Working scientifically skills are explored in detail in the activity at the end of most chapters. Work through these activities on your own or in groups. You will develop skills such as Dealing with data, Scientific thinking and Experimental skills.

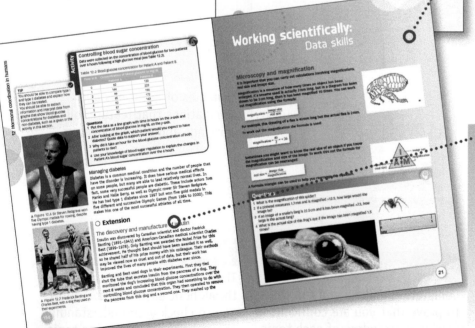

○ Extension

Occasionally we have included material that isn't in the AQA specification. You can use this for further reading and deepen your understanding of a topic. This may be especially useful for students hoping to study A Level science. This content is clearly marked with the green symbol seen here.

* AQA only approve the Student Book and Student eTextbook. The other resources referenced here have not been entered in the AQA approval process.

1 Cell structure

There are thousands of different types of cell found in millions of different species of life on Earth. These range from tiny bacteria that live around us to cells in birds that can fly over the Himalayas. Cells can be put into two broad groups: **prokaryotic cells** found in prokaryotic organisms (also called prokaryotes) and **eukaryotic cells** found in eukaryotic organisms (eukaryotes). Prokaryotic and eukaryotic cells have many features in common but also some key differences.

Specification coverage

This chapter covers specification points 4.1.1.1 to 4.1.1.5 and is called Cell structure.

It covers eukaryotic and prokaryotic cells, animal and plant cells in more detail, and microscopy.

Previously you could have learnt:

> Cells are the basic unit of living organisms.
> Each part of a cell has a particular function.
> Plant and animal cells have some similarities but also some differences.
> Unicellular organisms can be adapted for particular functions.

Test yourself on prior knowledge

1 What are the functions of plant cell walls?
2 Describe a difference between plant and animal cells.
3 Explain why plant leaves are green.

Eukaryotes and prokaryotes

Eukaryotes

KEY TERMS

Eukaryotic cells Cells that contain a nucleus.

Eukaryote An organism that is made of eukaryotic cells (those that contain a nucleus).

Prokaryotic cells Describe single-celled organisms that do not contain a nucleus.

Prokaryotes Prokaryotic organisms (bacteria).

All animal and plant cells are eukaryotic, which makes all plants and animals eukaryotes. Figure 1.1 shows examples of the huge diversity we can see in eukaryotic life on Earth.

You can see from Figure 1.1 that many eukaryotes are complex organisms. Organisms that are made from more than one cell are described as **multicellular**.

Prokaryotes (bacteria)

All bacterial cells are prokaryotic, which means that all bacteria are prokaryotes.

▼ **Figure 1.1** A range of different eukaryotic organisms.

Prokaryotes:

● are **single celled**
● do **not have a nucleus** containing their genetic material (DNA)
● are **smaller** than eukaryotic cells
● may also have small rings of DNA called **plasmids**.

Individual bacterial cells are usually between 1 μm and 10 μm in length. One million micrometres (μm) make up 1 metre (m), and 1000 make up 1 millimetre (mm). This means that between 100 and 1000 bacteria will fit in a straight line in a space of 1 mm. Groups of bacterial cells are called colonies.

$$1\,m\ (metre) = 100\,cm\ (centimetres)$$

$$1\,cm = 10\,mm\ (millimetres)$$

$$1\,mm = 1000\,\mu m\ (micrometres)$$

$$1\,\mu m = 1000\,nm\ (nanometres)$$

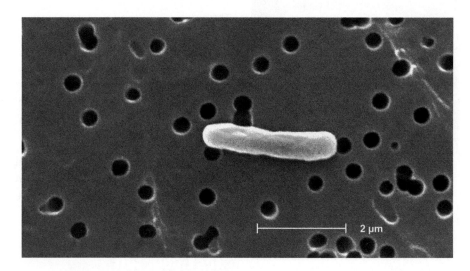

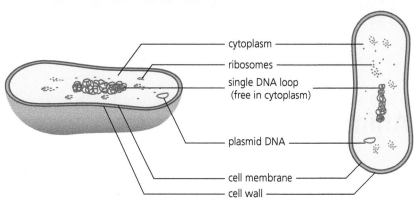

▲ **Figure 1.2** A bacterial cell as seen with a microscope (magnified ×20000) and as three- and two-dimensional diagrams.

A typical bacterial cell is shown in Figure 1.2. The functions of bacterial cell components are shown in Table 1.1.

TIP

It is important that you can explain what a chromosome is.

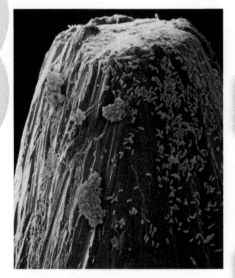

▲ **Figure 1.3** Prokaryotic bacterial cells seen on a pinhead.

Table 1.1 The components of bacterial cells and their functions.

Component	Structure and function
Cytoplasm	This fluid is the part of the cell inside the cell membrane. It is mainly water and it holds other components such as ribosomes. Here most of the chemical reactions in the cell happen (such as the making of proteins in ribosomes).
Cell wall	Like those of plants and fungi, bacterial cells have a cell wall to provide support. However, unlike plant cell walls this is not made of cellulose. The cell membrane is found on the inside surface of the cell wall.
Single DNA loop (DNA not in chromosomes)	DNA in prokaryotes is not arranged in complex chromosomes as in eukaryotic cells. It is not held within a nucleus.
Plasmids	These are small, circular sections of DNA. They have many functions, including giving bacterial cells resistance to some antibiotics.
Cell membrane	This controls what substances go in and out of a cell.
Ribosome	Proteins are made by ribosomes. Ribosomes are examples of cell organelles.

TIP

Copy out the headings in the first row and column of Table 1.1 and test yourself by filling in the rest of the table from memory. This will help you remember the details.

Figure 1.3 shows how small bacterial cells are. Typical eukaryotic cells are much larger than this. However, even eukaryotic cells are microscopic. This means you can't see a single cell without using a microscope.

Test yourself

1 Describe how DNA is arranged in bacteria.
2 Describe the function of ribosomes.

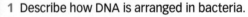

Animal and plant cells

◯ Generalised (typical) animal cells

Plant and animal cells are **eukaryotic**. Eukaryotic cells almost always have a **nucleus** and are generally **larger** than prokaryotic cells.

The structure of a generalised animal cell is shown in Figure 1.4.

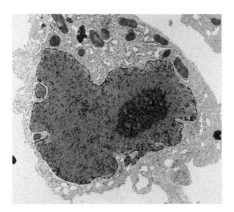
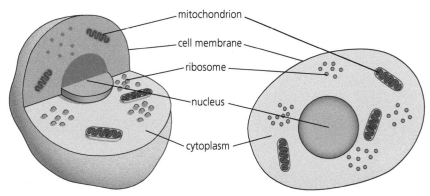

▲ **Figure 1.4** A generalised animal cell as seen with a microscope (magnified ×4800) and as three- and two-dimensional diagrams.

Components of animal cells

In the previous section we looked at bacterial cells. Animal cells, including human cells, have many components in common with these. The **cytoplasm** of animal cells is also mainly water and it holds other components such as ribosomes. In the cytoplasm most of the chemical reactions in the cell happen (such as the making of proteins using ribosomes).

The **cell membrane** of animal cells also surrounds the cell. There are **no cell walls** in animal cells and so the membrane is on the outside of these cells. The membrane controls what substances go in and out of the cell. Most cells need glucose and oxygen for respiration, and these substances move by diffusion or are transported into the cells from the blood, where they are found at a higher concentration. Carbon dioxide moves back into the blood capillaries through the membrane.

Mitochondria are small organelles found in the cytoplasm and are only present in eukaryotic cells. They are the site of aerobic respiration. Here the energy stored in glucose is released, using oxygen. More active cells, such as those in muscles or sperm cells, usually have more mitochondria because these cells need more energy.

Ribosomes are the site of protein synthesis. These organelles are present in the cytoplasm of animal cells.

Animal cells are unlike bacterial cells in that they usually possess a **nucleus**. This component is present in almost all eukaryotic cells. It is found in the cytoplasm and is surrounded by its own membrane. The cell's genetic material (DNA) is enclosed within it, arranged into chromosomes. The nucleus controls the activities of the cell.

KEY TERMS

Diffusion The net movement of particles from an area of high concentration to an area of lower concentration.

Aerobic respiration Respiration using oxygen.

Chromosome Structure containing DNA, found in the nucleus of eukaryotic cells.

TIP ✔

Mitochondria are found in eukaryotic cells but not in prokaryotic cells.

○ Generalised (typical) plant cells

Like animal cells, plant cells are eukaryotic. They have a nucleus and they are generally larger than prokaryotic (bacterial) cells.

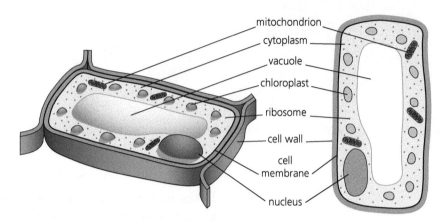

▲ **Figure 1.5** A generalised plant cell as three- and two-dimensional diagrams.

Components of plant cells

Plant cells have many components in common with animal cells, including a **nucleus** in which the organism's genetic material (DNA) is found. As in animal cells, the DNA is packaged into chromosomes. Plant cells also have **ribosomes** for protein synthesis and **mitochondria** for respiration, in their cytoplasm.

Plant cells have some components not present in animal cells. **Chloroplasts** are small organelles, full of a green pigment called **chlorophyll**, which absorb the light necessary for photosynthesis to occur. This reaction uses the light energy from the Sun to convert carbon dioxide and water into glucose and oxygen and so provides an energy source for the plant. It is the green chlorophyll in plants that gives some of their parts their green colour. Most roots are hidden from the Sun and so cannot photosynthesise. They do not have chloroplasts and so are often white, not green.

Plant cells also have a **cell wall**, unlike animal cells. This is made from **cellulose** and provides structure for the cell. Plants would not be able to stand upright to catch light energy from the Sun without cell walls. The **cell membrane** is found inside the cell wall.

Many plant cells also contain a **permanent vacuole**. This is filled with cell sap (water in which dissolved sugars and mineral ions are found). The pressure in the vacuole presses the cytoplasm against the wall to keep the cell turgid.

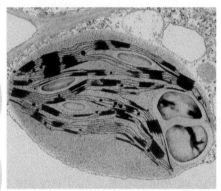

▲ Figure 1.6 The structure of a chloroplast (magnification about ×15000).

KEY TERMS

Photosynthesis A chemical reaction that occurs in the chloroplasts of plants and algae and stores energy as glucose or starch.

Turgid Describes cells that have a lot of water in their vacuole. The pressure created on the cell wall keeps the cell rigid.

TIP

Make a model of a bacterial, plant or animal cell, and label it with the cell components and their functions to help you remember them.

TIP

Most plant cells have chloroplasts, a permanent vacuole and a cell wall made of cellulose; these features are **not** present in animal cells.

Use a light microscope to observe, draw and label a selection of plant and animal cells

In this practical you will examine the structure and features of different animal and plant cells.

Your teacher may provide you with slides showing a range of cells from plants and animals. If this is the case, use Method 1 below.

Method 1

1 Place your slide on a microscope stage and observe using the lowest power objective lens.

2 Focus in on the image and then increase the magnification until you can clearly observe the cell's structure.

3 Make a drawing of what you observe, labelling any structures you recognise. Ensure that you record the magnification you used when making your observations.

Alternatively, your teacher may ask you to make up your own slides to examine the cells in a range of tissues. If this is the case, use Method 2 below.

Method 2

Examining plant cells

1 Wear eye protection.

2 Use tweezers to remove a thin sheet of cells (epidermal tissue) from the inner part of an onion layer.

3 Place this flat on a microscope slide, being careful not to fold it.

4 Place a drop of iodine onto the onion tissue.

5 Carefully lower a cover slip on top of the tissue, ensuring no air bubbles form (Figure 1.7).

6 Follow the steps in Method 1 above to examine and draw the cells present.

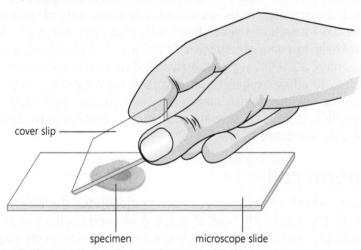

▲ **Figure 1.7** How to make a light microscope slide.

7 Repeat this process using a leaf from a piece of pondweed (*Elodea*), but add a drop of water rather than iodine.

Examining animal cells

1 Wear eye protection.

2 Using an interdental stick or flossing brush from a freshly opened pack, gently scrape the inside of your cheek.

3 Smear the cotton swab on the centre of the microscope slide in small circles.

4 Add a drop of methylene blue solution to the centre of the slide. This is an irritant and can be harmful, so avoid contact with the skin and wear eye protection.

5 Carefully lower a cover slip on top and remove any excess stain by allowing a paper towel to touch one side of the cover slip.

6 Follow the steps in Method 1 above to examine and draw the cells present.

7 Repeat this process using a single hair from your head. Place the base of the hair on a microscope slide and then stain it, and observe the cells using the microscope.

8 Put all slides in a solution of 1% Virkon.

Questions

1 Compare and contrast the structure of the cells you observed. Were any features missing from the animal cells you observed that were present in the plant cells?

2 Can you relate any of the structures or features of the cells you observed to their functions or position in the organisms they came from?

3 Order the cells you observed, from smallest to largest.

> **TIP**
> It is important that you can explain the function of the main components of animal and plant cells.

Show you can...

Describe the function of mitochondria.

Test yourself

3 Name two structures present only in plant cells.
4 In which types of cell would you find mitochondria?
5 Describe the function of the cytoplasm.

Cell specialisation

> **KEY TERMS** ⭐
>
> Biconcave Describes a shape with a dip that curves inwards on both sides.
>
> Ova (singular ovum) Eggs.

> **TIP** ✔
> If you are asked to explain how a cell is adapted to its function, don't forget to use a connecting phrase like 'so that it can' or 'to allow it to'.

The previous section looked at generalised animal and plant cells. Eukaryotic organisms like us are not usually made only of generalised cells. We have developed specialised cells that have adaptations to allow them to complete specific functions. Red blood cells, for example, have a biconcave shape (which dips in the middle on both sides) to allow oxygen to be absorbed more quickly. They also have no nucleus, which means they can absorb more oxygen. Some specialised cells in animals and plants, together with their adaptations, are listed below.

○ Sperm cell

In humans, about a teaspoon of semen is ejaculated during a male orgasm. In the semen are tens of millions of sperm cells, which must swim through the female reproductive system. Here one cell may fertilise an ovum (egg cell). Sperm cells have a **tail** to help them swim towards the ovum (Figure 1.8). They have a relatively **large number of mitochondria** to release the energy needed to help them swim. The nucleus of a human sperm contains the genetic material (DNA) of the father.

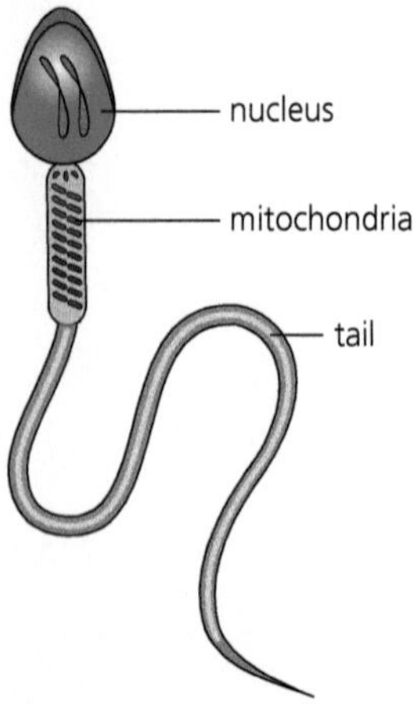

> **TIP**
> Sperm cells, nerve cells and muscle cells are only found in animals.

— nucleus

— mitochondria

— tail

▲ **Figure 1.8** The parts of a sperm cell.

○ Nerve cell

Our nervous system controls and coordinates all our actions. **Nerve impulses** are electrical signals that travel along nerve cells. Some of our nerve cells are the longest cells in our body. Their long extensions are called axons and these have a myelin sheath surrounding much of their length (Figure 1.9). This acts like the plastic coating on an electrical wire and insulates the electrical impulse. The cell body of the nerve cell also has smaller extensions which allow it to pick up signals from neighbouring cells.

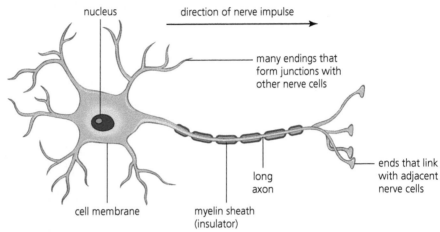

▲ **Figure 1.9** The parts of a nerve cell.

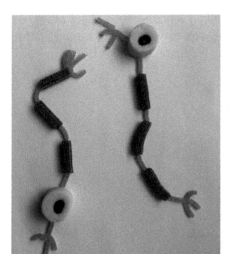

▲ **Figure 1.10** A model of nerve cells made from sweets.

> **TIP** ✅
> When muscle cells contract they get smaller.

> **TIP** ✅
> The reason why we can make large movements, such as bending our arms, is that many muscle cells contract at once in a muscle.

○ Muscle cell

Muscle cells are specialised cells that can **contract** and so move parts of the body. Muscle cells contain **large numbers of mitochondria**, as muscular contraction requires a lot of energy.

○ Root hair cell

Root hair cells in plants have a small thin extension which pokes out into the soil (Figure 1.11). The purpose of these hairs is to increase the **surface area** of the root that is in contact with the soil. This allows the plant to absorb more water and minerals from the soil. Without root hairs it is likely that the adult plant would not be able to absorb enough water to survive.

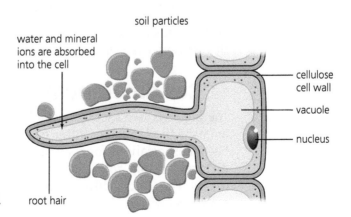

▶ **Figure 1.11** The parts of a root hair cell.

▲ **Figure 1.14** The veins of a leaf are made from xylem tubes transporting water to the leaf and phloem tubes transporting sugars away from it.

Xylem cell

Xylem cells form long tubes running along the roots and stems of plants. They carry water and some dissolved minerals from the roots upwards to other parts of the plant. This water evaporates and is lost from leaves as water vapour during the continual process of transpiration. Xylem cells also carry water to the green parts of plants for photosynthesis during the day. Xylem tubes are made from lots of individual cells that have died and have **no end walls** and **no contents**, leaving a hollow tube like a pipe (Figure 1.12). They have reinforced side walls to support the weight of the plant. The side walls are strengthened by a substance called **lignin**.

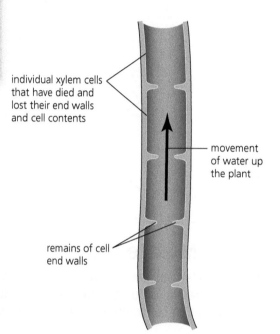

individual xylem cells that have died and lost their end walls and cell contents

movement of water up the plant

remains of cell end walls

▲ **Figure 1.12** The parts of a xylem tube.

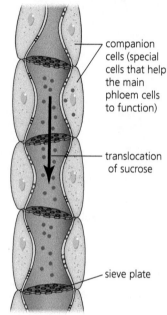

companion cells (special cells that help the main phloem cells to function)

translocation of sucrose

sieve plate

▲ **Figure 1.13** The parts of a phloem tube.

Phloem cell

Phloem cells carry the glucose (as sucrose) made in photosynthesis from the leaves of a plant to all other parts of the plant in cell sap. This process is called translocation. Unlike xylem, phloem cells are living. They have fewer cell organelles than many other types of cell, which allows the sugar to travel easily. Rather than having no end walls (as in xylem), phloem cells have specialised end walls called **sieve plates** that have small holes in them (Figure 1.13).

Phloem cells are located close to xylem cells in the plant, making up the veins you can see in a leaf (Figure 1.14).

Cell differentiation

The previous two sections have looked at generalised and specialised animal and plant cells. After generalised cells are formed, many become specialised for specific functions. This process is called **cell differentiation**. Your cells did this while you were in your mother's uterus. Part of cell differentiation involves cells developing specific structures within them to allow them to function. For example, muscle cells need to release lots of energy during respiration and so require a high number of mitochondria. Sperm cells and nerve cells are **very specialised** cells. Most types of animal cell differentiate very early in their development, but most plant cells can differentiate at any stage. This is why it is possible to take plant cuttings from different parts of a plant.

Test yourself ⚙

6 Give the function of nerve cells.

7 Describe how nerve cells are adapted for their function.

8 Describe how root hair cells are adapted to their function.

Microscopy

Light microscopes use objective and eyepiece lenses to magnify structures that allow light to pass through them. The light rays travel up through the specimen and are then magnified by the objective and eyepiece lenses.

eyepiece lens

coarse focus

fine focus

objective lens

specimen

stage

mirror

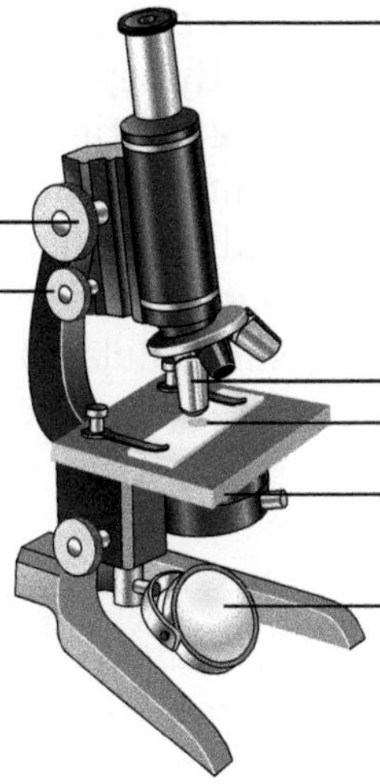

▲ Figure 1.15 A light microscope.

○ Light microscopes

The parts of a light microscope and their functions are shown in Table 1.2.

Table 1.2 The functions of the parts of a light microscope.

Part	Function
Eyepiece lens	You look through this lens to see your sample. This is often ×10.
Objective lens	Usually there are three to choose from (often ×5, ×10 and ×25). The smallest will be the easiest to focus, so select this first. When you have focused this lens try a different one with a greater magnification.
Stage	This holds the sample securely, often using two metal clips.
Specimen	This is usually placed in a drop of water or stain on a microscope slide under a very thin glass cover slip.
Mirror or light source	This sends light up through the specimen and then through the objective and eyepiece lenses into your eyes.
Coarse focus	This quickly and easily moves the stage up and down to focus on the sample.
Fine focus	This sensitively and slowly moves the stage up and down to allow you to make your image very sharp.

The total magnification of the image you are looking at is calculated by:

$$\text{total magnification} = \frac{\text{magnification of}}{\text{eyepiece lens}} \times \frac{\text{magnification of}}{\text{objective lens}}$$

▲ Figure 1.16 A scientist using a large modern electron microscope.

Electron microscopes

Electron microscopes use **electrons** in place of rays of light to make an image (Figure 1.16). The wavelength of electrons can be up to 100 000 times smaller than that of visible light. This means that electron microscopes can take images at **significantly higher magnifications** (Figure 1.17).

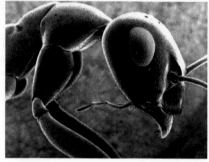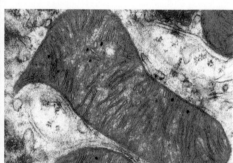

▲ **Figure 1.17** Images taken using an electron microscope: (a) a three-dimensional image of an ant's head; (b) a two-dimensional image of a mitochondrion. These have had colour added.

Electron microscopes can magnify much more than light microscopes, but the key thing is that they have a much greater resolution. The resolution of microscopes is the shortest distance between two parts of a specimen that can be seen as two distinctly separate points. An electron microscope can resolve points up to 2000 times closer than a light microscope, at a separation of just 0.1 nm. The greater resolution of an electron microscope means that sub-cellular structures can be seen in much finer detail. Electron microscopes were very important in enabling scientists to understand many sub-cellular structures. For example, many organelles cannot be seen in detail with a light microscope.

Test yourself

9 State how the resolution of light and electron microscopes differs.

Chapter review questions

1 Name two features of prokaryotic cells.

2 Name the substances present in a plant cell vacuole.

3 Explain why plant cells are often green.

4 Describe how the structure of sperm cells helps their function.

5 Describe how you would make a microscope slide to look at an onion cell.

6 Describe three differences between prokaryotic and eukaryotic cells.

7 Give the function of cytoplasm and what it is made from.

8 What is the function of ribosomes?

9 Describe three differences between plant and animal cells.

10 Describe how a nerve cell is adapted for its function.

11 Define the term 'resolution'.

12 a) Use the scale in Figure 1.18 to estimate the length of the sperm cell.

   b) Explain why the sperm cell will have a large number of mitochondria.

13 Explain why some cells have more mitochondria than other cells.

14 Define the term 'turgid'.

15 Explain how xylem cells are adapted for their function.

16 a) The root hair cell (X) in Figure 1.19 is 1.3 mm long. By how much has it been magnified?

   b) Explain how this cell is adapted.

17 Suggest why ribosomes are usually measured in nanometres (nm).

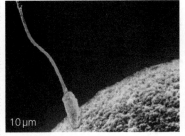

10 µm

▲ Figure 1.18

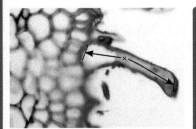

▲ Figure 1.19

Practice questions

1 Orchids are often found growing high up on other plants. They are unusual plants in that some species have green roots.

a) Choose the name of the structures that make parts of plants green:

A chloroplasts

B chlorophyll

C mitochondria

D ribosomes   [1 mark]

b) Suggest why orchids grow on other plants.   [1 mark]

c) Suggest why some orchids have green roots.   [3 marks]

d) Explain how root hair cells are adapted for their function.   [3 marks]

2 Life exists on Earth as single-celled or multicellular organisms. Bacteria are single-celled organisms that grow in many places.

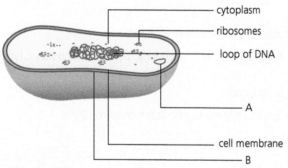

▲ Figure 1.20

a) Copy the diagram of a bacterial cell in Figure 1.20, and complete the missing labels.   [2 marks]

b) Which two of the following cell components are not found in prokaryotic organisms?

A cell wall

B DNA

C nucleus

D mitochondria   [2 marks]

3 Microscopes have been around since the end of the 16th century. Their invention allowed us to see single-celled organisms for the first time and also understand that multicellular organisms are made up of many cells.

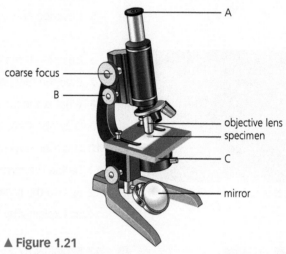

▲ Figure 1.21

a) Identify parts A, B and C in the diagram of a light microscope in Figure 1.21.   [3 marks]

b) Choose the part of the microscope that light first passes through:

A fine focus

B objective lens

C eyepiece lens

D slide   [1 mark]

c) How is the total magnification of a light microscope calculated?   [1 mark]

d) Describe two differences between a light microscope and an electron microscope.   [2 marks]

4 Describe the similarities and differences between prokaryotic cells and eukaryotic plant and animal cells.   [6 marks]

Working scientifically:
Dealing with data

Microscopy and magnification

It is important that you can carry out calculations involving magnifications, real size and image size.

Magnification is a measure of how many times an object has been enlarged. If a sesame seed is actually 3 mm long, but in a diagram has been drawn to be 3 cm long, then it has been magnified 10 times. You can work out magnification using the formula:

$$\text{magnification} = \frac{\text{image size}}{\text{actual size}}$$

For example, this drawing of a flea is 40 mm long but the actual flea is 2 mm.

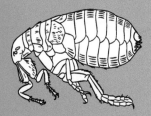

▲ Figure 1.22

To work out the magnification the above formula is used:

$$\text{magnification} = \frac{40 \text{ mm}}{2 \text{ mm}} = \times 20$$

Sometimes you might want to know the actual size of an object if you know the magnification and size of the image. To work this out the formula for magnification can be rearranged:

$$\text{actual size} = \frac{\text{image size}}{\text{magnification}}$$

Also, you might need to work out what image size would be produced if you were given the actual size of the image and its magnification:

$$\text{image size} = \text{actual size} \times \text{magnification}$$

A formula triangle can be used to help you rearrange the equation.

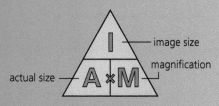

▲ Figure 1.23 A formula triangle.

> **TIP**
>
> It is really important to ensure that measurements are always in the same units. So if you have mixed units you will need to convert them all to the same format.

Questions

1 If a pinhead measures 1.8 mm and is magnified ×12.5, how large would the image be?

2 If an image of a snake's fang is 22.5 cm and it has been magnified ×7.5, how large is the actual fang?

3 What is the actual size of this frog's eye if the image has been magnified ×1.5?

▲ Figure 1.24

TIP

Cell lengths are usually measured in μm (micrometres). Sub-cellular structures can be measured in mm or nm (nanometres), depending on their size.

Extension

Often the actual object being studied is too small to be measured using a ruler, which means that a scale lower than a millimetre is needed. A micrometre (μm) is a thousandth of a millimetre and a millionth of a metre.

Using standard form, this can be written as:

$1 μm = 1 × 10^{-3} mm$ and

$1 μm = 1 × 10^{-6} m$.

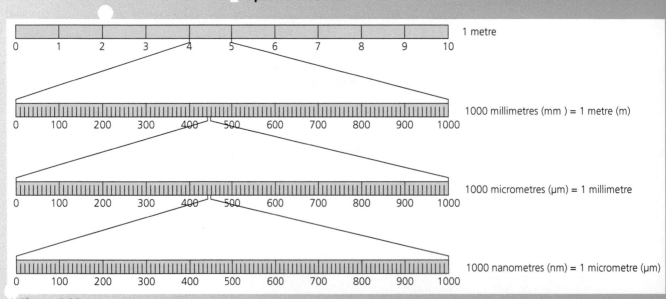

▲ Figure 1.25

Example

If the actual size of this cheek cell is 60 µm, by how much has it been magnified?

- First measure the size of the cell in mm. In this micrograph, the cell is 45 mm wide.

- Then convert this to µm by multiplying by 1000.

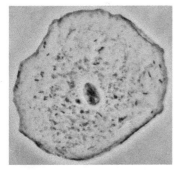

  So 45 × 1000 = 45 000 µm

- To work out the magnification:

$$magnification = \frac{image\ size}{real\ size}$$

$$magnification = \frac{45\,000}{60}$$

$$magnification = \times 750$$

▲ Figure 1.26

Question

4 What is the actual size of this red blood cell if it has been magnified ×6000?

▲ Figure 1.27

2 Cell division

When your father's sperm fused with your mother's ovum you were only a single stem cell. Years later, you are now made from thousands of billions of cells. These are arranged in a very specific way and specialised into several hundred different types. This chapter explains how your body cells grew from that single fertilised ovum by a process of cell division called mitosis and how the same process replaces your damaged tissues and those cells that die naturally. This process is key to life on Earth.

Specification coverage

This chapter covers specification points 4.1.2.1 to .3 and is called Cell division.

It covers the structure of chromosomes, mitosis, stem cells and cell differentiation.

Previously you could have learnt:

> Chromosomes, genes and DNA can be studied through the use of models.
> Cells are the fundamental unit of living organisms.

Test yourself on prior knowledge

1 Define the term 'gene'.
2 Describe the structure of DNA.
3 Put the following into size order starting with the largest: gene, chromosome, DNA.

Chromosomes

KEY TERMS

Gametes Sex cells, e.g. sperm, ova and pollen.

Haploid Describes a cell or nucleus of a gamete that has an unpaired set of chromosomes (i.e. only half the normal number).

Diploid Describes a cell or nucleus of a cell that has a paired set of chromosomes.

TIP

It is important that you can explain what a gamete is.

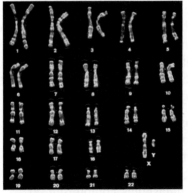

▲ **Figure 2.1** The 46 chromosomes present in a human body cell.

KEY TERM

Gene A section of a chromosome made from DNA that carries the code to make a protein.

Eukaryotic cells are those that contain a nucleus. These can either be single-celled organisms such as protozoa, or single cells of larger multicellular organisms such as trees, insects and you.

Almost all of the cells in your body have a nucleus in which your genetic material (DNA) is found. Half of this came from your father carried by his sperm, and the other half came from your mother in her ovum. Sex cells (sperm and ova) are called gametes, and these have only half of an organism's DNA in them. They are described as haploid cells. Apart from gametes and some cells such as red blood cells that have no nucleus, most of your cells contain two sets of DNA: half from your mother and half from your father. Any cell with these two copies is described as a diploid cell.

The DNA in any one of your body cells (not the gametes) stretches to approximately 2 metres long. Almost all the cells in your body are so small they can only be seen using a microscope. In order to fit all of this DNA into a cell this small, it is coiled into structures called **chromosomes**. Humans have **23 pairs** of chromosomes. We say chromosomes come in pairs to remind ourselves that half were inherited from each parent. This means that there are **46 chromosomes** in a diploid human cell. This is called the 'chromosome number'. Other eukaryotic animals and plants have different chromosome numbers. For example, mosquitos have a chromosome number of 2. This means they have one pair of chromosomes, one from each parent.

The 23 pairs of chromosomes present in a human body cell are numbered in order of their size (Figure 2.1), which varies considerably. Each of these chromosomes is divided into separate regions called genes. Each gene contains the genetic instructions to make a protein and therefore produce a characteristic. Because you have two copies of each chromosome, you also have two copies of each gene: one from your mother and one from your father.

Figure 2.1 also shows us that chromosomes are long, thin structures. They have a point towards the middle where they appear to pinch inwards. This is called the centromere, and no genes are present at this point.

Show you can...

Explain why we often say that we have 23 pairs of chromosomes, rather than 46 chromosomes.

Test yourself

1 What is the chromosome number of humans?
2 Define the term 'haploid'.
3 Define the term 'diploid'.
4 Describe the difference between a gene and a chromosome.

Mitosis and the cell cycle

House dust is mainly made from our dead skin cells, and those of our family or pets who live with us. Our houses are always becoming dusty, which means that our skin is continually dying and falling off. But we don't run out of skin. This means that we must be continually replacing our dead skin cells as they fall off. This replacement process is carried out by a type of cell division called mitosis. This process copies a diploid body cell, which contains all of your DNA, giving two new cells identical to the original cell and each other. Without this process we would not be able to **grow** from a fertilised ovum, **repair** ourselves from damage, or **replace** the cells that die naturally throughout our lives.

KEY TERM ★

Mitosis Cell replication that produces two identical copies of a diploid cell.

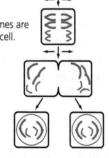

1 Chromosomes make copies of themselves and nucleus disappears.

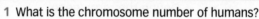
copied chromosome
original chromosome

2 Copied chromosomes line up.

3 Original and copied chromosomes are pulled to opposite ends of the cell.

4 Cell divides.

5 A new nucleus forms in each of the two new cells.

▲ **Figure 2.2** The main steps in mitosis for a cell with just two pairs of chromosomes. Note that the two new cells at the bottom of the diagram are identical to each other and also to the original 'parent' cell at the top in terms of chromosome number.

○ Steps in mitosis

Figure 2.2 shows the steps in mitosis. At the top of the diagram, you can see one cell with four chromosomes. The two red ones come from one parent and the two blue ones come from the second parent. The small blue and red chromosomes make one **pair**, and the large blue and red chromosomes make the other **pair**. If this were a human cell it would have 23 pairs, but this would be too confusing in a diagram.

The first step in mitosis is for the membrane around the nucleus to disappear and all the chromosomes to shorten and fatten. This helps make the following steps easier. The chromosomes have already copied themselves completely. At this point the cells contain 46 chromosomes and 46 copies. You can see in the second box in Figure 2.2 that each of the four chromosomes now looks like an X-shape. Each of these is a chromosome with its copy.

The chromosomes and their copies then migrate to the middle of each cell, which is shown in stage 2 in the diagram. The chromosomes and their copies split apart. The chromosomes are pulled to one side of the cell and the copies to the other. This is shown in stage 3 in the diagram. The cell membrane then starts to pinch inwards and eventually touches the other side, and splits into two identical cells. We call these daughter cells. Each new daughter cell is an exact copy of the original cell.

Before a cell can undergo mitosis, it needs to:

● grow and increase the number of sub-cellular structures such as ribosomes and mitochondria so that each daughter cell gets enough
● replicate (double) the amount of DNA through the duplication of each chromosome.

Mitosis is only one part of the sequence of cell growth, increase in the number of sub-cellular structures, duplication of DNA, and cell division. The whole sequence is referred to as the cell cycle.

Cell division by mitosis is important in the growth and development of multicellular organisms. Mitosis takes place where:

● new cells are being formed during growth
● parts of the body are being repaired or replaced.

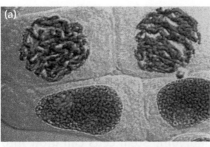
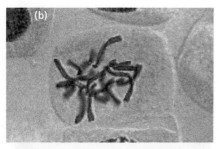

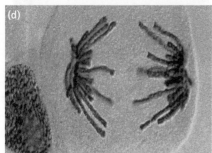
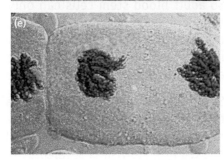

▲ **Figure 2.3** The steps in mitosis as seen in bluebell plant cells.

Show you can...

Describe the steps involved in mitosis.

Test yourself

5 Name the cells produced in mitosis.
6 Give two purposes of mitosis.

Activity

The stages of the cell cycle

Use a digital camera or mobile phone to make a stop-motion animation to show the stages of the cell cycle.

Before you start draw out a storyboard to show how you will animate:

1 the replication of DNA

2 the duplicate chromosomes being pulled to opposite ends of the dividing cell

3 the cytoplasm and cell membrane dividing to form two daughter cells.

Stem cells

KEY TERMS

Stem cell An undifferentiated cell that can develop into one or more types of specialised cell.

Differentiate To specialise, or adapt for a particular function.

TIP

It is important that you are able to explain the importance of cell differentiation for the examples given in this chapter.

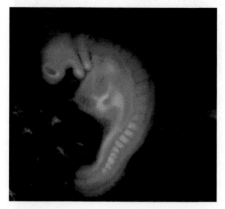

▲ **Figure 2.4** The stem cells in this embryo have already differentiated into a heart after only 26 days.

○ Stem cells in mammals

A **stem cell** is a cell that can **differentiate** into any other type of cell. In mammals there are two types of stem cell. The first type are **embryonic stem cells**, and these cells were present when you were a zygote and an embryo. They divide rapidly by mitosis and begin to differentiate within several days of the sperm fertilising the ovum. Within 21 to 22 days the human embryo has enough differentiated cells to form a beating heart.

Embryonic stem cells can grow into **most of the specialised** cells found in the adult organism. Once an embryonic stem cell has differentiated into a specialised cell it cannot change back or turn into any other type of cell.

We have a second type of stem cells, which are simply called stem cells or **adult stem cells** (although they are also found in children). These stem cells grow only in specific parts of the body, such as the **bone marrow.** They are used to repair the body when it is injured. Crucially they develop into the types of cell found in that location. So adult blood stem cells can only develop into red or white blood cells and some other types of cell. They cannot turn into any cell in the way that embryonic stem cells can. Scientists find these cells interesting to study, but they may not be as potentially useful as embryonic stem cells.

Some animals have stem cells that allow them to regenerate parts of their body. For example, lizards can shed, and later regrow, their tail if seized by a predator. Other animals can go even further than this. If one leg of a starfish is severed by a predator it will grow back.

▲ **Figure 2.5** Starfish can regenerate one or more legs they have lost.

○ Stem cells and differentiation in plants

Plants also have stem cells. However, those found in plants keep their ability to specialise into any type of cell. In a plant, the stem cells are located in a region called the meristem. This is where much of the plant's growth occurs. Meristems are found in shoot tips, where they encourage the shoots to grow towards the light. They are also found in the root tips, where they encourage the roots to grow downwards towards water.

The fact that plant stem cells can differentiate into other cells throughout the mature organism's life allows us to take cuttings. Here a small section of stem, usually with a few leaves, is removed. This is often dipped into rooting powder (Figure 2.6), which contains plant hormones to speed up differentiation. This cutting is then placed directly into the soil. The stem cells towards the bottom of the cutting will quickly grow into root cells and grow downwards. A little later we have a genetically identical copy of the parent plant, often described as a clone. Although there is no genetic variation, because of environmental variation the clone will not always look identical to the parent organism. Much of our food that comes from plants is grown from clones following this method. Many rare or valuable plants such as orchids and roses are grown in this way to stop them becoming too rare or becoming extinct. Also, stem cells can produce plants that all have good characteristics such as **disease resistance**.

▲ **Figure 2.6** A plant cutting being dipped into powder containing hormones to help the clone develop roots before being gently placed into compost.

TIP ✓

Plant cloning can produce plants with **good characteristics – quickly** and **economically**.

○ Stem cell research

Stem cell research is an ethical issue. This means that some people disagree with it for religious or moral reasons. Many scientists think that research into the medical uses of stem cells might help:

- treat **paralysed patients** by making new nerve cells to transplant into a severed spinal cord or damaged brain
- treat conditions such as **diabetes** by replacing the cells in the body that are no longer working properly
- replace injured or defective **organs**.

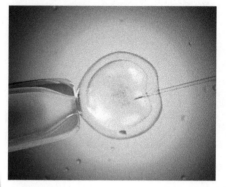

▲ **Figure 2.7** The cells in a fertilised ovum are living but does it constitute a life?

By using stem cells from an injured person's own body to repair damaged tissue in a process called **therapeutic cloning**, doctors can be sure that the cells will not be rejected in the way that some transplants are. In the future, treatments involving therapeutic cloning of stem cells might be used to treat many medical conditions.

The most useful stem cells for this research are **embryonic**, because they can develop into any type of cell. These are often collected from waste cells in the left-over umbilical cord after a mother has given birth. They are found in fertilised ova that are not selected to be put into a woman's uterus during in vitro fertilisation (IVF). It is with this that some people have an ethical issue. Some believe that a fertilised ovum is a life. Some believe that a fertilised ovum has rights and that its use in medical research amounts to murder. For this reason the regulations surrounding stem cell research are extremely tight, and some countries forbid it completely. Using stem cells can have other problems too, such as causing **viral infections** when infected stem cells are used.

Activity

Stem cell research

Stem cell research is one of the most hotly debated areas of medical science. Since the first isolation of embryonic stem cells from mice in the 1980s, there have been great advances in understanding of stem cells and their potential uses in medicine. Alongside this there has been much controversy about the ethical implications of stem cell use, particularly stem cells obtained from embryos.

The UK Government is funding stem cell research and wants the UK to be a world leader in such research.

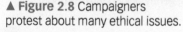

▲ **Figure 2.8** Campaigners protest about many ethical issues.

Questions

1 Working in small groups, come up with a list of views that different members of society might have for and against stem cell research.

2 Write each reason on a sticky note or piece of paper and rank the reasons based on which are the strongest and weakest arguments for and against stem cell research.

3 Write a letter to the Government expressing your views about stem cell research and whether you feel the Government should be funding it. Ensure you support your views with reasons. You may like to use the internet to find extra information from a range of sources to support your arguments.

Show you can...

Explain why stem cell research is of benefit.

Test yourself

7 Define the term 'stem cell'.

8 Name the two types of stem cell.

9 Describe why stem cell research is an ethical issue.

10 Describe the advantages of using embryonic stem cells in research.

Chapter review questions

1 Give the collective term for sperm and ova.
2 What is the number of chromosomes in a human skin cell?
3 How many cells are at the beginning and end of one mitotic division?
4 Give two examples of specialised cells in animals.
5 Define the term 'clone'.
6 Explain why plant cuttings are clones.
7 Give two examples of diploid human cells.

8 How many chromosomes are there in a human gamete?
9 Give the common name for ova.
10 Define the term 'gene'.
11 Give two purposes of mitosis.

12 Explain why gametes need to be haploid.
13 Define the term 'cell differentiation'.
14 Name the region of a plant in which cell differentiation occurs.
15 Suggest why some people protest against stem cell research.

Practice questions

1 How many chromosomes are there in a human diploid cell?
[1 mark]

  A 48

  B 46

  C 44

  D 42

2 Which of these cells are haploid? [1 mark]

  A Nerve cell

  B Epithelial cell

  C Sperm cell

  D White blood cell

3 Figure 2.9 shows the nucleus of a cell that is starting to divide.

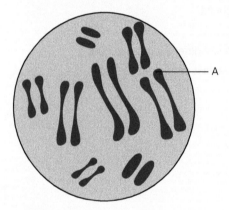

▲ Figure 2.9

a) Name structure A. [1 mark]

b) Draw a diagram to show the appearance of a nucleus from a cell produced by mitosis of the cell in Figure 2.9. [2 marks]

4 Figure 2.10 shows a section through human skin.

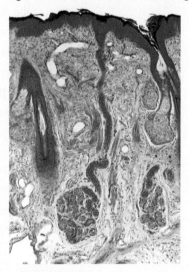

▲ Figure 2.10

a) Name the type of cell division that produces new skin cells. [1 mark]

b) Give a reason why it is important for skin cells to be able to divide. [1 mark]

c) Describe what must happen to the genetic material before a skin cell can divide. [1 mark]

5 Scientists believe that stem cells could have many potential uses in medicine.

a) Stem cells are described as being undifferentiated cells. What does this mean? [1 mark]

b) Stem cells found in liver tissue are called adult stem cells. These cells are often used to repair the body. Suggest another source of adult stem cells other than the liver. [1 mark]

c) Embryonic stem cells are useful in medicine. Give a reason why. [1 mark]

d) Many people have differing ethical views on the use of embryonic stem cells. Suggest two reasons why some people are against the use of these cells.
[2 marks]

Working scientifically:
Experimental skills

Hypotheses and predictions

Humans have a total of 46 chromosomes, arranged in 23 pairs. Not all living things have this many: some have more chromosomes and some have fewer. Since the 1900s, when scientists were beginning to observe chromosomes more closely, they have hypothesised and made predictions linked to chromosome number.

Questions

1 Using the data in Table 2.1, come up with a hypothesis that explains the reason for the chromosome number in the animals given.
2 Scientists use their hypotheses to make predictions. Predict the diploid number of chromosomes in an elephant.
3 Predict the diploid number of chromosomes in a hedgehog.
4 Predict the diploid number of chromosomes in a goldfish.

Your hypothesis is probably linked to the idea that more complex organisms will have more chromosomes and maybe even that because they are larger they need more DNA (genes) to code for the greater amount of proteins they need to produce.

This is exactly what scientists originally thought, and indeed if you predicted that an elephant would have more chromosomes than a human you would be right: it has 56 diploid chromosomes. However, your hypothesis and prediction would not be supported by the data for the hedgehog, which has a total of 90 chromosomes, or a goldfish, which has 94.

As more and more organisms have had their chromosome numbers determined it has become apparent that there is no link between the complexity of organisms and chromosome number. This means that scientists have rejected their original hypothesis and are trying to come up with new ideas to explain the variations seen.

Today, scientists still don't know the exact reason for differing chromosome numbers in organisms, but it is hypothesised that it is linked to their evolution and mutations that occurred in common ancestors.

For example, in some organisms two chromosomes can become fused together. This fusion of chromosomes is thought to explain the differences between chromosome number in humans and great apes. It is hypothesised that the ancestor of humans and apes had 48 chromosomes in 24 pairs but that in humans two of the chromosomes became fused so that we ended up with 46 chromosomes in 23 pairs, while the great apes still have 48 chromosomes in 24 pairs.

Scientists made a prediction that if two chromosomes had fused to make one, we should see similarities in the gene banding on the one human chromosome compared with that found on the two separate chromosomes.

Question

5 What other evidence from Figure 2.11 supports this hypothesis of two chromosomes fusing?

Table 2.1 Diploid chromosome number in different animals.

Animal	Diploid chromosome number
Jack jumper ant	2
Fruit fly	8
Red fox	34
Pig	38
Human	46

KEY TERMS

Hypothesis An idea that explains how or why something happens.

Prediction A statement suggesting what you think will happen.

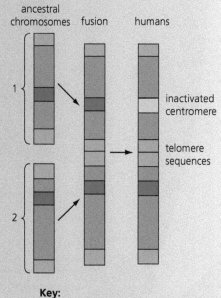

Key:
■ centromere
□ telomere (tip of chromosome)

▲ Figure 2.11 The banding pattern seen when comparing chromosomes.

3 Transport in cells

What does smelling your best friend's deodorant have in common with making a cup of tea? One thing is that they both involve the movement of particles. Particles spread out naturally from areas of high concentration to areas of low concentration. This movement is called diffusion and it is a key biological process. Without it your cells would not receive oxygen or glucose, and would quickly die.

Specification coverage

This chapter covers specification points 4.1.3.1 to 4.1.3.3 and is called Transport in cells.

It covers diffusion, osmosis and active transport.

Diffusion

KEY TERMS

Net Overall.

Concentration gradient A measurement of how a concentration of a substance changes from one place to another.

Diffusion is the process by which particles of gases or liquids spread out from an area where there are lots of them to areas where there are fewer of them. We say that areas with lots of particles have a high concentration and areas with fewer particles have a low concentration. This process happens naturally, and no additional energy is needed for it to occur. It is a **passive process**. Diffusion is defined as the net movement of particles from an area of high concentration to an area of lower concentration. Because the movement is from high to low concentration, we say it is down a concentration gradient.

TIP

We say 'net' movement, which means overall movement, because some of the particles may naturally diffuse back to the area of high concentration they have just come from. Far fewer will ever do this than diffuse away, however.

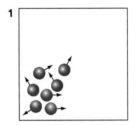 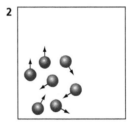 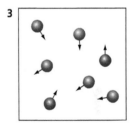

▲ Figure 3.1 Molecules in a gas spread out by diffusion.

Examples of diffusion

When you put your deodorant on in the morning the highest concentration is under your arm. But not all of the deodorant particles remain under your arm, or nobody would be able to smell them. They slowly move by diffusion (we say they diffuse) from a high concentration under your arm to the lower concentration found in the air. The same is true of tea particles when you add hot water to your teabag. The particles of tea don't remain in the high concentration within the bag – they spread out to the lower concentration found in the boiling water.

In the body, **diffusion occurs across cell membranes**. A good place to study diffusion is in the lungs.

▲ Figure 3.2 An everyday example of diffusion. The particles move from a high concentration in a scented candle to a low concentration in the air.

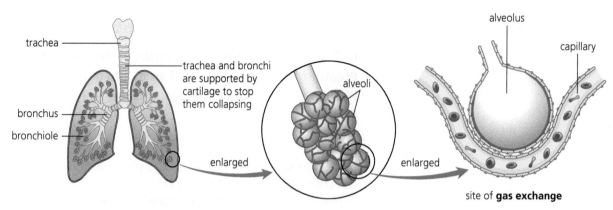

▲ **Figure 3.3** Lungs and alveoli.

Diffusion in the lungs

When we breathe in we take air that is relatively high in oxygen into our lungs. In the individual alveoli in our lungs the **oxygen** is in a higher concentration than in the blood, so the oxygen naturally diffuses from inside the alveoli into the blood. Because the blood in our body is always moving, the blood that now has a high concentration of oxygen is immediately moved away to the tissues and organs and is replaced by 'new' blood with lower levels of oxygen. This means more oxygen will always diffuse from the alveoli into the 'new' blood, keeping the blood rich in oxygen.

The reverse is also true when the blood that is now high in oxygen reaches our tissues and organs. Here it travels through the tiny capillaries that supply the tissues and cells. These cells have a low concentration of oxygen because they have just used their oxygen in aerobic respiration to release energy from glucose. So the oxygen moves from a high concentration in the blood to a lower concentration in these cells.

Carbon dioxide diffuses in the reverse direction. It is produced during respiration in the tissues and organs and so is in a higher concentration within them. The blood moving towards the tissues and organs has a low concentration of carbon dioxide because it has just come from the lungs, where it unloaded carbon dioxide and picked up oxygen. So in body cells carbon dioxide diffuses from a high concentration to a low concentration in the blood. The blood now has a high concentration of carbon dioxide. It is transported to the lungs, where the carbon dioxide diffuses from an area of high concentration in the blood to an area of lower concentration in the alveoli. You then breathe it out.

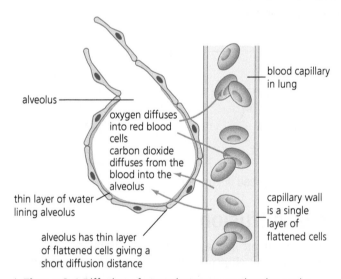

▲ **Figure 3.4** Diffusion of gases between an alveolus and a blood capillary in the lung.

Diffusion

TIP ✓

As a model for a group of alveoli, imagine a red mesh bag (the type that fruit comes in) full of balloons. What are the positives and negatives of this model?

Activity

Diffusion

Place some vapour rub or strong-smelling oil onto some cotton wool and place this inside a balloon. Blow the balloon up and time how long it takes for the smell to pass through the balloon so that you can smell it.

Questions
1 Explain how the scent got from inside the balloon to the outside.
2 Describe how you could speed up the process.
3 Explain how this is similar to the movement of oxygen in the lungs.

Adaptations of the lungs

The combined surface area of your lungs is the total area that is open to air inside your lungs and to blood on the other side. Your lungs have a surface area of about half the size of a tennis court. This means they have a huge area to allow oxygen to diffuse from the alveoli into the blood and carbon dioxide to diffuse from the blood into the alveoli.

TIP ✓

It is important that you can explain how the lungs in mammals are adapted to exchange materials.

KEY TERM ★

Ventilation Breathing in (inhaling) and out (exhaling).

In addition to having a **large surface area**, your lungs are adapted for effective gas exchange by:

- having **moist membranes** that allow substances to diffuse faster across them
- alveoli and capillaries having **thin linings** (usually one cell layer thick)
- having a **rich blood supply**
- **breathing** (ventilation), providing the lungs with a regular supply of fresh air and removing air low in oxygen and high in carbon dioxide.

TIP ✓

Some people confuse the process of breathing, called ventilation, with the release of energy from glucose in the chemical reaction called respiration.

Diffusion also occurs in other places in the body.

Activity

Surface area

Make a fist with one hand and use a piece of string and a ruler to measure around the outside of it from the base of your thumb to the base of your little finger. Now open your hand and measure from the base of your thumb around all your fingers, to end in the same place. What is this a model for?

KEY TERM ★

Excretion The removal of substances produced by chemical reactions inside cells.

Some of your cells make **urea** as a waste product. This is at high concentration in your cells and a lower concentration in your blood, so it diffuses from your cells to your blood. It is transported in the blood to your kidneys for excretion.

◯ **Diffusion in other organisms**

The size of many organisms is determined by the maximum distance that substances can diffuse quickly. Insects, for example, do not have lungs and therefore do not breathe. They simply have a number of small tubes that run into their bodies. Oxygen diffuses from these tubes into the cells of the insect because the cells are using it for aerobic respiration. So it moves from an area of higher concentration in the tubes to an area of lower concentration in the cells. The maximum size of insects is in part determined by the distance that oxygen can quickly diffuse into their cells.

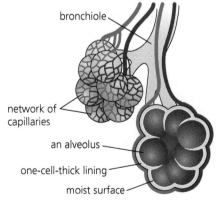

bronchiole

network of capillaries

an alveolus

one-cell-thick lining

moist surface

▲ **Figure 3.5** A cluster of alveoli showing their specialised features for gas exchange.

These smaller organisms do have an advantage over larger ones, however. **The smaller they are the greater the relative size of their surface area compared to their volume.** That is, they have a greater surface area to volume ratio. Large surface area to volume ratios in smaller organisms make it easier for them to get the oxygen they need (and get rid of carbon dioxide).

Large organisms like us need specialised exchange surfaces for exchange – such as alveoli in our lungs and villi in our intestines (see Chapter 4) – and a transport system such as the blood to transport substances around our bodies. This is because larger organisms have a smaller surface area to volume ratio.

Fish absorb dissolved oxygen into their blood by diffusion in their gills. These structures have a large surface area to maximise this.

○ Factors that affect diffusion

Concentration gradient

The steeper the **concentration gradient** (the bigger the difference in the number of particles between an area of high concentration and an area of lower concentration), the more likely the particles are to diffuse down the concentration gradient. For example, the more deodorant you put on, the more the particles of deodorant are likely to diffuse into the air, and so the more likely other people are to smell them.

Temperature

At **higher temperatures** all particles have more kinetic energy. They move faster as a consequence. This means that they are more likely to spread out from their high concentration to areas of lower concentration.

Surface area of the membrane

The **larger the surface area** of the membrane the more particles can diffuse at once. Many people who have smoked for long periods of time have a reduced surface area in their lungs. This is called emphysema. Because their lung surface area is reduced, they are less able to get oxygen into their blood. They therefore often find it harder to exercise.

Test yourself

1 Define the term 'diffusion'.
2 Describe four ways in which alveoli are adapted to their function.
3 Suggest why diffusion cannot happen in solids.

Show you can...

Explain where diffusion occurs to get oxygen to your cells.

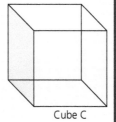

Activity

Investigating surface area to volume ratio and diffusion

A class carried out an investigation to examine the effect of surface area on the diffusion of dye. They were provided with three cubes of clear agar jelly that had been cut to different sizes (Figure 3.6).

Cube A was 1 × 1 × 1 cm.
Cube B was 2 × 2 × 2 cm.
Cube C was 4 × 4 × 4 cm.

Each cube was placed in a 200 cm³ beaker and the beaker was filled with 150 cm³ of blue dye. The cubes were left in the dye for 5 minutes. After this time the cubes were removed and any excess dye washed off before drying with a paper towel.

The cubes were then cut in half and observations were made on how far the dye had moved into the agar (Figure 3.7).

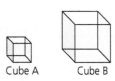

▲ Figure 3.6

Questions

1 Give two variables that were controlled in this investigation.

2 To work out the total surface area (SA) for cube A, first the surface area of one face needs to be calculated: this is 1 × 1. This then needs to be multiplied by the total number of faces (6), so the calculation is $1 \times 1 \times 6 = 6\,cm^2$. To work out the volume, all the dimensions should be multiplied: $1 \times 1 \times 1 = 1\,cm^3$. Copy and complete Table 3.1 by working out the surface area and volume for cubes B and C.

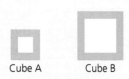

▲ Figure 3.7

Table 3.1

Cube	Total surface area in cm²	Total volume in cm³	Surface area/volume	SA : V
A	6	1	6	6:1
B				
C				

3 Calculate the surface area divided by volume for cells B and C and add this to the table.

4 Use the values you worked out for Question 2 to complete the surface area to volume ratios (SA:V) for cubes B and C.

5 In which cube had the greatest proportion been dyed blue?

6 Explain how the dye entered each cube.

To make sure diffusion rates are fast enough, multicellular organisms have many adaptations to increase diffusion. Examples can be found in the lungs in humans, root hair cells and leaves in plants, and gills in fish. Gills contain many finely divided sections of tissue that are rich in blood capillaries. All the finely divided sections added together give a very large surface area.

Osmosis

KEY TERMS

Osmosis The net diffusion of water from an area of high concentration of water to an area of lower concentration across a partially permeable membrane.

Partially permeable Allowing only substances of a certain size through.

We learnt in the previous section that particles of gases and liquids naturally move from areas of high concentration to areas of lower concentration by diffusion and that this can be across a membrane. Osmosis is the net diffusion of water from an area of high concentration of water (dilute solution) to an area of lower concentration of water (concentrated solution) across a partially permeable membrane. Water is the only substance that has a special name for diffusion. As in diffusion, no additional energy is used, and so this is a **passive process**. Because osmosis is from a high to a lower concentration of water, we say it is down a **concentration gradient**.

○ **Example of osmosis**

When it rains, water is present in a high concentration in the soil surrounding plant roots. The concentration of water inside the plant is lower, particularly if it hasn't rained for a while. So the water moves naturally from the soil into the plant cells across the membranes of the cells by diffusion. Because this is water and it moves across a membrane to get into the cells, we call this process **osmosis**.

The water will then be carried in the xylem up to the leaves, where most of it will evaporate into the air through stomata (tiny pores). This process is called transpiration. Because water is continuously evaporating, it will continuously be 'pulled up' from the roots, which means that the root cells almost always have a lower concentration of water than in the soil, therefore allowing water to continue to enter the plant by osmosis.

Comparing water concentrations

If two solutions have the same concentrations of water and solutes (substances in the water), then there is no net overall movement of water. The same volume of water will move in both directions if they are separated by a partially permeable membrane. We say these solutions are **isotonic**.

If one solution has a higher concentration of solute than another we describe the first one as being **hypertonic** to the other one. Crucially, because this has a higher solute concentration it has a lower water concentration. So if we took a red blood cell and put it into a very salty hypertonic solution (brine), the water from inside the blood cell would pass into the solution by osmosis. It would move from an area of high water concentration (inside the cell) to an area of lower water concentration (in the brine). The red blood cell would **shrivel up** as a result.

The reverse is also true. If one solution has a lower concentration of solute than another, we describe it as being **hypotonic** to the other solution. Crucially, because this has a lower solute concentration it has a higher water concentration. So if we took a red blood cell and put it into pure water (a hypotonic solution) the water from outside the blood cell would move into it. It would move from an area of high water concentration (outside the cell) to an area of lower water concentration (inside the cell). The red blood cell would **swell up** as a result.

▲ **Figure 3.8** The stomata of this sycamore leaf open and close to let water vapour evaporate during transpiration.

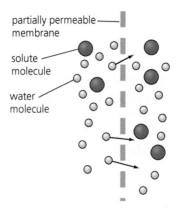

partially permeable membrane

solute molecule

water molecule

▲ **Figure 3.9** Osmosis: water moves from the left side of the partially permeable membrane to the right side because more water molecules are on the left. Note that the solute molecules are too big to pass through the partially permeable membrane.

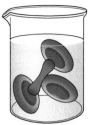

| correct concentration | low concentration | high concentration |
| of water | of water (brine) | of water |

▲ **Figure 3.10** This is what happens to red blood cells in solutions with different concentrations of water (not to scale).

Show you can...

Explain how osmosis leads to the uptake of water into a plant.

Test yourself

4 Define the term 'osmosis'.
5 Name the plant cell that is adapted to allow plants to absorb water.
6 Describe one key difference between osmosis and diffusion.

Required practical 2

Investigate the effect of a range of concentrations of salt or sugar solutions on the mass of plant tissue

Here is one way to investigate osmosis in potatoes, but your teacher may suggest you use another method or investigate different types of vegetables.

Method

1 Label six boiling tubes 0.0, 0.2, 0.4, 0.6, 0.8 and 1.0.

2 Using the volumes given in Table 3.2 and the 1.0 M solution of salt or sugar you have been provided with, make up the following concentrations in each boiling tube.

Table 3.2 Volumes used to make a range of concentrations.

Concentration in M	Volume of 1 M salt or sugar solution in cm³	Volume of distilled water in cm³	Total volume in cm³
0.0	0	25	25
0.2	5	20	25
0.4	10	15	25
0.6	15	10	25
0.8	20	5	25
1.0	25	0	25

3 Using a chipper or corer, remove tissue from the middle of a potato and cut it into six equal 1 cm-long pieces.

4 Pat the first tissue sample dry with a paper towel, then measure and record its starting mass in a table like the one shown on the right.

5 Place a 1 cm-long piece of potato in the tube labelled 0.0 M and start the stopwatch.

6 Repeat this for the other five concentrations.

7 After 30 minutes (or the time specified by your teacher), remove the potato piece from the tube and dry it gently using a paper towel.

8 Record the end mass for the potato piece.

9 Repeat for the other concentrations after each sample has been in the solution for 30 minutes.

Table 3.3 The results of an investigation into the effects of solute concentration on plant tissue.

Concentration in M	Starting mass in g	End mass in g	Change in mass in g	Change in mass in %
0.0				
0.2				
0.4				
0.6				
0.8				
1.0				

Questions

1 Work out the change in mass for each potato piece and record it in your table. The mass changes will be positive if the potato piece got heavier and negative if it became lighter. Ensure you have clearly indicated this.

2 Calculate the percentage mass change for each piece of potato using the equation:

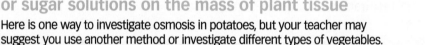

$$\% \text{ change in mass} = \frac{\text{change in mass}}{\text{starting mass}} \times 100$$

3 Now plot a graph of your data with the sugar or salt concentration on the *x*-axis and the percentage change in mass on the *y*-axis. Think carefully about how you will set up your axes to show both positive and negative values on the same graph.

4 Why did you have to pat dry the potato piece before and after each experiment?

5 Why did working out percentage change in mass give more appropriate results than simply recording the change in mass?

6 Write a conclusion for this investigation. You will need to describe the trend shown by your graph and consider how the rate of osmosis is affected by the concentration of solution.

7 Use your graph to predict what concentration of salt or sugar would have led to no change in mass in a piece of potato.

Active transport

KEY TERMS

Active transport The net movement of particles from an area of low concentration to an area of higher concentration using energy.

Mineral ions Substances that are essential for healthy plant growth, e.g. nitrates and magnesium.

On occasions organisms need to move particles from areas where they are in low concentration to areas of higher concentration across membranes. This is called active transport, and we say that the particles are moving **up (or against) a concentration gradient**. If this is the case, then **energy** must be used. This is not a passive process like diffusion and osmosis. The energy needed comes from respiration.

○ Examples of active transport

Mineral ions and plant roots

We learnt in the previous section that water moves from an area of high concentration in soil to a lower concentration in plant roots across a membrane and that this is called osmosis. But plants need to take up mineral ions from the soil as well. These exist in very low concentrations in the soil but in high concentrations in the plants. So we might expect the mineral ions to diffuse out from the plant roots into the soil. Because the plants need to move the mineral ions from low to high concentrations, against the concentration gradient, they need to use energy. This is an example of active transport.

TIPS

- It is important that you can explain the differences between diffusion, osmosis and active transport.
- It is important that you can explain how roots are adapted to absorb materials. You learnt about root hair cells in Chapter 1.

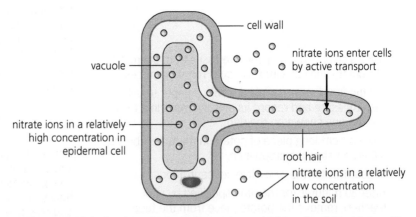

▲ **Figure 3.11** Active transport of nitrates in an epidermal cell (root hair cell).

Sugars and the digestive system

Following a sugary meal you will have high concentrations of sugars in your small intestine and lower concentrations in your blood. This means that sugars will naturally diffuse into your blood. But what happens if your last meal didn't have much sugar in it? The lining of the small intestine is able to use energy to move sugars from lower concentrations in your gut into your blood.

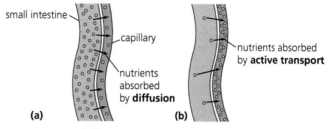

▲ **Figure 3.12** Look carefully at the nutrient concentrations in the intestine and the blood: (a) diffusion; (b) active transport.

Test yourself

7 Define the term 'active transport'.
8 Give one example of where active transport occurs in the human body.
9 Describe the key difference between diffusion and active transport in terms of the concentration gradient.

Show you can...

Explain where active transport occurs in a plant.

Chapter review questions

1 Define the term 'diffusion'.

2 Suggest an everyday example of diffusion of gases.

3 In which direction does oxygen diffuse in the lungs?

4 In which direction does carbon dioxide diffuse in the lungs?

5 Name the blood vessels from which oxygen diffuses into cells.

6 Give the scientific name for breathing.

7 Name the process by which oxygen moves into the blood from the lungs.

8 Define the term 'osmosis'.

9 Describe where and how osmosis occurs in a plant.

10 Name the tiny holes in leaves.

11 Define the term 'active transport'.

12 Is moving up a concentration gradient going from high to lower or from low to higher concentration?

13 Give an example of diffusion in a liquid.

14 Describe two ways in which your lungs are adapted for gas exchange.

15 Define the term 'partially permeable membrane'.

16 Explain why mineral ions moving into a plant root is not an example of osmosis.

17 Describe what would happen to the size of a red blood cell if it were placed into a solution with the same concentration of solutes.

18 Describe what would happen to the size of a red blood cell if it were placed into a solution with a higher concentration of solutes.

19 Describe what would happen to the size of a red blood cell if it were placed into a solution with a lower concentration of solutes.

20 Describe one place where active transport occurs in plants.

21 Describe one place where active transport occurs in humans.

22 Explain why we say 'net movement' in our definition of diffusion.

23 Explain, in terms of diffusion, why insects are small.

24 Describe how you could use your hand and a length of string to model increasing surface area.

25 Explain how temperature affects diffusion.

26 Explain how the surface area of the membrane affects diffusion through it.

27 Describe an experiment in which you could investigate osmosis in plants using pieces of potato.

Practice questions

1 Figure 3.13 shows an alveolus and blood capillary in the lung.

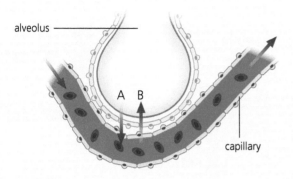

▲ Figure 3.13

a) During gas exchange, oxygen and carbon dioxide are exchanged between the alveolus and capillary. Which arrow (A or B) shows the net direction in which oxygen moves? [1 mark]

b) Gases move across cell membranes by diffusion.

i) Define the term 'diffusion'. [2 marks]

ii) Copy and complete the sentence using words from the box below:

active	passive	energetic	kinetic
oxygen	energy	carbon dioxide	

Diffusion is a _____ processes. This means it does not require additional _____. [2 marks]

2 Figure 3.14 shows three model cells (the pink areas) containing and surrounded by the same particles which can move freely into and out of the cell.

a) Which cell will have the greatest net movement of particles into it? [1 mark]

b) i) What effect would increasing the temperature have on the rate of movement of particles? [1 mark]

ii) Why would this occur? [1 mark]

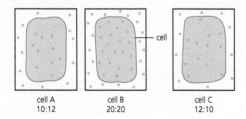

cell A	cell B	cell C
10:12	20:20	12:10

▲ Figure 3.14

c) i) Choose a cell which will have no net movement. [1 mark]

A Cell A   B Cell B
C Cell C

ii) Give the reason why you have chosen this cell. [1 mark]

d) Which cell will have the greatest rate of movement of particles into the cell? [1 mark]

A Cell A   B Cell B
C Cell C

3 Figure 3.15 shows a plant cell before and after it was placed in distilled water for 10 minutes.

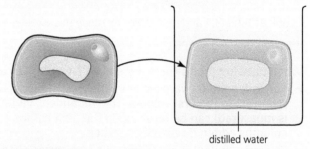

distilled water

▲ Figure 3.15

a) Describe one way in which the cell looks different after being left in distilled water. [1 mark]

b) i) Describe what would happen if an animal cell were placed in the distilled water. [1 mark]

ii) Give the reason why this is different to what happened to the plant cell. [1 mark]

c) i) Draw a diagram to show what the plant cell would look like if it had been placed in a concentrated salt solution. [3 marks]

ii) Explain why the cell would look this way. [3 marks]

4 A student investigated the effect of osmosis on potato pieces. This is the method used:

a) The student started out with three pieces of potato that each measured 2 cm in length.

b) They then placed one piece of potato into a concentrated salt solution and one piece of potato into a dilute salt solution.

c) They left the potato pieces for 10 minutes.

d) They then removed the potato pieces and re-measured their length.

Describe how this method could be improved to produce valid results. [6 marks]

Working scientifically:
Dealing with data

Presenting data in tables

Tables are an important part of most scientific investigations and are used to record the data collected. A good scientific table should present the data in a simple, neat way that is easy to understand.

Table 3.4

Time taken to smell the deodorant	Room temperature
1 minute and 45 seconds	10°C
54.0 seconds	25°C
1 minute and 30 seconds	15°C
1 minute	20°C
42.0 seconds	30°C

Questions

1 Look at Table 3.4 and note down as many mistakes as you can see.

When drawing tables there are some conventions (rules) to be followed:

▶ The independent variable (variable that is changed) should always be recorded in the first column. The dependent variable (variable that is measured) can be recorded in the next columns, with additional columns added if repeats are taken.
▶ The independent variable should be organised with an increasing trend.
▶ If a mean is calculated, this should be in the column furthest to the right.
▶ Column headers must have a clear title. If quantitative data are recorded, correct SI units must be given.
▶ Units must be given in the headers and not rewritten in the table body.
▶ All data in a column must be recorded to the same unit as the header, and mixed units should not be used.
▶ Data should be recorded to the same number of decimal places or significant figures.

Questions

2 Use this information to redraw Table 3.4 above so that it is correct.
3 Read the following instructions for an experiment examining osmosis in model cells. Draw a suitable table to record the volumes required in each beaker in order to prepare for the experiment.

Method for making up solutions

Collect five 100 cm³ beakers and label them A, B, C, D and E. In each beaker add the following amounts of a concentrated fruit squash: A 100 cm³, B 75 cm³, C 50 cm³, D 25 cm³ and none in E. Then use distilled water to bring the volume of beakers B–E up to 100 cm³. Stir the solutions to ensure they are mixed thoroughly.

Questions

4 Read through the method of the experiment on the next page and design a suitable table to record the results of the experiment. Ensure you identify the independent and dependent variables.

Method

Take five equal-sized pieces of Visking tubing that have been soaked in water. Tie one end of each securely. Using a pipette add 10 cm³ of the solution from beaker A into one piece of Visking tubing. Tie the other end of the tubing using string and ensure that no liquid can escape. Repeat this process for the other four solutions B–E. Use a balance to determine the starting mass of each tube.

Place each tube in a separate beaker containing 200 cm³ of distilled water. After 5 minutes remove the tubes and pat dry with a paper towel. Record the mass of each tube. Return to the beakers they came from and repeat, recording the mass at 10, 15 and 20 minutes.

▲ Figure 3.16 The liquid in the beaker is pure water. The red liquid is a very concentrated sugar solution with some red food dye added and is placed inside Visking tubing. This special tubing allows molecules of water through it but not larger sugar molecules. Water moves by osmosis through the Visking tubing from an area of high water concentration in the beaker to an area of lower water concentration (because of the added sugar) in the Visking tubing. This makes the red solution rise up the glass tube.

4 Animal tissues, organs and organ systems

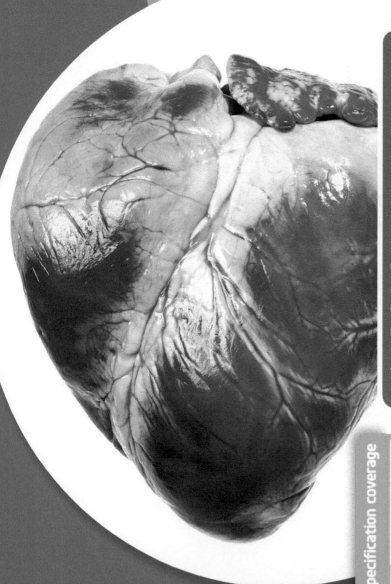

Have you ever thought of yourself as a complicated tube with your mouth at one end and your anus at the other? This is one way you could look at your digestive system. Your digestive system breaks down food into pieces that can be absorbed into your body and used by it. Your circulatory system transports this food, and other crucial substances such as oxygen, in the blood. This is only possible because a muscular organ called the heart is able to pump the blood around the body, through more than 50 000 miles of blood vessels.

Specification coverage

This chapter covers specification points 4.2.2.1 to 4.2.2.7 and is called Animal tissues, organs and organ systems.

It covers organisational hierarchy, the principles of organisation, the human digestive system and its enzymes, the heart and vessels, blood, related health issues, the effects of lifestyle, and cancer.

Previously you could have learnt:

> Multicellular organisms are organised in particular ways with a hierarchical arrangement in the sequence: cells, tissues, organs, organ systems and organisms.
> The human digestive system has a range of tissues and organs that are adapted for digesting and absorbing food.
> Bacteria are important organisms in the functioning of the human digestive system.
> There are consequences of having an unbalanced diet, including obesity, starvation and deficiency diseases.

Test yourself on prior knowledge

1 Name the part of the digestive system in which water is absorbed.
2 Describe the importance of bacteria in the human digestive system.
3 Explain why an imbalance in diet can lead to obesity.
4 Put the following into size order starting with the largest: organism, tissue, organ, organ system, cell.

Levels of organisation in living organisms

In multicellular organisms, there are a number of levels of organisation. For example, in animals there are:

- **cells**: the basic building blocks of all living organisms
- **tissues**: groups of cells with similar structure and functions
- **organs**: groups of tissues that perform a specific function
- **organ systems**: organs are organised into organ systems
- **organisms**: the different organ systems make up organisms.

Table 4.1 gives some examples of these terms.

Table 4.1 Examples of levels of organisation.

Organisational level	Examples
Cell	Nerve cell, muscle cell
Tissue	Nervous tissue, skin
Organ	Brain, heart
Organ system	Nervous system, digestive system
Organism	Human, frog

The human digestive system

KEY TERMS

Insoluble Cannot dissolve.

Soluble Can dissolve.

You eat large lumps of insoluble food. The breakdown products of this food must be transported around your body to reach the cells in which they are needed to complete life processes such as aerobic respiration. For this reaction alone each of your cells needs glucose (which comes from your food) and oxygen (which comes from your lungs). Digestion is the breakdown of food into smaller soluble pieces that can diffuse into your blood. The digestive system is the organ system that is responsible for doing this.

TIPS ✔

- It is important that you can identify the parts of the digestive system on a diagram like this.
- To help you remember facts about the digestive system, you could draw a table with the parts (in order), their functions and how each is adapted.

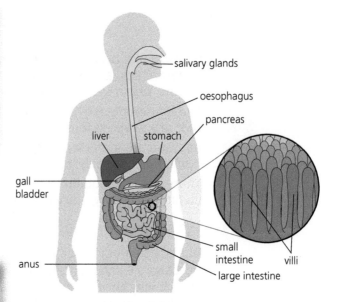

▲ **Figure 4.1** The digestive system, showing the location of the villi.

salivary glands
oesophagus
pancreas
liver
stomach
gall bladder
anus
small intestine
villi
large intestine

KEY TERM ★

Enzyme A biological molecule that speeds up a chemical reaction. It is called a **catalyst** for this reason.

TIP ✔

Amylase is the only enzyme for which you need to know a specific name.

TIP ✔

Amylase is an important enzyme, because it breaks down starch into glucose, which is used in respiration.

○ **Functions of the parts of the digestive system**

Salivary glands

Your salivary glands are found in your mouth and they make saliva, particularly when you are hungry and sense food. **Saliva** acts as a lubricant, making it easier to swallow food. It also contains an enzyme called **amylase**, which is an example of a **carbohydrase** enzyme as it breaks down starch (a carbohydrate) into simple sugar (glucose). When food mixes with saliva in your mouth the chemical breakdown (digestion) begins.

Oesophagus

Your mouth is connected to your stomach by a thin tube approximately 20 cm long. The only function of this tube, which is called the **oesophagus**, is to move food quickly and easily to your stomach. The saliva helps with this. The oesophagus is sometimes called the 'food tube' or 'gullet'.

▲ **Figure 4.2** A view looking down a patient's oesophagus towards his or her stomach.

KEY TERM ★

Pathogen A disease-causing microorganism (e.g. a bacterium or fungus).

Stomach

The stomach is a small organ found between the oesophagus and the small intestine. It releases a type of enzyme called **protease**. This starts the chemical breakdown of protein. The stomach also releases acid. This has a pH of about 2 to 3. Stomach acid does not break down food. Instead it reduces the pH of the stomach to the optimum (best) level for protease enzymes to work properly. It also is part of our first line of defence against infection, as it destroys any pathogens that may have entered the body with food or water.

▲ **Figure 4.3** A view looking down a patient's small intestine. You can see the folds in the intestine wall that increase the surface area. You cannot see villi as they are microscopic.

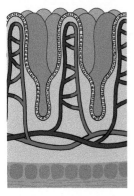

▲ **Figure 4.4** Villi are small, hair-like structures in your small intestine. They increase the surface area over which glucose is absorbed.

TIP ✓

The digestive system is an example of an organ system in which several organs work together to **digest** and **absorb** food.

TIP ✓

Imagine pushing a tennis ball through the leg of a pair of tights by squashing the bit immediately behind it. This is a good model for peristalsis.

Liver

The liver is a large organ found to the right of your stomach. It produces a green liquid called bile, which helps to break down fats. Bile is not an enzyme. After being made in the liver, bile is stored in the **gall bladder** before being released into the small intestine.

Food does not actually pass into the liver. It moves from the stomach to the small intestine.

Pancreas

The pancreas is an organ that produces the three types of enzyme found in the digestive system: **carbohydrase** enzymes, which break down carbohydrates; **protease** enzymes, which break down proteins; and **lipase** enzymes, which break down lipids (fats). The pancreas releases these into the top section of the small intestine, in an area called the duodenum, close to where the gall bladder releases bile.

As with the liver and the gall bladder, food does not actually pass into the pancreas. It moves from the stomach to the small intestine.

Small intestine

Despite its name, the small intestine is actually the longest single part of the digestive system. It is called the small intestine because it is narrower than the large intestine. It is about 7 metres long and is responsible for **absorbing** the products of digestion into the blood. They are then transported around the body in the blood to where they are needed. It is adapted by being **folded** and having **villi**, both of which increase the surface area.

Villi

Villi (Figure 4.4) are microscopic finger-like projections of the lining of your small intestine. In an area about the size of your thumbnail you will have about 4000 villi. They massively increase the surface area of the small intestine and allow much more digested food to be absorbed into your blood. Each tiny villus contains blood capillaries, providing a rich blood supply to move digested food molecules to other parts of your body.

Peristalsis

The walls of the digestive system have rings of muscle around them and all along their length. These contract to squash lumps of food called boluses along your digestive system. **Peristalsis** is the rhythmical contraction of this muscle behind a bolus to push it along. This happens in the oesophagus and the small intestine.

Large intestine

Food that enters the large intestine is mainly indigestible fibre and water. The large intestine is wider than the small intestine but also much shorter, at a length of about 1.5 metres long. The large intestine is responsible for **absorbing water** (and salts) from the remaining digested food.

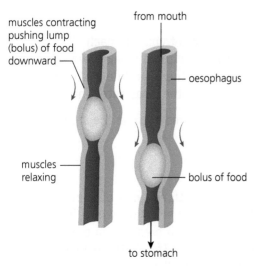

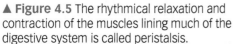

▲ **Figure 4.5** The rhythmical relaxation and contraction of the muscles lining much of the digestive system is called peristalsis.

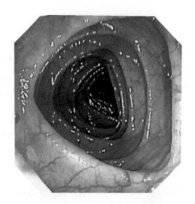

▲ **Figure 4.6** A view looking down a patient's large intestine.

> **TIP** ✔
> ....
> Infections of the large intestine can mean that water is not absorbed properly, resulting in too much liquid in the faeces. This is diarrhoea.

○ **Anus**

The anus is the opening at the end of your digestive system. This controls when you go to the toilet to remove solid waste (called faeces). This process is called defecation.

> **KEY TERM** ★
>
> Defecation Removing solid waste from the body.

Test yourself

1 What is stored in the gall bladder?
2 Describe the process of peristalsis in the small intestine.
3 Describe how villi are adapted.

Show you can...

Describe how food moves through the digestive system from mouth to anus.

• •

Human digestive enzymes

> **KEY TERMS** ★
>
> Substrate The molecule or molecules on which an enzyme acts.
>
> Product The substance or substances produced by an enzyme reaction.

We have seen that there are three types of digestive enzyme, which each act upon a different food group. The molecules that enzymes act upon are called substrates and the molecules that are produced are called products.

Table 4.2 The three types of enzyme found in the digestive system, the food groups they act upon and the molecules they are digested into.

Enzyme	Substrate	Products
Carbohydrase	Carbohydrate	Sugars
Protease	Proteins	Amino acids
Lipase	Lipids (fats and oils)	Fatty acids (three molecules) and glycerol (one molecule)

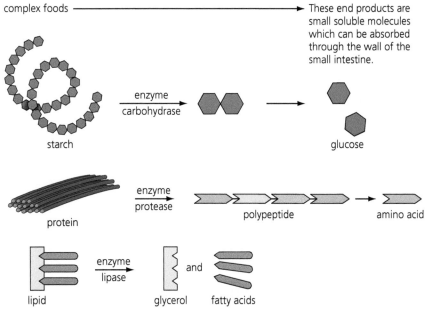

▲ **Figure 4.7** The breakdown of complex food molecules into small, soluble, usable molecules.

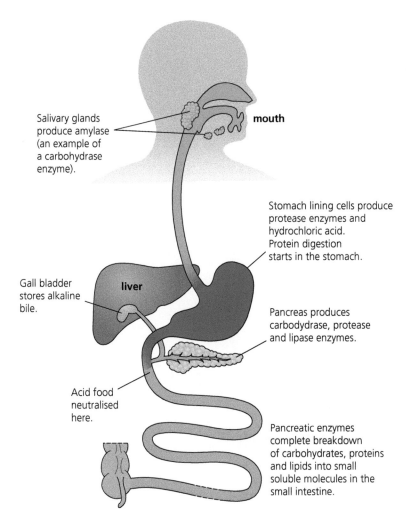

Salivary glands produce amylase (an example of a carbohydrase enzyme).

mouth

Stomach lining cells produce protease enzymes and hydrochloric acid. Protein digestion starts in the stomach.

Gall bladder stores alkaline bile.

liver

Pancreas produces carbodydrase, protease and lipase enzymes.

Acid food neutralised here.

Pancreatic enzymes complete breakdown of carbohydrates, proteins and lipids into small soluble molecules in the small intestine.

○ **Locations of enzymes in the digestive system**

Bile and lipase enzymes

As we have already seen, bile is produced by the liver and stored in the gall bladder before being released into the small intestine. It is not an enzyme but does help break down large globules of fat into tiny droplets. Any substance that does this is called an emulsifier, and the process is called **emulsification**. Bile does not actually break down the fat itself – instead it increases the **surface area** of the fat for the lipase enzymes to digest the fat into fatty acids and glycerol more quickly.

Bile is alkaline and so also **neutralises** hydrochloric acid entering the small intestine from the stomach. This increases the pH back towards neutral for the enzymes in the small intestine to work at their optimum.

◄ **Figure 4.8** Digestive enzymes control reactions that take place in the digestive system. No digestive enzymes are made or used in the oesophagus, liver (bile is not an enzyme), gall bladder, large intestine or anus.

Use qualitative reagents to test for a range of carbohydrates, lipids and proteins

In this practical you will carry out tests to identify starch, sugars, lipids and proteins.

Your teacher will provide you with five samples labelled A–E. You need to determine which is a starch solution, which is a glucose solution, which is a protein solution, which is a lipid oil and which is water.

Copy the table below and use to collect your results:

Table 4.3

Tube	Observation with starch test	Starch present?	Observation with Benedict's test	Glucose present?	Observation with Biuret test	Protein present?	Observation with emulsification test	Lipids present?
A								
B								
C								
D								
E								

Use eye protection throughout the experiment.

Testing for starch: method

1 Using a pipette, add two drops of solution A into a well of a spotting tile.

2 Add two drops of iodine solution to this and record the colour observed.

3 Repeat this for the other four solutions.

4 If starch is present, a blue-black colour will be produced. Use your results to determine which tube contained starch.

Testing for glucose: method

1 Add $1\,cm^3$ of solution A to a boiling tube.

2 Add 10 drops or $1\,cm^3$ of Benedict's reagent to this.

3 Place in a hot water bath (around $80\,°C$) and leave for 5 minutes.

4 Record the colour observed in your table.

5 Repeat this for the other four solutions.

6 If glucose is present, a brick-red precipitate will form. If it is not present, the solution will remain the blue of the Benedict's reagent. Use your results to determine which tube contained glucose.

Testing for protein (Biuret test): method

1 Add $2\,cm^3$ of solution A to a test tube.

2 Add $2\,cm^3$ of Biuret solution to this.

3 Record the colour change in your table.

4 Repeat this for the other four solutions.

5 If protein is present, the solution will turn a light lilac purple colour. If it is not present, the solution will be a cloudy blue. Use your results to determine which tube contained protein.

Testing for oil lipids: method

1 Half fill a test tube with water.

2 Add one drop of solution A to this.

3 Move the test tube from side to side to mix thoroughly.

4 Place your thumb over the top of the test tube and shake.

5 Repeat this for the other four solutions.

6 As oils do not dissolve in water, an emulsion will form. This will make the water go cloudy if lipids are present.

Questions

1 Which of your tubes contained water?

2 What are the negative results for each test?

3 Explain why food tests such as these can be considered subjective.

4 Produce a poster or pamphlet explaining how to do each food test and the positive results to expect.

○ **The lock and key hypothesis**

A hypothesis is a proposed explanation for scientific observations. The lock and key hypothesis explains how enzymes are **specific for their substrates**, just like a key is specific for the lock it fits. In the previous section you learnt that carbohydrase enzymes break down carbohydrate substrates and don't digest fats, for example. The lock and key hypothesis is a model that explains why.

Digestive enzymes are specific to the substrates they help break down because of their shape. Enzymes are proteins and all the **proteins** found in the body have a very specific shape to help them function.

A part of each enzyme is called its active site. This is the part that the substrate fits into. Any change in shape of an active site means that the breakdown will occur more slowly, or not at all. The enzyme and the substrate collide and become attached at the active site. The digestive enzyme then breaks the bonds holding the substrate together. Finally, the digestive enzyme releases the broken-down substrate.

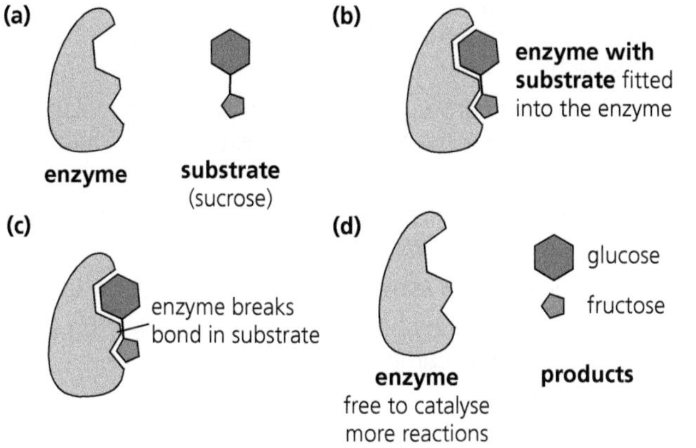

Denaturing of enzymes

Enzymes work at an optimum **temperature** and **pH**. This is simply the temperature and pH at which they are most effective. Here we would say they have the highest enzyme activity. This is when the largest number of successful collisions takes place between the enzyme's active site and the substrate molecules. Any movement away from these optimum conditions will reduce the effectiveness of the reaction and so lower the enzyme activity.

Enzymes are denatured by **high temperatures** and **extremes of pH**. However, at low temperatures enzyme activity falls because of the lower kinetic energy, and if re-warmed the enzyme would start to work again. When denatured, enzymes are permanently damaged and won't work any more. The lock and key hypothesis can be used to explain this as well. When an enzyme is denatured, the shape of the active site has been altered so that it will no longer fit the substrate. In other words, the 'key' will no longer fit the 'lock'.

(a) enzyme   substrate (sucrose)

(b) **enzyme with substrate** fitted into the enzyme

(c) enzyme breaks bond in substrate

(d) enzyme free to catalyse more reactions   products   glucose   fructose

▲ **Figure 4.9** How a digestive enzyme breaks down a substrate. Here, the substrate is sucrose and the products are glucose and fructose.

TIP

Avoid saying that enzymes have been 'killed'. Use the term 'denatured' instead.

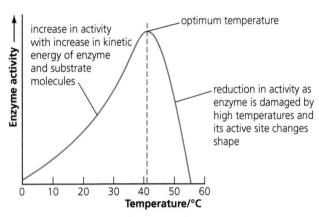

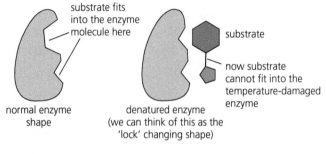

▲ **Figure 4.10** Graph showing the effect of increase in temperature on an enzyme reaction.

▲ **Figure 4.11** The effect of increase in temperature (and denaturation) on an enzyme molecule. High temperatures alter the shape of the active site and so the 'key' doesn't fit the 'lock'.

Required practical 4

Investigate the effect of pH on the rate of reaction of amylase enzyme

In this practical you will examine how pH affects the digestion of starch by amylase.

Method
1. Add 5 ml of 1% starch solution and 5 ml of 1% amylase solution into separate boiling tubes.
2. Add 5 ml of pH buffer to the amylase in the boiling tube.
3. Place each boiling tube in a water bath at 30°C for 5 minutes.
4. After the 5 minutes add all the amylase and buffer into the boiling tube containing the starch solution and stir using a glass rod. Make sure you keep the starch/amylase/buffer mixture in the water bath throughout the investigation. The time when the amylase/buffer is added to the starch is time 0.
5. After 30 seconds remove a sample of the starch/amylase mixture and add two drops of this to

iodine that had previously been added to a spotting tile.
6. Repeat every 30 seconds until the iodine solution remains yellow/brown.
7. Record the time taken.
8. Repeat at a range of pH values by using buffers of the chosen pH.

Questions
1. Why did you need to leave the tubes in the water bath for 5 minutes before mixing the amylase and the starch?
2. Plot a bar chart of your data with pH on the x-axis and time taken for the iodine to remain yellow-brown on the y-axis.
3. Use your bar chart to determine the optimum pH for amylase to work at.
4. Explain why the activity of amylase changes with pH.
5. Suggest how you could amend this investigation to improve the accuracy of the results.

TIPS ✔
- You could make a flicker-book of the lock and key hypothesis and denaturing to help you remember how they work.
- The term 'chemical scissors' is often used for enzymes, but this suggests that enzymes only break down substrates. As you know, they can join substrates too.

Show you can...
State which enzymes are produced in which parts of the digestive system.

○ Digestive (breakdown) and synthesis enzymes

So far you have learnt about the enzymes that are present in your digestive system. Digestive enzymes are those that break down (or digest) substances. Other enzymes, called synthesis enzymes, do the opposite (they help your body make complex molecules from simpler substances). Examples of this are the enzymes found in protein synthesis. They join amino acids to make proteins. Enzymes also build up absorbed sugars into carbohydrates in the body and glycerol and fatty acids into fats.

Test yourself
4. Give an example of a carbohydrase enzyme.
5. Name the products of the breakdown of protein.
6. Describe how breakdown and synthesis enzymes are different.
7. Describe two ways in which enzymes can be denatured.

The heart and blood vessels

Artery A large blood vessel that takes blood from the heart.

Vein A large blood vessel that returns blood to the heart.

Capillary Blood vessel that joins arteries and veins. Substances pass through capillary walls to and from the surrounding **cells**.

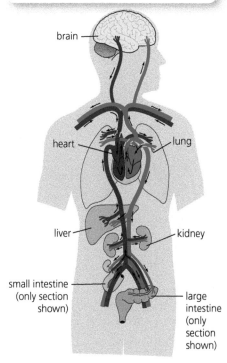

▲ **Figure 4.12** The circulatory system, showing the main organs with which the blood exchanges substances.
Dark red = blood without oxygen;
red = blood containing oxygen.

TIP ✓

If you are looking at a picture of a heart in a book, the left and right sides are always labelled the opposite way around to those on your body. Pick the diagram up and put it over your heart. Now the sides should make sense.

TIP ✓

The heart valves are between the atria and ventricles and at the base (start) of the arteries.

Many substances need to be moved around your body. For example, for respiration, you need oxygen and glucose to be taken to all your cells. You need the waste products of this reaction, carbon dioxide and water, to be removed. Other substances such as hormones are also needed in specific organs at specific times. All of these substances travel in the blood pumped through blood vessels by the heart. Transport is the function of your **circulatory system**. It is composed of:

- the **heart**, which pumps the blood around the body
- **blood**, which carries the blood cells and key molecules around your body
- **arteries**, which carry blood from your heart
- **veins**, which carry blood back to your heart
- **capillaries**, which join arteries and veins through tissues and organs.

○ **The heart and double circulation**

pulmonary artery – carries deoxygenated blood to lungs

aorta – carries oxygenated blood at high pressure around body

pulmonary vein – returns oxygenated blood from lungs

right atrium – receives blood from body

left atrium – receives blood from lungs

valves – prevent backflow of blood

vena cava – returns blood from body

left ventricle – pumps blood around body

right ventricle – pumps blood to lungs

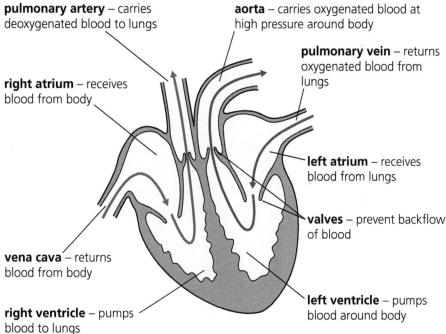

▲ **Figure 4.13** Diagram of a section through the heart. Note the positions of the valves and the wall thickness.

The heart is a pump which is responsible for pushing blood around your body. It is an **organ** made from muscle and nerve tissue. The muscle does the contracting and relaxing to push the blood around and the nerve tissue passes along electrical impulses to make sure the contractions happen correctly. The heart makes its own electrical impulses, which travel along its nervous tissue and cause the contractions. These electrical impulses are generated in the 'pacemaker' section, which is a small bunch of cells in the wall of the top right chamber (the right atrium). The pacemaker controls the rate of your heartbeat.

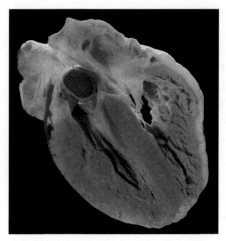

▲ **Figure 4.14** Cross-section through the heart.

There are four chambers in your heart. The top two chambers are called the left and right atria (singular **atrium**). These collect the blood as it returns from your body. The bottom two chambers are called the left and right **ventricles**. The blood is pumped from the atria into the ventricles and then from the ventricles to the rest of the body. The blood on the left and right sides of the heart never mixes. The right ventricle pumps blood to the lungs (where it is oxygenated), and the left ventricle pumps oxygenated blood around the body. This is called a **double circulation**. In effect, the two sides of the heart pump blood to different places.

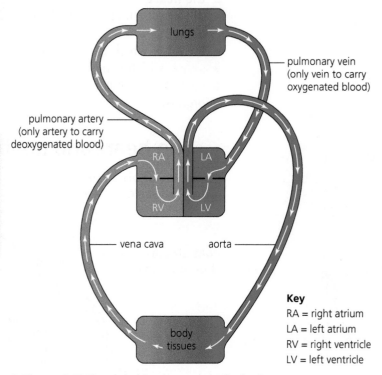

▲ **Figure 4.15** The main blood vessels in the body.

Blood flow through the heart

Blood returns from the lungs and is collected in the left atrium. Because it has come from the lungs it is high in oxygen and low in carbon dioxide. After entering the left ventricle, blood is pumped around the body. This pushes blood high in oxygen to all the tissues and organs that need it. The blood then is taken back to the right atrium by the vena cava. Because it has been to the tissues and organs, it now has low oxygen and high carbon dioxide levels. It enters the right ventricle and then is pumped to the lungs. Here diffusion removes the carbon dioxide and replenishes the oxygen. The blood then returns to where it began, the left atrium.

There are **valves** at the base of the arteries as they extend from the ventricles to prevent a backflow of blood. There are also valves between the atria and ventricles to stop blood being pumped backwards into the atria when the ventricles contract.

If you look closely at Figure 4.13 you will see that the walls of the left ventricle are thicker than in the right ventricle. This is because the left ventricle needs to pump the blood further to the extremities of your body, whereas the right ventricle only needs to pump it to the lungs, which are much closer.

○ The blood vessels

There are three types of blood vessel. Arteries move blood from the heart, veins take it back to the heart and capillaries carry it within tissues and organs. Capillaries link arteries and veins.

Arteries

Arteries must carry blood at high pressure, as it has just been pumped from the ventricles. Because of this they have thick walls made from elastic tissue and muscle tissue. This allows them to stretch. You can feel the surges of blood moving along your main arteries when you feel your pulse. You can do this in your wrist and neck, where arteries are particularly near the body surface.

The main artery coming from the left side of the heart, taking blood to the tissues and organs, is called the **aorta**. The main artery coming from the right side of the heart, taking blood to the lungs, is called the pulmonary artery. This is the only artery to carry deoxygenated blood.

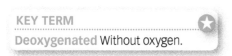

thick muscular and elastic wall

outer protective wall

thick muscle and elastic layer which can stretch to receive a 'pulse' of blood

lumen: round shape in section

smooth lining (in healthy individuals)

narrow lumen (hole)

— 0.85 cm —

▲ **Figure 4.16** Note the thickness of the artery wall.

Veins

Veins carry blood back to the heart at low pressure. This pressure has been lost as the blood travels through the arteries and capillaries. Veins also have to carry blood back to the heart against gravity from the lower parts of your body. Veins are wider than arteries but have much thinner walls. They have one-way valves to keep blood flowing in the correct direction. These are not present in arteries.

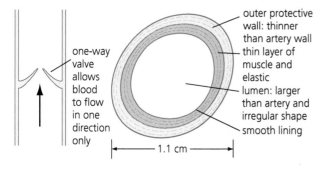

one-way valve allows blood to flow in one direction only

outer protective wall: thinner than artery wall

thin layer of muscle and elastic

lumen: larger than artery and irregular shape

smooth lining

— 1.1 cm —

▲ **Figure 4.17** Note the irregular shape and the thinner muscle and elastic layer of the vein.

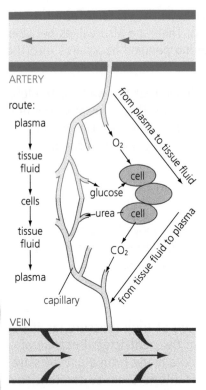

ARTERY

route:

plasma
↓
tissue fluid
↓
cells
↓
tissue fluid
↓
plasma

O₂

from plasma to tissue fluid

cell

glucose

cell

urea

CO₂

from tissue fluid to plasma

capillary

VEIN

▲ **Figure 4.18** Exchange between the blood and tissue cells in a capillary network.

KEY TERMS

Blood plasma The straw-coloured liquid that carries our blood cells and dissolved molecules.

Oxygenated Rich in oxygen.

Tissue fluid The liquid that surrounds ('bathes') cells in the body tissues. It is formed from plasma that diffuses through the capillary walls.

The main vein returning blood to the left side of the heart from the lungs is called the **pulmonary vein**. This is the only vein to carry oxygenated blood. The main vein returning blood to the right side of the heart from the tissues and organs is called the **vena cava**.

Capillaries

Capillaries are tiny blood vessels that spread out like the roots of a plant through your tissues and organs (including the heart muscle). You have billions and billions of these. They are extremely thin to allow as much oxygen as possible to diffuse from the blood into the cells and as much carbon dioxide as possible to diffuse the opposite way.

Blood plasma passes through capillary walls into the tissues (carrying oxygen and glucose with it), where it is called tissue fluid. This bathes the cells and helps provide them with the oxygen, glucose and other molecules they need. Waste products, such as carbon dioxide and urea, enter the capillaries as tissue fluid diffuses back into the blood. Other substances carried by the plasma include products from the digestive system such as amino acids and also hormones.

Test yourself

8 In which direction does blood flow in arteries?
9 Name the two types of chamber in the heart.
10 Describe how veins are adapted for their function.
11 Describe why the left side of the heart is bigger than the right side.

Show you can...

Explain how blood moves around your body from the left ventricle.

● ●

Blood

TIP ✓

Blood is never blue, even if some books show it as being this colour in diagrams. It sometimes looks blue when the walls of vessels are looked at through your skin. If blood is oxygenated it is bright red and if not it is a darker red.

◯ **Components of blood**

Red blood cells

The red blood cells are what give our blood its red colour. In a cubic centimetre of blood there are approximately 5000 million red blood cells.

Red blood cells contain a substance called haemoglobin. This binds with the oxygen that diffuses into your blood in the alveoli. When it is carrying oxygen, it is called oxyhaemoglobin, and it turns the colour of the red blood cells from dark red to a brighter red. These red blood cells then move through arteries and capillaries to the organs and tissues that need the oxygen. Here oxygen diffuses from the red blood cell in a reverse of the reaction in the lungs.

Red blood cells are adapted for carrying oxygen in many ways. Their biconcave shape gives a high surface area to volume ratio, and having no nucleus means there is more room for haemoglobin.

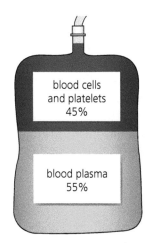

▲ **Figure 4.19** About 55% of blood is a pale yellow liquid called blood plasma. The other 45% is made up of red and white blood cells and platelets.

> **KEY TERMS** ⭐
>
> **Haemoglobin** The protein in red blood cells that can temporarily bind with oxygen to carry it around your body.
>
> **Oxyhaemoglobin** The name given to the substance formed when haemoglobin in your red blood cells temporarily binds with oxygen.

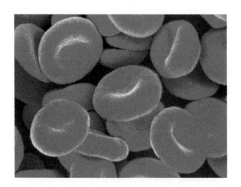

▲ **Figure 4.20** The biconcave shape of red blood cells maximises their surface area to volume ratio.

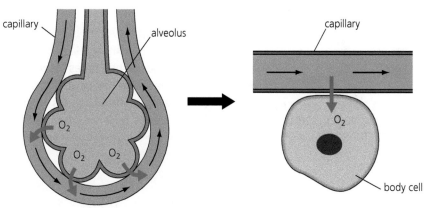

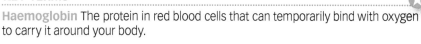

haemoglobin + oxygen → oxyhaemoglobin   oxyhaemoglobin → oxygen + haemoglobin

▲ **Figure 4.21** Haemoglobin transports oxygen from the lungs to other organs as oxyhaemoglobin in the blood.

> **TIP** ✅
>
> It is important that you can recognise images of different blood cells and explain how they are adapted to their function.

> **KEY TERMS** ⭐
>
> **Phagocyte** A type of white blood cell that engulfs pathogens.
>
> **Lymphocyte** A type of white blood cell that produces antibodies to help clump pathogens together to make them easier to destroy.
>
> **Antibody** A protein produced by lymphocytes that recognises pathogens and helps to clump them together.
>
> **Antigen** A marker on a pathogen that antibodies recognise and attach to.

White blood cells

White blood cells are part of your immune system, and they fight off invading pathogens (disease-causing **microorganisms**, such as bacteria). In a cubic centimetre of blood there are approximately 7.5 million white blood cells. Unlike red blood cells, white blood cells have a nucleus. There are two types of white blood cell. Phagocytes **engulf** pathogens and use enzymes to break them down. Lymphocytes produce antibodies to help clump pathogens together for phagocytes to destroy. Antigens are markers on the pathogens that the antibodies recognise and attach to.

▲ **Figure 4.22** A white blood cell among many red blood cells.

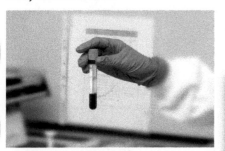

▲ **Figure 4.23** Blood after it has been spun in a centrifuge to separate the cells at the bottom from the plasma at the top.

Platelets

Platelets are **cell fragments**. In a cubic centimetre of blood there are approximately 350 million platelets. They are small structures that join together to form a scab when you cut yourself. Shortly after your skin is cut **platelets start the clotting process**. They do this by releasing chemicals called clotting factors. These turn a chemical called fibrinogen, which is found in your blood plasma, into fibrin. This forms a mesh and acts as a glue to help stick platelets together to form a scab.

Blood plasma

Plasma is a straw-coloured liquid that red and white blood cells and platelets are suspended in. It makes up about 55% of your blood and in turn is made from over 92% water. You have about 3 litres of blood plasma in the 5 litres of total blood. Many molecules that your cells need, such as glucose and amino acids, and those that are waste, such as carbon dioxide and urea, dissolve in your plasma.

Test yourself

12 Name the two types of white blood cell.
13 What is the colour of plasma?
14 Describe what happens when red blood cells meet oxygen in the lungs.
15 Describe how red blood cells are adapted to their function.

Show you can...

Describe the functions of the components of your blood.

Coronary heart disease: a non-communicable disease

The heart is a large muscle that contracts to push blood through the blood vessels all around your body. But the muscle (and nerve) cells that make up the heart organ need to respire themselves to keep on contracting and relaxing. In order to do this they must be supplied with oxygenated blood. This comes through the coronary arteries. Glucose and oxygen diffuse from the blood in these arteries and their capillaries into the cells of the heart.

In **coronary heart disease**, layers of fatty material build up inside the coronary arteries, narrowing them. This reduces blood flow to the heart cells, resulting in a lack of oxygen for the heart muscle. High levels of **cholesterol** in the blood can lead to this happening. The fatty deposits slow or stop the oxygenated blood reaching the cells of the heart. Lack of oxygen can cause cells to die and eventually lead to a heart attack.

KEY TERMS

Heart bypass A medical procedure in which a section of less important artery is moved to allow blood to flow around a blockage in a more important artery.

Stent A small medical device made from mesh that keeps arteries open.

Statin Drug that reduces blood cholesterol.

In recent years, coronary heart disease has become one of the major causes of death in the world. The traditional treatment for this was a heart bypass operation, in which a small section of artery is moved from another part of the patient's body to short-circuit the blockage in the coronary artery. Now more patients are being treated by using less invasive stents. These are small devices made from mesh that are inserted into the arteries to keep them open. This operation is less dangerous and faster to recover from than a heart bypass.

Eating a balanced diet, stopping (or not starting) smoking, reducing alcohol intake, maintaining a healthy weight and regular exercise all reduce the risk of coronary heart disease. Drugs such as statins are also prescribed by doctors to reduce blood cholesterol, which in turn reduces the risk of coronary heart disease. Statins slow the rate at which fatty material is deposited in the arteries.

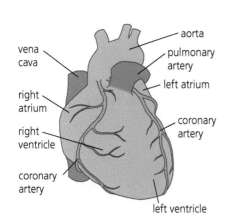

▲ **Figure 4.24** The coronary arteries serve the heart muscle tissue.

○ Faulty valves

The heart has four valves inside it to stop blood flowing backwards. Any flow of blood backwards reduces the efficiency at which blood flows around the body. This means that fewer glucose, oxygen and other molecules reach the cells that need them.

Valves that are faulty might not open properly or not close completely to stop backflow. This may cause breathlessness, tiredness, dizziness and chest pain for the patient. The most common form of treatment for severe cases is **replacement of the valves** during open-heart surgery. These valves can be replaced by valves from donors (biological valves) or artificial mechanical valves.

○ Other heart problems

Heart contractions are controlled by a bundle of cells called the pacemaker in the lining of the right atrium. These send electrical impulses down the heart's nerve cells to regulate the contractions. Some people are born with or develop problems with these cells, which affect the timing of the electrical impulses. An **artificial pacemaker** can be fitted to take over the generation of electrical impulses.

○ Transplants

Heart failure is when a person's heart stops beating. If this happens, or is likely to happen, a person usually requires a **heart** or **heart-and-lungs transplant**. These are the most serious of all operations described in this section. As with all transplants, a match between the donor and the patient must be found to stop the transplant being rejected by the patient's immune system. Often patients are on long waiting lists until a suitable donor is found.

An **artificial heart machine** can sometimes be used in hospital to keep people alive when waiting for a heart transplant or when recovering from a heart operation. Essentially these are mechanical pumps that are connected to the patient's blood supply.

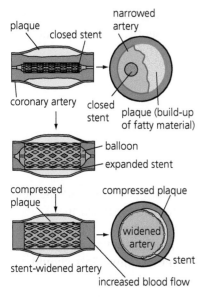

▲ **Figure 4.25** This stent allows blood to flow freely again.

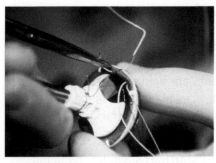

▲ **Figure 4.26** An artificial carbon-fibre heart valve.

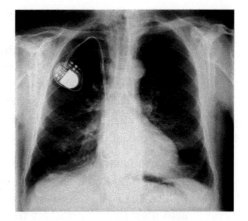

▲ **Figure 4.27** An X-ray showing a pacemaker connected to a patient's heart.

KEY TERM ⭐

Non-communicable disease A disease that is not passed from person to person.

The heart disease discussed in this section is an example of a **non-communicable disease** – a disease that is not infectious and cannot be passed from person to person.

TIPS ✔

- It is important that you appreciate that some people have religious or ethical objections to the use of human or animal tissue for transplants.
- It is important that you can evaluate the advantages and disadvantages of treating coronary heart disease with drugs, mechanical devices or transplants.

Test yourself ⚙

16 Describe the function of the pacemaker.
17 Describe how to reduce the chances of coronary heart disease.

Show you can...

Explain the difference between using stents and heart bypass operations, including which one doctors prefer to use and why.

Health issues

Health is defined as the state of **physical and mental wellbeing**. So being healthy means you are mentally as well as physically fit. Both physical and mental health can be maintained or improved by:

- a well-balanced diet
- regular exercise
- reducing stress
- seeking medical help for mental or physical difficulties.

◯ Well-balanced diet

A well-balanced diet means that you have the correct amount of the key food groups. This is often shown in a food pyramid, as shown in Figure 4.28. Vegetables are low in fat, high in fibre and provide your body with key vitamins. Fruits have more natural sugar than vegetables do, but are also low in fat and high in fibre and vitamins. Fats should only be consumed in lower quantities and are found in fish and nuts as well as many processed foods. Dairy products include milk, yoghurt and cheese. These are high in protein and some vitamins but also high

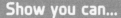

▲ **Figure 4.28** A balanced diet represented as a food pyramid.

in fats and cholesterol. Recent research suggests that the negative effects of dairy foods can outweigh the benefits such as strengthening bones. Meat and beans are a good source of protein as well as vitamins and minerals. Some scientists think that the food pyramid is an oversimplification, whilst others think that it is a useful guide for the public.

Regular exercise

The National Health Service (NHS) in the UK recommends that young people (aged 5 to 18) undertake at least 1 hour of physical activity every day. Some of this should be moderate intensity such as cycling and playground activities. Other activity should be vigorous, such as fast running and tennis. On 3 days a week this should involve muscle-strengthening exercise such as push-ups, and bone-strengthening activities such as running. Exercise also improves the effectiveness of your circulatory system.

Physical and mental ill health

Diseases can cause ill health. Some **different types of disease can interact** to cause health problems. Problems with a person's **immune system** might mean they are more likely to suffer from communicable diseases. A small number of specific virus infections can lead to the development of cancer. The reactions of a person's immune system to infection from a pathogen can trigger allergies such as skin rashes and asthma. Severe physical ill health can lead to mental ill health, such as stress, anxiety and depression.

Stress is the feeling of being under too much mental or emotional pressure. This can affect how you feel, think and behave. It is common for people who are stressed to sleep badly, lose their appetite and have difficulties concentrating. **Anxiety** is a feeling of unease, which might be worry or fear. This can be mild or severe depending upon the situation and the person. **Depression** affects different people in many different ways. Some people feel sad or hopeless, others lose interest in things they used to enjoy. Depression can also affect your physical health. It can make you feel tired and also lose your appetite. Severe depression can make people feel suicidal. People who feel stressed, anxious or depressed should speak to their doctor as soon as possible.

▲ **Figure 4.29** The NHS recommends that young people do an hour of exercise each day.

TIP ✓

Severe physical ill health or having a long-term condition can lead to depression and other mental illness.

TIP ✓

If reading this section makes you worry about your health it is extremely important that you speak to an adult (preferably your parents, a teacher or your doctor).

Test yourself

18 Name the two types of wellbeing that make up health.
19 What should you do if you are worried about your health?
20 Describe the ways in which you can improve your physical or mental health.

Show you can...

Describe the differences between stress, anxiety and depression.

Cancer

Sometimes cell differentiation or division goes wrong and cancerous cells are produced. These cells can divide rapidly by mitosis and quickly cause a lump or tumour. This rapid growth of cells is called **uncontrolled cell division**. There are two types of tumour. The first is described as malignant and causes cancer. These tumours divide rapidly and grow out of control. They can spread from one part of the body to another. If this happens it is called metastasis and forms secondary tumours. Prompt medical treatment is often needed to remove a malignant tumour or destroy the individual cells to stop the cancer spreading. The second type of tumour is described as benign and is medically less serious. Benign tumours are not cancerous, do not spread to other parts of the body and are usually contained within a membrane. They are often removed like malignant tumours, however. Nearly one in two people born after 1960 will suffer from cancer at some point in their life.

Signs of cancer include a lump formed by the tumour, unexplained bleeding, a long-term cough and a loss of weight without dieting. There are, however, over 200 different cancers in humans, and so there are many other symptoms not described here. The most important advice for anyone who thinks they might have cancer is to seek professional medical help as soon as they possibly can.

○ Screening

When doctors look for cancer it is called screening. This can be feeling a bump to see if it is a tumour, and taking blood tests, urine tests or X-ray images. Screening can also be undertaken before a person develops any symptoms, if they have a family history of developing cancer, for example.

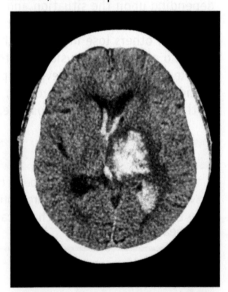

▲ **Figure 4.30** An X-ray shows a tumour (central lighter zone) as being different to the surrounding tissue.

○ Causes of cancer

More than 20% of cancers are caused by **smoking**, and too much **alcohol** can also lead to cancer. Cancers can also be caused by **infections** such as hepatitis B and C, and the human papillomavirus (HPV). Cancer can also occur as a result of **genetic disorders** inherited from parents. Other factors such as **ionising radiation** (including the Sun's ultraviolet rays) and **environmental pollutants** from industry are other causes. **Obesity** is a lifestyle factor that can contribute to cancer. The risk of cancer also increases as we **age**.

The most common cancers in the UK are breast cancer, lung cancer, prostate cancer (men only) and bowel (large intestine) cancer. It is very important to begin treatment of cancers as soon as possible. Sadly many cancers that are detected late become life-threatening. It is likely that some of these could have been treated if they were detected earlier.

○ Treating and preventing cancer

The two most common methods of treating cancer are chemotherapy and radiotherapy. **Chemotherapy** uses very powerful drugs to kill cancer cells. **Radiotherapy** uses X-rays to do the same thing. Both chemotherapy and radiotherapy cannot differentiate cancer cells from the other healthy cells around the tumour, so they can kill other nearby cells too.

Many cancers can be prevented from developing by leading a healthy lifestyle. This includes not smoking, not becoming over or underweight, not drinking too much alcohol, and eating healthily, including fresh fruit and vegetables.

Test yourself

21 What are the most common cancer types in the UK?
22 Name a cancer that can only occur in men and another that can only occur in women.
23 Describe the difference between malignant and benign tumours.
24 Describe how a number of cancers can be prevented.

The effect of lifestyle on some non-communicable diseases

KEY TERMS

Risk factor Any aspect of your lifestyle or substance in your body that increases the risk of a disease developing.

Causation The act of making something happen.

Correlation When an action and an outcome show a similar pattern but the action does not necessarily cause the outcome.

The risk of coronary heart disease increases with high blood pressure, smoking and excessive alcohol, high cholesterol and poor diet. Any aspects of your **lifestyle** or any **substances found in your body or environment** that are linked to the development of a disease are called risk factors. Some risk factors are proved to cause diseases (causation), while others are only linked to a higher chance of developing them (correlation). Examples of risk factors and their associated conditions (diseases) are shown in Table 4.4.

TIPS ✓

- It is important that you can explain the human and financial cost of non-communicable diseases.
- Copy out the table headings in the first row and column and test yourself by filling in the rest of the table from memory. This will help you remember the detail.

KEY TERM ★

Carcinogen A cancer-causing substance.

Table 4.4 Risk factors, conditions (diseases) and their effects.

Risk factor	Condition (disease)	Effects
Diet, smoking and lack of exercise	Cardiovascular disease	Layers of fat build up inside coronary arteries, narrowing them.
Obesity and lack of exercise	Type 2 diabetes	Body cells do not respond to the hormone insulin, which helps control the glucose level in the blood.
Alcohol	Liver function	Long-term alcohol use causes liver cirrhosis. The cells in the liver stop working and are replaced by scar tissue. This stops the liver from removing toxins, storing glucose as glycogen and making bile.
Alcohol	Brain function	Excessive use of alcohol can also alter the chemicals in the brain (neurotransmitters), that pass messages between nerve cells. This can cause anxiety, depression and reduced brain function.
Smoking	Lung disease and cancer	Smoking can cause cancer in many parts of the body, including the lungs, mouth, nose, throat, liver and blood. It also increases the chances of having asthma, bronchitis and emphysema.
Smoking and alcohol	Underdevelopment of unborn babies	Alcohol and chemicals from cigarettes in the mother's blood pass through the placenta to her baby. Without a fully developed liver the baby cannot detoxify these as well as the mother can. This can lead to miscarriage, premature birth, low birth weight and reduced brain function.
Carcinogens and ionising radiation	Cancer	Chemicals and radiation that cause cancer are called **carcinogens**. Tar in cigarettes, asbestos, ultraviolet from sunlight and X-rays are examples.

Many diseases are caused by the interaction of a number of factors. For example, people who smoke and drink excessive amounts of alcohol are more likely to be unfit and put on weight.

▲ **Figure 4.31** This liver tissue was taken from a heavy drinker. The darker areas are dead tissue.

Test yourself

25 Name the disease caused by carcinogens.
26 Name the two organs most likely to be damaged by long-term alcohol abuse.
27 Describe the effects of type 2 diabetes.

Show you can...

Explain the difference between causation and correlation, giving an example in your answer.

Chapter review questions

1 Describe the function of saliva.

2 Where is bile stored?

3 Name the types of enzyme produced in the pancreas.

4 Give the function of the large intestine.

5 Name the organ that pumps your blood.

6 In what direction do arteries carry blood?

7 Explain why the term 'double pump' is used for the heart in mammals.

8 Explain why malignant tumours are more serious than benign ones.

9 Explain why doctors often use stents rather than transplants.

10 Explain why we must digest our food.

11 Define the term 'enzyme'.

12 Name the enzyme that breaks down proteins, and the products.

13 Name the enzyme that breaks down carbohydrates, and the products.

14 Name the enzyme that breaks down fats, and the products.

15 Describe how you could use boiled and unboiled amylase to show that enzymes denature.

16 What are the two conditions that can denature an enzyme?

17 Define the term 'optimum' in relation to the effect of pH on enzyme activity.

18 Describe the pathway of the blood from the left atrium.

19 Describe how capillaries are adapted for their function.

20 Describe how phagocytes protect us from pathogens.

21 Name the blood vessels that provide the heart cells with glucose and oxygen.

22 Describe the effects of having faulty heart valves.

23 Describe one way in which doctors screen for cancer.

24 Define the term 'anxiety'.

25 Explain how the lock and key hypothesis models enzyme action.

26 Name the part of an enzyme that is specific to the substrate.

27 Explain denaturing of enzymes using the lock and key hypothesis.

28 Suggest the effects of having a reduced platelet count.

29 Describe the causes of coronary artery disease.

30 Explain why doctors prefer to use stents than complete bypass operations.

Practice questions

1 Figure 4.32 below shows the main organs in the human digestive system.

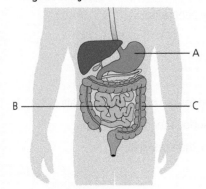

▲ Figure 4.32

a) Name the following organs:

   i) organ A [1 mark]

   ii) organ B [1 mark]

b) Organ C is the large intestine. What is its role? [1 mark]

c) Figure 4.33 shows several villi. Villi are found within the digestive system.

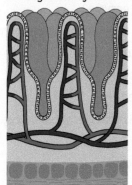

▲ Figure 4.33

   i) In which organ in Figure 4.32 would you expect to find the most villi? [1 mark]

      A Organ A   C Organ C
      B Organ B

   ii) From Figure 4.33, state one way the villus is adapted to maximise the absorption of the products of digestion. [2 marks]

d) i) Give the term used to describe how food is moved through a digestive system. [1 mark]

   ii) Explain how this movement is brought about. [2 marks]

2 Which of the following would be the least invasive method of treatment for coronary heart disease? [1 mark]

   A Stent   C Pacemaker
   B Bypass   D Transplant

3 Figure 4.34 shows amylase speeding up the breakdown (digestion) of a large molecule.

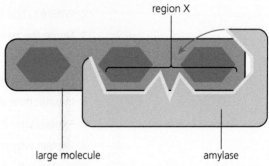

▲ Figure 4.34

a) Why do large molecules need to be digested? [1 mark]

b) What is region X in Figure 4.34? [1 mark]

4 Figure 4.35 below shows three types of blood vessel.

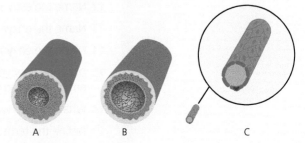

▲ Figure 4.35

a) Name the three blood vessels. [3 marks]

b) Explain how the build-up of fatty material in the blood vessels that supply the heart can lead to a heart attack. [2 marks]

c) Copy and complete Table 4.5 to show which part of blood fits the different descriptions listed. Choose your answers from the box. [4 marks]

plasma   platelets   red blood cells   white blood cells

Table 4.5

Description	Part of blood
Contain a substance called haemoglobin, which binds with oxygen	Red blood cells
Fight off invading pathogens	
Small structures that can join together to prevent blood loss	
Carries many molecules such as glucose and amino acids and carbon dioxide	
Have a shape designed to maximise their surface area	

5 A student carried out food tests to look for the presence of glucose, starch and protein. Describe how they would have carried out the food tests.

You should include:

a) what reagents you would use

b) what positive results would look like. [6 marks]

Working scientifically:
Scientific thinking

Understanding and evaluating models

The word 'model' is used a lot in science but what does it actually mean? Take a minute to try and define your understanding of the term 'model' and note down as many models you can think of that you have encountered in your science lessons.

Scientific models can take many forms, but their main purpose is to represent a process or feature in a way that is easier to predict, understand, visualise or test. Some models are scaled-up versions of real things, such as the model of DNA or a model cell. Others are scaled down, like the model of the solar system. Other models are much more abstract, explaining phenomena we can't see and simplifying the details of them, such as the particle model. Although models are designed to help us understand by simplifying versions of reality, they can also be misleading.

As you have seen, the lock and key hypothesis is a model used to explain how enzymes work. This is shown in Figure 4.36.

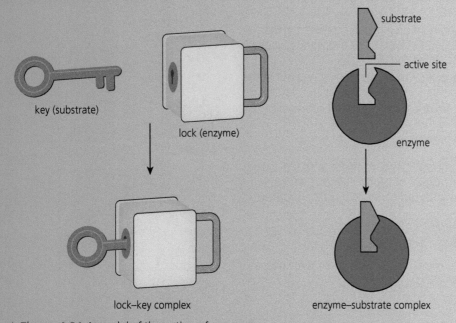

▲ Figure 4.36 A model of the action of enzymes.

Questions

1 What are the strengths of this model in representing how enzymes work?
2 What are the weaknesses of this model?
3 What does the model fail to explain or represent about enzymes?

Visking tubing is a membrane that has small holes in it. These holes are large enough to allow small molecules such as water and glucose to pass through but are too small to allow larger molecules such as starch to pass through. We can use this model to demonstrate how large molecules are digested into smaller molecules, which can leave the digestive system and enter the blood.

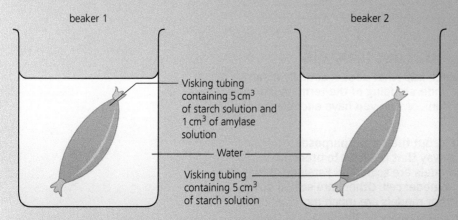

beaker 1

beaker 2

Visking tubing containing 5 cm^3 of starch solution and 1 cm^3 of amylase solution

Water

Visking tubing containing 5 cm^3 of starch solution

▲ Figure 4.37 An experiment to model the digestive system.

Questions

4 What does the Visking tubing, and the water surrounding it, represent in this model?

5 Explain why at the start of the experiment no glucose is present inside the Visking tubing or in the water in either beaker, but over time glucose is detected in the tubing and the water in beaker 1.

6 Explain why starch was only found inside the Visking tubing and not in the water.

7 What features of the digestive system does this model not represent appropriately?

Using Plasticine or other modelling materials, create models to show the key differences between the different types of blood vessel. Ensure you can explain how each blood vessel is adapted to carry out its function.

Questions

8 Evaluate your models and identify three strengths and three weaknesses.

5 Plant tissues, organs and organ systems

Plants are truly amazing. Without them and other photosynthesising species it is possible that the only life on Earth would be a few organisms surrounding volcanic vents on the bottom of the ocean. Plants support almost all life on Earth including you and me. Plants have far more in common with us than you might think. Like us, they are complex organisms that are arranged into tissues, organs and systems. And they are highly adapted to live and reproduce in their natural environment, just like us.

Specification coverage

This chapter covers specification points 4.2.3.1 and 4.2.3.2 and is called Plant tissues, organs and organ systems.

It covers the structure and organisation of plant tissues, and transportation in plants.

Prior knowledge

Previously you could have learnt:

> All multicellular organisms including plants are organised in particular ways with a hierarchical arrangement in the sequence: cells, tissues, organs, organ systems and organisms.

> Plants make carbohydrates in their leaves by photosynthesis and gain mineral nutrients and water from the soil via their roots.

> Leaf stomata are important structures for gas exchange in plants.

> As in animals, plants also reproduce to produce young.

> Leaves are adapted for photosynthesis.

Test yourself on prior knowledge

1 Where do plants absorb water and carbon dioxide?
2 Describe the role of stomata in gas exchange.
3 Explain how leaves are adapted for photosynthesis.

Plant tissues

KEY TERMS

Epidermis The outermost layer of cells of a plant.

Palisade mesophyll Tissue found towards the upper surface of leaves with lots of chloroplasts for photosynthesis.

Spongy mesophyll Tissue found below the palisade layer(s) of leaves with spaces between them to allow gases to diffuse.

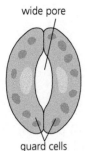

wide pore   narrow pore

guard cells

▲ **Figure 5.1** Pores in stomata become smaller if a plant needs to reduce the amount of water being lost by transpiration.

TIP

Guard cells of stomata are the only epidermal cells to have chloroplasts.

Epidermal tissue

The epidermis is a tissue made up of single layer of cells that forms the outer layer of a plant. It has many functions, including protecting against water loss, regulating the gases that are exchanged between the plant and the air (especially in the leaves), and water and mineral uptake (especially in the roots). The epidermis is usually transparent, possessing fewer green chloroplasts in its cells than other plant tissues. Because the epidermis is transparent, light can pass through it and reach the palisade mesophyll in leaves.

Palisade mesophyll

Immediately below the upper epidermis of plant leaves is the palisade mesophyll tissue. The cells in this tissue are often **more tightly packed** and have a more regular shape than the cells in the spongy mesophyll tissue below them. Palisade mesophyll cells have **more chloroplasts** than all other plant cells and so are the major site of photosynthesis. The palisade cells are highly adapted for photosynthesis.

Spongy mesophyll

Spongy mesophyll tissue is found below the palisade mesophyll tissue towards the lower surface of plant leaves. The cells in this tissue are more spherical in shape than palisade mesophyll cells and have many spaces between them. Gases enter leaves through tiny pores called **stomata** (singular, stoma; see Figure 5.1), which have guard cells around them. These cells control the size of the opening. More stomata are normally found on the underside of leaves than on the upper surface. Spongy mesophyll cells have a **large surface area** in contact with the **air spaces** in the leaf to maximise gas exchange.

Observing stomata on leaves

Stomata are the pores in a leaf that allow gas exchange with the atmosphere, which is required for photosynthesis and respiration. The numbers and arrangement of stomata vary on the upper and lower surfaces of plants and between plant species.

Method

1 Select the plant you want to study and use a small paintbrush to coat a small section of the upper and lower epidermis of a leaf with a water-based varnish or nail polish.

2 Leave the varnish to dry fully.

3 Using tweezers and sticky tape remove the varnish impressions from the leaves and mount each onto a labelled slide.

4 Observe the stomata using a microscope.

5 Draw and label a sketch of a stoma and the surrounding guard cells.

Questions

1 On which side of the leaf was there a higher density of stomata? Suggest why this is the case.

2 Explain how you would expect the stomata to look in a plant that is wilted compared with one that has been recently watered.

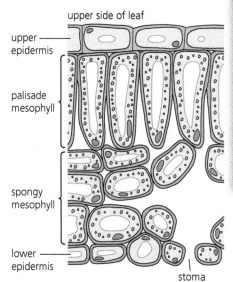

▲ **Figure 5.2** Part of a leaf in cross-section showing the epidermis (very top and bottom cell layers), palisade mesophyll (long, thin upright cells below the upper epidermis) and spongy mesophyll (more circular cells found towards the bottom).

> **TIP**
>
> It is important that you can explain how the structure of xylem and phloem are adapted to their functions. (For more detail on xylem and phloem cells, see Chapter 1.)

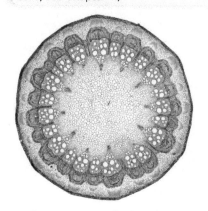

▲ **Figure 5.4** Vascular bundles in a stem.

Xylem and phloem

Water flows up through **xylem** tissue from the roots to the leaves during transpiration. **Phloem** cells carry the glucose (in the form of sucrose) made in photosynthesis from the leaves of a plant to all other parts of the plant during translocation. Xylem and phloem tissues are often found together in vascular bundles.

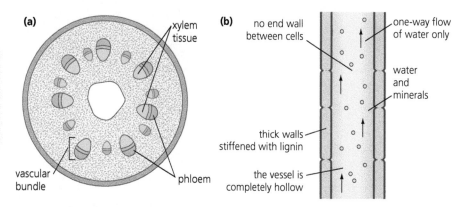

▲ **Figure 5.3** (a) Transverse section of a stem; (b) longitudinal section of xylem vessel from a vascular bundle (the arrows show the direction of water flow).

Meristem

The meristem is the region of plant tissue in which stem cells are produced and so where much of the plant's growth occurs. They are found in shoot tips reaching for the sunlight and root tips following gravity downwards.

Plant organs

TIP ✔

Remember that organs are parts of an organism that have a particular function and are formed of different types of tissue.

◯ Root

Roots are plant organs that are usually found below the soil. As a result they are white because they don't contain green chloroplasts for photosynthesis. Roots absorb water by osmosis and minerals by active transport from the soil. They also anchor the plant into the soil. In addition, in some plants, roots can store the glucose made during photosynthesis, usually as starch.

The meristem is found at the very tip of the root. Here new cells are produced to allow the root to grow deeper into the soil. On the outside of roots are root hair cells to absorb water by osmosis. These are specialised epidermal cells. In the middle of the root are the xylem and phloem tissues.

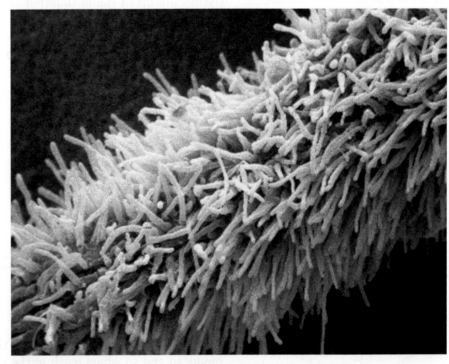

▲ **Figure 5.5** Look how many root hairs are on this one tiny root.

○ Shoot

Scientists define a shoot as the stem, its leaves, and its buds (not just the very tip of a young plant).

The meristem is found at the very tip of the shoot. Here new cells are produced to allow the shoot to grow towards the light. On the outside of shoots are epidermal cells.

TIP

You can see the bundles of xylem and phloem in the leaf 'skeleton' of a partially decayed leaf in autumn.

○ Leaf

The leaf is a plant organ and is the major site of photosynthesis. It also controls the flow of water through the plant. Previously you learnt that water is absorbed by osmosis from the soil into the roots. It is then 'pulled' through the plant by the **transpiration stream** because it is continuously being released from the leaves through stomata, which open and close to regulate this process.

Show you can...

Explain why animals often prefer to eat shoots rather than roots.

Test yourself

4 Name the process by which water is absorbed into the roots.
5 What are the two functions of roots?
6 Where are meristems of shoots found?
7 Explain why roots are often white.

Plant organ systems

○ Transportation organ system

You have already learnt that xylem and phloem are tissues, and that roots, shoots and leaves are plant organs. These combine to make the plant transportation organ system, which transports all substances around a plant.

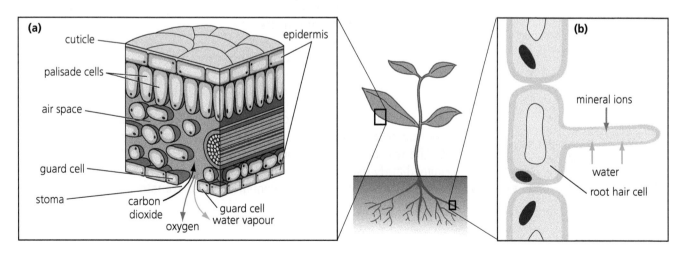

▲ **Figure 5.6** (a) A cross-section of a leaf showing the large surface area for gas exchange provided by the internal air spaces, and the movement of materials through the stomata. (b) Root hair cells give the roots a large surface area for absorption.

Transpiration and the transpiration stream

Water enters root hair cells in plant roots by osmosis. It then travels by osmosis through the cells of the root and then enters xylem cells. It travels up through the root and stem in long continuous columns of xylem cells. Eventually the xylem branches to form veins that carry the water to the leaves, where it enters the leaf cells.

Much of this water evaporates out of the leaf cells (mainly the spongy mesophyll cells) and enters the leaf air spaces as water vapour. This then diffuses out of the leaf through the air spaces and stomata. This is a continuous process, and the loss of water from a plant through the leaves is called transpiration. The constant evaporation of water from the leaves pulls, or 'sucks', the water up through the rest of the plant in a long, unbroken transpiration stream.

Transpiration has a number of functions, including:

- providing water for leaf cells and other cells (e.g. to keep them **turgid**)
- providing water to cells for the process of **photosynthesis**
- transporting **minerals** to the leaves.

Diffusion of any substance happens faster if the concentration gradient is greater (that is, the difference between the high and low concentrations is bigger). If the air surrounding a leaf is very humid (like just before a thunderstorm) then the water vapour gradient will be less steep so the rate of transpiration will be lower. On windy days the air surrounding the leaves is continually replaced. This keeps the concentration gradient steep and the rate of transpiration high. When temperatures are higher the rate of evaporation of water is higher and so transpiration occurs more rapidly. Water is also used up more rapidly during the daylight hours as some of it is used to make glucose by photosynthesis, so transpiration is increased. Also, the stomata are more likely to be open during the day.

In summary, high rates of transpiration are achieved when:

- there is **more wind**
- there is a **high temperature**
- the air is **less humid**
- the **light intensity is high** (during the day).

Translocation

Phloem tissue is also part of the transport organ system. Phloem transports dissolved sugars that are made in the leaves by photosynthesis to the rest of the plant. The transported sugar is usually either immediately used in respiration or stored as starch. The movement of dissolved food through the phloem is called **translocation**.

Activity

Investigating transpiration

A class were investigating water loss from plants and wanted to compare the amount of water lost from the upper and lower surfaces of a leaf. Four leaves of similar sizes were selected from a bush and their surface areas estimated by drawing around them on squared paper.

A thin layer of petroleum jelly was used to cover the stalks of the leaves and applied to the epidermises of some of the leaves, as shown in Table 5.1. The leaves were weighed and their starting mass recorded. They were then each hung on a piece of string using a paper clip attached to the stalk and left undisturbed in the light on a windowsill in the classroom. After an hour the leaves were re-weighed.

Table 5.1

Treatment	Estimated surface area in cm²	Starting mass in g	End mass in g	Change in mass in g
A No petroleum jelly	50	1.34	1.14	
B Petroleum jelly on the upper surface	55	1.46	1.38	
C Petroleum jelly on the lower surface	52	1.42	1.38	
D Petroleum jelly on both sides		1.56	1.55	

Questions

1 Copy and complete Table 5.1 by estimating the surface area of leaf D in Figure 5.7.
2 Determine the change in mass for each leaf.
3 Give an explanation for the class's results.
4 Calculate the water loss for leaf A in g per cm².
5 What was the independent variable in this experiment?
6 What was the dependent variable?
7 Name a variable that should have been controlled in this experiment but was not.

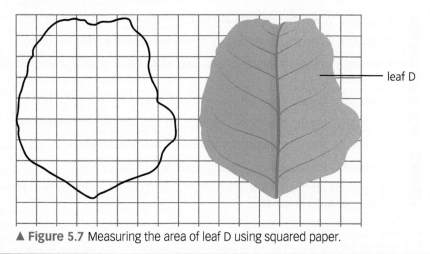

▲ **Figure 5.7** Measuring the area of leaf D using squared paper.

Show you can...

Describe the conditions in which transpiration increases.

Test yourself

8 Name the plant organ from which water evaporates.
9 Name a plant organ system.
10 Describe the diffusion of oxygen from plant leaves.

Chapter review questions

1 Describe where in a leaf the spongy mesophyll layer is found.

2 Describe an adaptation of the palisade mesophyll cells.

3 Name the tissue in which water moves up a plant from its roots.

4 Explain why most roots are white.

5 Describe two functions of roots.

6 Name the tissue in a plant that produces stem cells.

7 Name the process by which plants absorb mineral ions from the soil.

8 Give an example of a plant organ.

9 Name the type of cell in which most chloroplasts are found.

10 Name the organs that make up the plant transportation organ system.

11 Name the process by which water enters a plant root.

12 Describe the difference in structure of palisade mesophyll and spongy mesophyll tissues.

13 Describe the location and function of guard cells.

14 Describe an experiment in which you could use nail varnish to investigate the number of stomata on different plant leaves.

15 Name two tissues involved in transportation in a plant.

16 Suggest where you might find a plant with green roots.

17 Describe the process of transpiration.

18 Explain why more stomata are found on the lower surface of leaves.

19 Explain when you might expect all the stomata of a plant to be open.

20 Explain why plants must continuously allow water to evaporate from their leaves.

21 Explain how increasing humidity affects the rate of transpiration.

22 Explain how decreasing temperature affects the rate of transpiration.

23 Suggest when during the day transpiration is most likely to be highest.

24 Explain why more transpiration happens on a windy day.

25 Describe an experiment in which you could use petroleum jelly to investigate transpiration in leaves.

Practice questions

1 Figure 5.8 shows a cross-section part of a leaf.

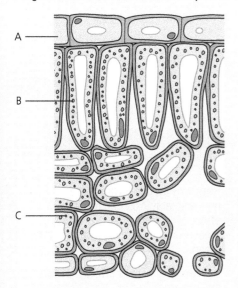

▲ **Figure 5.8**

a) Copy and complete Table 5.2 by identifying
   tissues A–C. [3 marks]

Table 5.2

Tissue	Letter
Spongy mesophyll	
Epidermis	
Palisade mesophyll	

b) Two other tissues found in plants are xylem and
   phloem. These are often found together in bundles
   and have an important role in transporting substances
   around plants. Name something that is transported by:

   i) the xylem [1 mark]

   ii) the phloem. [1 mark]

2 Figure 5.9 shows the arrangement of stomata on the
   underside of the leaves of two species of plant. Each
   square represents 0.02 mm² of leaf.

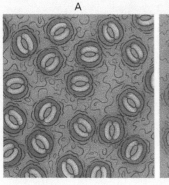
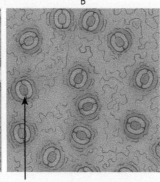

▲ **Figure 5.9**

a) i) Name cell X. [1 mark]

   ii) What is the role of cell X? [1 mark]

b) Do you think species A or B is adapted to live in a
   drier habitat? Explain your reason. [2 marks]

c) Suggest another adaptation that the leaves might have
   to help them survive in a dry habitat. [1 mark]

d) Calculate the number of stomata per 1 mm² of leaf
   epidermis for species B. Show your working. [2 marks]

3 Figure 5.10 shows an experiment involving water loss
   in plants.

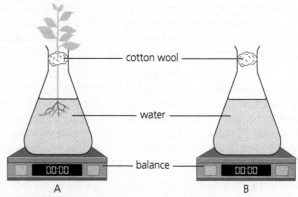

▲ **Figure 5.10**

a) Flasks A and B were weighed and shown to have the same
   mass at the start of the experiment. After 10 minutes they
   were re-weighed. Suggest what you would expect the mass
   of flask A to be compared to that of flask B at the end of
   the experiment. [1 mark]

b) Why was flask B needed? [1 mark]

c) Suggest why cotton wool was placed in each flask.
   [1 mark]

d) What is the name of the process by which leaves lose
   water vapour? [1 mark]

   A) Transportation

   B) Transformation

   C) Translocation

   D) Transpiration

e) Describe how this process occurs in a plant. [3 marks]

f) In which of these conditions would water loss from a
   plant be greatest? [1 mark]

   A) Hot and humid conditions

   B) Cold and humid conditions

   C) Hot and dry conditions

   D) Cold and dry conditions

Working scientifically:
Dealing with data

Understanding error

There are often differences in the results obtained in an experiment caused by different types of error. A random error is usually caused by a mistake being made by the person carrying out the experiment, a change in the measuring instrument or a change in the environment that was not controlled. Random errors cause the result to vary in an unpredictable way, spreading around the true value. We can reduce the effect of random error by carrying out more repeats and calculating a mean.

A systematic error causes the readings to differ from the true value by a consistent amount each time a measurement is made. These types of error usually come from the measuring instrument, either because it is incorrectly calibrated or because it is being used incorrectly by the experimenter. Systematic errors cannot be dealt with by more readings; instead the whole data collection needs to be repeated using a different technique or a different set of equipment.

Four students were asked to examine error by using a piece of equipment called a potometer. A potometer measures the rate of transpiration in plants. As water is lost from the leaves the plant draws up water to replace it. By measuring the distance moved by an air bubble over a set period of time the rate of transpiration can be measured. This allows transpiration rates under different conditions to be compared.

The students were first asked to set up their potometers and make a mark where the air bubble was at the start of the experiment. After 10 minutes they recorded the distance their bubble had moved. They repeated this process three times to determine a mean result.

Their results are shown in Table 5.3.

Table 5.3 The distance moved by an air bubble in a class potometer experiment.

Student	Distance moved by the air bubble in cm			
	Repeat 1	Repeat 2	Repeat 3	Mean
Amy	2.1	2.0	2.2	2.1
Chris	1.4	1.4	1.6	1.5
PJ	2.2	2.3	2.1	2.2
Kelly	1.4	2.8	2.4	2.2

Questions

1 Which set of data contains more errors? How do you know this from the data?

2 Chris's data were lower than those of the other members of the group, so the teacher asked him to show how he made his measurements. Figure 5.11 is a picture of how Chris recorded the distance moved, and how Amy did. What did Chris do wrong? What type of error is this? What should he now do?

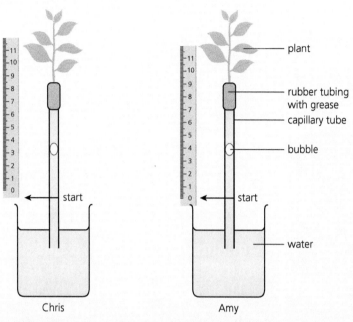

Chris   Amy

▲ **Figure 5.11** The results of an experiment using a potometer.

3 The teacher asked Kelly to explain how she took her readings. She explained that she took the first reading sitting down, the second standing up, and the third level with the bubble. Why did her method for measuring the bubble introduce errors? What should she have done? Draw a diagram to explain why changing her position would have led to errors in her readings.

4 Why did the rubber tube have grease smeared around it? If this was not present how could it lead to errors in the data?

5 Identify any other potential sources of error in the experiment.

6 The students repeated their experiment with three different conditions:

   a) on a bench with a lamp shining on the plant

   b) on a bench with a fan blowing onto the plant

   c) on a bench with a plastic bag around the plant, creating a humid environment.

   For each condition, explain how you think the result would be different and why.

We catch communicable diseases from infected people. They are contagious. HIV/AIDS currently kills over a million people per year and estimates suggest that it has killed about 39 million people in total so far. History reveals much higher numbers of deaths from other communicable diseases. Smallpox is likely to have killed more than 300 million people in the 20th century alone. Spanish flu killed between 50 and 100 million in the same timescale. The Black Death killed around 75 million, reducing the world's population to around 350 million in the 14th century.

Specification coverage

This chapter covers specification points 4.3.1.1 to 4.3.1.9 and is called Infection and response.

It covers communicable diseases caused by viruses, bacteria, fungi and protists. It also covers human defence systems, vaccination, and the discovery and development of drugs, including antibiotics and painkillers.

Communicable (infectious) diseases

KEY TERMS

Communicable A disease that can be transmitted from one organism to another.

Infectious Describes a pathogen that can easily be transmitted, or an infected person who can pass on the disease.

TIPS ✔

- It is important that you can explain how diseases caused by viruses, bacteria, fungi and protists are spread in animals and plants and how this can be prevented.
- Protista is a large group (kingdom) of eukaryotic organisms, each of which is called a protist.

▲ **Figure 6.1** Signs like this warn people to stay at home if they catch the highly contagious norovirus.

A **pathogen** is any microorganism that passes a communicable disease from one organism to another. There are four main types of microorganism that cause disease:

1 **viruses,** e.g. measles

2 **bacteria,** e.g. salmonella

3 **fungi,** e.g. rose black spot in plants

4 **protists,** e.g. malaria.

All four types of pathogen have a simple life cycle. They infect a host, reproduce (or replicate in the case of viruses), spread from their host and infect other organisms. This process then repeats. Many pathogens can survive without their host for a short period of time, but these are unable to reproduce without their host.

Pathogens are highly adapted to their role. They are very easily passed from one organism to another. We call these highly infectious. An example of a highly infectious pathogen is the measles virus. This is particularly easily spread because it is transmitted in the air. Other pathogens, such as the norovirus (winter vomiting bug), can reproduce very quickly. Others can survive for long periods without a host. An example of this is the *Staphylococcus* bacterium.

Test yourself

1 Give an example of a viral pathogen.
2 Give an example of a fungal pathogen.
3 Define the expression 'highly infectious'.
4 Explain the advantage to a pathogen of reproducing quickly.

Show you can...

Describe the life cycle of an infecting pathogen.

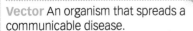

Spread of pathogens

▲ **Figure 6.2** Tiny droplets produced when we sneeze can spread pathogens through the air.

Pathogens have evolved many different ways of passing from one organism to another.

- **Airborne:** the common cold virus is often spread in tiny droplets of water propelled through the air when an infected person sneezes.
- **Through dirty water:** the cholera bacterium is often spread in unsterilised water.
- **By direct physical contact:** this can be sexual or non-sexual. Chlamydia is a bacterial pathogen that is one of the most common sexually transmitted diseases (STDs) in the world. Without treatment with antibiotics this can lead to serious reproductive problems.
- **Through contaminated food:** the *Escherichia coli* bacterium is often spread in uncooked or reheated food. It causes food poisoning.
- **Passed by another animal:** some farmers in the UK believe that badgers can catch the tuberculosis bacterium and pass it to their cattle. We call any organism that does this a vector.

Test yourself

5 Name the method by which cholera is often spread.
6 Describe how tuberculosis is spread.

Show you can...

Describe how the four different types of pathogen could be spread, giving an example of each in your answer.

Viral diseases

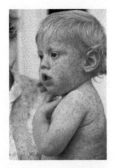

▲ **Figure 6.3** This small boy has measles.

Viral diseases are those caused by a virus. They are the smallest pathogens and the most simple, made from a strand of genetic material (DNA or RNA) surrounded by a protein coat. They infect a cell in a host and use it to copy their genetic material and protein coats. These are then assembled into new virus particles. At this point cells are often full to bursting with new viruses, and they then split open to release new infecting particles to repeat the cycle.

○ Measles

Measles is a highly infectious, common viral disease usually transmitted between young children. Its transmission is airborne, so it is passed in **tiny droplets** when an infected person sneezes. These are breathed in by those around them, who then may become infected. Its symptoms include a **fever** and a **red skin rash**. In developed countries babies are immunised (vaccinated) against this infection. It can infect adults that were not immunised or did not catch it as a child. More serious **medical complications** can then occur including sterility in adults and foetal abnormalities in pregnant women.

○ HIV/AIDS

HIV stands for human immunodeficiency virus. This is transmitted when **body fluids are shared**, often during sex, or by shared use of needles by drug users. It can be passed from mother to child in the uterus, during birth or in breast milk. Immediately after infection, the symptoms are like those of **flu**. After this, infected people usually show no symptoms. In fact many may not know they are infected at all. Months or years after infection, the virus attacks the body's immune cells. The virus enters the **lymph nodes** and destroys a particular type of immune cell. This means infected people are then less able to fight off **infections** such as the tuberculosis bacterium, and they are also more prone to a range of cancers. At this point the disease is called late-stage HIV or acquired immune deficiency syndrome **(AIDS)**. There is currently no cure for HIV/AIDS. Infected people are given **antiretroviral drugs,** which can slow the development of the disease.

○ Tobacco mosaic virus

The tobacco mosaic virus is a virus that affects the tobacco plant and many other related species including tomatoes.

The leaves of infected plants develop a mottled or **mosaic** appearance with the patches of normal tissue alternating with patches that have a light green or yellow appearance, as shown in Figure 6.4. Leaves are often wrinkled and curly.

The virus enters and damages chloroplasts in parts of the leaf, and this explains the loss of green colour and the yellowing of leaves.

As the chloroplasts and leaves of infected plants are damaged, the plants **cannot photosynthesise** as effectively and poor growth results. This obviously affects the profit of the crop growers.

Tobacco mosaic virus infects plants by entering through parts that have been damaged. The virus is easily spread from plant to plant on tools and on the clothes or hands of farm workers. It can also spread through contact between an uninfected and an infected plant (e.g. an infected plant being blown against a neighbouring plant in the wind). Occasionally, insects feeding on the plants may cause infection.

There is no cure, therefore the emphasis is on prevention. Measures to reduce infection include:

● using strains that are **resistant** to the virus
● **removing infected plants** immediately the infection is spotted
● **removing weeds** in close proximity to the crops as they can harbour the virus.

▲ **Figure 6.4** Leaves infected with tobacco mosaic virus.

Bacterial diseases

Bacterial diseases are caused by pathogenic bacteria. It is worth remembering that there are many useful bacteria that are not pathogens, including those that live in your digestive system and help you digest your food. All bacteria are prokaryotes. They are larger than viruses but are still only visible using a microscope. They live inside their hosts, often in mouths, noses and throats, but not inside cells as viruses do. Many pathogens produce toxins (poisons) as they grow, which irritate the surrounding cells of the host.

KEY TERM

Toxin A poison that damages tissues and makes us feel ill.

Salmonella

Salmonella is normally spread in food that has been prepared in unhygienic conditions, that has not been cooked well enough or that has been kept too long. We become infected if we eat food containing salmonella bacteria. In the UK, poultry are vaccinated against salmonella, to control its spread. Salmonella and its toxins cause **fever**, **abdominal cramps**, **vomiting** and **diarrhoea** in humans. Prevention measures include **vaccinating poultry** against salmonella, **cooking food thoroughly** and preparing food in **hygienic conditions**.

▲ Figure 6.5 This person is suffering from food poisoning caused by salmonella.

Gonorrhoea

Gonorrhoea is a sexually transmitted disease (STD) caused by a species of bacterium. Symptoms include a painful burning sensation when urinating and the production of a **thick yellow or green fluid** (discharge) from the vagina or penis. It is spread by **sexual contact**. Prevention measures include the use of a **barrier method of contraception** such as a condom, because this prevents the bacteria passing from person to person. Gonorrhoea can be treated with **antibiotics**, although some **resistant** strains of bacteria have evolved, making treatment more difficult.

Activity

Using a key to identify bacterial species

Use the statement key below to identify the six species of bacteria.

1 It is made of more than one cell .... Go to question 2.

It is a single cell .... Go to question 4.

2 It is made of two cells .... *Streptococcus pneumoniae.*

It is made of more than two cells .... Go to question 3.

3 Bacteria arranged in chains .... *Streptococcus pyogenes.*

Bacteria arranged in a cluster............................. *Staphylococcus aureus.*

4 It has no flagella .... *Treponema pallidum.*

It has flagella .... Go to question 5.

5 Its flagella are spread all around the cell .... *Salmonella typhi.*

The flagella are located to one side of the cell .... *Helicobacter pylori.*

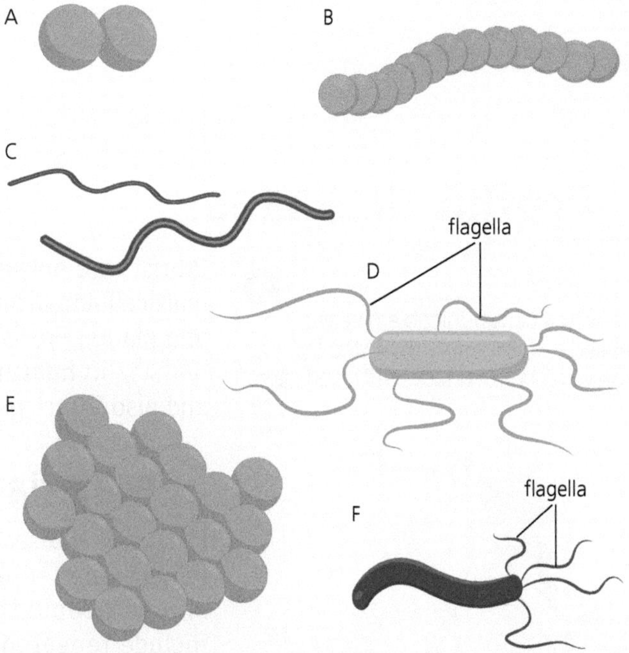

▲ **Figure 6.6** Bacterial species.

Fungal diseases

Fungi are eukaryotic, like animals and plants, but unlike bacteria. They have evolved a huge range of appearances, from the single-celled yeast fungus to much larger multicellular mushrooms. Fungi can cause many diseases, including athlete's foot in humans and rose black spot in roses.

○ Rose black spot

The fungus that causes this plant disease infects the leaves of the rose plant, causing them to develop **purple** or **black spots** which often turn yellow and fall off. Because the leaves are damaged, there is reduced photosynthesis and less growth. Fungal spores can travel by **wind** or **water** (e.g. rain splash) from plant to plant, causing the infection to spread. Rose black spot can be treated by using **fungicides** (chemicals that kill fungi) and **removing and destroying infected leaves** once they are first noticed.

Protist diseases

Protists are eukaryotic microorganisms. They are always unicellular or multicellular without tissues. This separates them from fungi, animals and plants. Protists are perhaps best known as pathogens for causing malaria in humans, but they also cause similar diseases in other animals and also infect plants.

○ Malaria

Malaria is a disease caused by *Plasmodium* protists. Approximately 200 million cases of malaria occur each year the world over. About half a million people then die from this disease each year. Its symptoms include repeated episodes of **fever**, and can eventually result in death. The *Plasmodium* (protist) pathogens are transmitted from one individual to another by **mosquitos**. Because the mosquito transmits the protist (which causes malaria) we call the mosquito a **vector**. Mosquitos bite and suck blood from organisms infected by *Plasmodium* and then pass the pathogen from their saliva to the blood of all other organisms they bite.

▲ **Figure 6.7** Is this mosquito about to transmit malaria to the person it bites?

Prevention is usually by avoiding being bitten. **Mosquito nets** and insect repellent sprays containing insecticides are often used. Mosquitos lay eggs in water that does not move, such as stagnant pools or puddles. To help stop the spread of malaria these can be filled in. No vaccination currently exists. Draining marshy areas can also destroy mosquito breeding sites.

Human defence systems

Your defence against pathogens can be divided into two key areas. We call these 'lines of defence' against infection by pathogens. Your first line of defence is preventing the pathogen entering, and your second line of defence is controlling pathogens once they have entered your bloodstream or tissues.

○ The first line of defence

The first line of defence is your body's natural barriers to infection. These are not specific to the infecting pathogen, and so we describe them as **non-specific**.

Skin

Your skin is an amazing organ that almost completely covers any outer part of you that is prone to attack from pathogens. As well as this, it insulates you, helps you regulate your temperature and is involved in the way your senses provide you with information. When we get a cut, the blood clots and the wound quickly seals over to restore the protective barrier of the skin.

In vulnerable parts of our bodies, we can produce an antimicrobial secretion that prevents pathogens entering. For example, our tears contain antimicrobial agents to protect our eyes.

Hairs and cilia

Pathogens that are breathed in through your nose and mouth are often stopped before they reach your lungs. Your **nose** has hairs and produces mucus, which acts as a filter, stopping larger particles containing pathogens. You blow your nose or sniff and swallow, moving this mucus and pathogens either out of your body or into your stomach. Hairs are physical barriers against infection.

If pathogens pass your saliva or hairs in your nose, they are often stopped by the ciliated cells lining the inside of your **trachea** and **bronchi,** the tubes that reach down to your lungs. Ciliated cells possess cilia, which are tiny hair-like projections that protrude into the airway. In between the ciliated cells are cells that produce mucus, which they pump into the airway. Many pathogens and other particles that have been breathed in get stuck in this mucus. The ciliated cells beat (or waft) their cilia in a rhythmical pattern, which propels the mucus back up the airway to the back of the throat. Cilia are physical barriers against infection.

Stomach acid

Your stomach acid does not actually digest your food. It provides the correct pH for protease enzymes to start digesting protein, but it also has a crucial role in the first line of defence. It is hydrochloric acid and is strong enough to kill many bacterial pathogens that enter your body through your mouth or nose. It is a chemical barrier against infection.

○ The second line of defence

The second line of defence is your defence against pathogens that have entered your bloodstream or tissues.

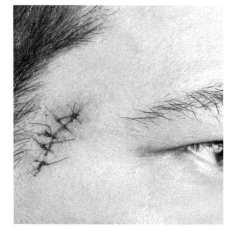

▲ **Figure 6.8** This cut is held together by stitches. It will soon scab over and eventually leave a small scar.

KEY TERMS ⭐

Antimicrobial secretions Chemicals that destroy pathogens.

Cilia Tiny hair-like projections from ciliated cells that waft mucus out of the gas exchange system.

Mucus A sticky substance that traps pathogens.

TIP ✓

The movement of cilia is like a Mexican wave in a football match.

TIPS ✓

- To help you revise this subject, draw an outline of a person on an A4 page and add the details of the first line of defence around the outside.
- It is important that you can explain the role of the immune system in defence against disease.

Lymphocytes

Lymphocytes are a type of white blood cell. There are between 1.5 and 4.5 million per litre of blood.

Almost all cells have markers (proteins) called antigens on their surface. Pathogens have different antigens from those found in your own cells. Your lymphocytes recognise the foreign antigens present on invading pathogens. When the lymphocyte identifies foreign antigens, it produces large numbers of antibodies. This takes several days, during which time you may feel ill. The antibodies are highly specific for the antigen present on the pathogen, which means that the shape of the antibody fits perfectly with the shape of the antigen. Antibodies help clump pathogens together and stop them moving around the body and causing harm. Several days later enough of the pathogen will normally have been destroyed and you are likely to be feeling better.

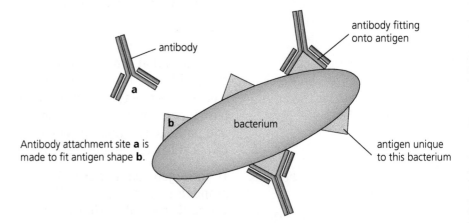

▲ **Figure 6.9** How an antibody fits onto the antigen of a pathogen.

If you are infected a second time with the same pathogen, a special type of lymphocyte will recognise its antigens and be able to produce larger numbers of antibodies more quickly. We call these white blood cells '**memory**' lymphocytes, and we produce these after we have been infected by a 'new' pathogen. This is likely to mean you won't fall ill from the same pathogen twice. How do we get colds every winter, then? This is because there are several hundred different strains of common cold viruses that all have different antigens. You are not likely to have been infected by the same one, but lots of different ones that have similar symptoms.

Phagocytes

Phagocytes are another a type of **white blood cell**. They take in or **engulf** pathogens, as well as your own dead or dying cells. Phagocytes are attracted to any area of your body in which an infection is present. Antibodies cause pathogens to clump together. When a phagocyte comes into contact with a pathogen it binds to it. The membrane of the phagocyte then surrounds the pathogen and absorbs it into a vacuole within its cytoplasm. Enzymes are added to the vacuole to break down the pathogen. This process in which phagocytes break down pathogens clumped together by antibodies is called **phagocytosis**.

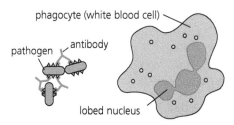

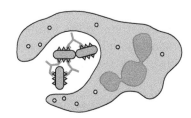

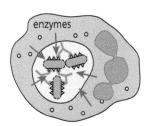

(a) Antibodies cause pathogens to clump together

(b) Phagocyte flows around pathogens to engulf them in a vacuole

(c) Enzymes added to vacuole to break down pathogen cell walls and membranes

▲ **Figure 6.10** A white blood cell engulfing pathogens.

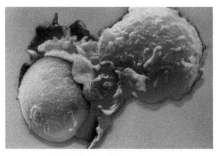

▲ **Figure 6.11** A phagocyte (white blood cell) engulfs a yeast cell before destroying it.

> **TIP** ✔
>
> Some people get the terms 'antibody', 'antigen' and 'antitoxin' mixed up. Make sure you can tell the difference between them.

Antitoxins

Many pathogens produce toxins that also make you ill (as well as the pathogen itself). To defend against these specific toxins, your lymphocytes can produce a special type of antibody called an antitoxin. This will bind with and neutralise the toxin helping you feel better.

> **KEY TERM**
>
> **Antitoxin** A protein produced by your body to neutralise harmful toxins produced by pathogens.

> **Test yourself**
>
> 20 Give an example of the first line of defence.
> 21 Name the two types of white blood cell.
> 22 Describe how stomach acid stops infection.
> 23 Describe the difference between chemical and physical barriers.

Vaccination

> **KEY TERM** ⭐
>
> **Vaccine** A medicine containing an antigen from a pathogen that triggers a low level immune response so that subsequent infection is dealt with more effectively by the body's own immune system.

> **TIPS** ✔
>
> - It is important that you can explain the use of vaccinations to prevent disease. You do not need to know specific times or dates for getting vaccines, though, or their side effects.
> - The term 'immunisation' means 'becoming immune to a pathogen'. This can be caused by vaccination but also naturally if you become infected by a pathogen and subsequently recover.

It is likely that you will have been vaccinated against a number of diseases since you were born. Typically you may have had the measles, mumps and rubella (MMR) vaccine at about 12 months old, and a second combined vaccine for diphtheria, tetanus, whooping cough and polio at about 3 years. These are all life-threatening diseases, and so vaccination is important.

If the vast majority of people in a population have a vaccination, then even if a small number of people become infected the disease is not likely to spread. This is called **herd immunity**. The reverse is also true. If few people have a vaccine and a small number become infected the disease will spread much more quickly.

A vaccine is a small quantity of a dead, inactive or genetically modified version of a pathogen. Crucially it must have the same antigens as the pathogen, or your body would not recognise the pathogen later. You are injected with this and your immune response begins. Lymphocytes produce antibodies and antitoxins. Because this process takes several days, you may feel slightly unwell after a vaccination.

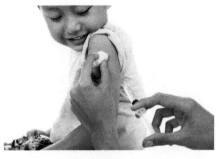

▲ **Figure 6.12** It is likely that you will have already had a number of vaccinations.

TIP ✓

Remember a primary immune response can also occur if you are infected by a pathogen for the first time.

However, if the same pathogen were to get past your first line of defence and infect you in the future, your 'memory' lymphocytes would respond by producing antibodies and antitoxins fast enough for you not to fall sick. This is your **secondary immune reponse**. Thus, vaccinations prepare your immune system in case you are infected later in life.

For many vaccines, you are likely to have had booster injections several years after the first injections. These serve as a timely reminder for your immune system and 'refresh' your memory lymphocytes.

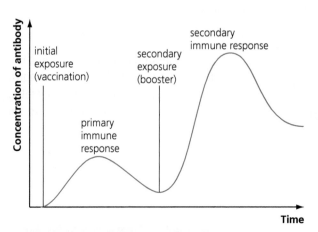

▲ **Figure 6.13** This graph shows the rate at which antibodies are produced after the first and second exposures to a pathogen.

Show you can...

Look at the graph in Figure 6.13 and explain the differences in the primary immune response and the secondary immune response.

Test yourself

24 What does 'MMR' in the MMR vaccine stand for?
25 Which cells produce antibodies?
26 Define the term 'vaccine'.
27 Describe the difference between antigens and antibodies.

Antibiotics and painkillers

TIP ✓

It is important that you are aware that it is difficult to develop drugs that kill viruses without harming our own tissues.

TIP ✓

It is important that you can explain the use of antibiotics and other medicines in treating disease.

KEY TERM ★

Antibiotic A medicine that kills bacteria (but not viruses) inside the body.

A drug is any substance that has a biological effect on the organism taking it. We do not normally include foods in this definition, even though many do affect people. Some drugs are natural, such as nicotine in tobacco. Others are manufactured, such as Viagra. Some have a positive effect on the taker, such as medicines, while others have negative side effects, such as cocaine. Some drugs are recreational. They are taken to alter a person's mood or emotions. Legal recreational drugs include caffeine and nicotine. Illegal recreational drugs include cannabis and cocaine.

○ Antibiotics

Antibiotics are a very important group of medicines which **kill bacteria**. They have **no effect on viruses** and so should not be prescribed for the common cold or other viral diseases.

Different antibiotics attack bacteria in different ways. Penicillin (see below) is a commonly used antibiotic. It makes the cell walls of the bacteria weaker, and so they burst and are killed. Other antibiotics alter bacterial enzymes, and others stop bacteria from reproducing.

Penicillin and Alexander Fleming

Sir Alexander Fleming first discovered penicillin in 1928. He won the Nobel Prize for this discovery and penicillin has saved countless numbers of lives ever since. Fleming returned to his laboratory where he was studying bacterial growth. Legend has it that one of his Petri dishes was mistakenly left open and had accidently been contaminated by the fungus *Penicillium notatum*. Where the fungus grew the bacteria did not. Instead of simply throwing the Petri dish away, Fleming realised that the fungus was naturally producing a chemical that killed bacteria. This chemical was eventually refined to become the first antibiotic drug, penicillin.

Antibiotic resistance

Since Fleming's discovery scientists have developed a large number of other antibiotics. These have saved many, many lives in recent years and led to the near removal of some major diseases such as tuberculosis. However, until very recently, we had not discovered any new antibiotics in 30 years. During this time some pathogens have been evolving to be resistant to our antibiotics. Antibiotic-resistant bacteria are difficult to treat, which is a major worry for doctors and scientists.

Drug companies are working extremely hard to find new antibiotics or alternatives to them.

Although antibiotics are used to kill bacteria, the fact that they do not kill viruses makes it hard to treat viral diseases. Also, because pathogenic viruses live in **body cells** and **tissues**, it is difficult to kill the viruses without damaging the cells.

▲ **Figure 6.14** This World War 2 advert shows that gonorrhoea can be treated by antibiotics.

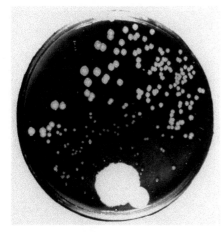

▲ **Figure 6.15** This is the original Petri dish that was contaminated by the *Penicillium notatum* fungus. Would you have looked at this and made one of the world's most important medical discoveries?

> **KEY TERM** ⭐
>
> **Antibiotic-resistant bacterium**
> A bacterium that cannot be killed by antibiotics.

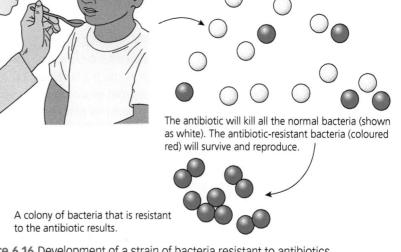

A sick child is being given an antibiotic.

The antibiotic will kill all the normal bacteria (shown as white). The antibiotic-resistant bacteria (coloured red) will survive and reproduce.

A colony of bacteria that is resistant to the antibiotic results.

▲ **Figure 6.16** Development of a strain of bacteria resistant to antibiotics.

KEY TERM ★

Painkiller A drug that treats the symptoms of disease, such as a headache, but does not kill any pathogens that may be causing the disease.

◯ Painkillers

Painkillers are drugs that **relieve pain** but do **not** kill the pathogens. Some painkillers are naturally found in plants. **Aspirin** comes from the bark of the **willow tree**. The naturally occurring compound is called salicin and was first discovered in 1763. It was not manufactured as aspirin until 1897. It is now one of the most widely used medicines in the world, with over 40 000 tonnes consumed each year. As well as relieving pain, it can reduce fever, swelling and inflammation. It is also used as a preventative drug for reducing the risk of heart attacks.

Other painkillers have been manufactured. A second common painkiller is **paracetamol**. Like aspirin, this is a mild painkiller often used to stop headaches or minor pain in other parts of the body. It is a major ingredient in many cold and flu remedies. Unfortunately, it is easy to take too much paracetamol at a time, which can cause fatal liver damage. This is why it is very important not to take more than the stated dose.

Both aspirin and paracetamol are 'over-the-counter' medicines, which means you can buy them in small numbers in chemists and supermarkets. Other stronger painkillers exist, such as **tramadol** and **morphine**. The use of these can only be prescribed by a doctor or used in accident and emergency situations.

▲ **Figure 6.17** Aspirin occurs as salicin in the bark of willow trees.

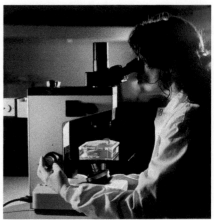

▶ **Figure 6.18** Looking at the effect of a drug on live cells.

Show you can...

Explain what was significant about the manner in which Fleming made his discovery.

Test yourself

28 Name the first antibiotic.
29 From which organism does aspirin come?
30 Describe another medical use for aspirin.

Discovery and development of drugs

In the course of our history we have found naturally occurring drugs and manufactured new ones. As we have seen, the discovery of some was by accident and others were designed.

Traditionally, drugs were extracted from plants and microorganisms, for example the **heart drug digitalis** originates from **foxgloves**. Digitalis can be used to treat people with irregular heartbeats. Now scientists in the pharmaceutical industry make new drugs, although the starting point is often a plant compound.

○ Modern drug development

Drug development is the process of identifying a new drug, testing it and then manufacturing it for sale. This is a very costly and time-consuming process. To get a drug to the stage at which it can be tested, the development is likely to take many years and hundreds of millions of pounds. Drug companies are some of the largest companies in the world. Many potential drugs don't pass though the stages described below. From perhaps ten thousand possible drugs only 10 (0.1%) may ever get to be tested on humans.

The first stage of drug development involves **computer modelling**. The structure of the drug and the interactions it might have on naturally occurring substances in the body are looked at on a computer.

This initial stage of drug development also involves testing in the laboratory. This may include testing on **live cells** taken from an organism and grown in a Petri dish. Once it appears that the drug may be able to work on live cells in a laboratory, it is then tested on **animals**. This is a very important stage, because it allows the scientists to test the drug on a **living organism** without risking testing it on humans at this stage. These first stages of drug testing are called **preclinical** testing.

The final stage involves **clinical trials** on humans. There are three phases here. In the first, small amounts of the drug are tested on a small number of **healthy volunteers** to determine whether it is toxic and has any side effects, and to identify safe dosing volumes. In the second phase the drug is given to **small numbers of sick patients** to test how well it works (its efficacy). Finally, in phase three trials it is given to a **large number of patients** to finalise safe doses and efficacy. If a drug passes all of these stages it will be given a **licence** and then can be manufactured and sold.

Some of these trials are described as double blind. This is because the patients are randomly allocated to receive either the drug or a placebo (which looks like the drug but does not contain it) and the doctors also don't know which patients are receiving which. This means those patients who receive the placebo are in effect a control group.

Test yourself

31 What is the first stage of drug development?

Show you can...

Describe how drug development occurs.

Chapter review questions

1 Define the term 'communicable disease'.
2 Name the four types of pathogen that can cause communicable diseases.
3 Give an example of a disease caused by a fungus.
4 Give an example of a disease caused by a protist.
5 What does HIV/AIDS stand for?
6 Name the scientist who first discovered antibiotics.
7 Name the microorganisms upon which antibiotics don't work.
8 Explain why medicines are often expensive.

9 How is salmonella often transmitted?
10 What are the symptoms of gonorrhoea?
11 What are the symptoms of malaria?
12 Describe how the transmission of malaria is reduced.
13 Describe how stomach acid acts in the first line of defence.
14 Explain how your immune system helps you 'remember' your previous infections.
15 Describe what a vaccine is.
16 Explain the idea of herd immunity.
17 Explain the purpose of booster injections.
18 Describe how antibiotics were first discovered.

19 Describe how cilia and mucus-producing cells prevent infection.
20 Describe how phagocytes prevent infection.
21 Describe how lymphocytes prevent infection.
22 What is the first step in drug development?
23 Define the term 'efficacy'.
24 Explain what a double blind trial is.

Practice questions

1 Figure 6.19 shows data from a drug trial involving a new asthma medicine called Breathrite.

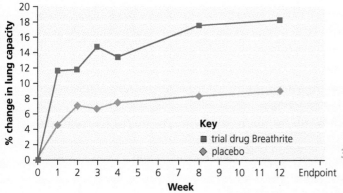

▲ Figure 6.19

a) Before the drug was trialled using people it was tested in the laboratory. What is the purpose of these tests? [1 mark]

b) The drug trial used a double blind test. Doctors tested the effectiveness of Breathrite against a placebo using two groups of volunteers over 12 weeks.

   i) What is a placebo? [1 mark]

   ii) Why is a placebo used? [1 mark]

   iii) Describe what a double blind trial means. [1 mark]

c) It is important that the two groups of volunteers are similar. Give one factor that should be similar in both groups. [1 mark]

d) Over the 12-week period, how much did the lung capacity of the volunteers increase in those who were given Breathrite? [1 mark]

2 In 2014 the largest Ebola epidemic in history broke out and affected a number of countries in West Africa. Ebola is caused by the *Ebola* virus.

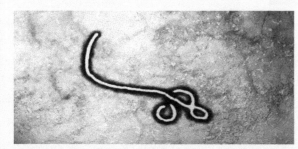

▲ Figure 6.20 The *Ebola* virus.

a) Give the term given to organisms like *Ebola* that cause disease. [1 mark]

b) Describe how viruses like *Ebola* cause illness. [1 mark]

c) Currently there are no cures for Ebola.

   i) Why don't antibiotics get rid of Ebola? [1 mark]

   ii) Some of the symptoms of Ebola are headaches and muscle pain. Suggest what might be given to someone suffering from Ebola to relieve the symptoms. [1 mark]

d) Scientists are trying to create a vaccine for Ebola by using an inactive form of the *Ebola* virus. Explain how this would allow a person to become immune to the disease. [3 marks]

3 a) Malaria is a disease cause by what type of pathogen?

   A Virus   B Bacteria
   C Fungi   D Protist   [1 mark]

b) Which two of the following are symptoms of this disease?

   A Fever   B Vomiting
   C Flaking skin   D Swollen feet   [1 mark]

4 The influenza virus is a microorganism that can cause the flu. The flu is an infectious disease, which can be spread quickly from person to person.

a) i) Name two other microorganisms that can cause disease. [2 marks]

   ii) Describe two ways in which microorganisms like influenza can be passed on from one person to another. [2 marks]

   iii) Suggest a simple hygiene measure that can be taken to reduce the spread of the flu. [1 mark]

b) The body has several non-specific ways of preventing infection from microorganisms. Copy and complete Table 6.1 by naming the part of the body described. [3 marks]

Table 6.1

How entry of microorganisms is prevented	Part of the body
Contains acid to destroy microorganisms	
Acts as a barrier	
Contain ciliated cells to trap bacteria	

c) If microorganisms make it past our body's defences, specific cells are involved in protecting us from harm.

   i) What is the name given to these types of cell? [1 mark]

   ii) Describe the ways in which these cells can protect us from infectious disease. [3 marks]

Working scientifically:
Scientific thinking

Evaluating the risks and benefits

The World Health Organization (WHO) has the goal of eliminating measles in WHO regions between 2015 and 2020. Measles is a highly infectious disease caused by the measles virus and is spread through droplet inhalation and contact with infected people and surfaces. Measles is not just a condition that causes spots; about one in five children infected with it experience complications and one in ten can end up in hospital. In rare cases measles can cause death. Anyone can catch measles, and there is no specific treatment for it. The most effective way of preventing it is to have two doses of the combined MMR (measles, mumps and rubella) vaccination, which gives almost total protection from the disease.

In recent years there has been a global decline in MMR vaccinations, especially in Western Europe and the USA. This has contributed to over 22 000 cases of measles worldwide and has raised concerns that, far from being eliminated, cases of measles are actually increasing. According to WHO, a growing number of parents are refusing to vaccinate their children. This is sparking a global debate over whether vaccinations should be compulsory.

So what do you think?

In making any decision a number of factors need to be examined. These can be split into advantages, disadvantages and risks. These need to be considered on a personal level for the person making the decision, as well as in terms of the impact on society as a whole.

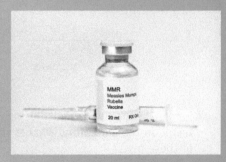

▲ **Figure 6.21** The MMR vaccine.

Questions

1 Copy the table below and use the internet and other sources of evidence, such as medical leaflets and newspaper articles, to help you determine the advantages, disadvantages and risks of compulsory vaccination programmes.

Table 6.2

	Advantages	Disadvantages	Risks
Vaccination compulsory			
Vaccination voluntary			

2 Make a decision on what you think and write a letter to the Government expressing your views on compulsory vaccinations. Ensure you support your views with reasons and evidence.

3 Using diagrams and your knowledge of the immune system, explain how the MMR vaccination creates immunity to measles.

7 Photosynthesis

Imagine being able to make your own food. Not make it from ingredients in your kitchen, but actually make the ingredients themselves. This is what plants and other photosynthesising species can do. In fact, almost all life on Earth depends upon their ability to do this. They feed themselves by storing the Sun's light as glucose during photosynthesis. Interestingly we could not live here without them, but they would grow equally well without us. In fact they might grow a little better!

Specification coverage

This chapter covers specification points 4.4.1.1 to 4.4.1.3 and is called Photosynthesis.

It covers photosynthesis, its rate of reaction and the uses of the glucose that it produces.

Prior knowledge

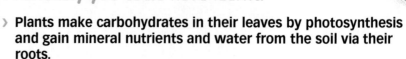

Previously you could have learnt:

> Plants make carbohydrates in their leaves by photosynthesis and gain mineral nutrients and water from the soil via their roots.

> Photosynthesis is an important process in plants that can be summarised by a word equation.

> Almost all life on Earth depends on the ability of photosynthetic organisms, such as plants and algae, to use sunlight in photosynthesis to build organic molecules that are an essential energy store, and also to regulate atmospheric levels of oxygen and carbon dioxide.

Test yourself on prior knowledge

1 Name the products of photosynthesis.
2 Explain why almost all life depends on photosynthetic organisms.

Photosynthetic reaction

TIP ✓

Algae are 'plant-like' protists: they have cell walls, chloroplasts and other features of plant cells.

TIPS ✓

● It is important that you can recognise the symbols for carbon dioxide (CO_2), water (H_2O), oxygen (O_2) and glucose ($C_6H_{12}O_6$).
● In photosynthesis, light transfers energy from the environment to chloroplasts.

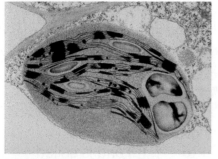

▲ **Figure 7.1** The structure of a chloroplast (magnification about ×15000).

KEY TERM ⭐

Endothermic reaction A reaction that requires energy to be absorbed to work.

Photosynthesis is a chemical reaction that occurs in the green chloroplasts of plants and algae. It needs light from the Sun and converts the reactants carbon dioxide and water into glucose and the by-product oxygen. The word equation for photosynthesis is:

$$\text{carbon dioxide} + \text{water} \xrightarrow{\text{light}} \text{glucose} + \text{oxygen}$$

The fully balanced symbol equation for photosynthesis is:

$$6CO_2 + 6H_2O \xrightarrow{\text{light}} C_6H_{12}O_6 + 6O_2$$

○ Reactants and products

Because it requires light and the green chlorophyll pigment in chloroplasts, photosynthesis mainly occurs in the leaves of plants. The palisade mesophyll cells (the cells near the top of leaves) have the highest number of chloroplasts and so most photosynthesis occurs here.

Gases, including carbon dioxide, are absorbed from the air and enter the leaves via the stomata, before diffusing through the air spaces to reach the palisade cells. Water is absorbed by osmosis into the root hair cells before being transported to the leaves (and rest of the plant) through xylem vessels in transpiration.

Plants complete photosynthesis to produce glucose. They require light to do this, so photosynthesis cannot occur in the dark. Because it requires energy to work, it is an endothermic reaction. These are

TIP

Plants produce oxygen as a by-product. They don't produce it to help animals live. It is handy for animals that they do, however, for without them there would be no oxygen to sustain their life on Earth.

reactions that require energy from their surroundings to work (as opposed to exothermic reactions, which release energy into their surroundings). The glucose produced in the leaves is then transported in phloem tissue to all other parts of the plant that require it during translocation.

Show you can...

Describe how oxygen levels would fluctuate during a 24-hour period.

Test yourself

1 Name the chemical pigment required for photosynthesis to occur.
2 Give the chemical formulae of the products of photosynthesis.
3 Explain why photosynthesis cannot happen in the dark.

Rate of photosynthesis

KEY TERM ⭐

Limiting factor Anything that reduces or stops the rate of a reaction.

TIP ✅

It is important that you can interpret and explain graphs of the rate of photosynthesis involving limiting factors.

A limiting factor is anything that reduces the rate of a reaction. If you were making cakes but ran out of eggs, you couldn't make any more cakes. So eggs would be the limiting factor. There are four possible limiting factors in photosynthesis:

1 **low temperatures**
2 **shortage of carbon dioxide**
3 **shortage of light**
4 **shortage of chlorophyll.**

At lower temperatures all chemical reactions occur more slowly because the reactant molecules have less kinetic energy, so they collide less, and therefore react less. Carbon dioxide is a reactant and so the reaction occurs slowly if carbon dioxide is in short supply. Light provides the energy necessary for this reaction so reduced levels mean reduced photosynthesis, and no light means no reaction. Plants absorb magnesium by active transport from the soil. They use this to make chlorophyll, so plants with a magnesium deficiency cannot photosynthesise as rapidly as those that have plenty of magnesium.

KEY TERM ⭐

Deficiency A lack or shortage.

(a)

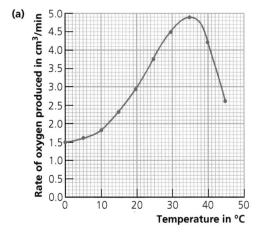

(b)

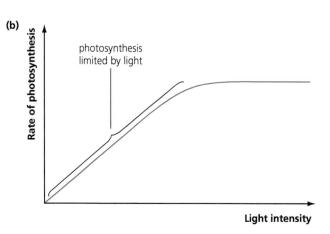

▲ **Figure 7.2** (a) Effect of temperature on the rate of photosynthesis at a fixed light intensity and fixed carbon dioxide concentration.
(b) Effect of increasing light intensity on the rate of photosynthesis at a fixed temperature and fixed carbon dioxide concentration.

The four limiting factors described on the previous page often interact. Any of them may, individually or together with another factor, be responsible for limiting the rate of photosynthesis.

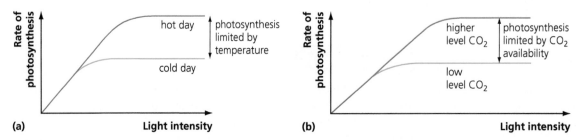

▲ **Figure 7.3** How the rate of photosynthesis is affected by increasing light intensity: (a) on a hot and a cold day; (b) at higher and lower CO_2 levels.

KEY TERM

Yield The amount of an agricultural product.

▲ **Figure 7.4** Farmers ensure their plants have optimum conditions for maximum growth.

Food production is an incredibly important job in all countries of the world. Farmers and growers of crop plants will do as much as they can to maximise the useful part of their crop (the yield) or make their plants grow as quickly as possible. To do this they ensure that plant growth is not limited by a limiting factor. They maintain their plants in greenhouses or polytunnels, which allow them to:

- be grown at the **optimum temperature** (using heaters if necessary)
- have burners near their plants to produce **extra carbon dioxide** for optimal growth
- provide their plants with **maximum light** (using artificial lighting if necessary).

Farmers and growers must balance these three conditions without making their crops too expensive to sell at a profit.

TIPS

- It is important that you can relate limiting factors to the cost effectiveness of raising the temperature, increasing the light intensity and adding carbon dioxide to greenhouses.
- You should be able to explain graphs of photosynthesis rate involving two or three factors and decide which is the limiting one (as in Figure 7.3).
- You should understand and use the inverse square law in the context of photosynthesis. This means that if the distance between plant and light is doubled, its rate of photosynthesis is quartered.

Test yourself

4 Define the term 'limiting factor'.
5 Name the limiting factors for photosynthesis.
6 Describe why plants need magnesium.
7 Describe how farmers can increase the rate of photosynthesis of plants inside a glasshouse.

Investigate the effect of light intensity on the rate of photosynthesis using an aquatic organism such as pondweed

In this practical you will investigate the effect of light intensity on the rate of photosynthesis.

Method

1 Take a piece of pondweed (*Cabomba* or *Elodea*) and cut it underwater so that the stalk is cut at a 45° angle.

2 Keeping it under the water, attach a paper clip at the opposite end to the cut stalk and transfer to a boiling tube.

3 Set up a lamp, metre ruler and tank or beaker of water as shown in Figure 7.5.

4 Position the boiling tube with pondweed in it so it is 50 cm away from the light source.

5 Turn off the classroom lights, turn the lamp on and wait for 2 minutes.

6 Ensure that the temperature of the water in the boiling tube with the pondweed remains constant throughout the experiment by monitoring with a thermometer.

7 Count the number of bubbles produced in 1 minute. Repeat this twice more so that a mean rate of bubbles produced can be determined.

8 Move the clamped boiling tube so that it is 40 cm away from the light source and repeat steps 5 and 6.

9 Repeat the experiment for a distance of 30, 20 and 10 cm.

10 Record your results in a table and draw a graph of mean number of bubbles against distance of lamp.

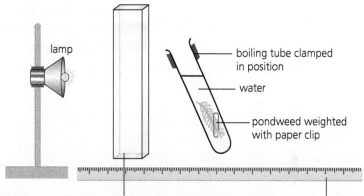

▲ **Figure 7.5** The equipment used to investigate the effect of light intensity on the rate of photosynthesis.

Questions

1 Use the equation 1/distance² to determine the light intensity for each distance used. For example, if the distance were 60 cm the light intensity would be $1/60^2$ or $1/(60 \times 60)$.

2 Plot a line graph of your data with light intensity on the *x*-axis and mean rate of bubbles on the *y*-axis.

3 What can you conclude from your data about how light intensity affects photosynthesis?

4 Why was a beaker of water placed between the lamp and the pondweed?

5 Why did you have to wait 2 minutes before counting the number of bubbles?

6 Suggest how you could modify this experiment to measure the rate of oxygen production more accurately.

Uses of glucose from photosynthesis

It often helps to think of photosynthesis in terms of where the energy is. The two reactants, carbon dioxide and water, have a relatively small amount of energy within them. The energy found within the two products, glucose and oxygen, is greater. You will have learnt that energy cannot ever be created or destroyed, only converted from one form to another. So how can the products of this reaction have more energy than the reactants? Where does this energy come from?

The answer is energy transferred by the light. This energy is used to power the reaction. It breaks apart the chemical bonds in the reactants and re-forms them with some extra energy stored in the bonds of the products. It is this extra energy that supports almost all life on Earth. When you think like this, it is obvious that photosynthesis can't happen in the dark.

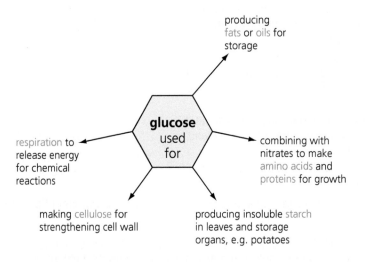

▲ **Figure 7.6** How plants make use of the glucose produced during photosynthesis.

The energy that plants have stored in the formation of glucose during photosynthesis has five general uses. It is:

1 used in **respiration**
2 converted into insoluble **starch** and **stored**
3 converted into **fats** and **oils** and **stored**
4 used to make **cellulose** which strengthens the **cell walls** in plants
5 used with **nitrate ions** absorbed from the soil to make **amino acids** for **protein synthesis**.

Show you can...

State whether there is more energy in the reactants or products of photosynthesis.

Chapter review questions

1 Name the two reactants in photosynthesis.

2 Name the plant organs in which most photosynthesis occurs.

3 Name the plant cell organelle in which photosynthesis occurs.

4 What is the source of the energy required for photosynthesis to occur?

5 Explain why there would be almost no life on Earth without plants.

6 Give the word equation for photosynthesis.

7 Name the green chemical that must be present for photosynthesis to occur.

8 Name the plant tissue in which most photosynthesis occurs.

9 Name the plant tissue with the highest concentration of chloroplasts.

10 Name the type of specialised cell in a plant that absorbs the most water.

11 Give the chemical formula for glucose.

12 Define the term 'limiting factor'.

13 Name the process by which plants absorb mineral ions from the soil.

14 Name the key metal element in making chlorophyll.

15 Explain why farmers often grow crops in polytunnels and greenhouses.

16 Give the five uses of glucose formed during photosynthesis.

17 Name the compound produced by plants to strengthen their cell walls.

18 Describe why photosynthesis is an endothermic reaction.

19 Give the balanced symbol equation for photosynthesis.

20 Name the four limiting factors in photosynthesis.

21 Explain why less photosynthesis occurs at lower temperatures.

22 Explain why more photosynthesis occurs with more carbon dioxide.

23 Explain why less photosynthesis occurs under lower light conditions.

24 Describe an experiment in which you investigate the effect of light intensity on photosynthesis.

25 Define the term 'yield'.

26 Describe what farmers can do to greenhouses or polytunnels in which they are growing plants to improve their yield.

27 Explain how we know there is more energy in glucose and oxygen combined than in carbon dioxide and water combined.

28 Name the type of ions absorbed by roots which are used by plants to make proteins.

Practice questions

1 a) Copy and complete the word equation for photosynthesis. [2 marks]

   carbon dioxide + X → glucose + Y

b) Describe how the carbon dioxide needed for photosynthesis gets into the plant. [1 mark]

c) Copy and complete the sentences by choosing the correct words from the box.

> light root mitochondria respiration chemicals flower
> chloroplasts haemoglobin chlorophyll ribosome leaf

i) The plant organ that is specialised to carry out photosynthesis is the
   _____. [1 mark]

ii) The energy needed for photosynthesis to occur comes from _____. [1 mark]

iii) Energy is absorbed by a green pigment called _____. [1 mark]

iv) This green pigment is found in small organelles called _____. [1 mark]

2 After a plant had been kept in the dark for 48 hours it was set up as shown in Figure 7.7. After another 24 hours the leaves from inside bags A, B and C were removed and a test was carried out to see if starch was present.

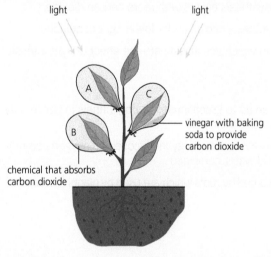

light   light

A
C
B

vinegar with baking soda to provide carbon dioxide

chemical that absorbs carbon dioxide

▲ Figure 7.7

a) Copy and complete Table 7.1 to show the likely results of the experiment by placing the letters in the correct column. [2 marks]

Table 7.1

Starch present	Starch not present

b) i) Which chemical would have been used to test the leaves for starch? [1 mark]

   A Biuret solution   C Iodine solution
   B Benedict's solution D Ethanol solution

ii) What colour would indicate a positive result for starch? [1 mark]

   A Purple   C Reddish orange
   B Blue-black   D Green

3 A farmer was growing tomatoes in a greenhouse. The graph shows the effect of temperature and concentration of carbon dioxide on the rate of photosynthesis of the tomato plants.

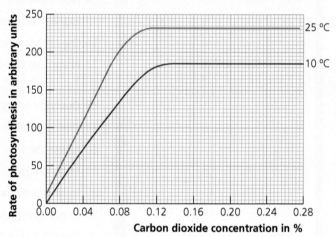

▲ Figure 7.8

a) From the graph conclude:

i) the best temperature at which to grow the tomato plants [1 mark]

ii) the best concentration of carbon dioxide to use to grow the tomato plants [1 mark]

iii) the maximum rate of photosynthesis recorded. [1 mark]

b) Apart from temperature and carbon dioxide concentration, what other factor could the farmer change to increase the rate of photosynthesis? [1 mark]

4 Outline a plan to investigate how the rate of photosynthesis in pondweed changed when the intensity of light was changed.

a) Describe how you would do the investigation and the measurements you would take.

b) Describe how you would make it a fair test. [6 marks]

Working scientifically:
Experimental skills

Planning and variables

A variable is a quantity or characteristic. In investigations three types of variable are discussed: independent variables, dependent variables and control variables.

By knowing what the variables are, a clear method can be written that outlines the procedure and the steps taken in order to make it a fair test.

A group of students were planning to investigate the effect of carbon dioxide concentration on the rate of photosynthesis. They decided they would measure the rate of photosynthesis by hole-punching small discs out of spinach leaves, and placing them in a 10 cm³ syringe of sodium hydrogencarbonate solution ($NaHCO_3$). By pressing on the plunger of the syringe they could increase the pressure. This forced any trapped air out of the spongy mesophyll of the leaf discs and made them sink. Because gas is produced in photosynthesis, the time taken for the leaf discs to rise (inflate again) could be used as a measure of the rate of photosynthesis.

Before they started, the students wrote up their experiment method.

1 Read through the method below and write down any parts where it is not clear how a control variable will be kept constant. Explain how the method could be improved to overcome this.

2 Then write down a clear method to explain how the above experiment could be modified to investigate the effect of temperature on the rate of photosynthesis. Ensure you clearly detail what the variables would be.

Method

1 Wearing eye protection, cut out the spinach leaves using a hole-punch and add these to the syringe by removing the plunger.

2 Add to this one drop of washing-up liquid and replace the plunger.

3 Draw up a 0.5% sodium hydrogencarbonate solution into the syringe.

4 Hold the syringe with the tip pointing upward and expel any air that remains in the syringe by depressing the plunger carefully. Stop before any of the sodium hydrogencarbonate solution comes out.

5 Place a finger over the tip of the syringe and hold firmly in place. Pull back on the plunger and hold for 10–15 seconds to create a partial vacuum inside the syringe. Look to see signs of bubbles escaping the edges of the leaf discs.

6 Release your finger and the plunger at the same time, tap gently on the side of the tube and the leaf discs should start to sink.

7 Repeat steps 5 and 6 until all the leaf discs have sunk.

8 Place the syringe under a lamp and time how long it takes for the first leaf disc to reach the surface.

9 Repeat the experiment using a 0.4, 0.3, 0.2 and 0.1% concentration of sodium hydrogencarbonate solution.

KEY TERMS

Independent variable This is the variable that is changed in an experiment or selected by the investigator.

Dependent variable This is the variable that is measured or recorded for each change of the independent variable.

Control variable This is the variable that can affect the outcome of an investigation and therefore must be kept constant or monitored. If all control variables are kept constant then the experiment is a fair test.

▲ Figure 7.9 The equipment used to investigate the effect of carbon dioxide concentration on the rate of photosynthesis.

Questions

1 What is the dependent variable in this experiment?

2 What is the independent variable in this experiment?

3 What variables must be controlled in order to make the experiment a fair test?

8 Respiration

Every one of the thousands of billions of cells that make up your body is respiring now and will continue to do so. If a cell stops it will die. Respiration is therefore an extremely important reaction. It is one of the seven life processes. Just like you, all other life on our planet undergoes respiration or a chemically similar reaction to release the energy it needs to survive. When we discover life elsewhere in the Universe it will probably do something similar.

Specification coverage

This chapter covers specification points 4.4.2.1 to 4.4.2.3 and is called Respiration.

It covers aerobic and anaerobic respiration, response to exercise and metabolism.

Previously you could have learnt:

> Aerobic and anaerobic respiration are important processes in living organisms, that involve the breakdown of organic molecules produce energy.
> Aerobic respiration is an important process in living organisms that can be described by a word equation.
> Anaerobic respiration occurs in humans and microorganisms (e.g. fermentation), and can be described by a word equation.
> Aerobic and anaerobic respiration have differences in terms of reactants used, products formed, and implications for the organism.

Test yourself on prior knowledge

1 Name the products of aerobic respiration.
2 Describe the difference between aerobic and anaerobic respiration.
3 Explain how we use organisms that respire anaerobically. Give an example in your answer.

Aerobic respiration

TIPS

● Some people confuse the process of breathing, called ventilation, with the release of energy from glucose in the chemical reaction called respiration.
● Respiration occurs continuously in all living cells.

KEY TERMS

Aerobic In the presence of oxygen.

Oxidation A reaction that uses oxygen.

TIP

It is important that you can recognise the symbols for glucose ($C_6H_{12}O_6$), oxygen (O_2), carbon dioxide (CO_2) and water (H_2O).

Respiration is a chemical reaction that occurs in the **mitochondria** of your cells. This reaction releases the energy stored in glucose to allow your cells to complete chemical reactions and to allow animals to move and keep warm. The word equation for aerobic respiration is:

$$glucose + oxygen \xrightarrow{energy} carbon\ dioxide + water$$

The fully balanced symbol equation for aerobic respiration is:

$$C_6H_{12}O_6 + 6O_2 \xrightarrow{energy} 6CO_2 + 6H_2O$$

Respiration is an oxidation reaction, because it uses oxygen.

Reactants and products

Plants and algae photosynthesise to store energy in glucose. Respiration releases this energy by reversing the process. Glucose reacts in the presence of oxygen to form carbon dioxide and water.

▲ **Figure 8.1** Water is produced during respiration and condenses on windows when you exhale.

It is easy to remember the reactants by answering these questions:

1 Why do I eat? Answer: To get glucose (and other key nutrients).

2 Why do I breathe? Answer: To get oxygen.

It is easy to remember the products by picturing yourself breathing out on to a cold window. What two substances do you breathe out? Carbon dioxide and water (vapour which condenses on the window).

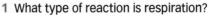

> **TIP** ✔
> Aerobic respiration does not mean 'in the presence of air', even though it sounds like it does. It means 'in the presence of oxygen'.

Unlike photosynthesis, which only occurs in the light, your cells must respire continuously through the day and night. Respiration does not require energy to drive the reaction like photosynthesis does. In fact, respiration releases energy to allow respiring cells to live. This makes it an **exothermic reaction**.

> **KEY TERM** ⭐
> **Exothermic reaction** A reaction that gives out thermal energy.

> **TIP** ✔
> **Ex**o sounds like **ex**it. You leave through an exit, so energy leaves in an exothermic reaction.

Show you can...

Explain why some cells, such as sperm and muscle, contain more mitochondria than others.

Test yourself

1 What type of reaction is respiration?

2 Name the type of cellular organelle in which respiration occurs.

3 Define the term 'aerobic'.

4 State the balanced symbol equation for respiration.

Activity

Investigating respiration in invertebrates

Students used a simple respirometer to investigate how much oxygen is used by different invertebrate species. This is shown in Figure 8.2.

Questions

1 a) Which gas is produced by both invertebrates during the experiment?

b) What happens to this gas during the experiment?

2 a) What happens to the water drops in the capillary tubes over the course of the experiment in both respirometers?

b) Explain the reason for this movement.

3 Which respirometer gives the more accurate reading? Explain why.

Note: the grasshopper and cricket were released from the respirometers before they ran out of oxygen, and no insects were harmed during this experiment!

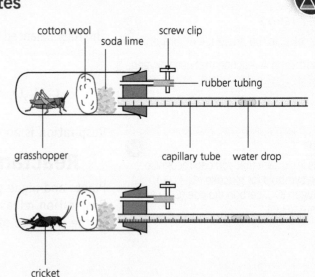

▲ **Figure 8.2** The equipment used to investigate how much oxygen is used by different invertebrate species.

Conversion of energy in respiration (and photosynthesis)

So the reactants and products in respiration are the opposite of those in photosynthesis.

Photosynthesis:

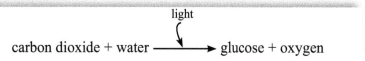

carbon dioxide + water $\xrightarrow{\text{light}}$ glucose + oxygen

Respiration:

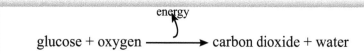

glucose + oxygen $\xrightarrow{\text{energy}}$ carbon dioxide + water

However, these two crucial chemical reactions are not simply the opposite of each other when we focus on the flow of energy. Photosynthesis is an **endothermic** reaction, which requires energy from its surroundings to occur. The arrow on the photosynthesis equation shows energy in. Respiration is an **exothermic** reaction, which releases energy. The arrow on the respiration equation shows energy out.

These two equations work beautifully together for us and many other organisms on Earth. They show how:

- energy transferred by light (mainly from the Sun) is converted into a chemical store of energy in glucose by photosynthesis
- energy for life processes is released from glucose by respiration.

The energy released from glucose has two main functions. It is converted into:

- **thermal energy** to keep an organism warm (especially in warm-blooded birds and mammals)
- a chemical store of energy that is available for **reactions** and processes, such as **movement**, in an organism.

▲ **Figure 8.3** When this 400-metre runner starts the race, his muscles will need more glucose and oxygen.

Investigating the temperature rise caused by respiration in yeast

To investigate the temperature rise produced by respiring organisms, a student heated 200 cm³ of a 10% glucose solution to 35 °C and then stirred in 20 g of baker's yeast (*Saccharomyces cerevisiae*). The mixture was then poured into a thermos flask and a tightly fitting bung with two bore holes was used to stopper the flask. A thermometer was inserted in one of the holes so the temperature could be monitored over the course of an hour.

Questions

1 Explain why a second hole was needed in the bung.

2 a) Describe how you would expect the temperature readings to vary over the course of an hour.

  b) Explain why this is.

3 Describe a suitable control that could be used in this experiment and explain the purpose of having one.

— flask containing glucose solution plus yeast

▲ **Figure 8.4** The equipment used to investigate the heat produced by respiring organisms.

Test yourself

5 Name the reactants in photosynthesis.
6 Name the products in respiration.
7 Describe the energy flow through photosynthesis and respiration.
8 Describe the main functions of the energy released from glucose.

Anaerobic respiration

TIP

It is important that you can compare the processes of aerobic and anaerobic respiration in terms of the need for oxygen, the products formed and the amount of energy released.

KEY TERM

Anaerobic In the absence of oxygen.

TIP

The 5% in the word equation means that in anaerobic respiration, one glucose molecule releases 5% of the energy it would do in aerobic respiration.

○ Response to exercise

When we exercise, we use up energy more quickly. To produce the extra energy, we respire more quickly and therefore use up oxygen and glucose faster. We **breathe faster** and **more deeply** to take in more oxygen. Our **hearts beat faster** to deliver glucose and oxygen more quickly to our rapidly respiring muscles. There will be times in your life when you cannot breathe quickly or deeply enough to supply all of your cells with all the oxygen they need to keep on respiring aerobically. This might be the last time you had to run the cross-country at school or when you were last out of breath. At this point your cells, particularly your muscles, start to run out of oxygen. They can only respire anaerobically. The word equation for this is:

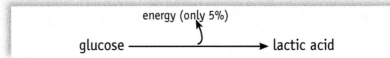

energy (only 5%)

glucose ⟶ lactic acid

○ Reactants and products

Cells respiring anaerobically are missing oxygen and so cannot make carbon dioxide and water. Instead they make an intermediary substance called **lactic acid**. A build-up of lactic acid in a muscle causes cramp. This causes muscle **fatigue** and stops the muscle contracting so efficiently.

Because the reaction has not been fully completed (because of the lack of oxygen) the total energy released from anaerobic respiration is much less than during aerobic respiration. Only about 5% of the energy (or $\frac{1}{20}$) is released. This means that when you are respiring anaerobically, your body is releasing less energy and producing a substance that hurts you (lactic acid). Is it trying to tell you something?!

○ Paying your oxygen debt

When you have finished exercising vigorously it is likely that you will sit down and relax. We say you have an **oxygen debt** to your body at this point. (You owe it oxygen.) For the next few minutes you will continue to **breathe deeply** and **quickly** to replenish the oxygen you have used up. You are repaying your oxygen debt. Your **pulse rate** will remain **high** to pump the newly oxygenated blood and glucose as quickly as possible to your muscles for more aerobic respiration.

After a few minutes your breathing will return to normal. Now you probably feel tired but not nearly as tired as you did a few minutes ago.

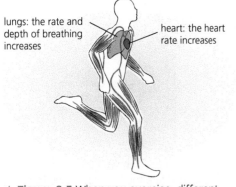

lungs: the rate and depth of breathing increases

heart: the heart rate increases

▲ **Figure 8.5** When you exercise, different parts of your body respond to increase the rate at which energy is released in your muscles.

▲ **Figure 8.6** These sprinters have produced a lot of lactic acid in their muscles during the race. What happens to the lactic acid?

The lactic acid that builds up during anaerobic respiration diffuses from a high concentration in your muscle cells to a low concentration in your blood. By the time it reaches your liver it is at a high concentration in your blood and so diffuses again into the low concentration in your **liver**. Here it is converted back into glucose by an **oxidation** reaction. This then diffuses into the blood for use in either aerobic or anaerobic respiration.

When you convert the lactic acid to glucose in the liver, you are paying back your oxygen debt and making the energy available that was locked up in the lactic acid.

KEY TERM

Oxygen debt The amount of extra oxygen the body needs after exercise to break down the lactic acid.

Test yourself

9 Give the percentage of energy released in anaerobic respiration compared to that released in aerobic respiration.
10 Where is lactic acid broken down?
11 Describe where the majority of the energy is stored in anaerobic respiration.
12 Define the term 'anaerobic'.

Show you can...

Explain what happens from when your muscle cells start to respire anaerobically during a run until you recover completely afterwards.

Anaerobic respiration in plants and microorganisms

▲ **Figure 8.7** All life in this rainforest respires, including the plants.

It is a common misunderstanding that plants only photosynthesise and animals only respire. If plants only photosynthesised they would all be mass-producing glucose but not be able to use this energy to complete the life processes (and so die). Just like each and every one of your cells, every plant cell in every plant on our planet must respire or it will die. So plants photosynthesise during the day, and respire during the day and also the night. Just like animals, plants respire anaerobically when they do not have enough oxygen. This occurs in root cells in waterlogged soils and can cause roots, and eventually the whole plant, to die. Some microorganisms such as yeast, a single-celled fungus, also respire in this way. The equation for their anaerobic respiration is:

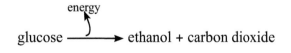

$$\text{glucose} \xrightarrow{\text{energy}} \text{ethanol} + \text{carbon dioxide}$$

KEY TERM

Fermentation The chemical breakdown of glucose into ethanol and carbon dioxide by respiring microorganisms such as yeast.

If this reaction happens in yeast cells it is called fermentation. This microorganism and this reaction are economically important in the manufacture of **alcoholic drinks**. The compound ethanol that is produced is commonly known as alcohol.

Yeast is also economically important in the manufacture of **bread**. In bread making, the carbon dioxide produced by the yeast causes the dough (mixture of flour and water) to rise. If the same reaction is occurring, why does our bread not taste alcoholic? Bread is baked in an oven and so the yeast is killed before it can make too much ethanol. Any that it does make is evaporated away by the warmth of the oven.

Test yourself

13 Give the word equation for fermentation.
14 Describe why bread isn't alcoholic even though yeast ferments in its making.

Show you can...

Explain why some organisms such as yeast produce alcohol.

Metabolism

KEY TERM

Metabolism The sum of all the chemical reactions that happen in an organism.

Metabolism is the sum of all the chemical reactions that happen in a cell or in your body. These reactions include digestion of food, aerobic and anaerobic respiration which you have just learnt about, and protein synthesis. For plant cells, metabolism also includes photosynthesis. Most of these reactions are controlled by enzymes.

○ Breakdown and synthesis reactions

Some metabolic reactions make more, smaller, often less complicated molecules. These are breakdown reactions.

Other reactions make larger, more complex molecules. These are synthesis reactions. To occur, these reactions need energy from respiration.

Table 8.1 Breakdown reactions and synthesis reactions.

	Type of reaction		Comment
	Breakdown reactions	Synthesis reactions (building up)	
Carbohydrates	Complex carbohydrates, e.g. starch, are broken down into simple sugars, e.g. glucose. Glucose is broken down in respiration in plants and animals to release energy.	Simple sugars, e.g. glucose, are built up into starch (in plants) and glycogen (in animals) for storage. Cellulose for plant cell walls is also made from simple sugars.	
Proteins	Proteins broken down into amino acids.	Amino acids are built up into protein. Amino acids are formed from glucose and nitrate ions.	In animals, excess protein (or amino acids) cannot be stored. The excess is broken down into urea, which is excreted via the kidneys.
Lipids (fats)	Lipids are broken down to glycerol and fatty acids. Each lipid molecule breaks down to form one glycerol molecule and three fatty acid molecules.	Glycerol and fatty acids build up into lipids.	
Note	The above breakdown reactions take place in the gut in animals to produce smaller, soluble sub-units that can be absorbed into the bloodstream.	The above synthesis reactions take place in the body cells in animal and also in plants.	

Chapter review questions

1 Explain the difference between respiration and ventilation.

2 Name the reactants in aerobic respiration.

3 Define the term 'aerobic'.

4 Give the word equation for respiration.

5 Describe the conditions under which anaerobic respiration occurs.

6 Give the word equation for anaerobic respiration.

7 What are the effects of producing too much lactic acid?

8 Name the process by which oxygen moves into the cells for respiration.

9 Give the word equation for anaerobic respiration in yeast.

10 Explain the economic importance of anaerobic respiration in yeast.

11 Name the organelle in which respiration occurs.

12 What are the uses of the energy released during respiration?

13 State why respiration is an exothermic reaction.

14 Give the proportion of energy released in anaerobic respiration compared with aerobic respiration.

15 Define the term 'oxygen debt'.

16 Explain why your heart rate and breathing rate increase during exercise.

17 Define the term 'fermentation'.

18 Define the term 'metabolism'.

19 Describe an experiment in which you investigate how respiring yeast raises the temperature of the substrate it is growing in.

20 Explain the flow of energy through photosynthesis and aerobic respiration.

21 Name the main source of energy for the vast majority of the reactions in your cells.

22 Where is the unreleased energy stored in anaerobic respiration?

23 What happens to the lactic acid produced during anaerobic respiration?

24 Define the term 'oxidation'.

25 Describe an experiment in which you investigate the oxygen consumption for respiration in two invertebrate species.

Practice questions

1 Respiration occurs in living organisms.

a) i) What is the purpose of respiration?   [1 mark]

ii) Using your knowledge of the word equation for aerobic respiration, identify which of the following are chemical products.   [2 marks]

A Carbon dioxide   C Glucose

B Energy   D Water

iii) Glucose is a reactant in respiration. Which of the following is the correct chemical formula for glucose?   [1 mark]

A $C_{12}H_6O_{12}$ C $C_6H_6O_{12}$

B $C_{12}H_6O_6$ D $C_6H_{12}O_6$

b) Figure 8.8 shows how a respirometer was set up to investigate aerobic respiration in germinating peas. Apparatus A shows the starting point of the water in the respirometer.

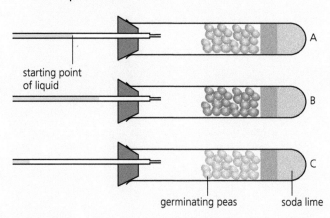

▲ Figure 8.8

i) Predict which tube, B or C, shows the direction of liquid movement in the capillary tube after 10 minutes.   [1 mark]

ii) Explain the role of the soda lime in this experiment.   [2 marks]

2 A student was investigating the effect of exercise on an athlete's body. She collected data on their heart rate before and after exercise. Table 8.2 shows the data collected.

Table 8.2

	Resting	During exercise
Heart rate in beats per minute	65	125
Volume of blood pumped out of the heart in each beat in cm³	90	145
Cardiac output in cm³ per minute	5850	

a) By how much did the athlete's heart rate increase during exercise?   [1 mark]

b) Explain why the athlete's heart rate increased.   [3 marks]

c) Calculate the cardiac output in cm³ per minute for the athlete during exercise. Show your working.   [2 marks]

d) Give two other changes that occur in the body during exercise.   [2 marks]

3 Yeast cells can respire anaerobically. This means they can be grown in anaerobic conditions inside a fermenter.

a) What does anaerobic mean?   [1 mark]

b) i) Which of the following is the correct word equation for anaerobic respiration in yeast?   [1 mark]

A Glucose → lactic acid + carbon dioxide

B Glucose + oxygen → lactic acid + carbon dioxide

C Glucose → ethanol + carbon dioxide

D Glucose + oxygen → ethanol + carbon dioxide

ii) Give one way anaerobic respiration in yeast is different to anaerobic respiration in human cells.   [1 mark]

c) Figure 8.9 shows the changes that occur in a fermenter over the course of 24 hours. Describe and explain how ethanol production changed over the 24-hour period.   [2 marks]

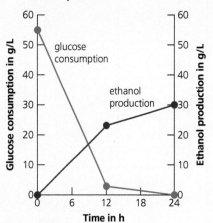

▲ Figure 8.9

4 A woman is taking part in an exercise class. Initially her body cells are carrying out aerobic respiration.

a) Write a balanced equation for aerobic respiration.   [2 marks]

b) As the exercise becomes more vigorous, her cells switch to anaerobic respiration.

i) Name the chemical produced in anaerobic respiration that is not produced in aerobic respiration.   [1 mark]

ii) Explain how the substance produced is broken down after the exercise has finished.   [2 marks]

5 Describe the similarities and differences between aerobic and anaerobic respiration in humans.   [6 marks]

Working scientifically:
Dealing with data

Means and ranges

Two students were investigating the effect of temperature on respiration in yeast. Terry set up his experiment using the apparatus shown in Figure 8.10a and Afreen as shown in Figure 8.10b.

Both students added the same amount of glucose solution and yeast to their apparatus and covered it using $20\,cm^3$ of liquid paraffin. They then placed their equipment in a water bath at 20 °C and left it for 5 minutes. After the 5-minute period, they measured the volume of gas produced in 10 minutes: Terry by counting the number of bubbles and Afreen by recording the movement of the gas syringe. They repeated their experiment five times and then repeated it over a range of temperatures, from 20 to 80 °C.

(a)

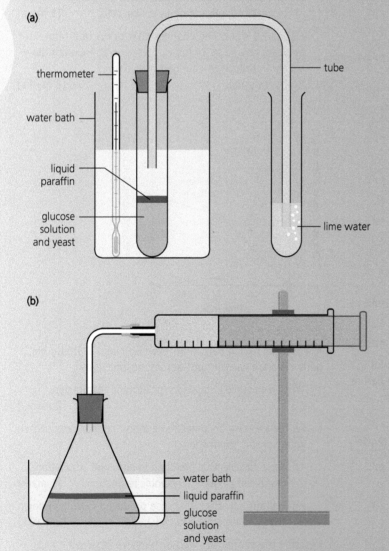

(b)

▲ **Figure 8.10** The equipment used by Terry (a) and Afreen (b) to investigate the growth of yeast.

Their results are shown in Table 8.3 and Table 8.4.

Table 8.3 Terry's results.

Temperature in °C	Number of bubbles produced over 10 minutes				
	Trial 1	Trial 2	Trial 3	Trial 4	Trial 5
20	400	356	362	345	345
35	623	645	682	654	652
50	520	519	302	512	518
65	105	115	186	203	186
80	2	1	4	1	2

Table 8.4 Afreen's results.

Temperature in °C	Volume of gas produced over 10 minutes in cm³				
	Trial 1	Trial 2	Trial 3	Trial 4	Trial 5
20	15	14	16	18	20
35	45	44	43	44	45
50	38	41	4	42	55
65	25	28	24	27	25
80	4	1	5	2	1

Both of the students' data show variation in results. This is to be expected with repeated results. The range of the results is the highest and lowest value recorded for each set of data. The narrower the range, the closer the results are to each other and the more repeatable **the data. For Terry's results at 20 °C the range is 345–400.**

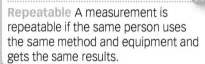

KEY TERM

Repeatable A measurement is repeatable if the same person uses the same method and equipment and gets the same results.

Questions

1 Calculate the range for Terry's other temperature results and determine which of his data sets has the most repeatable data.

Example

For Terry's results at 20 °C the mean is (400 + 356 + 362 + 345 + 345)/5 = 361.6.

The more repeatable the data, the more confident you can be that the true value of a measurement lies within the range of your data. To estimate a true value from a range you should calculate a mean. This is done by adding up the repeats and dividing the number by the total number of repeats taken. The result should be rounded to the same number of decimal places or one more than the raw data.

Questions

2 Calculate the means for Terry's results at 35 °C, 50 °C, 65 °C and 80 °C.

The more repeats taken, the easier it is to spot anomalies. An anomaly is a value that is not in line with the other data and is therefore not likely to be caused by random variation.

Questions

3 Identify any anomalous results in Afreen's data.

When data can be identified as being anomalous they should be left out of the ranges and means, as they can skew the data and make them less accurate.

Questions

4 Work out the ranges and means for Afreen's data.
5 What is the trend shown in both students' data? Can you explain this?
6 Whose equipment allowed them to gather more accurate data? Explain why.
7 Why in both experiments was the yeast and glucose solution left for 5 minutes before the measurements were started?
8 Why in both experiments was the yeast and glucose solution covered in liquid paraffin?
9 What gas were both students collecting, and how could they test it to prove what it is?

9 Atomic structure and the periodic table

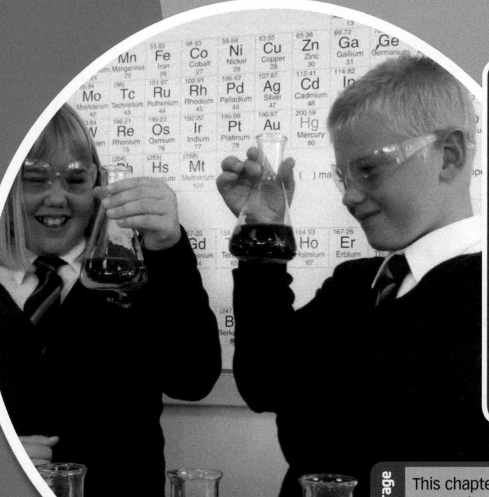

Until you reached GCSE, Chemistry was studied at the particle level. In order to take chemistry further, you now need to understand what is inside atoms. The elements in the periodic table are ordered by what is inside their atoms. An understanding of the periodic table allows you to explain and/or work out a lot of chemistry even if you have never studied it.

Specification coverage

This chapter covers specification points to 5.1.2.6 and is called Atomic structure and the periodic table.

It covers the structure of atoms, reactions of elements, the periodic table and mixtures.

Writing formulae and equations is covered separately in Chapter 14.

Previously you could have learnt:

> Elements are made of particles called atoms.
> Elements are substances containing only one type of atom – this means they cannot be broken down into simpler substances.
> Each element has its own symbol and is listed in the periodic table.
> Elements are either metals or non-metals.
> Compounds are substances made from atoms of different elements bonded together.
> Compounds have different properties from the elements from which they are made.
> Compounds are difficult to break back down into their elements.
> Substances in mixtures are not chemically joined to each other.
> Substances in mixtures can be separated easily by a range of techniques.

Test yourself on prior knowledge

1 What is an element?
2 What is a compound?
3 Why do compounds have different properties from the elements from which they are made?
4 List some differences between metals and non-metals.
5 Why is it easy to separate the substances in a mixture but not to break apart a compound?
6 Name four methods of separating mixtures.

Structure of atoms

KEY TERMS

Atom The smallest part of an element that can exist. A particle with no electric charge made up of a nucleus containing protons and neutrons surrounded by electrons in energy levels.

Proton A positively charged particle found in the nucleus of an atom.

Neutron A neutral particle found in the nucleus of an atoms.

Electron Negatively charged particle found in energy levels (shells) surrounding the nucleus inside atoms.

Nucleus Central part of an atom containing protons and neutrons.

Energy level (shell) The region an electron occupies surrounding the nucleus inside an atom.

○ Protons, neutrons and electrons

Atoms are the smallest part of an element that can exist. Atoms are made up of smaller particles called protons, neutrons and electrons. The table below shows the relative mass and electric charge of these particles. The mass is given relative to the mass of a proton. Protons and neutrons have the same mass as each other while electrons are much lighter (Table 9.1).

Table 9.1 Relative mass and relative charge of each part of an atom.

	Proton	Neutron	Electron
Relative mass	1	1	very small
Relative charge	+1	0	−1

○ The structure of atoms

Atoms are very small. Typical atoms have a radius of about 0.1 nm (0.000 000 000 1 m, that is 1×10^{-10} m). Atoms have a central nucleus which contains protons and neutrons (Figure 9.1). The nucleus is surrounded by electrons. The electrons move around the nucleus in energy levels, or shells.

TIP ✓

The charge of a proton can be written as + or +1. The charge of an electron can be written as – or –1.

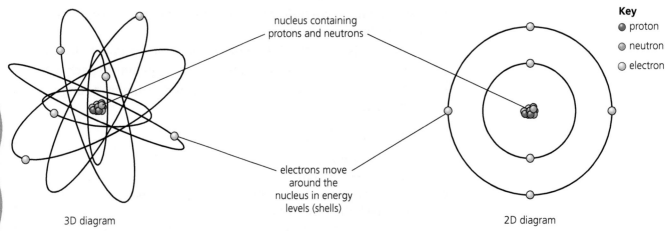

nucleus containing protons and neutrons

electrons move around the nucleus in energy levels (shells)

Key
● proton
● neutron
○ electron

3D diagram

2D diagram

▲ **Figure 9.1** 3D and 2D diagram of an atom.

The nucleus is tiny compared to the size of the atom as a whole. The radius of the nucleus is less than 1/10000th of that of the atom (1×10^{-14} m). This difference in size between a nucleus and an atom is equivalent to a pea placed in the middle of a football pitch (Figure 9.2).

The nucleus contains protons and neutrons. These are much heavier than electrons. This means that most of the mass of the atom is contained in the tiny nucleus in the middle.

TIP ✓

$1\,nm = 1 \times 10^{-9}$ m (0.000000001 m)

TIP ✓

The SI unit for length is metres (m).

▲ **Figure 9.2** The size of the nucleus compared to the atom is like a pea compared to a football pitch.

Test yourself

1 Carbon atoms have a radius of 0.070 nm. Write this in standard form in the units of metres.

2 The radius of a hydrogen atom is 2.5×10^{-11} m. Write this in nanometres.

3 The radius of a chlorine atom is 1×10^{-10} m and the radius of a silicon atom is 0.060 nm. Which atom is bigger?

4 Sodium atoms have a radius of 0.180 nm. The nucleus of an atom is about 10000 times smaller. Estimate the radius of the nucleus of a sodium atom. Write your answer in both nanometres and metres.

5 A copper atom has a diameter of 0.256 nm. A copper wire has a diameter of 0.0440 cm.

   a) Write the diameter of the atom and the wire in metres.

   b) How many times wider is the copper wire than a copper atom? Give your answer to 3 significant figures.

6 A gold atom has a diameter of 0.270 nm. The largest gold bar in the world is 45.5 cm long. How many gold atoms fit into 45.5 cm? Give your answer to 3 significant figures.

○ Atomic number and mass number

The number of protons that an atom contains is called its atomic number. Atoms of different elements have different numbers of protons. It is the number of protons that determines which element an atom is. For example, all atoms with 6 protons are carbon atoms, while all atoms with 7 protons are nitrogen atoms.

All **atoms** are **neutral**, which means they have **no overall electric charge**. This is because the number of protons (which are positively charged) is the same as the number of electrons (which are negatively charged).

Most of the mass of an atom is due to the protons and neutrons. Protons and neutrons have the same mass as each other. The mass number of an atom is the sum of the number of protons and neutrons in an atom. For example, an atom of sodium has 11 protons and 12 neutrons and so has a mass number of 23.

> **ATOMIC NUMBER** = number of protons
>
> **MASS NUMBER** = number of protons + number of neutrons

The atomic number and mass number of an atom can be used to work out the number of protons, neutrons and electrons in an atom:

- **number of protons = atomic number**
- **number of neutrons = mass number – atomic number**
- **number of electrons = atomic number** (*only for atoms, not ions*).

The mass number and atomic number of atoms can be shown as in Figure 9.3.

KEY TERM

Atomic number Number of protons in an atom.

TIP ✓

A carbon atom is neutral because it contains 6 protons (charge + 6) and 6 electrons (charge – 6) and so has no overall charge.

KEY TERM ★

Mass number Number of protons plus the number of neutrons in an atom.

mass number = 23

atomic number = 11

mass number: protons + neutrons
atomic number: protons
number of protons = 11
number of neutrons = 23 – 11 = 12

▲ **Figure 9.3** Working out the number of neutrons in an atom.

TIP ✓

Atoms are sometimes shown in the form ^{23}Na. As the atom is a sodium atom, it must have an atomic number of 11 and so it is not necessary to include the atomic number.

Example

How many protons, neutrons and electrons are there in an atom of $^{81}_{35}Br$?

Answer

Number of protons: 35 (we could also find this by looking at the atomic number in the periodic table if it was not shown).

Number of neutrons: 81 – 35 = 46 (the mass number minus the number of protons).

Number of electrons: 35 (the same as the number of protons).

○ Isotopes

For most elements there are atoms with different numbers of neutrons. Atoms with the same number of protons but a different number of neutrons are called isotopes. This means that **isotopes** have the **same atomic number** but a **different mass number**.

For example, carbon has three isotopes and so there are three different types of carbon atoms. These are shown in the table below. These three isotopes are all carbon atoms because they all contain 6 protons, but they each have a different number of neutrons (Table 9.2).

KEY TERM ★

Isotopes Atoms with the same number of protons, but a different number of neutrons.

TIP ✓

The three isotopes of carbon are all **atoms** because they have the same number of electrons as **protons**. They are neutral.

TIP
Relative atomic mass is not the same as mass number. Mass numbers are integers because they are the number of protons plus the number of neutrons. Relative atomic mass is an average mass that takes into account all the isotopes of an element and so is not an integer.

KEY TERM
Relative atomic mass The average mass of atoms of an element taking into account the mass and amount of each isotope it contains (on a scale where the mass of a ^{12}C atom is 12).

TIP
When you calculate relative atomic mass, the answer should have a value somewhere between the mass of the lightest isotope and the heaviest isotope.

Table 9.2 The three isotopes of carbon.

Atom	$^{12}_{6}C$	$^{13}_{6}C$	$^{14}_{6}C$
Protons	6	6	6
Neutrons	6	7	8
Electrons	6	6	6

Isotopes have a different mass and mass number, but their chemical properties are the *same* because they contain the same number of electrons.

Relative atomic mass

The relative atomic mass (A_r) of an element is the average mass of atoms of that element taking into account the mass and amount of each isotope it contains.

This can be calculated as shown:

$$\text{Relative atomic mass } (A_r) = \frac{\text{total mass of all atoms of element}}{\text{total number of atoms of that element}}$$

Example

Find the relative atomic mass of chlorine which is found to contain 75% of atoms with mass number 35, and 25% of atoms with mass number 37. Give the answer to one decimal place.

Answer

$$\text{Relative atomic mass } (A_r) = \frac{[(75 \times 35) + (25 \times 37)]}{75 + 25} = \frac{3550}{100} = 35.5$$

Test yourself

7 List the three particles found inside atoms.

8 Identify the particle found inside the nucleus of atoms that has no charge.

9 Atoms contain positive and negative particles. Explain why atoms are neutral.

10 How many protons, neutrons and electrons are there in an atom of $^{31}_{15}P$?

11 What is it about the atom $^{39}_{19}K$ that makes it an atom of potassium?

12 Describe the similarities and differences between atoms of the isotopes $^{35}_{17}Cl$ and $^{37}_{17}Cl$.

13 The element copper contains 69% ^{63}Cu and 31% ^{65}Cu. Find the relative atomic mass of copper to one decimal place. Show your working.

14 The element magnesium contains 79% ^{24}Mg, 10% ^{25}Mg and 11% ^{26}Mg. Find the relative atomic mass of magnesium to one decimal place. Show your working.

15 Explain why mass number is an integer, but relative atomic mass is not.

Show you can...

Copy and complete the table for each of the elements listed.

Table 9.3

Element	Atomic number	Mass number	Number of protons	Number of electrons	Number of neutrons
$^{7}_{3}Li$					
$^{24}_{12}Mg$					
$^{27}_{13}Al$					
$^{39}_{19}K$					
$^{107}_{47}Ag$					

○ Electron arrangement

The **electrons** in an atom are in **energy levels**, also known as **shells**. Electrons occupy the lowest available energy levels. The lowest energy level (the first shell) is the one closest to the nucleus and can hold up to two electrons. Up to eight electrons occupy the second energy level (the second shell) with the next eight electrons occupying the third energy level (third shell). If there are two more electrons they occupy the fourth energy level.

The arrangement of electrons in some atoms are shown in Table 9.4. The electronic structure can be drawn on a diagram or written using numbers. For example, the electronic structure of aluminium is 2,8,3 which means that it has two electrons in the first energy level, eight electrons in the second energy level and three electrons in the third energy level (Table 9.4).

> **KEY TERM**
>
> **Electronic structure** The arrangement of electrons in the shells (energy levels) of an atom.

Table 9.4 The electronic structures of some atoms.

Atom	He	F	Al	K
Atomic number	2	9	13	19
Number of electrons	2	9	13	19
Electronic structure (written)	2	2,7	2,8,3	2,8,8,1
Electronic structure (drawn)				

○ Ions

Ions are particles with an electric charge because they do **not** contain the same number of protons and electrons. Remember that protons are positive and electrons are negative. Positive ions have more protons than electrons. Negative ions have more electrons than protons.

> **KEY TERM**
>
> **Ion** An electrically charged particle containing different numbers of protons and electrons.

For example:

● An ion with 11 protons (total charge 11+) and 10 electrons (total charge 10−) will have an overall charge of 1+.
● An ion with 16 protons (total charge 16+) and 18 electrons (total charge 18−) will have an overall charge of 2−.

Test yourself

16 Write the electronic structure of the following atoms:
$^{16}_{8}O$, $^{23}_{11}Na$, $^{40}_{20}Ca$.

17 Lithium atoms contain 3 electrons and have the electronic structure 2,1. State why the electrons are not all in the first shell.

Show you can...

The diagram shows an atom of an element X, where: **e** represents an electron; **n** represents a neutron; and **p** represents a proton.

a) Name the element X.
b) Write the electronic structure of X.
c) What is the mass number of this atom of element X?
d) Name the part of the atom shaded red.

▲ Figure 9.4

Table 9.5 shows some common ions.

Table 9.5 The charges and electronic structures of some atoms.

Ion	Li⁺	Al³⁺	Cl⁻	O²⁻
Atomic number	3	13	17	8
Number of protons	3 (charge 3+)	13 (charge 13+)	17 (charge 17+)	8 (charge 8+)
Number of electrons	2 (charge 2−)	10 (charge 10−)	18 (charge 18−)	10 (charge 10−)
Overall charge	1+	3+	1−	2−
Electronic structure (written)	2	2,8	2,8,8	2,8
Electronic structure (drawn)				

Simple ions (those made from single atoms) have the same electronic structure as the elements in Group 0 of the periodic table (Table 9.6). The elements in Group 0 are called the noble gases. The noble gases have very stable electronic structures.

Table 9.6 Common ions with the same electronic structure as Group 0 elements.

Group 0 element	He	Ne	Ar
Electronic structure	2	2,8	2,8,8
Common ions with the same electronic structure	Li⁺, Be²⁺	O²⁻,F⁻, Na⁺, Mg²⁺, Al³⁺	S²⁻, Cl⁻, K⁺, Ca²⁺

The hydrogen ion (H⁺) is the only simple ion that does not have the electronic structure of a noble gas. It does not have any electrons at all. This makes it a very special ion with special properties, and it is the H⁺ ion that is responsible for the behaviour of acids.

Show you can...

Table 9.7 gives some information about six different particles, A, B, C, D, E and F. Some particles are **atoms** and some are **ions**. (The letters are not chemical symbols).

Table 9.7 Features of some atoms and ions.

Particle	Atomic number	Mass number	Number of protons	Number of neutrons	Number of electrons	Electronic structure
A	18	40				2,8,8
B		27	13			2,8
C			20	20	20	
D		35	17			2,8,7
E	16	32			18	
F	17			20	17	

a) Copy and complete the table.
b) Particle C is an atom. Explain, using the information in the table, why particle C is an atom.
c) Particle E is a negative ion. What is the charge on this ion?
d) Which two atoms are isotopes of the same element?

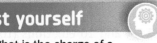

Test yourself

18 What is the charge of a particle with 19 protons and 18 electrons?

19 What is the charge of a particle with 7 protons and 10 electrons?

20 What is the electronic structure of the P³⁻ ion?

21 How many protons, neutrons and electrons are there in the ¹⁹₉F⁻ ion?

22 What is the link between the electronic structure of ions and the Group 0 elements (the noble gases)?

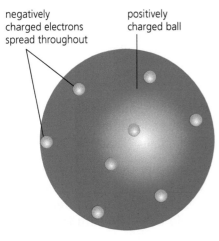

negatively
charged electrons
spread throughout

positively
charged ball

▲ **Figure 9.5** Plum pudding model of the atom (1897).

○ Development of ideas about the structure of atoms

The idea that everything was made of particles called **atoms** was accepted in the early 1800s after work by John Dalton. At that time, however, people thought that atoms were the smallest possible particles and the word *atom* comes from the Greek word *atomos* which means something that cannot be divided.

However, in 1897 the electron was discovered by J.J. Thomson while carrying out experiments on the conduction of electricity through gases. He discovered that electrons were tiny, negatively charged particles that were much smaller and lighter than atoms. He came up with what was called the '**plum pudding**' model of the atom. In this model, the atom was a ball of positive charge with the negative electrons spread through the atom (Figure 9.5).

A few years later in 1911, this model was replaced following some remarkable work from **Hans Geiger** and **Ernest Marsden** working with Ernest Rutherford. They fired alpha particles (He^{2+} ions) at a very thin piece of gold foil. They expected the particles to pass straight through the foil but a tiny fraction were deflected or even bounced back. This did not fit in with the plum pudding model. **Rutherford** worked out that the scattering of some of the alpha particles meant that there must be a tiny, positive nucleus at the centre of each atom. This new model was known as the **nuclear model** (Figure 9.6).

plum pudding model (1897)

nuclear model (1911, but revised since)

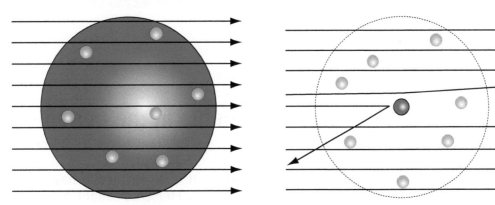

The alpha particles would all be expected to travel straight through the gold foil according to the plum pudding model

A tiny fraction of alpha particles were deflected or bounced back. Rutherford worked out that there must be a tiny, positive nucleus to explain this

▲ **Figure 9.6** The plum pudding model and an early nuclear model of the atom.

In 1913, **Neils Bohr** adapted the nuclear model to suggest that the **electrons** moved in stable orbits at **specific distances from the nucleus** called **shells**. Bohr's theoretical calculations agreed with observations from experiments.

Further experiments led to the idea that the positive charge of the nucleus was made up from particles which were given the name protons.

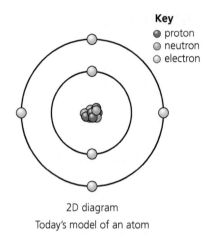

Key
- proton
- neutron
- electron

2D diagram
Today's model of an atom

▲ **Figure 9.7** Today's model of an atom.

Scientists realised that there was some mass in atoms that could not be explained by this model, and in 1932 **James Chadwick** discovered a new particle inside the nucleus that had the same mass as a proton but had no electric charge. This particle was given the name **neutron**.

The model has been developed further since then, but the basic idea of atoms being made up of a tiny central nucleus containing protons and neutrons surrounded by electrons in shells remains (Figure 9.7).

The development of ideas about atomic structure shows very well how scientific models and theories develop over time. When new discoveries are made, models and theories may have to be altered or sometimes completely replaced if they do not fit in with the new discoveries.

Test yourself

23 What was discovered that led to scientists realising that atoms were made up of smaller particles?
24 Why was the plum pudding model replaced?
25 Why would a nucleus deflect an alpha particle?

Show you can...

Use a table to compare and contrast the plum pudding model, the nuclear model and today's model of an atom.

Reactions of elements

KEY TERM

Element A substance containing only one type of atom; a substance that cannot be broken down into simpler substances by chemical methods.

TIP

When writing symbols you must be very careful to ensure your capital letters cannot be mistaken for small letters and vice versa.

◯ Elements in the periodic table

An **element** is a substance containing only one type of atom. For example, in the element carbon all the atoms are carbon atoms meaning that all the atoms have 6 protons and so have the atomic number 6. Elements cannot be broken down into simpler substances.

Atoms are known with atomic numbers ranging from 1 to just over 100. All the elements are listed in the periodic table. The elements are listed in order of atomic number (Figure 9.8).

Atoms of each element are given their own symbol, each with one, two or three letters. The first letter is always a capital letter with any further letters being small letters. For example, carbon has the symbol C while copper has the symbol Cu.

Group 1	Group 2											Group 3	Group 4	Group 5	Group 6	Group 7	Group 0
						1 **H** hydrogen 1											4 **He** helium 2
7 **Li** lithium 3	9 **Be** beryllium 4											11 **B** boron 5	12 **C** carbon 6	14 **N** nitrogen 7	16 **O** oxygen 8	19 **F** fluorine 9	20 **Ne** neon 10
23 **Na** sodium 11	24 **Mg** magnesium 12											27 **Al** aluminium 13	28 **Si** silicon 14	31 **P** phosphorus 15	32 **S** sulfur 16	35.5 **Cl** chlorine 17	40 **Ar** argon 18
39 **K** potassium 19	40 **Ca** calcium 20	45 **Sc** scandium 21	48 **Ti** titanium 22	51 **V** vanadium 23	52 **Cr** chromium 24	55 **Mn** manganese 25	56 **Fe** iron 26	59 **Co** cobalt 27	59 **Ni** nickel 28	63.5 **Cu** copper 29	65 **Zn** zinc 30	70 **Ga** gallium 31	73 **Ge** germanium 32	75 **As** arsenic 33	79 **Se** selenium 34	80 **Br** bromine 35	84 **Kr** krypton 36
85 **Rb** rubidium 37	88 **Sr** strontium 38	89 **Y** yttrium 39	91 **Zr** zirconium 40	93 **Nb** niobium 41	96 **Mo** molybdenum 42	[98] **Tc** technetium 43	101 **Ru** ruthenium 44	103 **Rh** rhodium 45	106 **Pd** palladium 46	108 **Ag** silver 47	112 **Cd** cadmium 48	115 **In** indium 49	119 **Sn** tin 50	122 **Sb** antimony 51	128 **Te** tellurium 52	127 **I** iodine 53	131 **Xe** xenon 54
133 **Cs** caesium 55	137 **Ba** barium 56	139 **La*** lanthanum 57	178 **Hf** hafnium 72	181 **Ta** tantalum 73	184 **W** tungsten 74	186 **Re** rhenium 75	190 **Os** osmium 76	192 **Ir** iridium 77	195 **Pt** platinum 78	197 **Au** gold 79	201 **Hg** mercury 80	204 **Tl** thallium 81	207 **Pb** lead 82	209 **Bi** bismuth 83	[209] **Po** polonium 84	[210] **At** astatine 85	[222] **Rn** randon 86
[223] **Fr** francium 87	[226] **Ra** radium 88	[227] **Ac**** actinium 89	[261] **Rf** rutherfordium 104	[262] **Db** dubnium 105	[266] **Sg** seaborgium 106	[264] **Bh** bohrium 107	[277] **Hs** hassium 108	[268] **Mt** meitnerium 109	[271] **Ds** darmstadtium 110	[272] **Rg** roentgenium 111	[285] **Cn** copernicium 112	[284] **Uut** ununtrium 113	[289] **Fl** flerorium 114	[288] **Uup** ununpentium 115	[293] **Lv** livermorium 116	[294] **Uus** ununseptium 117	[294] **Uuo** ununoctium 118

Key
relative atomic mass **atomic symbol** name atomic (proton) number

*

140 **Ce** cerium 58	141 **Pr** praseodymium 59	144 **Nd** neodymium 60	(145) **Pm** promethium 61	150 **Sm** samarium 62	152 **Eu** europium 63	157 **Gd** gadolinium 64	159 **Tb** terbium 65	162 **Dy** dysprosium 66	165 **Ho** holmium 67	167 **Er** erbium 68	169 **Tm** thulium 69	173 **Yb** ytterbium 70	175 **Lu** lutetium 71
232 **Th** thorium 90	231 **Pa** protactinium 91	238 **U** uranium 92	237 **Np** neptunium 93	(244) **Pu** plutonium 94	(243) **Am** americium 95	(247) **Cm** curium 96	(247) **Bk** berkelium 97	(251) **Cf** californium 98	(252) **Es** einsteinium 99	(257) **Fm** fermium 100	(258) **Md** mendelevium 101	(259) **No** nobelium 102	(260) **Lr** lawrencium 103

**

▢ Metal   ▢ Non-metal   ▢ Difficult to classify

▲ **Figure 9.8** The periodic table.

TIP ✓

The periodic table shows relative atomic mass and not mass number. This is the average mass of the isotopes of each element.

TIPS ✓

- Metals are found to the left and towards the bottom of the periodic table.
- Non-metals are found towards the right and top of the periodic table.

⭕ Metals and non-metals

Over three-quarters of the elements are metals, with most of the rest being non-metals. Typical properties of metals and non-metals are shown in Table 9.8, although there are some exceptions.

Table 9.8 Properties of metals and non-metals.

	Metals	Non-metals
Melting and boiling points	High	Low
Conductivity	Thermal and electrical conductor	Thermal and electrical insulator (except graphite)
Density	High density	Low density
Appearance	Shiny when polished	Dull
Malleability	Can be hammered into shape	Brittle as solids
Reaction with non-metals	React to form positive ions in ionic compounds	React to form molecules
Reaction with metals	No reaction	React to form negative ions in ionic compounds
Acid-base properties of oxides	Metal oxides are basic	Non-metal oxides are acidic

There are a few elements around the dividing line between metals and non-metals, such as silicon and germanium, that are hard to classify as they have some properties of metals and some of non-metals.

Test yourself

26 Is each of the following elements a metal or non-metal?
  a) Element **1** is a dull solid at room temperature that easily melts when warmed.
  b) Element **2** is a dense solid that is a thermal conductor.
  c) Element **3** reacts with oxygen to form an oxide which dissolves in rain water to form acid rain.
  d) Element **4** reacts with chlorine to form a compound made of molecules.
  e) Element **5** reacts with sodium to form a compound made of ions.

Show you can...

Figure 9.9 shows magnesium and oxygen reacting to form a single product.

a) **State two differences in the physical properties of magnesium and oxygen.**
b) **Suggest the name of the product of this reaction.**
c) **Is the product acidic or basic?**
d) **Does the product consist of ions or molecules?**

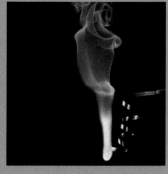

▲ Figure 9.9

KEY TERM

Compound Substance made from different elements chemically bonded together.

○ Reactions between elements

When elements react with each other they form compounds. **Compounds** are substances made from different elements **bonded** (chemically joined) together. A chemical reaction takes place when elements combine to form compounds. Chemical reactions always involve the formation of one or more new substances and there is usually a detectable energy change.

When elements react with each other, electrons are either **shared** with other elements or **transferred** from one element to another. This is done so that atoms obtain the stable electronic structure of the noble gases (Group 0 elements). Bonding will be covered in more detail in the next chapter.

Table 9.9 shows what happens in general when elements react with each other.

Table 9.9 Overview of reactions involving metals and/or non-metals.

Elements reacting	What happens to the electrons to obtain noble gas electronic structures	Type of particles formed	Type of compound formed	Example
Non-metal + non-metal	Electrons shared	Molecules (where atoms are joined to each other by covalent bonds)	Molecular compound	Hydrogen reacts with oxygen by sharing electrons and forming molecules of water
Metal + non-metal	Electrons transferred from metal to non-metal	Positive and negative ions	Ionic compound	Sodium reacts with chlorine by transferring electrons from sodium to chlorine to form sodium chloride which is made of ions
Metal + metal	No reaction as both metals cannot lose electrons			

Test yourself

27 Do the following elements react with each other, and if they do, what type of compound is formed?
   a) potassium + oxygen
   b) bromine + iodine
   c) oxygen + sulfur
   d) sulfur + magnesium
   e) calcium + potassium
   f) nitrogen + hydrogen

Show you can...

The electronic structures of the atoms of 5 different elements, A, B, C, D and E, are shown below.

A 2,8,8   B 2,8,8,1   C 2,6
D 2,1   E 2,8,7

Using the letters A, B, C, D or E choose:

a) An unreactive element.
b) Two elements found in the same Group of the periodic table.
c) An element whose atoms will form ions with a charge of 2–.
d) Two elements that react to form an ionic compound.
e) Two elements that react to form a covalent compound.

The periodic table

TIP
The columns in the periodic table are **groups** and the rows are **periods**.

○ Electronic structure and the periodic table

The elements are placed in the periodic table in **order of increasing atomic number** (the number of protons). Figure 9.10 shows the first 36 elements in the periodic table.

Group 1	Group 2										Group 3	Group 4	Group 5	Group 6	Group 7	Group 0
														H 1		He 2
Li 3	Be 4										B 5	C 6	N 7	O 8	F 9	Ne 10
Na 11	Mg 12										Al 13	Si 14	P 15	S 16	Cl 17	Ar 18
K 19	Ca 20	Sc 21	Ti 22	V 23	Cr 24	Mn 25	Fe 26	Co 27	Ni 28	Cu 29	Zn 30	Ga 31	Ge 32	As 33	Se 34	Br 35	Kr 36

▲ **Figure 9.10** The first 36 elements in the periodic table, with their atomic (proton) number.

KEY TERM

Period The name given to a row in the periodic table.

The table can be seen as arranging the elements by electronic structure. At the end of each **period**, a noble gas stable electronic structure is reached and then a new energy level (shell) starts to be filled at the start of the next period. The electronic structure of the first 20 elements is shown in Figure 9.11. Note the increasing number of electrons in the elements going from left to right in each row (period).

Group 1	Group 2											Group 3	Group 4	Group 5	Group 6	Group 7	Group 0
								H 1									He 2
Li 2,1	Be 2,2											B 2,3	C 2,4	N 2,5	O 2,6	F 2,7	Ne 2,8
Na 2,8,1	Mg 2,8,2											Al 2,8,3	Si 2,8,4	P 2,8,5	S 2,8,6	Cl 2,8,7	Ar 2,8,8
K 2,8,8,1	Ca 2,8,8,2																

▲ **Figure 9.11** The electronic structures of the first 20 elements in the periodic table.

TIP

The chemical properties of the elements in the periodic table repeat at regular (periodic) intervals. This is why it is called the periodic table.

Elements in the **same group** (column) have the **same number of electrons in their outer shell.** The number of electrons in the outer shell equals the Group number. For example, all the elements in Group 1 have 1 electron in their outer shell (Li = 2,1; Na = 2,8,1; K = 2,8,8,1) (Figure 9.11). All the elements in Group 7 have 7 electrons in their outer shell (F = 2,7; Cl = 2,8,7). The only exception to this is Group 0 where all the elements have 8 electrons in their outer shell except helium which has 2 electrons (but the first shell can only hold 2 electrons).

All the elements in the same group have similar chemical properties because they have the same number of electrons in their outer shell.

Elements in the same period (row) have the same number of shells (electron levels) – see Figure 9.11.

Show you can...

Element A has electronic structure 2,8,1.

a) Explain why element A is not found in Group 5.
b) Determine the atomic number of A.

Test yourself

28 In what order are the elements in the periodic table?
29 In which group of the periodic table do the elements with these electron structures belong?
a) 2,8,4 b) 2,8,8,1 c) 2,8,18,3
30 Explain why the periodic table has the word 'periodic' in its name.

○ Group 0 – the noble gases

KEY TERM

Noble gases The elements in Group 0 of the periodic table (including helium, neon and argon).

The main elements of Group 0 are helium, neon, argon, krypton, xenon and radon (Figure 9.12 and Table 9.10). They are known as the **noble gases** (Table 9.11). These atoms all have stable electronic structures. Helium's outer shell is full with 2 electrons while the others have 8 electrons in their outer shells.

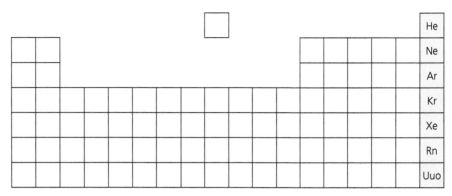

▲ **Figure 9.12** Group 0 – the noble gases.

Table 9.10 The noble gases.

Element	Formula	Appearance at room temperature	Number of electrons in outer shell	Relative mass of atoms	Boiling point in °C
Helium	He	Colourless gas	2	4	−269
Neon	Ne	Colourless gas	8	20	−246
Argon	Ar	Colourless gas	8	40	−190
Krypton	Kr	Colourless gas	8	84	−157
Xenon	Xe	Colourless gas	8	131	−111
Radon	Rn	Colourless gas	8	222	−62

Table 9.11 Properties of the noble gases.

Metals or non-metals?	All the elements are **non-metals**.
Boiling points	The noble gases are all colourless gases with low boiling points. The **boiling points increase** as the **atoms get heavier going down the group**.
Reactivity	The Group 0 elements are very **unreactive** and do not easily react to form molecules or ions because their atoms have stable electron arrangements.

Test yourself

31 Why are the noble gases unreactive?
32 Suggest a reason why the noble gases are refered to as being in Group 0 rather than Group 8.
33 Some atoms of element 118 (Uuo) have been produced. Element 118 is in Group 0. Predict the chemical and physical properties of this element.

○ Group 1 – the alkali metals

The main elements of Group 1 are lithium, sodium, potassium, rubidium and caesium (Table 9.13). They are known as the alkali metals (Figure 9.13 and Table 9.14). The Group 1 elements have similar chemical and physical properties because they all have one electron in their outer shell. They are all soft metals that can be cut with a knife (Figure 9.14). They are very reactive and so are stored in bottles of oil to stop them reacting with water and oxygen. They are very reactive because they have only one electron in their outer shell, which can easily be lost.

▲ **Figure 9.14** The alkali metals are all soft and can be cut with a knife.

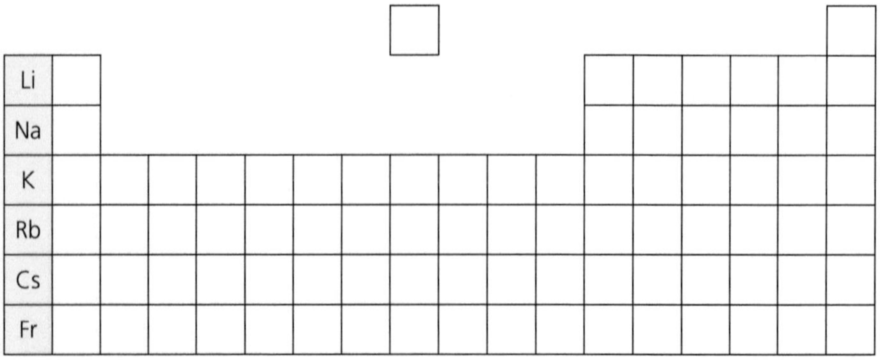

| Li |
| Na |
| K |
| Rb |
| Cs |
| Fr |

▲ **Figure 9.13** Group 1 – the alkali metals.

Table 9.13 The alkali metals.

Element	Formula	Appearance at room temperature	Number of electrons in outer shell	Relative mass of atoms	Melting point in °C	Density in g/cm^3
Lithium	Li	Silvery-grey metal	1	7	180	0.53
Sodium	Na	Silvery-grey metal	1	23	98	0.97
Potassium	K	Silvery-grey metal	1	39	63	0.89
Rubidium	Rb	Silvery-grey metal	1	85	39	1.53
Caesium	Cs	Silvery-grey metal	1	133	28	1.93

Table 9.14 Properties of the alkali metals.

Metals or non-metals?	All the elements are **metals**.
Melting points	The alkali metals are all solids with relatively low melting points at room temperature. The **melting points decrease** as the atoms get bigger **going down** the group.
Density	The alkali metals have **low densities for metals**. Lithium, sodium and potassium all float on water as they are less dense than water.
Reaction with non-metals	The metals all **react easily** with non-metals by the transfer of electrons from the metal to the non-metal forming compounds made of ions. Alkali metals always form **1+ ions** (e.g. Li$^+$, Na$^+$, K$^+$, Rb$^+$, Cs$^+$) as they have one electron in their outer shell which they lose when they react to obtain a noble gas electronic structure.
Reaction with oxygen	The alkali metals all burn in oxygen to form metal oxides which are white powders: **metal + oxygen → metal oxide** sodium + oxygen → sodium oxide e.g. $4Na + O_2 \rightarrow 2Na_2O$ The metals burn with different colour flames. For example lithium burns with a crimson-red flame, sodium with a yellow-orange flame and potassium with a lilac flame.
Reaction with chlorine	The alkali metals all burn in chlorine to form metal chlorides which are white powders: **metal + chlorine → metal chloride** e.g. sodium + chlorine → sodium chloride e.g. $2Na + Cl_2 \rightarrow 2NaCl$
Reaction with water	The alkali metals all react with water, releasing hydrogen gas and forming a solution containing a metal hydroxide: **metal + water → metal hydroxide + hydrogen**: e.g. sodium + water → sodium chloride + hydrogen e.g. $2Na + 2H_2O \rightarrow 2NaOH + H_2$ The solution of the metal hydroxide that is formed is **alkaline**.
Compounds made from Group 1 metals	Compounds made from alkali metals: are **ionic** are **white solids** dissolve in water to form **colourless solutions**

Reactivity trend of the alkali metals

The alkali metals get **more reactive** the **further down the group** (Figure 9.15). This can be seen when the alkali metals react with water.

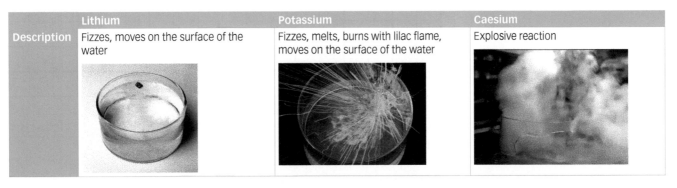

	Lithium	Potassium	Caesium
Description	Fizzes, moves on the surface of the water	Fizzes, melts, burns with lilac flame, moves on the surface of the water	Explosive reaction

lithium (2, 1)

sodium (2, 8, 1)

potassium (2, 8, 8, 1)

▲ **Figure 9.15** The further down the group, the more shells are added and the further the outer electron is from the nucleus.

TIP ✓

A molecule is atoms joined together by covalent bonds.

KEY TERMS ★

Halogens The elements in Group 7 of the periodic table (including fluorine, chlorine, bromine and iodine).

Diatomic molecule A molecule containing two atoms.

When the alkali metals react they are losing their outer shell electron in order to get a stable electronic structure. The further down the group, the further away the outer electron is from the nucleus as the atoms get bigger. This means that the outer electron is less strongly attracted to the nucleus and so easier to lose. The easier the electron is to lose, the more reactive the alkali metal.

Test yourself

34 Why are the alkali metals reactive?

35 Write a word equation for the reaction of potassium with water.

36 Explain why the solution formed when potassium reacts with water has a high pH.

37 Potassium reacts with chlorine to form an ionic compound. Explain why this reaction happens.

38 Explain why potassium is more reactive than sodium.

39 Francium is the last element in Group 1. Predict the chemical and physical properties of francium.

○ Group 7 – the halogens

The main elements of Group 7 are fluorine, chlorine, bromine and iodine (Figure 9.16 and Table 9.15). They are known as the halogens (Figure 9.16 and Table 9.16). The Group 7 elements have similar chemical and physical properties because they all have seven electrons in their outer shell. The halogens are reactive because they only need to gain one electron to gain a noble gas electronic structure. The particles in each of the elements are **molecules** containing **two atoms** (diatomic molecules), such as F_2, Cl_2, Br_2 and I_2.

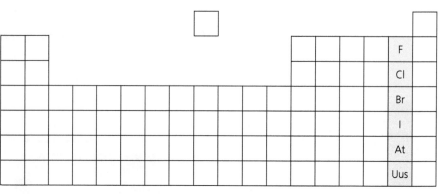

▲ **Figure 9.16** Group 7 – the halogens.

Table 9.15 The halogens.

Element	Formula	Appearance at room temperature	Number of electrons in outer shell	Relative mass of molecules	Melting point in °C	Boiling point in °C
Fluorine	F_2	Pale yellow gas	7	38	−220	−188
Chlorine	Cl_2	Pale green gas	7	71	−102	−34
Bromine	Br_2	Dark brown liquid	7	160	−7	59
Iodine	I_2	Grey solid	7	254	114	184

Table 9.16 Properties of the halogens.

Metals or non-metals?	Fluorine, chlorine, bromine and iodine are all **non-metals**.
Toxicity	Each of the halogens is toxic.
Melting and boiling points	The halogens have low melting and boiling points. The **melting and boiling points increase** as the molecules get heavier going down the group.
Reaction with non-metals	The halogens react with other non-metals by sharing electrons to form **compounds made of molecules** (molecular compounds).
Reaction with metals	The halogens all react easily with metals by the transfer of electrons from the metal to the halogen forming compounds made of ions (ionic compounds). Halogens always form **1− ions** (e.g. F⁻, Cl⁻, Br⁻, I⁻, all known as **halide ions**) as they have seven electrons in their outer shell and gain one more electron when they react to get a noble gas electronic structure.

▲ **Figure 9.17** Bromine is a dark brown liquid that easily vaporises to give an orange gas.

Reactivity trend of the halogens

The halogens get less reactive the further down the group. This can be seen by looking at which halogens can displace each other from compounds. Compounds containing halogens, such as sodium chloride and potassium bromide are often called halides or halide compounds.

A more reactive element can displace a less reactive element from a compound. You have seen this (before GCSE) with metals when a more reactive metal can displace a less reactive metal from a compound. For example, aluminium can displace iron from iron oxide because aluminium is more reactive than iron.

aluminium + iron oxide → aluminium oxide + iron

In a similar way, a more reactive non-metal can displace a less reactive non-metal from a compound. This means that a more reactive halogen can displace a less reactive halogen from a halide compound.

▲ **Figure 9.18** Colourless chlorine water reacts with colourless potassium bromide solution to form a yellow solution of bromine.

This can be seen when aqueous solutions of the halogens react with aqueous solutions of halide compounds (aqueous means dissolved in water) (Table 9.17 and Figure 9.18).

Table 9.17 Displacement reactions involving halogens and halide compounds.

	Chlorine (aq)	Bromine (aq)	Iodine (aq)
Potassium chloride (aq)		**No reaction** Bromine cannot displace chlorine	**No reaction** Iodine cannot displace chlorine
Potassium bromide (aq)	chlorine + potassium bromide → potassium chloride + bromine $Cl_2 + 2KBr \rightarrow 2KCl + Br_2$ ($Cl_2 + 2Br^- \rightarrow 2Cl^- + Br_2$) Yellow solution formed (due to production of bromine) Chlorine displaces bromine		**No reaction** Iodine cannot displace bromine
Potassium iodide (aq)	chlorine + potassium iodide → potassium chloride + iodine $Cl_2 + 2KI \rightarrow 2KCl + I_2$ ($Cl_2 + 2I^- \rightarrow 2Cl^- + I_2$) Brown solution formed (due to production of iodine) Chlorine displaces iodine	bromine + potassium iodide → potassium bromide + iodine $Br_2 + 2KI \rightarrow 2KBr + I_2$ ($Br_2 + 2I^- \rightarrow 2Br^- + I_2$) Brown solution formed (due to production of iodine) Bromine displaces iodine	

It can be seen from these reactions that the trend in reactivity for these three halogens is:

Most reactive   Chlorine

↑   Bromine

Least reactive   Iodine

In general, the **further down the group the less reactive the halogen** (Figure 9.19). The higher up the group, the more reactive the halogen. This means that fluorine is the most reactive halogen. You will not do experiments with fluorine because it is very reactive and toxic.

When the halogens react they gain one electron in order to get a noble gas electronic structure. The further down the group, the electron gained will enter an energy level further away from the nucleus as the atoms get bigger. This means that the electron gained is less strongly attracted to the nucleus and so harder to gain. The harder the electron is to gain, the less reactive the halogen.

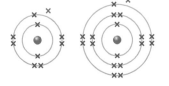

fluorine (2,7)   chlorine (2,8,7)

▲ **Figure 9.19** The further down the group, the further the electron gained is from the nucleus.

TIP ✓

The explanation for the reactivity trend in Group 7 is about the distance between the nucleus and the electron gained which is from **outside** the atom – it is not about the outer shell electrons.

Test yourself

40 Why are the halogens reactive?
41 All the halogens are made of diatomic molecules. What are diatomic molecules?
42 Predict what would happen, and why, if fluorine and sodium chloride were mixed.
43 Bromine reacts with chlorine to form a molecular compound. Explain why this reaction happens.
44 Explain why chlorine is more reactive than bromine.

Reactions of the halogens

The diagram shows chlorine gas being passed through a dilute solution of potassium iodide. The upper layer is a hydrocarbon solvent. A colour change occurs in the potassium iodide solution due to the displacement reaction that occurs.

1 a) What is the most important safety precaution, apart from wearing safety glasses, which must be taken when carrying out this experiment?

b) Explain why a displacement reaction occurs between chlorine and potassium iodide.

c) Name the products of the displacement reaction which occurs.

d) What is the colour change that occurs in the potassium iodide solution?

e) Write a balanced symbol equation for the reaction between chlorine and potassium iodide.

f) If this experiment was repeated using potassium bromide, instead of potassium iodide, explain if the observations would be different.

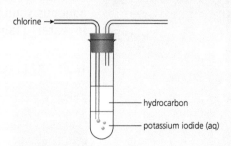

chlorine

hydrocarbon

potassium iodide (aq)

▲ Figure 9.20

2 The halogens are more soluble in hydrocarbon solvents than in water and produce coloured solutions. After the reaction of chlorine with aqueous potassium iodide, the aqueous layer is shaken with the hydrocarbon solvent and most of the displaced halogen dissolves in the upper layer.

a) Explain the meaning of the word solvent.

b) Use the information in the table to suggest what happens to the hydrocarbon solvent, after shaking.

Table 9.18

Halogen	Colour of hydrocarbon when halogen dissolves
Chlorine	Pale green
Bromine	Orange
Iodine	Purple

Show you can...

The diagram shows part of the periodic table divided into five sections A, B, C, D and E.

Select from A to E the section in which you would find:

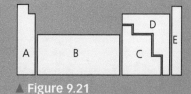

▲ Figure 9.21

a) A gas that forms no compounds.
b) A metal that is used as a catalyst in industrial reactions.
c) A metal that reacts rapidly with cold water.
d) A coloured gas.
e) A metal that forms coloured compounds.

○ History of the periodic table

As more elements were discovered, scientists tried to classify the elements into some sort of order and pattern. This was originally done before the discovery of protons, neutrons and electrons. Scientists' first attempts were based on the use of the atomic weights of elements which we now know as relative atomic mass.

John Newlands spotted that the properties of elements seemed to repeat every eighth element when placed in order of atomic weight. He called this the '**law of octaves**' as it was similar to notes in musical scales. One of the successes of his table was that he had lithium, sodium and potassium in the same group, each of which has very similar properties (Figure 9.22).

	No.		No.		No.		No.		No.		No.		No.		No.		No.
H	1	F	8	Cl	15	Co & Ni	22	Br	29	Pd	36	I	42	Pt & Ir	50
Li	2	Na	9	K	16	Cu	23	Rb	30	Ag	37	Cs	44	Os	51
G	3	Mg	10	Ca	17	Zn	24	Sr	31	Cd	38	Ba & V	45	Hg	52
Bo	4	Al	11	Cr	19	Y	25	Ce & La	33	U	40	Ta	46	Tl	53
C	5	Si	12	Ti	18	In	26	Zr	32	Sn	39	W	47	Pb	54
N	6	P	13	Mn	20	As	27	Di & Mo	34	Sb	41	Nb	48	Bi	55
O	7	S	14	Fe	21	Se	28	Ro & Ru	35	Au	43	Au	49	Th	56

▲ **Figure 9.22** Newlands' table (1865).

At the time though, only about 50 elements were known and his table was not accepted because it only worked for the first 20 or so of the known elements. After that there were problems, such as copper being in the same group as lithium, sodium and potassium. Copper has very different properties to those metals. For example, copper does not react with water but the other three react vigorously with water.

A few years later in 1869, a Russian chemist called **Dmitri Mendeleev** devised a table which has become the basis for the periodic table today (Figure 9.23). He also placed elements in **order of atomic weight**, but crucially he did two things differently to Newlands.

1 Mendeleev **left gaps for elements** he predicted had yet to be discovered. He also predicted the properties of these elements.

2 Mendeleev was prepared to **alter slightly the order of the elements** if he thought it fitted the properties better. For example, he swapped around the order of iodine (atomic weight 127) and tellurium (atomic weight 128). He placed iodine after tellurium so that it was in the same group as fluorine, chlorine and bromine which have very similar properties. Mendeleev actually believed that the atomic weights must have been measured incorrectly.

Tabelle II.

Reihen	Gruppe I. — R^2O	Gruppe II. — RO	Gruppe III. — R^2O^3	Gruppe IV. RH^4 RO^2	Gruppe V. RH^3 R^2O^5	Gruppe VI. RH^2 RO^3	Gruppe VII. RH R^2O^7	Gruppe VIII. — RO^4
1	H=1							
2	Li=7	Be=9,4	B=11	C=12	N=14	O=16	F=19	
3	Na=23	Mg=24	Al=27,3	Si=28	P=31	S=32	Cl=35,5	
4	K=39	Ca=40	—=44	Ti=48	V=51	Cr=52	Mn=55	Fe=56, Co=59, Ni=59, Cu=63.
5	(Cu=63)	Zn=65	—=68	—=72	As=75	Se=78	Br=80	
6	Rb=85	Sr=87	?Yt=88	Zr=90	Nb=94	Mo=96	—=100	Ru=104, Rh=104, Pd=106, Ag=108.
7	(Ag=108)	Cd=112	In=113	Sn=118	Sb=122	Te=128	J=127	
8	Cs=133	Ba=137	?Di=138	?Ce=140	—	—	—	— — — —
9	(—)		—	—	—	—	—	
10	—	—	?Er=178	?La=180	Ta=182	W=184	—	Os=195, Ir=197, Pt=198, Au=199.
11	(Au=199)	Hg=200	Tl=204	Pb=207	Bi=208	—	—	
12	—	—	—	Th=231	—	U=240	—	— — — —

der chemischen Elemente.

▲ **Figure 9.23** Mendeleev's table (1869).

Over the next few years, elements were discovered that Mendeleev had predicted would exist (Table 9.19). These included gallium (1875), scandium (1879) and germanium (1886). In each case the properties of the element closely matched Mendeleev's predictions. Table 9.19 shows some of the properties that Mendeleev predicted for the element he called eka-silicon and that we call germanium.

Table 9.19 Mendeleev's predictions for eka-silicon (germanium).

	Element	Appearance	Atomic weight	Melting point in °C	Density in g/cm³	Formula of oxide	Formula of chloride
Mendeleev's predictions	Eka-silicon (Es)	Grey metal	72	high	5.5	EsO_2	$EsCl_4$
Actual properties	Germanium (Ge)	Grey-white metal	73	947	5.4	GeO_2	$GeCl_4$

TIP ✔
Make sure you know the difference between atomic number (the number of protons in an atom) and atomic weight.

As these elements were discovered, Mendeleev's ideas and table became well accepted and formed the basis of the periodic table as we know it today. We now know that Mendeleev placed the elements in **order of atomic number** (the number of protons in an atom) even though he did not know about the existence of protons. It is the atomic number rather than atomic weight that matters, because elements are made of a mixture of isotopes and, depending on the isotope, the atomic weight will differ.

The story of Mendeleev illustrates how strong support can come for a scientific idea if predictions made using that theory are later found to be correct.

Show you can...

State some features of the periodic table developed by Mendeleev which are different from today's modern periodic table.

Test yourself

45 In what order did Mendeleev put the elements?
46 Why did Mendeleev not stick to this order for some elements?
47 Why did Mendeleev leave some gaps in his table?
48 Why did Mendeleev's ideas become accepted?

Mixtures

○ Mixtures compared to compounds

KEY TERM ★
Mixture More than one substance that are not chemically joined together.

A mixture consists of two or more substances that are mixed together and not chemically combined. In a mixture, each substance has its own properties. Mixtures are very different from compounds (Table 9.20).

Table 9.20 Differences between compounds and mixtures.

	Compound	Mixture
Description	A substance made from two or more elements chemically bonded together. A compound is a single substance with its own unique properties	Two or more substances each with their own properties (the different substances are not chemically joined to each other)
Proportions	Each compound has a fixed proportion of elements (so each compound has a fixed formula)	There can be any amount of each substance in the mixture
Separation	Compounds can only be separated back into elements by chemical reaction because the elements are chemically joined.	No chemical reaction is needed as the substances in the mixture are not chemically joined. They can be separated by physical methods (e.g. filtration, distillation).

Sodium is a very reactive, dangerous, grey metal that reacts vigorously with water. Chlorine is a pale green, toxic gas that is very reactive. In a mixture of sodium and chorine each substance keeps its own properties as a grey metal and green gas, respectively. It is easy to separate the sodium and chlorine because they are not chemically joined together.

However, if heated together sodium reacts with chlorine to make the compound sodium chloride. Sodium chloride is very different from both sodium and chlorine. Sodium chloride is a white solid that is not very reactive and is safe to eat. It is very difficult to break sodium chloride back down into the elements because the sodium and chlorine are chemically joined together.

Table 9.21

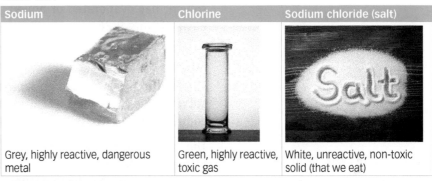

Sodium	Chlorine	Sodium chloride (salt)
Grey, highly reactive, dangerous metal	Green, highly reactive, toxic gas	White, unreactive, non-toxic solid (that we eat)

Show you can...

For each of the substances A, B, C, D decide if it is an element, compound or mixture.

If any substance is a mixture decide if it is a mixture of elements, a mixture of elements and compounds, or a mixture of compounds.

▶ **Figure 9.24**

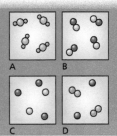

○ Separating mixtures

The substances in a mixture are quite easy to separate because the substances are not chemically joined to each other. Different methods are used depending on what type of mixture there is (Table 9.22).

Table 9.22 Different methods of separating mixtures.

Type of mixture	Insoluble solid and liquid	Soluble solid dissolved in a solvent	Soluble solids dissolved in a solvent	Two miscible liquids (liquids that mix)	Two immiscible liquids (liquids that do not mix)
Method of separation	Filtration	Evaporation (to obtain solid) Crystallisation (to obtain solid) Simple distillation (to obtain solvent)	Chromatography	Fractional distillation	Separating funnel

TIP ✓

Some definitions of key words:
- **solute**: the solid substance that dissolves in a solvent
- **solvent**: the liquid that a solute dissolves in
- **solution**: a solute dissolved in a solvent
- **soluble**: when a substance will dissolve in a solvent
- **insoluble**: when a substance does not dissolve in a solvent.

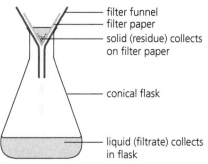

▲ **Figure 9.25** Filtration.

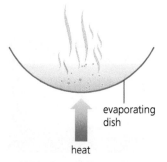

▲ **Figure 9.26** Evaporation.

▲ **Figure 9.27** Crystallisation.

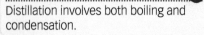

KEY TERMS ⭐

Filtrate Liquid that comes through the filter paper during filtration.

Residue Solid left on the filter paper during filtration.

Saturated (in the context of solutions) A solution in which no more solute can dissolve at that temperature.

Condenser The part of the apparatus that causes the solvent to condense to a liquid.

TIP ✔

Distillation involves both boiling and condensation.

Filtration

This method is used to separate an **insoluble solid** from a **liquid**. For example, it could be used to separate sand from water.

The mixture is poured through a funnel containing a piece of filter paper. The liquid (called the filtrate) passes through the paper and the solid (called the residue) remains on the filter paper (Figure 9.25).

Evaporation

This method is used to separate a **dissolved solid** from the **solvent** it is dissolved in. For example, it could be used to separate salt from water.

The mixture is placed in an evaporating dish and heated until all the solvent has evaporated or boiled, leaving the solid in the evaporating basin (Figure 9.26).

Crystallisation

This method is also used to separate a **dissolved solid** from the **solvent** it is dissolved in. For example, it could be used to separate copper sulfate crystals from a solution of copper sulfate (Figure 9.27).

The mixture is heated to boil off some of the solvent to create a hot, saturated solution. A saturated solution is one in which no more solute can dissolve at that temperature. As it cools down, the solute becomes less soluble and so cannot remain dissolved, so some of the solute crystallises out of the solution as crystals. The crystals can then be separated from the rest of the solution by filtration.

Simple distillation

This method is used to separate the **solvent** from a **solution**. For example, it could be used to separate pure water from sea water.

The mixture is heated and the solvent boils. The vaporised solvent passes through a water-cooled condenser where it cools and condenses. The condenser directs the condensed solvent into a container away from the original solution (Figure 9.28).

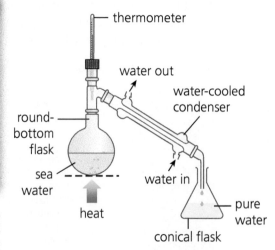

▲ **Figure 9.28** Distillation of sea water.

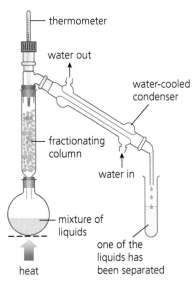

thermometer

water out

water-cooled condenser

fractionating column

water in

mixture of liquids

one of the liquids has been separated

heat

▲ **Figure 9.29** Fractional distillation.

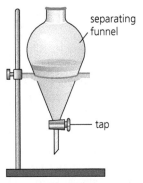

separating funnel

tap

▲ **Figure 9.30** Separating funnel.

▲ **Figure 9.31** Substances separate as they move up the paper with the solvent at different speeds.

Fractional distillation

Liquids that mix together are called miscible liquids. Water and alcohol are examples of miscible liquids. Fractional distillation is used to **separate mixtures of miscible liquids**. It works because the liquids have different boiling points.

The apparatus used is similar to that for simple distillation, but a long column (called a fractionating column) is used to help separate different liquids as they boil. The fractionating column often contains glass beads.

In industry, such as in the fractional distillation of crude oil, the whole mixture is vaporised and then condensed in a fractionating column which is hot at the bottom and cold at the top. The liquids will condense at different heights in the fractionating column (Figure 9.29).

Separating funnel

Liquids that do not mix together are called immiscible liquids. Oil and water is an example of liquids that are immiscible with each other. They can be separated in a separating funnel. The liquids form two layers and the bottom layer can be removed using the tap at the bottom of the funnel. The liquid with the greater density is the lower layer (Figure 9.30).

Chromatography

There are many forms of chromatography. Paper chromatography is used **to separate mixtures of substances dissolved in a solvent**.

A piece of chromatography paper, with the mixture on, is placed upright in a beaker so that the bottom of the paper is in the solvent. Over time, the solvent soaks up the paper. The substances move up the paper at different speeds and so are separated (Figure 9.31).

Test yourself

49 How would you separate the following mixtures?
   a) alcohol from a mixture of alcohol and water
   b) magnesium hydroxide from a suspension of insoluble magnesium hydroxide in water
   c) pure dry cleaning solvent from waste dry cleaning solvent containing dirt that dissolved in the solvent from clothes
   d) sunflower oil and water
   e) food colourings in a sweet.

Show you can...

Three common methods of separation include filtration, distillation and fractional distillation.

For each of these separation methods pick **two** words or phrases from the list and insert them into a copy of the table with an explanation of their meaning. Also include the type of mixture separated by each method:

condenser, distillate, fractionating column, filtrate, miscible liquids, residue.

Table 9.23

	Filtration	Distillation	Fractional distillation
Type of mixture separated			
Important word and definition			
Important word and definition			

Rock salt

Common salt is sodium chloride and is found naturally in large amounts in seawater or in underground deposits. Sodium chloride can be extracted from underground by the process of solution mining.

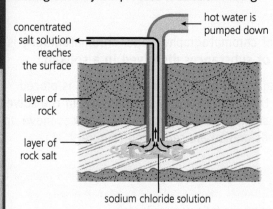

▲ Figure 9.32

1 a) On what physical property of sodium chloride does this process depend?

b) Suggest one reason why solution mining uses a lot of energy.

c) Suggest one negative effect which solution mining has on the environment.

d) Suggest how sodium chloride is obtained from the concentrated salt solution.

2 Rock salt is a mixture of salt, sand and clay. To separate pure salt from rock salt, the method listed below can be used in the laboratory.

Method:

i Place 8 spatulas of rock salt into a mortar and grind using a pestle.

ii Place the rock salt into a beaker and quarter fill with water.

iii Place on a gauze and tripod and heat, stirring with a glass rod. Stop heating when the salt has dissolved – the sand and clay will be left undissolved.

iv Allow to cool and then filter.

v Heat until half the volume of liquid is left.

vi Place the evaporating basin on the windowsill to evaporate off the rest of the water slowly. Pure salt crystals should be left.

Choose one step of the method (i to vi) which is best represented in each photograph A–C.

3 a) Why is rock salt considered to be a mixture?

b) What was the purpose of grinding the rock salt?

c) Why was the mixture heated and stirred?

d) State what the filtrate contains.

e) State what the residue contains.

f) Explain why the salt obtained may still be contaminated with sand and suggest how you would improve your experiment to obtain a purer sample of salt.

Chapter review questions

1 Choose from the following list of elements to answer the questions below:

   bromine   calcium   krypton   nickel   nitrogen   potassium   silicon

   a) Which element is most like lithium?

   b) Which element is most like iron?

   c) Which element is most like helium?

   d) Which element is most like fluorine?

   e) Which element is most like carbon?

2 In which group or area of the periodic table would you find these elements?

   a) Element **A** has 7 electrons in its outer shell.

   b) Element **B** reacts vigorously with water to give off hydrogen gas and an alkaline solution.

   c) Element **C** is a metal with 4 electrons in its outer shell.

   d) Element **D** is a colourless gas that does not react at all.

   e) Element **E** forms coloured compounds.

   f) Element **F** is toxic and is made of diatomic molecules.

   g) Element **G** forms 1– ions when it reacts with metals to form ionic compounds

   h) Element **H** can form both 1+ and 2+ ions

   i) Element **I** is a metal that floats on water

   j) Element **J** has the electronic structure 2,8,8,6

   k) Element **K** has 12 protons

   l) Element **L** has a full outer shell

   m) Element **M** can act as a catalyst

3 Identify a mixture that could be separated by each of the following methods.

   a) simple distillation

   b) filtration

   c) crystallisation

   d) evaporation

   e) chromatography

   f) fractional distillation

4 Look at the following atoms and ions.

   $^{12}_{6}C$ $^{14}_{6}C$ $^{16}_{8}O^{2-}$ $^{19}_{9}F^{-}$ $^{20}_{10}Ne$

   Which of these atoms and ions, if any,

   a) are isotopes?

   b) have 9 protons?

   c) have 10 electrons?

   d) have 10 neutrons?

   e) have more protons than electrons?

5 Caesium atoms are among the largest atoms. A caesium atom has a radius of 0.260 nm. Write this in metres in standard form.

6 Predict whether each of the following pairs of elements will (i) react by sharing electrons; (ii) react by transferring electrons; or (iii) not react:

a) magnesium + oxygen

b) sulfur + hydrogen

c) aluminium + magnesium

d) argon + oxygen

e) bromine + phosphorus

f) fluorine + lithium

7 Sodium (Na) reacts with bromine (Br_2) to form the ionic compound sodium bromide (NaBr). Sodium is in Group 1 of the periodic table. Bromine is in Group 7 of the periodic table.

a) What names are often given to Groups 1 and 7?

b) Describe in detail what happens in terms of electrons when sodium reacts with bromine.

c) Potassium reacts with bromine more vigorously than sodium. Explain why potassium reacts more vigorously than sodium.

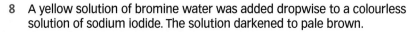

8 A yellow solution of bromine water was added dropwise to a colourless solution of sodium iodide. The solution darkened to pale brown.

a) Explain why the solution darkened.

b) Write an ionic equation for the reaction that took place.

c) Explain, in terms of electrons, why this reaction took place.

9 The following four substances are mixed together: salt; water; cyclohexane; diethyl ether. Cyclohexane and diethylether are liquids made of organic molecules. Salt is soluble in water but not in cyclohexane or diethyl ether. Cyclohexane and diethyl ether are miscible, but water is not miscible with cyclohexane or diethyl ether. Describe how the four substances could be separated.

Practice questions

1 How many electrons are there in a potassium ion (K^+)?

A 18

B 19

C 20

D 39   [1 mark]

2 In which of the following atoms is the number of protons greater than the number of neutrons?

A ^2_1H

B ^3_2He

C $^{10}_5B$

D $^{16}_8O$ [1 mark]

3 An aluminium atom contains three types of particle.

a) Copy and complete the table below to show the name, relative mass and relative charge of each particle in an aluminium atom.   [4 marks]

Table 9.24

Particle	Relative charge	Relative mass
Proton		1
		Very small
Neutron	0	

b) Complete the sentences below about an aluminium ion by choosing one of the words or numbers in bold.   [4 marks]

i) In an aluminium atom, the protons and neutrons are in the **nucleus/shells.**

ii) The number of protons in an aluminium atom is the **atomic number/group number/mass number.**

iii) The sum of the number of protons and neutrons in an aluminium atom is the **atomic number/group number/mass number.**

iv) The number of electrons in an aluminium atom is **13/14/27.**

4 The structure of the atom has caused debate for thousands of years. In the late 19th century the 'plum pudding model' of the atom was proposed. This was replaced at the beginning of the 20th century with the nuclear model of the atom which is the basis of the model we use today.

a) Describe the differences between the 'plum pudding' model of the atom and the model of the atom we use today.   [5 marks]

b) The diagram represents an atom of an element. The electrons are missing from the diagram.

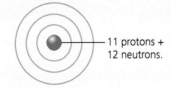

11 protons + 12 neutrons.

▲ Figure 9.33

i) State the atomic number of this element. [1 mark]

ii) State the mass number of this element.   [1 mark]

iii) Name the part of the atom in which the protons and neutrons are found.   [1 mark]

iv) Copy and complete the diagram to show the electron configuration of the atom, using x to represent an electron.   [1 mark]

c) The table shows some information for several atoms and simple ions. Copy and complete the table.   [6 marks]

Table 9.25

Atom/Ion	Number of protons	Electronic structure
	7	2,5
S^{2-}		
Ca^{2+}		
	12	2,8

5 Mixtures may be separated in the laboratory in many different ways. Three different methods of separating mixtures are shown below.

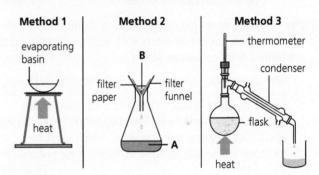

▲ Figure 9.34

a) Name each method of separation.   [3 marks]

b) Which method (**1, 2** or **3**) would be most suitable for obtaining water from potassium chloride solution?   [1 mark]

c) Which method would be most suitable for removing sand from a mixture of sand and water?   [1 mark]

d) What general term is used for liquid **A** and solid **B** in method 2?   [2 marks]

e) State why method 2 would **not** be suitable to separate copper(II) chloride from copper(II) chloride solution.   [1 mark]

6 To determine if two different orange drinks **X** and **Y** contained the food colourings E102, E101 or E160 a student put a drop of each orange drink and a drop of each food colouring along a pencil line on filter paper.

The filter paper was placed in a tank containing 1 cm depth of solvent. The solvent soaked up the paper and carried different components with it. After 5 minutes, the filter paper was removed and allowed to dry. The results are shown.

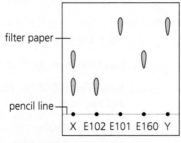

filter paper

pencil line

X E102 E101 E160 Y

▲ Figure 9.35

a) What is the name of the process used by the student to analyse the two orange drinks? [1 mark]

b) i) Orange drink X contains the food colouring E102. How do the results show this? [1 mark]

ii) What other food colouring does orange drink X contain? [1 mark]

iii) Re-draw the diagram and add a spot to show that orange drink Y also contained food colouring E160. [1 mark]

iv) The line across the bottom of the filter paper was drawn with a pencil not with ink. Why should the line not be drawn with ink? [1 mark]

7 When Group 1 elements react, the atom forms an ion. For example when potassium reacts with water, potassium ions are formed from potassium atoms.

a) Why is potassium stored under oil in the laboratory? [1 mark]

b) Before reacting Group 1 elements with water a risk assessment is carried out. Give two safety precautions, apart from wearing safety glasses, which must be included in the risk assessment for reacting potassium with water. [2 marks]

c) Equal sized pieces of three Group 1 metals are added to separate troughs of water which contain universal indicator. The observations made are recorded in the table. Use the information in the table to answer the questions that follow.

Table 9.26

Group 1 metal	Observation on reacting with water	Colour of universal indicator solution
Potassium	Melts Burns with a lilac flame Moves on the surface of the water Disappears quickly	Changes colour from green to purple
Lithium	Floats Moves on the surface of the water Eventually disappears	Changes colour from green to purple
Sodium	Melts Moves on the surface of the water Disappears	Changes colour from green to purple

i) What happens to the reactivity of the Group 1 elements as the Group is descended? [1 mark]

ii) Explain fully why the universal indicator changed colour from green to purple. [2 marks]

iii) Give one more observation which could be added to the table for all three reactions. [1 mark]

iv) Write a word equation to describe the reaction between sodium and water. [1 mark]

8 The modern periodic table has been in use for over 100 years. Its development included the work of several chemists including that of Dmitri Mendeleev.

a) Fill in the blanks in the following passage.

The modern periodic table arranges the elements in order of increasing atomic _____ whereas early versions of the periodic table arranged them in order of increasing atomic _____. [2 marks]

b) State one other difference between the modern periodic table and Mendeleev's table. [1 mark]

c) Elements in the periodic table are arranged in groups. The table gives details of some of the groups of the periodic table. Copy and complete the table. [6 marks]

Table 9.27

Group number	Name of group	Number of electrons in the outer shell of an atom of this group	Reactive or non-reactive group?
1			
		7	

d) Many trends in reactivity and physical properties are apparent as a group is descended.

i) State the trend in reactivity as Group 7 is descended. [1 mark]

ii) Name the least reactive element in Group 1. [1 mark]

iii) Astatine is found at the bottom of Group 7. Predict its state at room temperature and pressure. [1 mark]

9 Dmitri Mendeleev produced a table that is the basis of the modern periodic table. Describe the key features of Mendeleev's table and explain why his table came to be accepted over time by scientists. [6 marks]

Working scientifically:
How theories change over time

A version of the periodic table hangs on the wall of almost every chemistry laboratory across the world – it is a powerful icon and a single document that summarises much of our knowledge of chemistry. The history of the periodic table can be traced back over centuries and illustrates how scientific theories change over time.

▲ Figure 9.36

After the discovery of the new element phosphorus in 1649, scientists began to think about the definition of an element. In 1789 Antoine-Laurent de Lavoisier produced a table similar to that below of simple substances, or elements, which could not be broken down further by chemical reactions.

Table 9.28

Acid-making elements	Gas making elements	Metallic elements	Earth elements
Sulfur	Light	Cobalt, mercury, tin	Lime (calcium oxide)
Phosphorus	Caloric (heat)	Copper, nickel, iron	Magnesia (magnesium oxide)
Charcoal (carbon)	Oxygen	Gold, lead, silver, zinc	Barytes (barium sulfate)
	Azote (nitrogen)	Manganese, tungsten	Argila (aluminium oxide)
	Hydrogen	Platina (platinum)	Silex (silicon dioxide)

In addition to many elements which form the basis of our modern periodic table, Lavoisier's list also included 'light' and 'caloric' (heat) which at the time were believed to be material substances. Lavoisier incorrectly classified some compounds as elements because high temperature smelting equipment or electricity was not available to break down these compounds. The incorrect classification of these compounds as elements was due to a lack of technology as much as a lack of knowledge.

1 What is an element?

2 Which elements in Lavoisier's table also appear in today's periodic table?

3 Which group of elements did Lavoisier classify correctly?

4 Why do you think sulfur, phosphorus and charcoal are described as 'acid-making' elements?

5 Which substances in Lavoisier's list, from your own modern knowledge, are compounds? Why do you think Lavoisier thought these were elements?

Following on from the work of Lavoisier, in the early 19th century Johann Döbereiner noted that certain elements could be arranged in groups of three because they have similar properties. For example

▶ lithium, sodium and potassium – very reactive metals which produce alkalis with water

▶ calcium, strontium and barium – reactive metals but with higher melting points and different formulae of their oxides

▶ chlorine, bromine and iodine – low melting point, coloured, reactive non-metals.

He also noted that the 'atomic weight' of the middle element was close to the average of the other two:

Li = 7; Na = 23; K = 39; $\frac{7 + 39}{2} = 23$

Ca = 40; Sr = 88; Ba = 137; $\frac{40 + 137}{2} = 88.5$

These groups were called triads and were the first partial representation of a group of elements with similar properties.

6 State the group represented by each of the three Döbereiner triads.

7 Does the final triad listed above follow Döbereiner's atomic weight rule? Show your working.

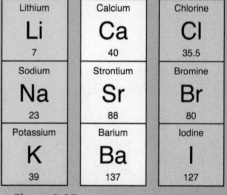

▲ Figure 9.37

You have already learned of the work of Newlands and Mendeleev in the development of the periodic table. It is important to realise that their work was built on the theory and tables suggested by Lavoisier, Döbereiner and others. Use the following questions to think about the ways in which scientific theories and methods develop over time.

8 In what ways is Newlands' periodic table superior to Lavoisier's classification of the elements?

9 Why is Newlands' classification superior, yet building on Johann Döbereiner's work?

10 State as many features as you can think of in which the modern periodic table is superior to Mendeleev's periodic table.

In the 1940s Glenn Seaborg was part of a research team working on 'nuclear synthesised' elements with atomic masses beyond the naturally occurring limit of uranium. When isolating the elements americium and curium, he wondered if these elements belonged to a different series which would explain why their chemical properties were different from what was expected. In 1945, against the advice of colleagues, he proposed a significant change to Mendeleev's table – the actinide series. Today this series is well accepted and included in the periodic table.

Through the history of the periodic table we can easily see

▶ the ways in which scientific methods and theories develop over time

▶ how a variety of concepts and models are used to develop scientific explanations and understanding.

10 Bonding, structure and the properties of matter

Atoms are so small that we cannot see them. We cannot see them even using the most powerful light microscope because atoms are much smaller than the wavelength of light. However, being able to picture what the particles are like in a substance and how they are bonded to each other is vital to understand chemistry. In this chapter we will examine what the particles are and how they bond together in different substances to help us understand the properties of these different substances.

Specification coverage

This chapter covers specification points 5.2.1.1 to 5.2.3.3 and is called Bonding, structure and the properties of matter.

It covers ionic, molecular, giant covalent and metallic substances, as well an overview of types of bonding and structures, nanoscience and the different forms of carbon.

Related work on writing formulae and equations can be found in Chapter 14.

Prior knowledge

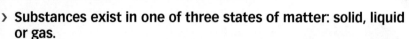

Previously you could have learnt:

> Substances exist in one of three states of matter: solid, liquid or gas.
> In solids the particles are vibrating and packed close together in a regular pattern; in liquids the particles are close together but moving about randomly; in gases the particles are a long way apart and moving about randomly.
> Substances change state at their melting and boiling points.
> How easily substances melt or boil depends on the strength of the forces or bonds between the particles.
> Different substances have different properties; e.g. some have high melting points but others have low melting points, some conduct electricity but many do not.

Test yourself on prior knowledge

1 What are the three states of matter?
2 In general terms, what is the name of the temperature at which a substance changes from:
   a) A solid to a liquid.   b) A gas to a liquid.   c) A liquid to a solid.
3 Identify two substances in each case that:
   a) Have a high melting point.   b) Have a low melting point.
   c) Conduct electricity as a solid. d) Do not conduct electricity as a solid.
4 The particles in liquids and gases can move around, but those in a solid only vibrate and cannot move around. Explain why the particles in a solid cannot move around.

Ionic substances

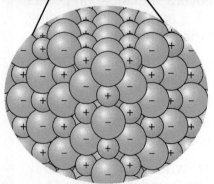
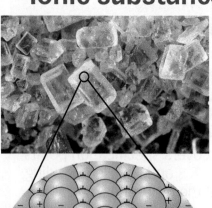

▲ **Figure 10.1** Ionic substances are made up of a giant lattice of positive and negative ions.

○ What are ionic substances?

Many substances are made of ions. **Ions** are **electrically charged** particles which have a different number of protons (which are positively charged) and electrons (which are negatively charged).

Most compounds made from a combination of metals and non-metals have an ionic structure. For example, sodium chloride is made from sodium (metal) and chlorine (non-metal) and is ionic. Copper sulfate is made from copper (metal), sulfur (non-metal) and oxygen (non-metal) and is also ionic.

○ The structure of ionic substances

In substances made of ions, there are lots of positive and negative ions in a giant lattice. A giant lattice contains a massive number of particles in a regular structure that continues in all directions throughout the substance (Figure 10.1).

KEY TERMS

Giant lattice Ionic substances are made up of a giant lattice of positive and negative ions in a regular structure.

Ionic bonding The electrostatic attraction between positive and negative ions.

The positive and negative ions are attracted to each other by electrostatic attraction because opposite charges attract. Ionic bonding is the attraction between positive and negative ions. Each ion is attracted to all the ions of opposite charge around it. This attraction is strong and so all ionic substances are solids at room temperature.

Four alternative ways of representing the ions in a lattice are shown in Table 10.1 (using sodium chloride as an example).

Table 10.1 Methods of representing ions in a lattice.

	Dot and cross diagram	2D space-filling structure	3D space-filling structure	Ball and stick structure
Diagram ⊖ Cl⁻ ion ⊕ Na⁺ ion	$[\text{Na}]^+$ (2,8) $[\text{Cl}]^-$ (2,8,8)			The ions are shown with gaps between them Lines are drawn between the ions to show how they are arranged
Advantages of this representation	Shows the electronic structure of the ions	Very easy to draw	Gives very good representation of how the ions are packed together	Helps to show how the ions are arranged relative to each other
Disadvantages of this representation	Can give the impression that the structure is made of pairs of ions rather than being a continuous structure containing a massive number of ions	Can give the impression that the structure is limited to a few ions rather than being a continuous structure with a massive number of ions		May make you think there are covalent bonds between the ions (there are **NO** covalent bonds in an ionic lattice) May make you think the ions are a long way apart (but they are packed close together)
		Only shows the structure in 2D		

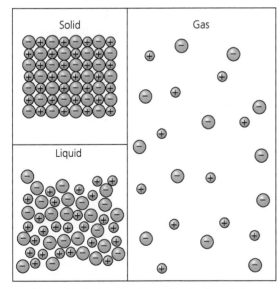

▲ **Figure 10.2** The structure of an ionic substance as a solid, liquid and gas.

○ The properties of ionic substances

Melting and boiling points

In order to melt and boil ionic substances, the strong attraction between the positive and negative ions has to be overcome (Figure 10.2). This is difficult and requires a lot of energy and so ionic substances have high melting and boiling points. For example, sodium chloride melts at 801 °C and aluminium oxide melts at 2072 °C.

Electrical conductivity

An electric current is the flow of electrically charged particles such as ions or electrons. Ionic substances are made of ions, but as a solid the ions cannot move so they cannot conduct electricity. However, when melted, the ions can move and carry charge, so ionic substances will conduct electricity when molten. Many ionic substances dissolve in water and

► **Figure 10.3** When an ionic substance dissolves in water the ions separate, mix in with the water molecules and move around.

Show you can...

B²⁺ and A⁻ form a compound in which the bonding is ionic.

a) What is the formula of the compound formed between these two ions?
b) Explain fully what is meant by an ionic bond.
c) What type of structure does this compound have? State two of its properties.

will conduct electricity if they dissolve because the ions can move (Figure 10.3).

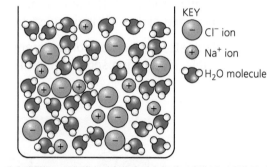

KEY
○⁻ Cl⁻ ion
○⁺ Na⁺ ion
H₂O molecule

Test yourself

1 What are ions?
2 Explain why ionic substances have high melting points.
3 Explain why ionic substances conduct electricity when molten or dissolved.
4 Explain why ionic substances do not conduct electricity as solids.
5 Ionic substances are made of a giant lattice of ions. What is a giant lattice?
6 Which of the following substances are likely to be ionic: CO_2, PH_3, Fe_2O_3, CH_4O, SiO_2, $MgBr_2$?

○ The formula of ionic substances

The charge on ions

You can work out the charge on some ions easily. For example, all the elements in

● **Group 1** have one electron in their outer shell and so **lose** one electron when they react forming 1+ ions (e.g. Na^+, K^+).
● **Group 2** have two electrons in their outer shell and so **lose** two electrons when they react forming 2+ ions (e.g. Ca^{2+}, Mg^{2+}).
● **Group 6** have six electrons in their outer shell and so **gain** two electrons when they react forming 2– ions (e.g. O^{2-}, S^{2-}).
● **Group 7** have seven electrons in their outer shell and so **gain** one electron when they react forming 1– ions (e.g. Cl^-, Br^-).

These charges and those of other common ions are shown in the Tables 10.2 and 10.3.

Table 10.2 Positive ions.

Group 1 ions (form 1+ ions)		Group 2 ions (form 2+ ions)		Group 3 ions (form 3+ ions)		Others	
Li^+	lithium	Mg^{2+}	magnesium	Al^{3+}	aluminium	NH_4^+	ammonium
Na^+	sodium	Ca^{2+}	calcium			H^+	hydrogen
K^+	potassium	Ba^{2+}	barium			Cu^{2+}	copper(ɪɪ)
						Fe^{2+}	iron(ɪɪ)
						Fe^{3+}	iron(ɪɪɪ)
						Ag^+	silver
						Pb^{2+}	lead
						Zn^{2+}	zinc

Table 10.3 Negative ions.

Group 6 ions (form 2– ions)		Group 7 ions (form 1– ions)		Others	
O^{2-}	oxide	F^-	fluoride	CO_3^{2-}	carbonate
S^{2-}	sulfide	Cl^-	chloride	OH^-	hydroxide
		Br^-	bromide	NO_3^-	nitrate
		I^-	iodide	SO_4^{2-}	sulfate

> **TIP**
>
> You need to be able to work out the charge on ions of elements in Groups 1, 2, 6 and 7. You will be provided with the charges on other ions.

What the formula of an ionic substance means

The formula of an ionic substance represents the ratio of the ions in the lattice. For example, in sodium chloride the formula NaCl means that the ratio of sodium (Na^+) ions to (Cl^-) chloride ions in the lattice is 1 : 1. In aluminium oxide, the formula Al_2O_3 means that the ratio of aluminium (Al^{3+}) ions to oxide (O^{2-}) ions in the lattice is 2 : 3.

Working out the formula of an ionic substance

In an ionic substance the total number of positive charges must equal the total number of negative charges. This allows us to work out the formula of ionic substances.

> **TIP**
>
> The charge on the nitrate (NO_3^-) ion is 1– not 3–. The charge on the ammonium (NH_4^+) ion is 1+ not 4+.

Some ions contain atoms of different elements. Examples include sulfate (SO_4^{2-}), hydroxide (OH^-) and nitrate (NO_3^-). If you need to write more than one of these in a formula, then those ions should be placed in a bracket (Table 10.4).

Table 10.4 Ionic substances and their ions.

Name	Positive ions		Negative ions		Formula
Sodium chloride	Na^+	(1+ charge)	Cl^-	(1– charge)	NaCl
Magnesium chloride	Mg^{2+}	(2+ charges)	Cl^- Cl^-	(2– charges)	$MgCl_2$
Magnesium sulfide	Mg^{2+}	(2+ charges)	S^{2-}	(2– charges)	MgS
Copper(II) sulfate	Cu^{2+}	(2+ charges)	SO_4^{2-}	(2– charges)	$CuSO_4$
Sodium carbonate	Na^+ Na^+	(2+ charges)	CO_3^{2-}	(2– charges)	Na_2CO_3
Ammonium sulfate	NH_4^+ NH_4^+	(2+ charges)	SO_4^{2-}	(2– charges)	$(NH_4)_2SO_4$
Calcium nitrate	Ca^{2+}	(2+ charges)	NO_3^- NO_3^-	(2– charges)	$Ca(NO_3)_2$
Aluminium oxide	Al^{3+} Al^{3+}	(6+ charges)	O^{2-} O^{2-} O^{2-}	(6– charges)	Al_2O_3
Iron(III) hydroxide	Fe^{3+}	(3+ charges)	OH^- OH^- OH^-	(3– charges)	$Fe(OH)_3$

> **TIP**
>
> In nitrate (NO_3^-), the 3 represents the number of oxygen atoms.

Test yourself

7 What will be the charge on ions of strontium, astatine, selenium and rubidium?

8 What does the formula K_2O mean?

9 Write the formula of the following substances: sodium sulfide, calcium fluoride, magnesium hydroxide, potassium carbonate, barium nitrate, caesium oxide.

Show you can...

Copy and complete the table to give information about some copper compounds.

Table 10.5

Name of compound	Formula of positive ion in compound	Formula of negative ion in compound	Formula of compound	What the numbers in the formula represent
Copper(II) carbonate				
Copper(II) hydroxide		OH^-		
Copper(II) nitrate	Cu^{2+}			
Copper(II) oxide				
Copper(II) sulfate			$CuSO_4$	1 copper, 1 sulfur, 4 oxygen
Copper(II) sulfide				

TIP

When metal atoms lose electrons they form positive ions. When non-metal atoms gain electrons they form negative ions.

TIP

When atoms gain electrons to form ions, the name changes to end in -ide. For example, oxygen atoms gain electrons to form oxide ions, chlorine atoms gain electrons to form chloride ions, sulfur gains electrons to form sulfide ions.

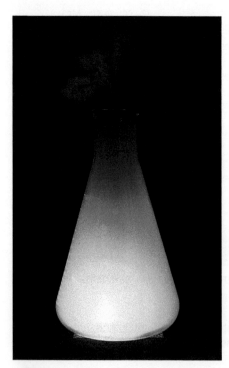

▲ **Figure 10.5** Sodium reacting with chlorine.

○ The reaction between metals and non-metals

Ionic compounds can be formed when **metals** react with **non-metals.** In these reactions, **electrons are transferred** from the outer shell of the metal atom to the outer shell of the non-metal atom, to produce ions. These ions have the electronic structure of the noble gases (Group 0 elements).

For example, when sodium reacts with chlorine, each sodium atom loses one electron and each chlorine atom gains one electron (Figures 10.4–10.6). This produces sodium ions and chloride ions which have noble gas electronic structures (i.e. sodium chloride).

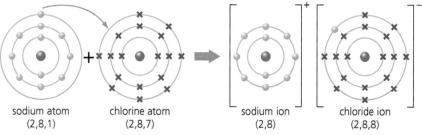

+
sodium atom
(2,8,1)
chlorine atom
(2,8,7)
sodium ion
(2,8)
chloride ion
(2,8,8)

▲ **Figure 10.4** Electron transfer during the formation of sodium chloride.

This can be drawn as a 'dot and cross' diagram which only shows the outer shell electrons. Electrons from one atom are shown as dots (●) and electrons from the other atom as crosses (×) (Figure 10.6).

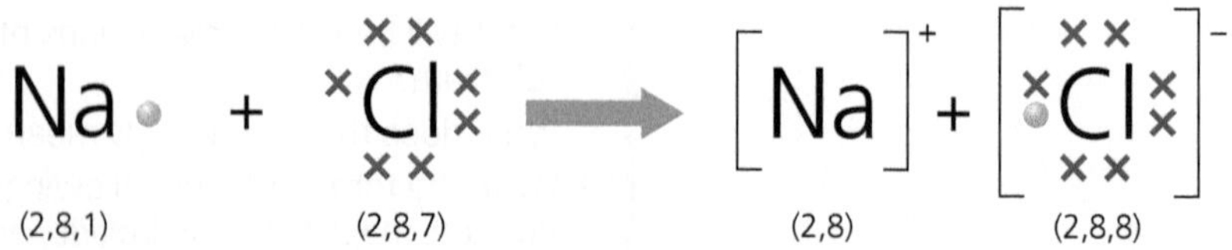

Na
+
Cl
Na
+
Cl
(2,8,1)
(2,8,7)
(2,8)
(2,8,8)

▲ **Figure 10.6** Sodium atoms react with chlorine atoms in the ratio 1 : 1 to form NaCl.

TIP

Note in these reactions that to form ionic compounds, electrons are transferred from the metal to the non-metal.

Some other examples of these reactions are shown in Figure 10.7.

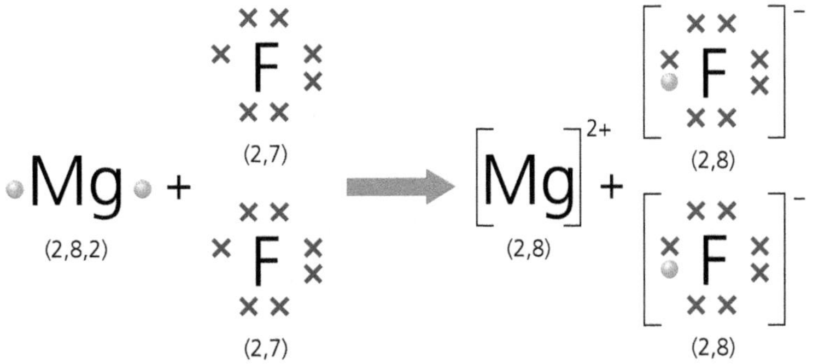

(a) Magnesium atoms react with fluorine atoms in the ratio $1:2$ to form MgF_2.

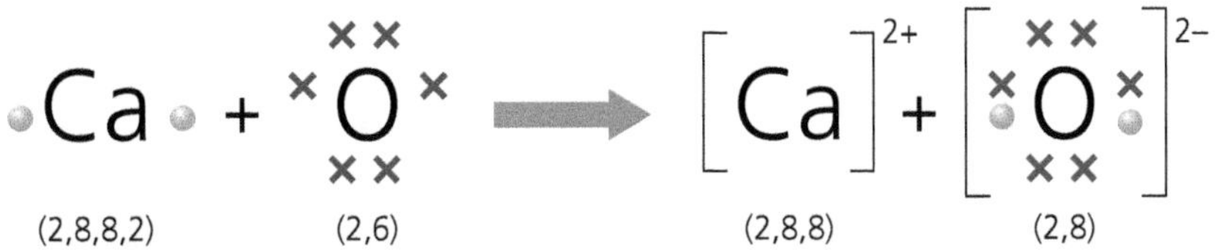

(b) Calcium atoms react with oxygen atoms in the ratio $1:1$ to form CaO.

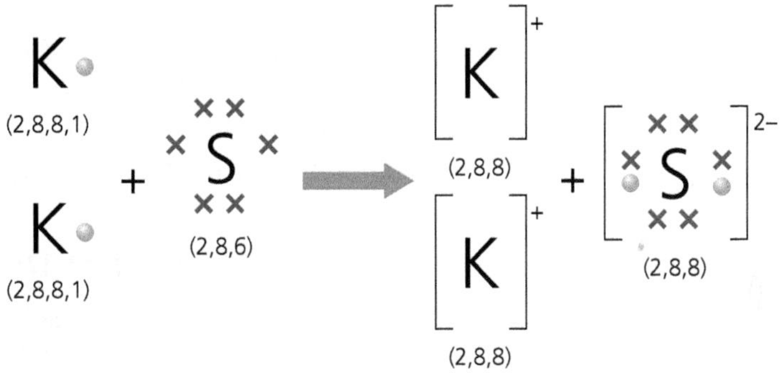

(c) Potassium atoms react with sulfur atoms in the ratio $2:1$ to form K_2S.

▲ Figure 10.7

Test yourself

10 Explain why ions are formed when metals react with non-metals.
11 Draw a diagram to show what happens in terms of electrons when lithium reacts with oxygen.

Show you can...

a) Explain fully what happens when magnesium atoms react with oxygen atoms to produce magnesium oxide.
b) Explain fully what happens when magnesium atoms react with fluorine atoms to produce magnesium fluoride.
c) Highlight any similarities and differences between the reaction of magnesium with oxygen and the reaction with fluorine.

Molecular substances

○ What are molecular substances?

Many substances are made of molecules. Molecules are atoms joined together by covalent bonds. A covalent bond is two **shared** electrons that join atoms together. How covalent bonds form is described later in this chapter.

Many substances are made of molecules. Some non-metal elements are made of molecules. The most common ones are listed in the periodic table as shown in Figure 10.8.

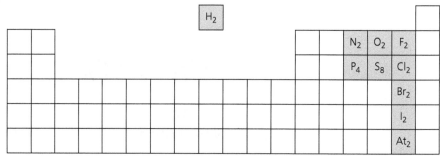

▲ Figure 10.8 Common elements that are formed of molecules.

Most compounds are made from a combination of non-metals. For example, water (H_2O) is made from hydrogen (non-metal) and oxygen (non-metal); glucose ($C_6H_{12}O_6$) is made from carbon (non-metal), hydrogen (non-metal) and oxygen (non-metal).

○ The structure of molecular substances

A molecular substance is made of many identical molecules that are not joined to each other (Figure 10.9). Within each molecule, the atoms are joined together by the very strong covalent bonds. However, the molecules are not bonded to each other. There are only some weak forces between the molecules – these weak forces are intermolecular forces.

Molecules are often quite small, containing just a few atoms, but some substances are made of big molecules (e.g. wax and many polymers).

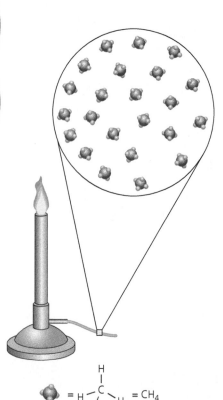

= H $-\overset{\overset{\displaystyle H}{|}}{\underset{\underset{\displaystyle H}{|}}{C}}-$ H = CH_4

▲ Figure 10.9 In molecular substances such as methane (natural gas, CH_4) there are many separate molecules. The molecules are not bonded to each other.

○ The properties of molecular substances

Melting and boiling points

Molecules are not bonded to each other. The intermolecular forces (forces between molecules) are only weak and so are easy to overcome. This means that molecular substances have low melting and boiling points. Many molecular substances with small molecules are gases and liquids at room temperature. For example, methane boils at −162°C and water boils at 100°C.

Generally, the bigger the molecules, the stronger the forces between the molecules and so the higher the melting and boiling points. Molecules of glucose are quite large and it melts at 146°C.

When molecular substances change state, the covalent bonds do **not** break. For example, water molecules are identical as H_2O whether it is steam, water or ice (Figure 10.10). No covalent bonds are broken when water changes state.

TIP

The covalent bonds in molecules do NOT break when molecules change state.

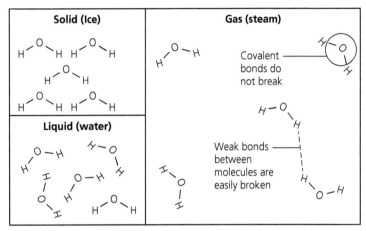

▲ **Figure 10.10** The structure of water, a molecular substance as a solid, liquid and gas. No covalent bonds are broken when it changes state.

Electrical conductivity

Molecules are electrically neutral which means that molecular substances do not conduct electricity at all. They do not contain delocalised electrons or charged ions.

TIP

Pure water does not conduct electricity because it is made of molecules. If there are any ionic substances dissolved in water, those ionic substances will conduct electricity.

Test yourself

12 What are molecules?

13 What is a covalent bond?

14 Explain why molecular substances have low melting and boiling points.

15 What happens to the covalent bonds in a molecular substance when it melts and boils?

16 Explain why molecular substances do not conduct electricity.

17 Which of the following substances are likely to be molecular: H_2S, Na_2O, KNO_3, $ZnBr_2$, CO, N_2H_4, C_2H_6O?

Show you can...

Metal oxides and non-metal oxides have different properties. Sulfur dioxide, a non-metal oxide has a melting point of –72°C and calcium oxide, a metal oxide, has a melting point of 2613°C. Explain why the melting point of sulfur dioxide is low but that of calcium oxide is high.

○ **The formula of molecular substances**

Molecular substances have two formulae, the empirical formula and the molecular formula. The molecular formula is the one that is normally used.

Table 10.6 Molecular and empirical formulae of molecules.

	Diagram of molecule	Molecular formula (gives the number of atoms of each element in each molecule)	Empirical formula (gives the simplest ratio of the atoms of each element in the substance)
Butane	H H H H │ │ │ │ H—C—C—C—C—H │ │ │ │ H H H H	**C_4H_{10}** There are 4 C atoms and 10 H atoms in each molecule	**C_2H_5** Simplest ratio of C:H = 2:5
Water	H—O—H	**H_2O** There are 2 H atoms and 1 O atom in each molecule	**H_2O** Simplest ratio of H:O = 2:1
Glucose	(structural diagram of glucose)	**$C_6H_{12}O_6$** There are 6 C atoms, 12 H atoms and 6 O atoms in each molecule	**CH_2O** Simplest ratio of C:H:O = 1:2:1

Test yourself

18 Benzene has the molecular formula C_6H_6. What does this tell us about benzene?

19 What is the empirical formula of benzene? What does this tell us about benzene?

Show you can...

a) Copy and complete the table:

Table 10.7

Molecular formula	Empirical formula
C_2H_6	
$C_{21}H_{22}N_2O_2$	
$C_2H_4O_2$	

b) Are the empirical formula and the molecular formula of a substance always different? Using an example, explain your answer.

○ Drawing molecules

When atoms join together to form molecules they share electrons in order to obtain noble gas electronic structures. A **covalent bond** is two shared electrons joining atoms together. Table 10.8 shows how many covalent bonds atoms typically form.

Table 10.8 The relationship between number of covalent bonds and electronic structure.

Atoms	H	Group 4 atoms (e.g. C, Si)	Group 5 atoms (e.g. N, P)	Group 6 atoms (e.g. O, S)	Group 7 atoms (e.g. F, Cl, Br, I)
Number of electrons in their outer shell	1	4	5	6	7
Number of electrons needed to obtain a noble gas electronic structure	1	4	3	2	1
Number of covalent bonds formed	1	4	3	2	1

There are several ways to show how atoms join together by sharing electrons in covalent bonds to form molecules (Table 10.9).

Table 10.9 Different ways of representing covalent bonding.

Dot and cross diagram showing all electrons and shell circles	Dot and cross diagram showing only outer shell electrons and shell circles	Dot and cross diagram showing only outer shell electrons	Stick diagram	Ball and stick diagram	Space-filling diagram
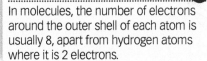					
Note that the ● and ✗ represent electrons that came from different atoms			Each stick (or line) represents one covalent bond (i.e. 2 shared electrons)	A good representation of a molecule showing how atoms merge into each other, but the covalent bonds are not visible	

When drawing stick diagrams, each atom makes the number of covalent bonds shown in Table 10.8. When drawing dot-cross diagrams, each single covalent bond is made up of two electrons. Atoms can make **double** and **triple** covalent bonds (Figure 10.11). A double covalent bond contains four electrons (two from each atom), while a triple bond contains six electrons (three from each atom).

Any outer shell electrons that are not used up in making covalent bonds are found in non-bonding electron pairs, often called lone pairs.

> **TIP**
>
> In molecules, the number of electrons around the outer shell of each atom is usually 8, apart from hydrogen atoms where it is 2 electrons.

> **TIP**
>
> In a **double** covalent bond, four electrons are shared. In a **triple** covalent bond, six electrons are shared.

	H_2	Cl_2	O_2	N_2
Stick diagram	H—H	Cl—Cl	O=O	N≡N
Dot and cross diagram				

	HCl	H_2O	NH_3	CH_4	CO_2
Stick diagram	H—Cl	H—O—H	H—N—H \n | \n H	H—C—H (with H above and below)	O=C=O
Dot and cross diagram					

▶ **Figure 10.11** Examples of covalent bonding, shown as stick diagrams and dot and cross diagrams.

Test yourself

20 What does each stick represent in a stick diagram?

21 What do the dots and crosses represent in dot and cross diagrams?

22 Draw a stick diagram and a dot and cross diagram for H_2S.

23 Draw a stick diagram and a dot and cross diagram for CS_2.

Show you can…

Phosphorus bonds with hydrogen to form phosphine. PH_3 is a colourless gas which has an unpleasant, rotting fish odour. Phosphorus also bonds with chlorine to form phosphorus trichloride which is a toxic colourless liquid.

a) Draw a dot and cross diagram to show the bonding in PH_3.
b) Suggest the formula of phosphorus trichloride.
c) Draw a dot and cross diagram to show the bonding in phosphorus trichloride.
d) Using your diagram from c), explain what is meant by a covalent bond and a non-bonding electron pair.

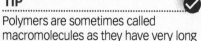

○ Polymers

There are many different types of polymer (plastics), including polythene, PVC, Perspex, Teflon and polystyrene. Polymers contain **very large molecules**, often with hundreds or thousands or atoms. Within each molecule, the atoms are joined to each other by **strong covalent bonds** and so are **solids** at room temperature (Figure 10.12). The basic sub-units of polymers are called monomers.

close-up view of a polymer molecule

long polymer molecules – these are not joined together

▲ **Figure 10.12** Polymers contain very large molecules, within which atoms are joined together by covalent bonds.

Polymers can be represented in the form:

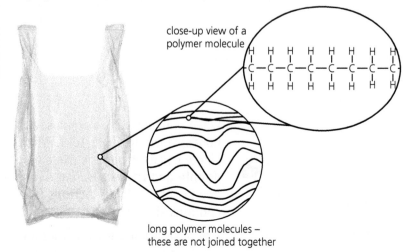

where:

the point where other units (monomers) join

the number of times the basic structure in brackets is repeated

polythene in made from many sub-units (**monomers**) of ethene

ethene — single covalent bond

double covalent bond

Practical

Testing the electrical conductivity of ionic and molecular covalent substances

To investigate the conduction of electricity by a number of compounds in aqueous solution.

The apparatus was set up as shown in the diagram.

Questions

1 Describe the experimental method which you would use to test the solutions using the apparatus shown.
2 Copy and complete the results table.
3 Using the results from column three and four of the table write a conclusion for this experiment stating and explaining any trends shown in the results.
4 Would the results be different if solid copper(II) sulfate was used in place of copper(II) sulfate solution? Explain your answer.
5 Predict and explain the results you would obtain for calcium nitrate solution.
6 Predict and explain the results you would obtain for bromine solution.

Table 10.10

Test solution	Does the bulb light?	Does the substance conduct electricity?	Does the substance contain ionic or covalent bonding?
Copper(II) sulfate	yes		
Ethanol (C_2H_5OH)	no		
Magnesium sulfate	yes		
Potassium iodide	yes		
Glucose ($C_6H_{12}O_6$)	no		
Sodium chloride	yes		

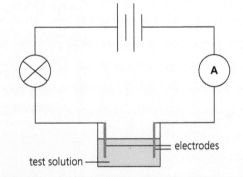

▶ Figure 10.13

test solution —
electrodes

Giant covalent substances

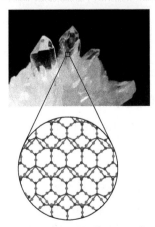

▲ Figure 10.14 Giant covalent substances are made up of a continuous network of atoms linked by covalent bonds.

○ What are giant covalent substances?

There are a few substances that have atoms joined by covalent bonds in a continuous network. Common examples are

● diamond, C (a form of carbon) – studied in detail later in the chapter ('The different forms of carbon' section)
● graphite, C (a form of carbon) – studied in detail later in the chapter ('The different forms of carbon' section)
● silicon, Si
● silicon dioxide, SiO_2 (also known as silica).

○ The structure of giant covalent substances

In a giant covalent substance all the **atoms** are in a **giant lattice**. They are all joined together by covalent bonds in a continuous network throughout the structure (Figure 10.14).

These substances are **not** molecules. In molecular substances, there are lots of separate molecules with the atoms in each molecule joined by covalent bonds but the molecules are not joined together.

○ The properties of giant covalent substances

Melting and boiling points

In order to melt a giant covalent substance, many covalent bonds have to be broken. Covalent bonds are very strong and so it takes a lot of energy to break them. Therefore, giant covalent substances are **solids** with **very high melting and boiling points**. For example, diamond melts at over 3500°C.

Electrical conductivity

Most giant covalent substances do not conduct electricity because they do not contain any delocalised electrons. However, graphite does as it does have some delocalised electrons. Delocalised electrons are able to move throughout the substance.

Test yourself

24 Describe the structure of a giant covalent substance.
25 Why do giant covalent substances have very high melting points?
26 Why do giant covalent substances, except graphite, not conduct electricity?

Show you can...

Carbon and silicon both form dioxides; carbon dioxide and silicon dioxide. Carbon dioxide is a gas at room temperature and silicon dioxide is a solid with melting point of 1610°C.

a) Copy and complete the table:

Table 10.11

Substance	Type of bonding	Structure
Carbon dioxide		
Silicon dioxide		

b) Explain why silicon dioxide is a solid with a high melting point and carbon dioxide is a gas with a low melting point.

Metallic substances

○ What are metallic substances?

Metals are metallic substances. Over three-quarters of all the elements are metals and have a metallic structure.

○ The structure of metallic substances

KEY TERM

Delocalised Free to move around.

Metals consist of a giant lattice of positively charged atoms arranged in a regular pattern. The outer shell electrons from each atom are delocalised which means they are free to move throughout the **whole structure** (Figures 10.15 and 10.16).

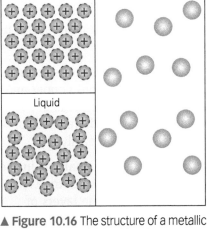
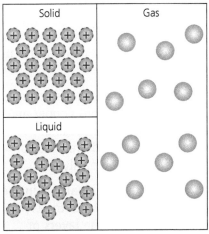

▲ **Figure 10.16** The structure of a metallic substance as a solid, liquid and gas.

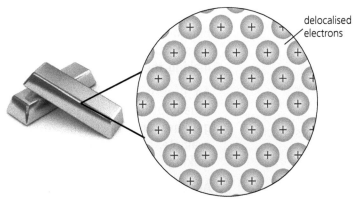

▲ **Figure 10.15** Metallic substances are made up of a giant lattice of positively charged atoms in a cloud of delocalised electrons.

There is a strong attraction between the positive nucleus of these atoms and the delocalised electrons. This attraction between the nucleus and the delocalised electrons is called metallic bonding.

○ The properties of metallic substances

Melting and boiling points

In metals, the metallic bonding is strong. This means that most metals are solids and have **high melting and boiling points**. For example, aluminium melts at 660°C and iron melts at 1538°C.

Electrical conductivity

Metals are good conductors of electricity because the delocalised electrons are able to move through the structure and carry electrical charge through the metal (Figure 10.17).

Thermal conductivity

Metals are also good thermal conductors. The thermal energy is transferred by the delocalised electrons.

Malleability

Metals are malleable, which means they can be bent and hammered into shape. This is because the layers of atoms can slide over each other while maintaining the metallic bonding (Figure 10.18). This makes metals soft.

▲ **Figure 10.17** Copper is used in electrical cables because it is an excellent electrical conductor, has a high melting point and can be shaped into wires easily.

KEY TERMS ⭐

Metallic bonding The attraction between the nucleus of metal atoms and delocalised electrons.

Malleable Can be hammered into shape.

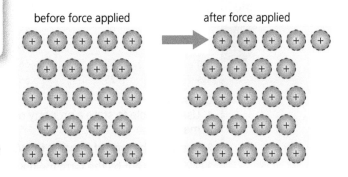

before force applied   after force applied

▶ **Figure 10.18** Metals are soft because atoms can slide over each other.

○ Alloys

Pure metals are very malleable. This can make them too soft for most uses as they lose their shape easily. Metals can be made more useful by making them into alloys.

An alloy is a mixture of a metal with small amounts of other elements, usually other metals. Pure metals such as aluminium, iron, copper and gold are rarely used, and alloys of these metals are used instead. For example, steels are alloys made from iron. Alloys of gold are used for making jewellery as pure gold would lose its shape too easily.

Alloys also have metallic structures. However, some of the atoms in the alloy are a different size to those of the metal. This distorts the layers in the structure and makes it much **more difficult** for the layers of atoms to **slide over each other** (Figure 10.19).

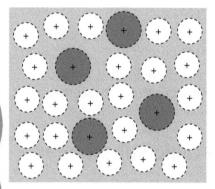

▲ **Figure 10.19** Alloys are harder than pure metals because atoms cannot slide over each other as easily due to the presence of some different sized atoms.

Test yourself

27 What is metallic bonding?
28 Why do metals have high melting points?
29 Why do metals conduct electricity?
30 Why are metals malleable?
31 What are alloys?
32 Why are alloys harder than pure metals?

Show you can...

Aluminium is used in overhead electricity cables and to make saucepans.

a) Explain why aluminium is a good conductor of electricity.
b) State two other reasons why aluminium is used in overhead electricity cables.
c) Explain in terms of structure why aluminium is malleable.
d) Aluminium oxide is a compound of aluminium. Compare and contrast how aluminium and aluminium oxide conduct electricity.

Activity

The physical properties of Group 1 elements

The table shows some physical properties of Group 1 elements.

Table 10.12

Element	Melting point in °C	Boiling point in °C	Density in g/cm³	Electrical conductivity
Li	180	1340	0.53	Good
Na	98	880	0.97	Good
K	63	766	0.89	Good
Rb	39	686	1.53	Good
Cs	28	669	1.93	Good

Questions

1 Use the table to state a property of Group 1 metals which is common to all metals.
2 Describe the type of bonding found in Group 1 metals.
3 What is meant by the term melting?

4 State the trend in melting point as the group is descended.

5 Write the electronic structure of Li, Na and K and state what happens to the distance of the outer electron from the nucleus as the group is descended. Now try and explain the trend in melting point down the group.

6 State the trend in density as the group is descended.

7 Use the data in the table to plot a bar chart to show the density of the different Group 1 metals.

8 State if the Group 1 metals will all float on water (the density of water is $1\,g/cm^3$).

Overview of types of bonding and structures

○ Types of bonding

There are three types of bonding that are summarised in the Table 10.13.

Table 10.13 The three types of bonding.

	Ionic bonding	Covalent bonding	Metallic bonding
Description	The electrostatic attraction between positive and negative ions	Atoms that are joined together by sharing pairs of electrons	The attraction between positive nucleus of metal atoms and delocalised outer shell electrons
Which substances have this bonding	Ionic compounds	Molecular substances Giant covalent substances	Metallic substances

○ Types of structure

There are five types of structure that are summarised in the Table 10.14. The elements in Group 0 of the periodic table (the noble gases) have a unique structure called monatomic which is very similar to molecules except that the particles are individual atoms and not molecules. The elements in Group 0 are stable and have no need to lose, gain or share electrons.

Table 10.14 Different types of structure.

	Monatomic	Molecular	Giant covalent	Ionic	Metallic
Which substances have this structure	Group 0 elements	Many non-metal elements Most compounds made from a combination of non-metals	Diamond Graphite Silicon Silicon dioxide	Most compounds made from a combination of metals with non-metals	Metals and alloys
Description of structure	Made up of many separate atoms. The atoms are not bonded to each other. There are very weak forces of attraction between the atoms	Made up of many separate molecules. The atoms within each molecule are joined by covalent bonds. There are no bonds between molecules. There are only weak forces of attraction between the molecules	Made of a giant lattice of atoms joined to each other by covalent bonds	Made of a giant lattice of positive and negative ions. There are strong electrostatic forces of attraction between the positive and negative ions	Made of a giant lattice of metal atoms with a cloud of delocalised outer shell electrons. There are strong forces of attraction between the positive nucleus of the metal atoms and the delocalised electrons

	Monatomic	Molecular	Giant covalent	Ionic	Metallic
Melting and boiling points	VERY LOW as it is requires little energy to overcome the very weak forces between atoms	LOW as it requires little energy to overcome the weak forces between molecules	VERY HIGH as it requires a lot of energy to break lots of strong covalent bonds	HIGH as it is requires a lot of energy to overcome the strong attraction between the positive and negative ions	HIGH as it requires a lot of energy to overcome the strong attraction between the positive nucleus of the metal atoms and delocalised electrons
Electrical conductivity	NON-CONDUCTOR as atoms are neutral and there are no delocalised electrons or ions	NON-CONDUCTOR as molecules are neutral and there are no delocalised electrons or ions	NON-CONDUCTOR (except graphite) as there are no delocalised electrons (graphite does have delocalised electrons)	Solid = NON-CONDUCTOR as ions cannot move Liquid/solution = CONDUCTOR as ions can move and carry the charge	CONDUCTOR as outer shell electrons are delocalised and can carry charge through the metal

Test yourself

33 For each of the following substances, state the type of bonding and the type of structure it is likely to have:

a) copper(II) oxide (CuO)

b) diamond (C)

c) lead carbonate ($PbCO_3$)

d) phosphorus oxide (P_4O_{10})

e) argon (Ar)

f) copper (Cu)

Show you can...

Substances may be classified in terms of their physical properties. Use the table to answer the following questions:

Table 10.15

Substance	Melting point in °C	Boiling point in °C	Electrical conductivity as solid	Electrical conductivity as liquid
A	3720	4827	Good	Poor
B	−95	69	Poor	Poor
C	327	1760	Good	Good
D	3550	4827	Poor	Poor
E	801	1413	Poor	Good

a) Which substance could be sodium chloride? Explain your answer.

b) Which substance consists of small covalent molecules? Explain your answer.

c) Explain why substance A could not be diamond.

d) Which substance is a metal?

○ States of matter

The three states of matter are solid, liquid and gas (Figure 10.20). Substances change state at their melting and boiling points. A substance is a:

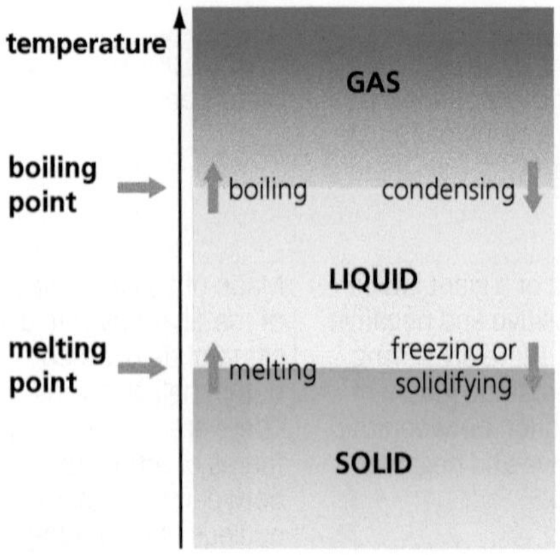

- solid at temperatures below its melting point
- liquid at temperatures between its melting and boiling point
- gas at temperatures above its boiling point.

temperature

GAS

boiling point → boiling   condensing

LIQUID

melting point → melting   freezing or solidifying

SOLID

▲ **Figure 10.20** Changes of state.

The amount of energy needed for substances to melt and boil depends on the strength of the forces or bonds between their particles. **The stronger the forces or bonds between the particles, the higher their melting and boiling points**. For example, giant covalent substances have very high melting and boiling points as strong covalent bonds have to be broken. Molecular substances have low melting and boiling points as there are only weak forces between the molecules that are easy to overcome.

You may have used a very simple model to represent the particles in solids, liquids and gases like the top row in Figure 10.21. It can help to understand how particles are arranged when substances are in each state.

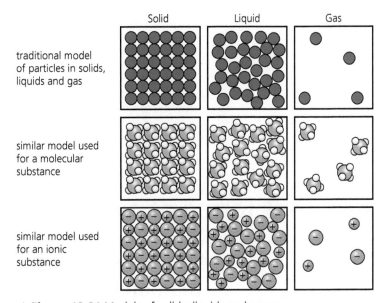

▲ **Figure 10.21** Models of solids, liquids and gases.

The top row in Figure 10.21 is particularly over-simplistic. The particles in most substances are actually molecules or ions and they are not solid spheres. These diagrams also do not show that there are forces or bonds between the particles. They also cannot show whether the particles are moving around or not.

Test yourself

34 Use the data in the table to answer the questions that follow.

Table 10.16

Substance	Melting point in °C	Boiling point in °C
A	45	137
B	595	984
C	−30	56
D	−189	−186
E	186	302

a) Which substance(s) is/are gases at room temperature?
b) Which substance(s) is/are liquids at 100°C?
c) Which substance(s) is/are solids at 100°C?
d) Which substance is a liquid over the widest temperature range?

The different forms of carbon

TIP

In diamond, each carbon atom has four covalent bonds with other carbon atoms.

There are several different forms of the element carbon which usefully illustrate several parts of this chapter.

○ Diamond

Diamond is probably the best known form of carbon. It has a giant covalent structure with all the carbons joined by covalent bonds in a giant lattice. This can be thought of as a continuous network of atoms linked by covalent bonds (Figure 10.22).

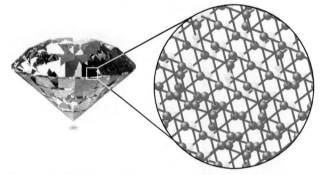

▲ **Figure 10.22** Diamond has a giant covalent structure with carbon atoms joined by covalent bonds in a giant lattice.

Close examination of the structure of diamond shows that each carbon atom is covalently bonded to **four** other carbon atoms. Figure 10.23 shows a very small part of the diamond lattice to help see how the carbon atoms are bonded together.

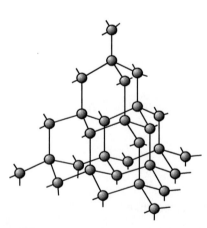

▲ **Figure 10.23** A tiny part of the diamond lattice. Each C atom is bonded to four others.

○ Graphite

Graphite is another form of carbon. It is the grey substance that runs through the inside of a pencil and rubs off onto the paper. Like diamond, graphite has a giant covalent structure with all the carbons joined by covalent bonds in a giant lattice (Figure 10.24). However, the carbon atoms are bonded together in **flat layers**. The layers of atoms are not bonded to each other and there are only very weak attractive forces between these layers of atoms.

Close examination of the structure of graphite shows that each carbon atom is covalently bonded to **three** other carbon atoms (Figure 10.25).

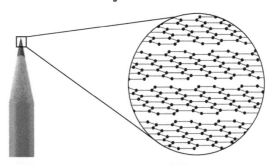 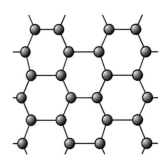

▲ **Figure 10.24** Graphite has a giant covalent structure with carbon atoms joined by covalent bonds within layers in a giant lattice.

▲ **Figure 10.25** A tiny part of the graphite lattice. Each C atom is bonded to three others.

The diagram shows a very small part of the graphite lattice to help see how the carbon atoms are bonded together.

This bonding leaves one outer shell electron on each carbon atom that is not used in bonding. These electrons become delocalised and are free to move along the layers.

Table 10.18 Comparison of the physical properties of diamond and graphite.

	Diamond	Graphite
Melting point	**Very high melting point** (over 3500°C) because lots of strong covalent bonds need to broken	**Very high melting point** (over 3500°C) because lots of strong covalent bonds need to be broken
Hardness	**Very hard** (the hardest natural substance) because the atoms are arranged in a very rigid continuous network held together by strong covalent bonds	**Soft** because the **layers of atoms** are **not** bonded together and so can easily slide over each other
Conductivity	**Does not conduct** because it contains no delocalised electrons	**Conducts** because it contains delocalised electrons (one from each atom) that move along the layers and carry charge through the graphite
Uses	Drill and saw tips – due to its hardness	Electrodes – as it conducts electricity In pencils – it rubs off easily onto the paper because layers can easily move over each other

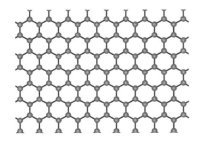

▲ **Figure 10.26** Graphene is a single layer of graphite.

○ Graphene

Graphene is a new substance. It is a **single layer of graphite** (Figure 10.26). Scientists at the University of Manchester won a Nobel Prize in 2010 for their work on graphene.

Graphene has some remarkable properties. It is extremely thin being just **one atom thick**, but is extremely strong due to its giant covalent structure. It is also semi see-through as it is so thin. In a similar way to graphite, it is a thermal and electrical conductor due to having some delocalised electrons.

Graphene is a very exciting new material and a lot of research is being done to make use of it. Its properties make it very useful in electronics (e.g. touchscreens) and composite materials (e.g. carbon fibres).

○ Fullerenes

The molecule C_{60} was identified in the 1980s as another form of carbon. The molecule has a shape that resembles a football (Figure 10.27). It was named **buckminsterfullerene** after the American architect Richard Buckminster Fuller who built domes that had similar structures. C_{60} is often referred to as a buckyball and was the first fullerene produced. Scientists at the University of Sussex won a Nobel Prize in 1996 for their work on fullerenes.

▲ **Figure 10.27** C_{60} has the shape that resembles a football.

A whole family of similar molecules, such as C_{70} and C_{84}, have been produced. These molecules are all called fullerenes. The structure of fullerenes is based on carbon atoms in **hexagonal rings** (rings of six atoms), but some rings have five or seven atoms. They all have a hollow part in the centre of the molecule.

Fullerenes are being used:

- For **delivery of drugs** into specific parts of the body and/or cells – the drugs are often carried inside the hollow centre of the fullerene molecule.
- In **lubricants** to reduce friction when metal parts of machines move past each other – the spherical shape of the molecule allows molecules to roll past each other.
- As catalysts – a lot of research is taking place into the use of fullerenes as catalysts and a wide range of potential applications are being found.

○ Carbon nanotubes

Carbon nanotubes are **cylindrical fullerenes**, sometimes called buckytubes (Figure 10.28). They have very high length to diameter ratios, significantly higher than for any other material. They can also be thought of as being tubes of graphene sheets.

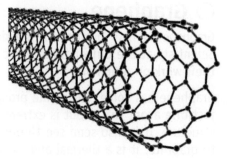

▲ **Figure 10.28** A carbon nanotube.

TIP

C_{60} is a molecule with a simple molecular structure. It does NOT have a giant covalent structure.

KEY TERMS

Fullerenes Family of carbon molecules each with carbon atoms linked in rings to form a hollow sphere or tube.

Catalyst Substance that speeds up a chemical reaction but is not used up in it.

These carbon nanotubes have some excellent properties making them very useful. They have:

- **High tensile strength** – in other words it is very strong when it is pulled – this is due to the many strong covalent bonds throughout its structure.
- **High thermal and electrical conductance** – this is due to some of the electrons being delocalised.

Carbon nanotubes have many uses, for example to **reinforce the materials** used to make sports equipment like tennis racquets (Figure 10.29) and golf clubs.

▲ **Figure 10.29** Carbon nanotubes are very strong and are used to reinforce tennis racquets.

Test yourself

35 Explain why diamond and graphite both have very high melting points.
36 Explain why diamond is hard but graphite is soft.
37 Explain why graphite conducts electricity but diamond does not.
38 What are fullerenes?
39 Carbon nanotubes are used to reinforce and strengthen tennis racquets. Explain why carbon nanotubes strengthen materials.
40 A typical carbon nanotube is 12 cm long and has a diameter of 1 nm. Calculate the length to diameter ratio of this carbon nanotube.

Show you can...

Copy and complete the table to give information about the different structures and uses of carbon.

Table 10.19

	Graphite	Diamond	Graphene	Fullerenes	Carbon nanotubes
Description of structure					
Example of use					

Chapter review questions

1   Four structure types are: ionic, metallic, molecular and giant covalent.

   a) Which of these structure types usually have low melting and boiling points?

   b) Which of these structure types usually conduct electricity as solids?

   c) Which of these structure types usually conduct electricity when melted?

2   Carbon dioxide is a molecular compound with the formula CO_2.

   a) Explain why carbon dioxide has a low boiling point.

   b) Explain why carbon dioxide does not conduct electricity.

3   Iron is a metal with the formula Fe.

   a) Explain why iron has a high melting point.

   b) Explain why iron conducts electricity.

   c) Explain why pure iron is soft.

   d) Steels are alloys of iron. Explain why steels are harder than pure iron.

4   Potassium fluoride is an ionic compound containing potassium (K^+) ions and fluoride (F^-) ions.

   a) Give the electronic structure of potassium (K^+) ions.

   b) Give the electronic structure of fluoride (F^-) ions.

   c) Give the formula of potassium fluoride.

   d) Explain why potassium fluoride has a high melting point.

   e) Explain why potassium fluoride does not conduct electricity as a solid.

   f) Explain why potassium fluoride conducts electricity when molten or dissolved.

5   Decide whether each of the following substances has an ionic, molecular, giant covalent or metallic structure.

   a) zinc (Zn)

   b) ethane (C_2H_6)

   c) diamond (C)

   d) magnesium oxide (MgO)

   e) iodine trifluoride (IF_3)

   f) potassium carbonate (K_2CO_3)

6   Decide whether each of the following substances has an ionic, molecular, giant covalent or metallic structure.

Table 10.20

Substance	Melting point in °C	Boiling point in °C	Electrical conductivity as solid	Electrical conductivity as liquid
A	838	1239	Does not conduct	Conducts
B	89	236	Does not conduct	Does not conduct
C	678	935	Conducts	Conducts
D	1056	1438	Does not conduct	Conducts
E	2850	3850	Does not conduct	Does not conduct
F	−39	357	Conducts	Conducts

7 Draw a stick and a dot-cross diagram for each of the following molecules.

a) phosphine (PH_3)

b) bromine (Br_2)

c) carbon dioxide (CO_2)

8 Calcium reacts with chlorine to form the ionic compound calcium chloride. Draw diagrams to show the electronic structure in calcium atoms, chlorine atoms, calcium ions and chloride ions.

9 Work out the formula of the following ionic compounds. The charge of some ions is given (sulfate = SO_4^{2-}, hydroxide = OH^-)

a) potassium oxide

b) magnesium fluoride

c) lithium sulfide

d) iron(III) sulfate

e) copper(II) hydroxide

10 Ethyne is a molecule with the molecular formula C_2H_2.

a) Explain what this formula tells us about ethyne.

b) Draw a stick diagram to show the covalent bonds in a molecule of ethyne.

c) Draw a dot-cross diagram to show the outer shell electrons in a molecule of ethyne.

11 The diagram shows part of a carbon nanotube. They are used to reinforce the materials used to make tennis racquets as they have high tensile strength.

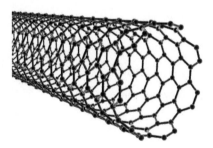

▲ Figure 10.30

a) How many covalent bonds does each carbon atom make in a carbon nanotube?

b) Explain clearly by considering your answer to (a) why nanotubes can conduct electricity.

c) Explain clearly why carbon nanotubes have high tensile strength.

Practice questions

1 The elements P and Q are in found in Groups 2 and 7 respectively, of the periodic table. Which one of the following shows the formula and the bonding type of the compound that they form? [1 mark]

A PQ$_2$ covalent

B PQ$_2$ ionic

C P$_2$Q covalent

D P$_2$Q ionic

2 Which one of the following does not have a giant covalent structure? [1 mark]

A diamond

B graphene

C graphite

D sulfur dioxide

3 A dot and cross diagram is given here:

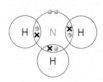

▲ Figure 10.31

a) Name and write the formula for this compound. [2 marks]

b) On a copy of the diagram above, use an arrow to label:

i) A covalent bond.

ii) A non-bonded pair of electrons. [2 marks]

c) Using a line to represent a single covalent bond, redraw the diagram shown above. [1 mark]

d) What is meant by the term single covalent bond? [2 marks]

4 The following diagram shows some changes between the states of matter.

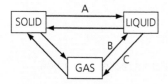

▲ Figure 10.32

a) What is the name for each of the changes labelled A, B and C? [3 marks]

b) Which of the changes A, B, or C is achieved by a decrease in temperature? [1 mark]

c) Draw a diagram to represent the arrangement of atoms in a solid using small solid spheres to represent atoms. [1 mark]

d) Explain the limitations of this simple particle theory in relation to change in state. [2 marks]

e) Explain why the melting points of some solids are greater than others. [2 marks]

5 The diagrams show two different structures of the Group 4 element carbon and two compounds of Group 4 elements.

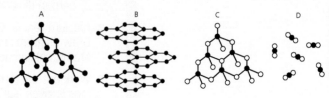

▲ Figure 10.33

a) Which two of the diagrams represent different structures of the element carbon? [1 mark]

b) The substances are carbon dioxide, diamond, graphite and silicon dioxide. Name each substance, A to D. [4 marks]

c) Name the type of bonding which occurs between the atoms in all of the substances A to D. [4 marks]

d) Name the type of structure for each substance A to D. [4 marks]

6 The photographs show two uses of graphite. It is used in pencils and for electrodes in the electrolysis of sodium chloride solution.

▲ Figure 10.34

a) With reference to the structure of graphite, explain why it is used in pencils. [3 marks]

b) Graphite electrodes conduct electricity. Explain why graphite is a good conductor of electricity. [2 marks]

c) Describe, as fully as you can, what happens when sodium atoms react with chlorine atoms to produce sodium chloride. You may use a diagram in your answer. [3 marks]

d) Explain why sodium chloride solution will conduct electricity, but sodium chloride solid will not. [2 marks]

7 The table gives some of the properties of the Period 3 element magnesium and one of its compounds, magnesium chloride.

Table 10.21

Property	Magnesium	Magnesium chloride
Melting point in °C	649	714
Electrical conductivity when solid	Conducts	Does not conduct
Electrical conductivity when molten	Conducts	Conducts

Use ideas about structure and bonding to explain the similarities and differences between the properties of magnesium and magnesium chloride. [6 marks]

8 a) Chlorine is a green gas which exists as diatomic molecules.

   i) Suggest what is meant by the term diatomic. [1 mark]

   ii) Use a dot and cross diagram to clearly show how atoms of chlorine combine to form chlorine molecules. [2 marks]

b) Chlorine can form a range of compounds with both metals and non-metals.

   i) Describe, as fully as you can, what happens when calcium atoms react with chlorine atoms to produce calcium chloride. You may use a diagram in your answer. [3 marks]

   ii) Name the type of bonding found in calcium chloride. [1 mark]

   iii) Name the type of structure for calcium chloride. [1 mark]

   iv) Use a dot and cross diagram to show how atoms of chlorine combine with atoms of carbon to form tetrachloromethane CCl_4. [2 marks]

   v) Name the type of bonding found in CCl_4. [1 mark]

c) The properties of compounds depend very closely on their bonding. Redraw the following table with only the correct words to show some of the properties of calcium chloride and tetrachloromethane. [2 marks]

Table 10.22

Compound	Solubility in water	Relative melting point
Calcium chloride	Soluble/insoluble	Low/high
Tetrachloromethane	Soluble/insoluble	Low/high

d) The bonding in the elements calcium and carbon is very different. Describe the bonding in calcium and in carbon (in the form of graphite). [6 marks]

e) Both calcium and graphite can conduct electricity. State two properties of calcium which are different from those of graphite. [2 marks]

f) Why might fullerenes be used in new drug delivery systems? Choose the correct statement A, B or C. [1 mark]

A They are made from carbon atoms.

B They are hollow.

C They are very strong.

g) How does the structure of nanotubes make them suitable as catalysts? Choose the correct statement A, B or C. [1 mark]

A They have a large surface area to volume ratio.

B They are made from reactive carbon atoms.

C They have strong covalent bonds.

Working scientifically:
Units: Using prefixes and powers of ten for orders of magnitude

Standard form

Standard form is used to express very large or very small numbers so that they are more easily understood and managed. It is easier to say that a speck of dust has a mass of 1.2×10^{-6} grams than to say it has a mass of 0.0000012 grams. It uses powers of ten.

Standard form must always look like this:

A must always be between 1 and 10	n is the number of places the decimal point moves (for numbers less than 1, the value of n will be negative)

$$A \times 10^n$$

▲ Figure 10.35 The mass of the Earth is 5972200000000000000000000 kg. This is more conveniently written in scientific form as 5.9722×10^{24} kg.

Example

Write 4600000 in standard form.

Answer

- Write the non-zero digits with a decimal place after the first number and then write × 10 after it:

  4.6×10

- Then count how many places the decimal point has moved and write this as the *n* value. The *n* value is positive as 4600000 is greater than 1.

  $4600000 = 4.6 \times 10^6$

Example

Write 0.000345 in standard form.

Answer

- Write the non-zero digits with a decimal place after the first number and then write × 10 after it:

  3.45×10

- Then count how many places the decimal point has moved and write this as the *n* value. The *n* value is negative as 0.000345 is less than 1.

  $0.000345 = 3.45 \times 10^{-4}$

▲ Figure 10.36 Make sure that you are familiar with how standard form is presented on your calculator. This calculator reads 1.23×10^{99} to 3 significant figures.

SI units

The International System of Units (SI) is a system of units of measurements that is widely used all over the world, and the one which you will use in your study of chemistry. It uses several base units of measure, for example metres, grams and seconds. When a numerical unit is very small or large, the units may be modified by using a **prefix**. Some prefixes are shown in the table.

Table 10.23

Prefix name	Prefix symbol	Scientific notation 10^n	Decimal
tera	T	10^{12}	1 000 000 000 000
giga	G	10^9	1 000 000 000
mega	M	10^6	1 000 000
kilo	k	10^3	1 000
centi	c	10^{-2}	0.01
milli	m	10^{-3}	0.001
micro	μ	10^{-6}	0.000 001

A prefix goes in front of a basic unit of measure to indicate a multiple of the unit. For example instead of writing 1000 grams we can add the prefix kilo (10^3) and write 1 kg.

For example using scientific notation:

▶ 1 cm = 1 centimetre $= 1 \times 10^{-2}$ m = 0.01 m

▶ 1 μl = 1 microlitre $= 1 \times 10^{-6}$ l = 0.000 001 l

▶ 1 μm = 1 micrometre $= 1 \times 10^{-6}$ m = 0.000 001 m

Questions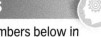

1 Write the numbers below in standard form.
   a) 0.00024
   b) 3 230 000 000
   c) 0.02
   d) 0.000 000 007
   e) 24 000

2 Write the numbers below as decimals.
   a) 2.3×10^{-3}
   b) 4.6×10^5
   c) 9.5×10^{-5}
   d) 5.34×10^4
   e) 3.3×10^3

Questions

3 Write the following quantities in units with the appropriate prefixes.
   a) 31 000 000 m
   b) 0.001 g
   c) 9700 m
   d) 0.000 000 002 s

11 Quantitative chemistry

Water companies regularly analyse samples of water supplied to homes to check that it is safe to drink. Their analytical chemists have to be able to carry out very accurate experiments to analyse the water. They also need to be able to carry out calculations using their results to work out how much of each substance is in the water. In this chapter you will learn how to perform a range of calculations.

Specification coverage

This chapter covers specification points 5.3.1.1 to 5.3.2.5 and is called Quantitative chemistry.

It covers relative mass and moles, conservation of mass, and reacting.

Related work on writing and balancing equations can be found in Chapter 14.

Relative mass and moles

▲ **Figure 11.1** The elephant has a relative mass of 500 compared to the child. This means that the elephant is 500 times heavier than the child.

KEY TERMS

Relative atomic mass The average mass of atoms of an element, taking into account the mass and amount of each isotope it contains.

Relative formula mass The sum of the relative atomic masses of all the atoms shown in the formula (often referred to as *formula mass*).

○ Relative atomic mass

Individual atoms have a tiny mass. For example, an atom of ^{12}C has a mass of about 2×10^{-23} g (that is 0.000 000 000 000 000 000 000 02 g). As numbers like this are awkward to use, scientists measure the mass of atoms relative to each other. They use a scale where the mass of a ^{12}C atom is defined as being exactly 12. On this scale, an atom of ^{24}Mg has a relative mass of 24 and is twice as heavy as a ^{12}C atom, whereas an atom of ^{1}H has a relative mass of 1 and is 12 times lighter than a ^{12}C atom. In effect, the relative mass of a single atom equals the mass number of that atom (mass number is the number of protons plus the number of neutrons).

Many elements are made up of a mixture of atoms of different isotopes. For example, 75% of chlorine atoms are ^{35}Cl with relative mass of 35 and the remaining 25% are ^{37}Cl atoms with relative mass of 37. The average relative mass of chlorine atoms is 35.5 as there are more chlorine atoms with relative mass of 35 than 37.

The relative atomic mass (A_r) of an element is the average mass of atoms of that element taking into account the mass and amount of each isotope it contains on a scale where the mass of a ^{12}C atom is 12.

○ Relative formula mass

The relative formula mass (M_r) of a substance is the sum of the relative atomic masses of all the atoms shown in the formula. It is often just called *formula mass*.

For example, the formula of water is H_2O and so the relative formula mass is the sum of the relative atomic mass of two hydrogen atoms (2×1) and one oxygen atom (16) which adds up to 18. This and other examples are shown in Table 11.1.

Table 11.1 Relative atomic mass and relative formula mass for some substances.

Name	Formula	A_r values	Sum	M_r (relative formula mass)
Water	H_2O	H = 1, O = 16	2(1) + 16 =	18
Copper	Cu	Cu = 63.5		63.5
Sodium chloride	NaCl	Na = 23, Cl = 35.5	23 + 35.5 =	58.5
Sulfuric acid	H_2SO_4	H = 1, S = 32, O = 16	2(1) + 32 + 4(16) =	98
Magnesium nitrate	$Mg(NO_3)_2$	Mg = 24, N = 14, O = 16	24 + 2(14) + 6(16) =	148
Ammonium sulfate	$(NH_4)_2SO_4$	N = 14, H = 1, S = 32, O = 16	2(14) + 8(1) + 32 + 4(16) =	132

Show you can...

Many compounds, for example carbon dioxide and calcium nitrate contain the element oxygen.

a) The formula of carbon dioxide is CO_2. How many oxygen atoms are present in one molecule? What is the relative formula mass of carbon dioxide?

b) Calcium nitrate is an ionic compound. Use ion charges to write the formula of calcium nitrate. How many oxygen atoms are shown in the formula? What is the relative formula mass of calcium nitrate?

c) A compound containing oxygen has the formula OX_2 and a relative formula mass of 54. Calculate the relative atomic mass of X and use your periodic table to identify X.

Test yourself

1 Calculate the relative formula mass of the following substances. You can find relative atomic masses on the periodic table in the Appendix of this book.
   a) ammonia, NH_3
   b) nickel, Ni
   c) butane, C_4H_{10}
   d) calcium hydroxide, $Ca(OH)_2$
   e) aluminium nitrate, $Al(NO_3)_3$

○ The mole

What is a mole?

Amounts of chemicals are often measured in moles. This makes it much easier to work out how much of a chemical is needed in a reaction.

One atom of ^{12}C has a relative mass of 12. The mass of 602 000 000 000 000 000 000 000 (6.02×10^{23}) atoms of ^{12}C is exactly 12 g. The number 6.02×10^{23} is a very special number and is known as the Avogadro constant.

When you have a pair of ^{12}C atoms, you have two of them. When you have a dozen ^{12}C atoms, you have 12 of them. When you have a mole of ^{12}C atoms, you have 6.02×10^{23} of them.

One mole of a substance contains the same number of the stated particles (atoms, molecules or ions) as one mole of any other substance. For example, the number of carbon atoms in one mole of carbon atoms (6.02×10^{23} atoms) is the same number of particles as there are molecules in one mole of water molecules (6.02×10^{23} molecules).

The value of the Avogadro constant was chosen so that the mass of one mole of that substance is equal to the relative formula mass (M_r) in grams. Table 11.2 shows some examples.

Table 11.2 Relative formula mass and the mole.

Name	Formula	Relative formula mass (M_r)	Mass of 1 mole of that substance
Water	H_2O	18	18 g
Copper	Cu	63.5	63.5 g
Sodium chloride	NaCl	58.5	58.5 g
Sulfuric acid	H_2SO_4	98	98 g
Magnesium nitrate	$Mg(NO_3)_2$	148	148 g
Ammonium sulfate	$(NH_4)_2SO_4$	132	132 g

How many moles?

$$mass (g) = M_r \times moles$$

This formula triangle may help you. Cover up the quantity you want to show the equation you need to use.

Thinking of **Mr Moles** with a mass on his head may help you remember this triangle

▲ **Figure 11.2** Formula triangle for working out mass and number of moles.

If one mole of ^{12}C atoms has a mass of 12 g, then it follows that the mass of two moles of ^{12}C atoms will be 24 g. There is a simple equation linking the mass of a substance to the number of moles (Figure 11.2).

When using this equation, the mass must be in grams. Table 11.3 shows conversion factors if the masses are not given in grams.

Table 11.3 Conversion factors to work out mass in grams.

Conversion factor	Example
1 tonne = 1 000 000 g	3 tonnes = 3 × 1 000 000 = 3 000 000 g
1 kg = 1000 g	0.5 kg = 0.5 × 1000 = 500 g
1 mg = 0.001 g	20 mg = 20 × 0.001 = 0.020 g

Table 11.4 gives some examples using the equation that links mass and moles. In calculations, the units of moles is usually abbreviated to mol.

Table 11.4 Examples of working out mass and number of moles.

How many moles in each of the following?	180 g of H_2O	M_r of H_2O = 2(1) + 16 = 18 Moles $H_2O = \dfrac{mass}{M_r} = \dfrac{180}{18} = 10$ mol
	4 g of CH_4	M_r of CH_4 = 12 + 4(1) = 16 Moles $CH_4 = \dfrac{mass}{M_r} = \dfrac{4}{16} = 0.25$ mol
	2 kg of Fe_2O_3	M_r of Fe_2O_3 = 2(56) + 3(16) = 160 Mass of Fe_2O_3 = 2 kg = 2 × 1000 g = 2000 g Moles $Fe_2O_3 = \dfrac{mass}{M_r} = \dfrac{2000}{160} = 12.5$ mol
	50 mg of NaOH	M_r of NaOH = 23 + 16 + 1 = 40 Mass of NaOH = 50 mg = $\dfrac{50}{1000}$ g = 0.050 g Moles NaOH = $\dfrac{mass}{M_r} = \dfrac{0.050}{40} = 0.00125$ mol
What is the mass of each of the following?	20 moles of CO_2	M_r of CO_2 = 12 + 2(16) = 44 Moles $CO_2 = M_r \times moles = 44 \times 20 = 880$ g
	0.025 moles of Cl_2	M_r of Cl_2 = 2(35.5) = 71 Moles $Cl_2 = M_r \times moles = 71 \times 0.025 = 1.78$ g
What is the M_r of the following?	3.6 g of a substance is found to contain 0.020 mol	$M_r = \dfrac{mass}{moles} = \dfrac{3.6}{0.020} = 180$

Show you can...

Use Figure 11.2 to write three different equations linking mass, moles and M_r. Use these equations to answer the questions below.

a) Calculate the number of moles in 9.8 g of H_2SO_4.

b) Calculate the mass in grams of 0.5 moles of $Ca(OH)_2$.

c) 6.9 g of a substance Y_2CO_3 contains 0.05 moles. What is the relative formula mass of the substance? Identify Y.

Test yourself

You can find relative atomic masses on the periodic table in the Appendix to help you answer these questions.

2 What is the mass of one mole of the following substances?
   a) iron, Fe
   b) oxygen, O_2
   c) ethane, C_2H_6
   d) potassium chloride, KCl
   e) calcium nitrate, $Ca(NO_3)_2$

3 Which one of the substances in each of the following pairs of substances contains the most particles or are they the same?
   a) 2 moles of water (H_2O) molecules and 2 moles of carbon dioxide (CO_2) molecules
   b) 10 moles of methane (CH_4) molecules and 10 moles of argon atoms (Ar)
   c) 10 moles of helium atoms (He) and 5 moles of oxygen molecules (O_2)

4 What are the following masses in grams?
   a) 20 kg
   b) 5 mg
   c) 0.3 tonnes

5 What is the mass of each of the following?
   a) 3 moles of oxygen, O_2
   b) 0.10 moles of ethanol, C_2H_5OH

6 How many moles are there in each of the following?
   a) 50 g of hydrogen, H_2
   b) 4 kg of calcium carbonate, $CaCO_3$
   c) 80 mg of bromine, Br_2
   d) 2 tonnes of calcium oxide, CaO

7 0.300 g of a substance was analysed and found to contain 0.0050 moles. Calculate the M_r of the substance.

◯ Significant figures

We often quote answers to calculations to a certain number of significant figures. In chemistry, we usually give values to 2, 3 or 4 significant figures (sf), but it can be more or less than this. Table 11.5 shows some numbers given to 2, 3 and 4 significant figures.

Table 11.5 Some numbers to 2, 3 and 4 significant figures.

Number	2 sf	3 sf	4 sf
2.7358	2.7	2.74	2.736
604531	600000	605000	604500
0.10836	0.11	0.108	0.1084
0.0042981	0.0043	0.00430	0.004298

We quote values to a limited number of significant figures because we cannot be sure of the exact value to a greater number of significant figures.

For example, if we measure the temperature rise in a reaction three times and find the values to be 21°C, 21°C and 22°C, then the mean temperature rise shown on a calculator would be 21.33333333°C. However, it is impossible for us to say that the temperature rise is exactly 21.33333333°C as the thermometer could only measure to ±1°C. Therefore, we should quote the temperature rise to 2 significant figures, i.e. 21°C, which is the same number of significant figures as the values we measured.

Test yourself

8 Give the following numbers to 2, 3 and 4 significant figures.
   a) 34.8226
   b) 28554210
   c) 0.0231876
   d) 0.000631947

9 Find the mean of these measurements and give your answer to 3 significant figures.
   a) 25.4 cm³, 25.1 cm³, 25.3 cm³
   b) 162 s, 175 s, 169 s, 173 s
   c) 1.65 g, 1.70 g, 1.69 g, 1.64 g, 1.71 g

Conservation of mass

N₂ + 3H₂ → 2NH₃

1 molecule of N₂	3 molecules of H₂	2 molecules of NH₃
2 N atoms	6 H atoms	2 N atoms & 6 H atoms

▲ **Figure 11.3** There are 2 N and 6 H atoms on both sides of this equation.

KEY TERM

Conservation of mass In a reaction, the total mass of the reactants must equal the total mass of the products.

N_2 $M_r = 28$ + H_2 $M_r = 2$ + H_2 $M_r = 2$ + H_2 $M_r = 2$ → NH_3 $M_r = 17$ + NH_3 $M_r = 17$

Sum of M_r of all reactants = 28 + 2 + 2 + 2 = 34

Sum of M_r of all products = 17 + 17 = 34

▲ **Figure 11.4** Conservation of mass: the mass of the products must equal the mass of the reactants.

○ Balanced equations

In a balanced chemical equation, the **number of atoms of each element is the same on both sides of the equation**. This is because atoms cannot be created or destroyed in chemical reactions. This means that you have the same atoms before and after the reaction, although how they are bonded to each other changes during the reaction.

For example, in Figure 11.3 one molecule of nitrogen (N_2) reacts with three molecules of hydrogen (H_2) to make two molecules of ammonia (NH_3). There are two N atoms and six H atoms on both sides of the equation.

There is a section on how to balance symbol equations later in this chapter.

○ Conservation of mass

As there are the same atoms present at the start and end of a chemical reaction, mass must be conserved in the reaction. In other words, the total mass of the reactants must equal the total mass of the products. This is known as the law of conservation of mass.

One way to look at this is using relative formula masses. The total of the relative formula masses of all the reactants in the quantities shown in the equation will add up to the total of the relative formula masses of all the products in the quantities shown in the equation (Figure 11.4).

All chemical reactions obey the law of conservation of mass. There are some reactions that may appear to break this law, but they do not as shown below.

○ Reaction of metals with oxygen

When metals react with oxygen the mass of the product is greater than the mass of the original metal. However, this does not break the law of conservation of mass. The 'extra' mass is the mass of the oxygen from the air that has combined with the metal to form a metal oxide.

In the example shown in Figure 11.5, some magnesium has been heated in a crucible. It may appear as though the products are 0.16 g heavier than the reactants, but 0.16 g of oxygen has reacted with the 0.24 g magnesium to make 0.40 g magnesium oxide.

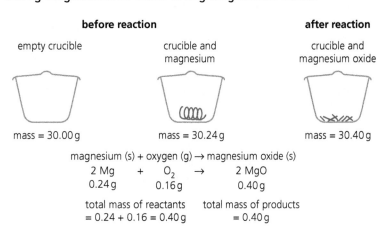

before reaction		after reaction
empty crucible	crucible and magnesium	crucible and magnesium oxide
mass = 30.00 g	mass = 30.24 g	mass = 30.40 g

magnesium (s) + oxygen (g) → magnesium oxide (s)

2 Mg + O₂ → 2 MgO
0.24 g   0.16 g   0.40 g

total mass of reactants = 0.24 + 0.16 = 0.40 g

total mass of products = 0.40 g

▶ **Figure 11.5** In this reaction, the total mass of products is the same as the total mass of reactants.

○ Thermal decomposition reactions

A thermal decomposition reaction is one where high temperature causes a substance to break down into simpler substances. In decomposition reactions, one or more of the products may escape from the reaction container into the air as a gas. These reactions may appear to lose mass, but the 'missing' mass is the mass of the gas that has escaped into the air.

In the example shown in Figure 11.6, some copper carbonate is heated in a crucible. It may appear as though the products are 0.22 g lighter than the reactants, but 0.22 g of carbon dioxide has been released into the air.

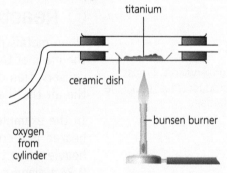

before reaction		after reaction
empty crucible	crucible and copper carbonate	crucible and copper oxide
mass = 30.00 g	mass = 30.62 g	mass = 30.40 g

copper carbonate (s) → copper oxide (s) + carbon dioxide (g)

$$CuCO_3 \rightarrow CuO + CO_2$$
$$0.62\,g \qquad 0.40\,g \qquad 0.22\,g$$

total mass of reactants = 0.62 g   total mass of products = 0.40 + 0.22 = 0.62 g

▲ Figure 11.6 Thermal decomposition and conservation of mass.

Oxidation of titanium

The metal titanium reacts with oxygen to form an oxide of titanium. In an experiment a sample of titanium metal was heated in a crucible with a lid. During heating the lid was lifted from time to time.

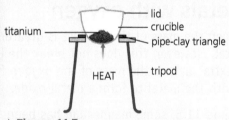

▲ Figure 11.7

The following results were obtained:
Mass of crucible = 16.34 g
Mass of crucible + titanium metal = 17.36 g
Mass of crucible + titanium oxide = 18.04 g

Questions

1 Use the results to calculate
   a) the mass of titanium used in this experiment
   b) the mass of titanium oxide formed in this experiment
   c) the mass of oxygen used in this experiment.

2 The equation for the reaction is $Ti + O_2 \rightarrow TiO_2$
   Explain using the masses calculated in (1) how this reaction follows the law of conservation of mass.

3 Suggest why it was necessary to lift the crucible lid during heating.

In a different experiment titanium metal was heated in a stream of oxygen:

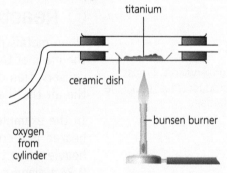

▲ Figure 11.8

Questions

4 What masses should be found before heating to determine the mass of titanium used in the experiment?

5 The ceramic dish and its contents are repeatedly weighed, heated, reweighed and heated until the mass is constant. State and explain if the mass increases or decreases during this experiment.

6 What safety precautions should be taken in this experiment?

7 How would the repeatability of this experiment be checked?

○ Balancing symbol equations

To balance symbol equations, you need to be able to work out:

● the number of **elements** in a chemical formula; for example, NaCl has two elements – Na (sodium) and Cl (chlorine); similarly, NaOH has three different elements – Na (sodium), O (oxygen) and H (hydrogen)
● the number of **atoms**; NaCl has one atom of sodium (Na), and one atom of chlorine (Cl); however, water (H_2O) has two atoms of hydrogen and one of oxygen.

When working out the number of atoms you should remember that if the number is subscript (the small number at the bottom), it refers to the element immediately before it; for example, in $2NaHCO_3$ the subscript 3 means there are three atoms of oxygen (O).

If the number is normal size, for example $2NaHCO_3$, it refers to all the elements after it. In this example there are two sodium (Na) atoms, two hydrogen (H), two carbon (c) and six oxygen (O) atoms (2 × 3).

The formula $Ca(HCO_3)_2$ has one atom of calcium, two of hydrogen, two of carbon and six of oxygen (the brackets indicates that everything inside the bracket is multiplied by the number in subscript outside).

Example

The symbol equation for the reaction between sodium hydroxide and hydrochloric acid is

$$NaOH + HCl \rightarrow NaCl + H_2O$$

1O \ 1Cl   1Cl   1O
1Na 2H   1Na   2H

This equation is balanced as there are the same numbers of each type of atom on either side.

Example

Magnesium burns in oxygen to give magnesium oxide.

$$Mg + O_2 \rightarrow MgO$$

In this equation there are two oxygen atoms on the left side and only one on the right.

We need to double the number of oxygens on the right by using the balancing number 2, i.e. 2MgO, but as we have also doubled the number of magnesium atoms we need to balance them too. We do this by doubling the Mg on the left.

$$2Mg + O_2 \rightarrow 2MgO$$

Example

When heated sodium hydrogencarbonate breaks down to form sodium carbonate, carbon dioxide and water.

The unbalanced symbol equation is:

$$NaHCO_3 \rightarrow Na_2CO_3 + CO_2 + H_2O$$

The balanced symbol equation for heating sodium hydrogencarbonate is:

$$\textbf{2NaHCO}_3 \rightarrow \textbf{Na}_2\textbf{CO}_3 + \textbf{CO}_2 + \textbf{H}_2\textbf{O}$$

We know it is balanced because:

- there are two Na atoms on the left (2Na) and two on the right (Na_2)
- two H atoms on the left and two on the right
- two C atoms on the left and two on the right
- six O atoms on the left and six on the right.

Example

Lithium reacts with water to give lithium hydroxide and hydrogen.

Using symbols the unbalanced equation is:

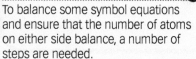

This equation does need to be balanced, as there are different numbers of atoms of hydrogen on each side.

To balance the number of atoms on either side, we need to balance the hydrogen atoms (currently two on the left and three on the right). We do this by using balancing numbers as shown below to give the balanced symbol equation:

$$2Li + 2H_2O \rightarrow 2LiOH + H_2$$

| 2Li 4H | 2Li 4H |

TIP

To balance some symbol equations and ensure that the number of atoms on either side balance, a number of steps are needed.

State symbols

Balanced equations sometimes state symbols to show the state of each substance: (s) solid, (l) liquid, (g) gas, (aq) aqueous. For example:

$$CaCO_3(s) + 2HCl(aq) \rightarrow CaCl_2(aq) + H_2O(l) + CO_2(g)$$

Test yourself

You can find relative atomic masses on the periodic table in the Appendix to help you answer these questions.

10 Hydrogen reacts with oxygen to make water as shown in this equation below.

$$2H_2 + O_2 \rightarrow 2H_2O$$

a) Describe in words what this tells you about the reaction of hydrogen with oxygen in terms of how many molecules are involved in the reaction.

b) Show that the sum of the relative formula masses of all the reactants equals the sum of the relative formula masses of all the products.

11 A piece of copper was heated in air. After a few minutes it was reweighed and found to be heavier.

a) Explain why the copper gets heavier.

b) What is the law of conservation of mass?

c) Explain why this reaction does not break the law of conservation of mass.

12 When 1.19 g of solid nickel carbonate is heated for several minutes, only 0.75 g of solid remains. Explain clearly why the mass decreases and what happens to the remaining mass.

Reacting masses

○ Molar ratios in equations

Chemical equations can be interpreted in terms of moles. For example, in the equation in Figure 11.9 one mole of nitrogen (N_2) molecules reacts with three moles of hydrogen (H_2) molecules to form two moles of ammonia (NH_3) molecules.

> **TIP**
> The Avogadro constant is the number of particles in one mole of a substance and has a value of 6.02×10^{23}.

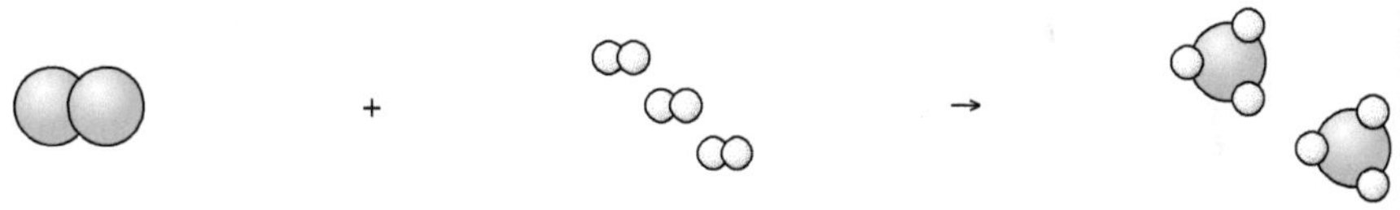

N_2	+	$3H_2$	→	$2NH_3$
1 molecule of N_2	reacts with	3 molecules of H_2	to make	2 molecules of NH_3
100 molecules of N_2	reacts with	300 molecules of H_2	to make	200 molecules of NH_3
602 molecules of N_2	reacts with	1806 molecules of H_2	to make	1204 molecules of NH_3
6.02×10^{23} molecules of N_2	reacts with	18.06×10^{23} molecules of H_2	to make	12.04×10^{23} molecules of NH_3
1 mole of molecules of N_2	reacts with	3 moles of molecules of H_2	to make	2 moles of molecules of NH_3

▶ **Figure 11.9** Balanced equations give ratios in which substances react.

These ratios can be used to calculate how many moles would react and be produced in reactions (Figures 11.10 and 11.11).

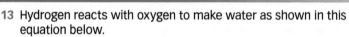

N₂	+	3H₂	→	2NH₃
N_2 + $3H_2$ → $2NH_3$				
1 mole N₂	reacts with	3 moles H₂	to make	2 moles NH₃
10 moles N₂	reacts with	30 moles H₂	to make	20 moles NH₃
2 moles N₂	reacts with	6 moles H₂	to make	4 moles NH₃
0.5 moles N₂	reacts with	1.5 moles H₂	to make	1.0 moles NH₃

▲ Figure 11.10 Examples of molar quantities for the reaction of nitrogen with hydrogen.

C_3H_8 + $5O_2$ → $3CO_2$ + $4H_2O$

C₃H₈	+	5O₂	→	3CO₂	+	4H₂O
1 mole C₃H₈	reacts with	5 moles O₂	to make	3 moles CO₂	and	4 moles H₂O
10 moles C₃H₈	reacts with	50 moles O₂	to make	30 moles CO₂	and	40 moles H₂O
2 moles C₃H₈	reacts with	10 moles O₂	to make	6 moles CO₂	and	8 moles H₂O
0.5 moles C₃H₈	reacts with	2.5 moles O₂	to make	1.5 moles CO₂	and	2.0 moles H₂O

▲ Figure 11.11 Examples of molar quantities for the reaction of propane with oxygen.

Show you can...

a) Select the correct words to complete the sentence below.

In the equation C + O_2 → CO_2 one mole of C *atoms/ molecules* reacts with one mole of O_2 *atoms/molecules* to form one mole of CO_2 *atoms/molecules*.

b) Write a similar sentence about each equation below
   i) $C_2H_4 + 2O_2 → CO_2 + 2H_2O$
   ii) $Be + Cl_2 → BeCl_2$
   iii) $C_xH_y + 3O_2 → 2CO_2 + 2H_2O$

c) What is the value of x and y in equation (ii)?

Test yourself

13 Hydrogen reacts with oxygen to make water as shown in this equation below.
   $2H_2 + O_2 → 2H_2O$
   a) How many moles of oxygen react with 10 moles of hydrogen?
   b) How many moles of hydrogen react with 3 moles of oxygen?
   c) How many moles of oxygen react with 0.3 moles of hydrogen?
   d) How many moles of hydrogen react with 2.5 moles of oxygen?

14 Titanium is made when titanium chloride reacts with sodium as shown in the equation below.
   $TiCl_4 + 4Na → Ti + 4NaCl$
   a) How many moles of sodium react with 4 moles of titanium chloride?
   b) How many moles of titanium are made from 2.5 moles of titanium chloride?
   c) How many moles of sodium react with 0.5 moles of titanium chloride?
   d) How many moles of titanium are made from 0.65 moles of titanium chloride?

15 Potassium chlorate ($KClO_3$) decomposes to form potassium chloride and oxygen as shown below.
   $2KClO_3 → 2KCl + 3O_2$
   a) How many moles of potassium chloride are formed when 10 moles of potassium chlorate decomposes?
   b) How many moles of oxygen are formed when 4 moles of potassium chlorate decomposes?
   c) How many moles of potassium chloride are formed when 0.5 moles of potassium chlorate decomposes?
   d) How many moles of oxygen are formed when 3 moles of potassium chlorate decomposes?

TIP

When doing reacting mass calculations, it is advised to stick to one method only. If you are thinking of studying chemistry beyond GCSE then you are probably best to use the moles method.

○ Calculating reacting masses

Scientists need to be able to calculate how much of each substance to use in a chemical reaction. There are two common ways to do these calculations. One method uses moles while the other uses ratios and relative formula masses.

Examples

Using moles

1 Work out the relative formula mass (M_r) of the substance whose mass is given and the one you are finding the mass of. (Remember that the balancing numbers in the equation are not part of the formulae).
2 Calculate the moles of the substance whose mass is given (using moles = mass / M_r).
3 Use molar ratios from the balanced equation to work out the moles of the substance the question asks about.
4 Calculate the mass of that substance (using mass = M_r × moles).

Example 1
Calculate the mass of hydrogen needed to react with 140g of nitrogen to make ammonia.
$N_2 + 3H_2 \rightarrow 2NH_3$
M_r N_2 = 2(14) = 28, H_2 = 2(1) = 2
Moles $N_2 = \dfrac{\text{mass}}{M_r} = \dfrac{140}{28} = 5$
Moles H_2 = moles of N_2 × 3 = 5 × 3 = 15
Mass H_2 = M_r × moles = 2 × 15 = 30g

Example 2
What mass of oxygen reacts with 4.6 g of sodium?
$4Na + O_2 \rightarrow 2Na_2O$
M_r Na = 23, O_2 = 2(16) = 32
moles Na = $\dfrac{\text{mass}}{M_r} = \dfrac{4.6}{23} = 0.20$
Moles O_2 = moles of Na ÷ 4 = 0.20 ÷ 4 = 0.05
Mass O_2 = M_r × moles = 32 × 0.05 = 1.6g

Example 3
What mass of iron is produced when 32kg of iron oxide is heated with carbon monoxide?
$Fe_2O_3 + 3CO \rightarrow 2Fe + 3CO_2$
M_r Fe_2O_3 = 2(56) + 3(16) = 160, Fe = 56
moles $Fe_2O_3 = \dfrac{\text{mass}}{M_r} = \dfrac{32\,000}{160} = 200$
Moles Fe = moles of Fe_2O_3 × 2 = 200 × 2 = 400
Mass Fe = M_r × moles = 56 × 400 = 22 400g

Using ratios

1 Work out the relative formula mass (M_r) of the substance whose mass is given and the one you are finding the mass of. (Remember that the balancing numbers in the equation are not part of the formulae).
2 Find the reacting mass ratio for these substances using the M_r values and the balancing numbers in the equation.
3 Scale this to find what happens with 1 g of the substance given.
4 Scale this up/down to the mass you were actually given.

Example 1
Calculate the mass of hydrogen needed to react with 140g of nitrogen to make ammonia.
$N_2 + 3H_2 \rightarrow 2NH_3$
M_r N_2 = 2(14) = 28, H_2 = 2(1) = 2
N_2 reacts with $3H_2$
28g of N_2 reacts with 6g (3 × 2) of H_2
1g of N_2 reacts with $\dfrac{6}{28}$ g of H_2
140g of N_2 reacts with 140 × $\dfrac{6}{28}$ = 30g of H_2

Example 2
What mass of oxygen reacts with 4.6 g of sodium?
$4Na + O_2 \rightarrow 2Na_2O$
M_r Na = 23, O_2 = 2(16) = 32
4Na reacts with O_2
92g (4 × 23) of Na reacts with 32g of O_2
1g of Na reacts with $\dfrac{32}{92}$ g of O_2
4.6g of Na reacts with 4.6 × $\dfrac{32}{92}$ = 1.6g of O_2

Example 3
What mass of iron is produced when 32kg of iron oxide is heated with carbon monoxide?
$Fe_2O_3 + 3CO \rightarrow 2Fe + 3CO_2$
M_r Fe_2O_3 = 2(56) + 3(16) = 160, Fe = 56
Fe_2O_3 makes 2Fe
160g of Fe_2O_3 makes 112g (2 × 56) of Fe
1g of Fe_2O_3 makes $\dfrac{112}{160}$ g of Fe
32kg of Fe_2O_3 reacts with 32 × $\dfrac{112}{160}$ = 22.4kg of Fe

Test yourself

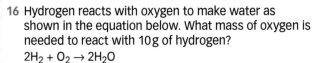

16 Hydrogen reacts with oxygen to make water as shown in the equation below. What mass of oxygen is needed to react with 10g of hydrogen?
$2H_2 + O_2 \rightarrow 2H_2O$

17 Calcium carbonate decomposes when heated as shown in the equation below. What mass of calcium oxide is formed from 25g of calcium carbonate?
$CaCO_3 \rightarrow CaO + CO_2$

18 Magnesium burns in oxygen to form magnesium oxide as shown in the equation below. What mass of magnesium oxide would be made from 3g of magnesium?
$2Mg + O_2 \rightarrow 2MgO$

19 Lithium reacts with water as shown in the equation below. What mass of hydrogen would be made from 1.4g of lithium?
$2Li + 2H_2O \rightarrow 2LiOH + H_2$

20 Aluminium is made by the electrolysis of aluminium oxide as shown in the equation below. What mass of aluminium can be formed from 1kg of aluminium oxide? Give your answer to 3 significant figures.
$2Al_2O_3 \rightarrow 4Al + 3O_2$

21 Hot molten copper can be produced for welding electrical connections in circuits by the reaction of copper oxide with aluminium powder shown below. What mass of aluminium is needed to react with 10g of copper oxide? Give your answer to 3 significant figures.
$3CuO + 2Al \rightarrow 3Cu + Al_2O_3$

○ Deducing the balancing numbers in an equation from reacting masses

The balancing numbers in a chemical equation can be calculated by calculating the moles of the substances in the reaction. In order to do this:

1 Calculate the moles of each substance (using moles $= \frac{\text{Mass}}{M_r}$)

2 Find the simplest whole number ratio of these mole values by dividing all the mole values by the smallest mole value.

3 If this does not give a whole number ratio, multiply up by a factor of 2 (where there is a value ending in approximately 0.5), of 3 (where there is a value ending in approximately 0.33 or 0.67), of 4 (where there is a value ending in approximately 0.25 or 0.75), etc.

Example

1.2 g of magnesium (Mg) reacts with 0.8 g of oxygen (O_2) to make 2.0 g of magnesium oxide (MgO). Use this information to deduce the equation for this reaction.

Answer

Substance	Magnesium (Mg)	Oxygen (O_2)	Magnesium oxide (MgO)
Calculate the moles of each substance	moles $= \frac{\text{mass}}{M_r}$ $= \frac{1.2}{24} = 0.05$	moles $= \frac{\text{mass}}{M_r}$ $= \frac{0.8}{32} = 0.025$	moles $= \frac{\text{mass}}{M_r}$ $= \frac{2.0}{40} = 0.05$
Find the simplest whole number ratio	$\frac{0.05}{0.025} = 2$	$\frac{0.025}{0.025} = 1$	$\frac{0.05}{0.025} = 2$

Therefore the reacting ratio is 2:1:2, and so the balanced equation is
$2Mg + O_2 \rightarrow 2MgO$

Example

0.81 g of aluminium (Al) reacts with 3.20 g of chlorine (Cl_2) to make 4.01 g of aluminium chloride ($AlCl_3$). Use this information to deduce the equation for this reaction.

Answer

Substance	Aluminium (Al)	Chlorine (Cl_2)	Aluminium chloride ($AlCl_3$)
Calculate the moles of each substance	moles $= \frac{\text{mass}}{M_r}$ $= \frac{0.81}{27} = 0.030$	moles $= \frac{\text{mass}}{M_r}$ $= \frac{3.2}{71} = 0.045$	moles $= \frac{\text{mass}}{M_r}$ $= \frac{4.01}{133.5} = 0.030$
Find the simplest whole number ratio	$\frac{0.030}{0.030} = 1$ × 2 to get rid of 0.5 values $1 \times 2 = 2$	$\frac{0.045}{0.030} = 1.5$ × 2 to get rid of 0.5 values $1.5 \times 2 = 3$	$\frac{0.030}{0.030} = 1$ × 2 to get rid of 0.5 values $1 \times 2 = 2$

Therefore the reacting ratio is 2:3:2, and so the balanced equation is
$2Al + 3Cl_2 \rightarrow 2AlCl_3$

Example

2.2g of propane (C_3H_8) reacts with 8.0g of oxygen (O_2). Calculate the molar ratio in which propane and oxygen react here.

Table 11.6

Substance	propane (C_3H_8)	oxygen (O_2)
Calculate the moles of each substance	moles $= \dfrac{mass}{M_r} = \dfrac{2.2}{44} = 0.050$	moles $= \dfrac{mass}{M_r} = \dfrac{8.0}{32} = 0.25$
Find the simplest whole number ratio	$\dfrac{0.050}{0.050} = 1$	$\dfrac{0.25}{0.050} = 5$

Therefore the reacting ratio is 1:5, and so the ratio they react in is **$C_3H_8 + 5O_2$**

Test yourself

22 6.8g of hydrogen peroxide (H_2O_2) decomposes into 3.6g of water (H_2O) and 3.2g of oxygen (O_2). By calculating molar ratios, deduce the balanced equation for this reaction.

23 2.0g of calcium (Ca) reacts with 1.9g of fluorine (F_2) to form 3.9g of calcium fluoride (CaF_2). By calculating molar ratios, deduce the balanced equation for this reaction.

24 1.0g of nickel (Ni) reacts with 0.27g of oxygen (O_2) to form 1.27g of nickel oxide (NiO). By calculating molar ratios, deduce the balanced equation for this reaction.

25 When 4.1g of calcium nitrate ($Ca(NO_3)_2$) is heated, 1.4g of calcium oxide (CaO), 2.3g of nitrogen dioxide (NO_2) and 0.4g of oxygen (O_2) are formed. By calculating molar ratios, deduce the balanced equation for this reaction.

26 11.7g of potassium (K) reacts with 24g of bromine (Br_2). Calculate the molar ratio in which potassium reacts with bromine.

27 11.4g of titanium chloride ($TiCl_4$) reacts with 5.52g of sodium (Na). Calculate the molar ratio in which titanium chloride reacts with sodium.

▲ **Figure 11.12** Some camping stoves use propane gas.

Practical

Experiment to find the equation for the action of heating sodium hydrogencarbonate

A student suggested that there are three possible equations for the thermal decomposition of sodium hydrogencarbonate.

Equation 1: $NaHCO_3 \rightarrow NaOH + CO_2$
Equation 2: $2NaHCO_3 \rightarrow Na_2CO_3 + H_2O + CO_2$
Equation 3: $2NaHCO_3 \rightarrow Na_2O + H_2O + 2CO_2$

In order to find out which is correct, she carried out the following experiment and recorded her results.

• Find the mass of an empty evaporating basin.
• Add approximately 8g of sodium hydrogencarbonate to the basin and find the mass.
• Heat gently for about 5 minutes.
• Allow to cool and then find the mass.
• Reheat, cool and find the mass.
• Repeat the heating and measurement of mass until constant mass is obtained.

Results

Mass of evaporating basin = 21.05g
Mass of basin and sodium hydrogencarbonate = 29.06g
Mass of sodium hydrogencarbonate = 8.01g
Mass of basin and residue after heating to constant mass = 26.10g
Mass of residue = 5.05g

Questions

1 Draw a labelled diagram of the assembled apparatus for this experiment.

2 Calculate the number of moles of sodium hydrogencarbonate used.

3 From the possible equations, and your answer to question 2, calculate:
   a) The number of moles of NaOH that would be formed in equation 1.
   b) The number of moles of Na_2CO_3 that would be formed in equation 2.
   c) The number of moles of Na_2O that would be formed in equation 3.

4 From your answers to question 3, calculate
   a) The mass of NaOH that would be formed in equation 1.
   b) The mass of Na_2CO_3 that would be formed in equation 2.
   c) The mass of Na_2O that would be formed in equation 3.

5 By comparing your answers in question 4 with the experimental mass of residue, deduce which is the correct equation for the decomposition of sodium hydrogencarbonate.

6 Why did the mass decrease?

○ Using an excess

In many reactions involving two reactants, it is very common for an excess of one of the reactants to be used to ensure that all of the other reactant is used up. This is often done if one of the reactants is readily available but the other one is expensive or is in limited supply. For example, when many fuels are burned an excess of oxygen is used. Fuels are expensive and in limited supply. The oxygen is readily available from the air and using an excess of oxygen ensures that all the fuel burns.

When one of the reactants is in excess, the other reactant is a limiting reactant that is completely used up. This is because it is the amount of this substance that determines the amount of product formed in a reaction, in other words it limits the amount of product made.

Example

Magnesium reacts with sulfuric acid as shown below: 5 moles of magnesium (Mg) is reacted with 7 moles of sulfuric acid (H_2SO_4). One of the reagents is in excess. Calculate the moles of the products formed.

$$Mg + H_2SO_4 \rightarrow MgSO_4 + H_2$$

Answer

As magnesium and sulfuric acid react with the ratio 1 to 1, only 5 moles of the sulfuric acid can react as there are only 5 moles of magnesium to react with. The rest of the sulfuric acid is in excess (2 moles). Therefore, the magnesium is the limiting reactant and determines how much of the products are made. In this case, 5 moles of $MgSO_4$ and 5 moles of H_2 would be made.

Table 11.7

	Mg	+	H_2SO_4	→	$MgSO_4$	+	H_2
Reacting ratio from the equation	1 mole of Mg	reacts with	1 mole of H_2SO_4	to make	1 mole of $MgSO_4$	and	1 mole of H_2
Amount provided	5 moles of Mg		7 moles of H_2SO_4				
	limiting reactant		in excess so it does not all react				
Reaction that takes place	5 moles of Mg	reacts with	5 moles of H_2SO_4	to make	5 moles of $MgSO_4$	and	5 moles of H_2

Example

Iron oxide reacts with carbon monoxide as shown below to produce iron. 10 moles of iron oxide (Fe_2O_3) is reacted with 50 moles of carbon monoxide (CO). Calculate the moles of products formed.

$$Fe_2O_3 + 3CO \rightarrow 2Fe + 3CO_2$$

Answer

In this reaction, the reacting ratio is one mole of iron oxide for every three moles of carbon monoxide. Therefore, 10 moles of iron oxide reacts with 30 moles of carbon monoxide. The rest of the carbon monoxide is in excess (20 moles). Therefore, the iron oxide is the limiting reactant and determines how much of the products are made. In this case, 20 moles of iron and 30 moles of carbon dioxide would be made.

Table 11.8

	Fe_2O_3	+	3CO	→	2Fe	+	$3CO_2$
Reacting ratio from the equation	1 mole of Fe_2O_3	reacts with	3 moles of CO	to make	2 moles of Fe	and	3 moles of CO_2
Amount provided	10 moles of Fe_2O_3		50 moles of CO				
	limiting reactant		in excess so it does not all react				
Reaction that takes place	10 moles of Fe_2O_3	reacts with	30 moles of CO	to make	20 moles of Fe	and	30 moles of CO_2

Example

Magnesium reacts with oxygen as shown below. 0.30 moles of magnesium (Mg) is reacted with 0.20 moles of oxygen (O_2). Calculate the moles of products formed.

$$2Mg + O_2 \rightarrow 2MgO$$

Answer

In this reaction, the reacting ratio is two moles of magnesium for every mole of oxygen. Therefore, 0.30 moles of magnesium can only react with 0.15 moles of oxygen. The rest of the oxygen is in excess (0.05 moles). Therefore, the magnesium is the limiting reactant and determines how much of the products is made. In this case, 0.30 moles of MgO would be made.

Table 11.9

	2Mg	+	O_2	→	2MgO
Reacting ratio from the equation	2 moles of Mg	reacts with	1 mole of O_2	to make	2 moles of MgO
Amount provided	0.30 moles of Mg		0.20 moles of O_2		
	limiting reactant		in excess so it does not all react		
Reaction that takes place	0.30 moles of Mg	reacts with	0.15 moles of O_2	to make	0.30 moles of MgO

Example

Tungsten oxide reacts with hydrogen as shown below to produce tungsten (Figure 11.13). 23.2 g of tungsten oxide (WO_3, $M_r = 232$) is reacted with 20.0 g of hydrogen (H_2, $M_r = 2$). Calculate the mass of tungsten ($M_r = 184$) formed.

$$WO_3 + 3H_2 \rightarrow W + 3H_2O$$

▲ **Figure 11.13** Tungsten.

Answer

In this reaction, the reacting ratio is one mole of tungsten oxide for every three moles of hydrogen. Therefore, 0.10 moles (calculated from 23.2 g) of tungsten oxide reacts with 0.30 moles of hydrogen, and so the hydrogen is in excess (there are 10 moles of hydrogen calculated from 20 g). Therefore, the tungsten oxide is the limiting reactant and determines how much of the products are made. In this case, 0.10 moles of tungsten and 0.30 moles of water would be made.

Table 11.10

	WO_3	+	$3H_2$	→	W	+	$3H_2O$
Reacting ratio from the equation	1 mole of WO_3	reacts with	3 moles of H_2	to make	1 mole of W	and	3 moles of H_2O
Amount provided	Moles = $\frac{23.2}{232}$ = 0.10		Moles = $\frac{20}{10}$ = 10				
	0.10 moles of WO_3		10 moles of H_2				
	limiting reactant		in excess so it does not all react				
Reaction that takes place	0.10 moles of WO_3	reacts with	0.30 moles of H_2	to make	0.10 moles of W	and	0.30 moles of H_2O
					Mass of W = 184 × 0.10 = 18.4 g		

Test yourself

28 Copper can be made by reacting copper oxide with hydrogen as shown below.
$$CuO + H_2 \rightarrow Cu + H_2O$$
How many moles of copper would be made if:
   a) 5 moles of copper oxide were reacted with 10 moles of hydrogen?
   b) 2 moles of copper oxide were reacted with 2 moles of hydrogen?
   c) 0.4 moles of copper oxide were reacted with 0.3 moles of hydrogen?

29 Copper can be also be made by reacting copper oxide with methane as shown below.
$$4CuO + CH_4 \rightarrow 4Cu + 2H_2O + CO_2$$
How many moles of copper would be made if:
   a) 2 moles of copper oxide were reacted with 2 moles of methane?
   b) 2 moles of copper oxide were reacted with 1 mole of methane?
   c) 10 moles of copper oxide were reacted with 2 moles of methane?

30 Hydrogen gas is formed when magnesium reacts with hydrochloric acid as shown below.
$$Mg + 2HCl \rightarrow MgCl_2 + H_2$$
How many moles of hydrogen would be made if:
   a) 3 moles of magnesium were reacted with 3 moles of hydrochloric acid?
   b) 0.2 moles of magnesium were reacted with 0.3 moles of hydrochloric acid?
   c) 0.5 moles of magnesium were reacted with 1.5 moles of hydrochloric acid?

31 Calcium reacts with sulfur as shown in the equation below. What mass of calcium sulfide can be made when 8 g of calcium reacts with 8 g of sulfur?
$$Ca + S \rightarrow CaS$$

32 Titanium is made when titanium chloride reacts with magnesium as shown in the equation below. What mass of titanium can be made when 1.9 g of titanium chloride reacts with 6 g of magnesium?
$$TiCl_4 + 2Mg \rightarrow Ti + 2MgCl_2$$

33 Aluminium bromide is made when aluminium reacts with bromine as shown in the equation below. What mass of aluminium bromide can be made when 0.81 g of aluminium reacts with 6.4 g of bromine?
$$2Al + 3Br_2 \rightarrow 2AlBr_3$$

Show you can...

Calculate the mass of calcium carbide (CaC_2) formed when 84 g of calcium oxide (CaO) is reacted with 48 g of coke (C).

$$CaO + 3C \rightarrow CaC_2 + CO$$

Use the following headings.

- Number of moles of calcium oxide
- Number of moles of coke
- The reactant in excess is
- Number of moles of calcium carbide formed
- Mass of calcium carbide formed in grams.

The concentration of solutions

○ Concentration of solutions in g/dm³

Figure 11.14 shows two solutions of copper sulfate. The one that is darker blue has much more copper sulfate dissolved in it. The darker blue one is more concentrated and the paler blue one is more dilute.

▲ **Figure 11.14** The darker copper sulfate solution is more concentrated than the lighter blue solution.

We can measure the concentration of a solution by considering what mass of solute is dissolved in the solution. This is usually found in g/dm³, which means the number of grams of solute dissolved in each dm³ of solution (Figure 11.15); 1 dm³ is the same volume as 1000 cm³ or 1 litre. For example, if 50 grams of copper sulfate is dissolved in 2 dm³ of solution, then the concentration is 25 g/dm³.

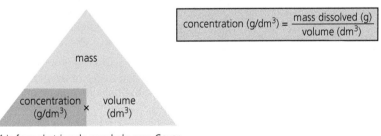

$$\text{concentration (g/dm}^3) = \frac{\text{mass dissolved (g)}}{\text{volume (dm}^3)}$$

mass

concentration (g/dm³) × volume (dm³)

This formula triangle may help you. Cover up the quantity you want, to show the equation you need to use.

▲ **Figure 11.15** Working out the concentration of a solution.

In the laboratory, we often use volumes measured in cm³ rather than dm³. As there are 1000 cm³ in 1 dm³, we should divide the volume in cm³ by 1000 to find the volume in dm³. For example, 25 cm³ is $\frac{25}{1000}$ = 0.025 dm³.

Chapter review questions

1 Calculate the relative formula mass (M_r) of the following substances.

   a) oxygen, O_2 b) propane, C_3H_8

   c) magnesium sulfate, $MgSO_4$ d) calcium nitrate, $Ca(NO_3)_2$

2 Sodium reacts with oxygen to make sodium oxide as shown in this equation:
   $4Na + O_2 \rightarrow 2Na_2O$

   a) Show that the sum of the relative formula masses of all the reactants equals the sum of all the relative formula masses of the products.

   b) In a reaction, 11.5 g of sodium reacts with 4.0 g of oxygen. What mass of sodium oxide is formed?

3 When calcium carbonate is heated it decomposes. In a reaction 10 g of calcium carbonate was heated. At the end of the reaction, only 5.6 g of solid was left. What has happened to the other 4.4 g?

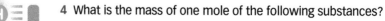

4 What is the mass of one mole of the following substances?

   a) potassium, K   b) nitrogen, N_2 c) sucrose, $C_{12}H_{22}O_{11}$

5 What is the mass of each of the following?

   a) 2.8 moles of chlorine, Cl_2 b) 0.05 moles of methanol, CH_3OH

6 How many moles are there in each of the following?

   a) 2.4 g of oxygen, O_2 b) 10 kg of iron oxide, Fe_2O_3

   c) 0.5 tonnes of ammonium nitrate, NH_4NO_3 d) 25 mg of platinum, Pt

7 a) What mass of water molecules is the same number of molecules as the number of molecules in 88 g of carbon dioxide molecules?

   b) What mass of oxygen molecules is the same number of molecules as the number of atoms in 10 g of calcium atoms?

8 Hydrogen can be made by the reaction of methane with steam. What mass of hydrogen is formed from 80 g of methane?
   $CH_4 + H_2O \rightarrow 3H_2 + CO$

9 Iron for welding railway lines together is produced from the reaction of iron oxide with aluminium. How much iron oxide will react with 1.00 kg of aluminium? Give your answer to 3 significant figures.
   $Fe_2O_3 + 2Al \rightarrow 2Fe + Al_2O_3$

10 When 1.70 g of sodium nitrate ($NaNO_3$) is heated, 1.38 g of sodium nitrite ($NaNO_2$) and 0.32 g of oxygen (O_2) are formed. By calculating molar ratios, deduce the balanced equation for this reaction.

11 The salt potassium sulfate can be made by reaction of sulfuric acid with potassium hydroxide.

$$H_2SO_4 + 2KOH \rightarrow K_2SO_4 + 2H_2O$$

How many moles of potassium sulfate would be made if:

a) 10 moles of sulfuric acid was reacted with 10 moles of potassium hydroxide?

b) 10 moles of sulfuric acid was reacted with 15 moles of potassium hydroxide?

c) 10 moles of sulfuric acid was reacted with 25 moles of potassium hydroxide?

12 Chromium metal can be made when chromium oxide reacts with aluminium as shown in the equation below. What mass of chromium can be made when 30.4 g of chromium oxide reacts with 13.5 g of aluminium? One of the reagents is in excess. Give your answer to 3 significant figures.

$$Cr_2O_3 + 2Al \rightarrow 2Cr + Al_2O_3$$

Practice questions

1 Which one of the following would contain the same number of moles as 6 g of magnesium? [1 mark]

A 3 g of carbon   B 6 g of carbon

C 20 g of calcium   D 40 g of calcium

2 What is the mass of one mole of calcium nitrate, $Ca(NO_3)_2$? [1 mark]

A 82 g   B 164 g

C 204 g   D 220 g

3 Most metals are found naturally in rocks called ores. Some examples are shown in the table:

Table 11.11

Metal ore	Formula of compound in ore
Galena	PbS
Haematite	Fe_2O_3
Dolomite	$CaMg(CO_3)_2$

a) Calculate the relative formula mass (M_r) of each compound in the metal ore. [3 marks]

b) The relative formula mass of another metal ore was calculated to be 102. The formula of this ore can be represented as X_2O_3. Use this information to calculate the relative atomic mass of metal X. Find the identity of metal X. [2 marks]

c) Iron is extracted from the Fe_2O_3 in haematite by reaction with carbon monoxide as shown in the equation below.

$Fe_2O_3 + 3CO \rightarrow 2Fe + 3CO_2$

i) What is meant by the law of conservation of mass? [1 mark]

ii) By working out the formula masses of the reactants and products in this equation show that the equation follows the law of conservation of mass. [2 marks]

4 Anaemia is a condition that occurs when the body has too few red blood cells. Anaemia often occurs in pregnant women. To prevent anaemia, iron(II) sulfate tablets can be taken to provide the iron needed by the body to produce red blood cells. One brand of iron(II) sulfate tablets contains 200 mg of iron(II) sulfate.

a) Calculate the relative formula mass (M_r) of iron(II) sulfate ($FeSO_4$). [1 mark]

b) What is the mass of one mole of iron(II) sulfate? [1 mark]

c) Calculate the number of moles of iron(II) sulfate present in one tablet. [2 marks]

5 Lead is extracted from the ore galena PbS.

a) The ore is roasted in air to produce lead(II) oxide PbO. Calculate the maximum mass of lead(II) oxide PbO produced from 4780 g of galena PbS. [3 marks]

$2PbS + 3O_2 \rightarrow 2PbO + 2SO_2$

b) The lead(II) oxide is reduced to lead by heating it with carbon in a blast furnace. The molten lead is tapped off from the bottom of the furnace.

$PbO + C \rightarrow Pb + CO$

Using your answer to part (a) calculate the maximum mass of lead that would eventually be produced. [2 marks]

6 Sodium hydrogencarbonate decomposes when it is heated. 3.36 g of sodium hydrogencarbonate were placed in a test tube and heated in a Bunsen flame for some time.

$2NaHCO_3 \rightarrow Na_2CO_3 + H_2O + CO_2$

a) Calculate the number of moles of sodium hydrogencarbonate used. [2 marks]

b) Calculate the number of moles of sodium carbonate formed. [1 mark]

c) Calculate the mass of sodium carbonate expected to be formed. [2 marks]

7 Zinc sulfate crystals are prepared in the laboratory by reacting zinc carbonate with sulfuric acid, as shown in the equation below.

$ZnCO_3 + H_2SO_4 \rightarrow ZnSO_4 + H_2O + CO_2$

What is the maximum mass of zinc sulfate which could be formed when 2.5 g of zinc carbonate are reacted with sulfuric acid? [3 marks]

8 Willow bark contains salicylic acid and was once used as a painkiller. Salicylic acid is now used to manufacture aspirin. A student reacted 4.00 g of salicylic acid with 6.50 g of ethanoic anhydride.

$C_6H_4(OH)COOH + (CH_3CO)_2O \rightarrow HOOCC_6H_4OCOCH_3 + CH_3COOH$

salicylic acid   ethanoic   aspirin
                anhydride

a) How many moles of salicylic acid were used? [2 marks]

b) How many moles of ethanoic anhydride were present? [2 marks]

c) What is the maximum number of moles of aspirin which could be formed? [1 mark]

d) Calculate the maximum mass of aspirin which could be formed. [2 marks]

Working scientifically:
Interconverting units

Units

Many of the calculations used in chemistry will require different units. It is important that you can convert between units.

Volume

Volume is measured in cm³ or dm³ (cubic decimetres) or m³.

1000 cm³ = 1 dm³

You need to be able to convert between these volume units. The flow scheme in Figure 11.16 will help you to convert between volume units.

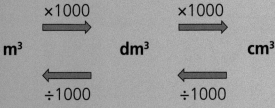

▲ **Figure 11.16** Converting between volume units.

Example

What is 15 cm³ in dm³?

Answer

To convert from cm³ to dm³ you need to divide by 1000

$15\,cm^3 = \dfrac{15}{1000} = 0.015\,dm^3$

Example

What is 0.4 dm³ in cm³?

Answer

To convert from dm³ to cm³ you need to multiply by 1000

$0.4\,dm^3 = 0.4 \times 1000 = 400\,cm^3$

Questions

1 Convert the following volumes to the units shown.
   a) 25 cm³ to dm³
   b) 100 cm³ to dm³
   c) 10 dm³ to cm³
   d) 20 dm³ to m³
   e) 24 000 cm³ to dm³

Mass

Mass can be measured in milligrams (mg), grams (g), kilograms (kg) and in tonnes.

1 tonne = 1000 kg

1 kilogram = 1000 g

1 gram = 1000 mg

The flow diagram in Figure 11.17 will help you to convert between mass units.

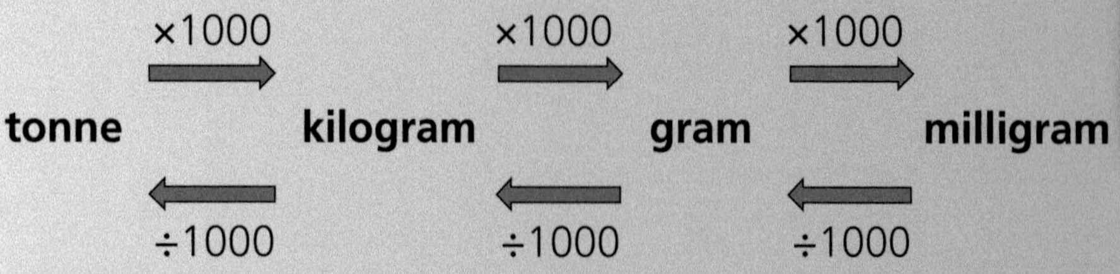

×1000   ×1000   ×1000

tonne   **kilogram**   **gram**   **milligram**

÷1000   ÷1000   ÷1000

▲ **Figure 11.17** Converting between mass units.

TIP ✓

Think logically when converting between units. A kilogram is bigger than a gram so when converting from kilograms to grams you would expect to get a bigger number.

Example

Convert 420 mg to grams.

Answer

To convert from mg to g you need to divide by 1000:

$$\frac{420}{1000} = 0.420\,g$$

Example

Convert 3.2 kg to grams.

Answer

To convert from kg to g you need to multiply by 1000:

$3.2 \times 1000 = 3200\,g$

Example

Convert 0.44 tonnes to grams.

Answer

First you need to convert from tonnes to kilograms by multiplying by 1000:

$0.44 \times 1000 = 440\,kg$

Then convert 440 kg to g by multiplying by 1000:

$440 \times 1000 = 440\,000\,g$

Example

Convert 250 mg to kg.

Answer

First you need to convert mg to g by dividing by 1000:

$$\frac{250}{1000} = 0.25\,g$$

Then convert 0.25 g to kg by dividing by 1000:

$$\frac{0.25}{1000} = 0.00025 = 2.5 \times 10^{-4}\,kg$$

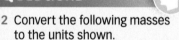

Questions

2 Convert the following masses to the units shown.
   a) 25 g to kg
   b) 1 032 kg to tonnes
   c) 10 tonnes to kg
   d) 43 mg to g
   e) 6.13 tonnes to g
   f) 0.3 kg to g

3 Carry out the following unit conversions.
   a) 50 cm³ to dm³
   b) 32 000 g to tonnes
   c) 22 000 cm³ to dm³
   d) 0.7 kg to g
   e) 2.45 tonnes to g
   f) 12 cm³ to dm³

12 Chemical changes

People have been using chemical reactions to produce metals since the Bronze Age and Iron Age. Chemistry is all about making useful substances from everyday resources and to do this we need to carry out chemical reactions. This chapter looks at some of the most common and important chemical reactions including reactions of metals, reactions of acids and electrolysis.

Specification coverage

This chapter covers specification points 5.4.1.1 to 5.4.3.5 and is called Chemical changes.

It covers reaction and extraction of metals, reactions of acids, making salts and electrolysis.

Related work on writing equations and half equations can be found in Chapter 14.

Prior knowledge

Previously you could have learnt:

> Metals can be listed in a reactivity series which compares metals in terms of their reactivity.
> Low reactivity metals include copper, silver and gold.
> High reactivity metals include sodium, calcium and magnesium.
> Metals are extracted from compounds in ores.
> Metals with low reactivity are extracted by heating metal compounds with carbon; high reactivity metals are extracted by electrolysis.
> Metals react with acids to produce hydrogen gas.
> Acids react with carbonates to form carbon dioxide gas.
> Acids have a pH less than 7; neutral solutions have a pH of 7; alkalis have a pH greater than 7.

Test yourself on prior knowledge

1 Name two reactive metals.
2 Name two unreactive metals.
3 Sodium is found in the ore halite which contains sodium chloride. How is sodium extracted from this ore?
4 What gas is made when acids react with:
   a) metals
   b) carbonates?
5 State whether each of the following solutions is acidic, neutral or alkaline.
   a) pH 11
   b) pH 2
   c) pH 7
   d) pH 6

Reactions of metals

○ The reactivity series of metals

Metals have many uses. For example, they are used in electrical cables, cars, aeroplanes, buildings, mobile phones and computers. Some metals, such as gold, are very unreactive. Other metals, such as sodium, are very reactive.

The reactivity series of metals shows the metals in order of reactivity. This order can be worked out by comparing how metals react with substances such as oxygen, water and dilute acids. The more vigorous the reaction, the higher the reactivity of the metal.

The reactivity series in Figure 12.1 shows some common metals. Carbon and hydrogen are included for comparison although they are non-metals.

KEY TERM

Reactivity series An arrangement of metals in order of their reactivity.

Metal		Reaction with oxygen	Reaction with water	Reaction with acids
Potassium	K	Burns to form oxide	Reacts and gives off H₂(g)	Reacts violently and gives off H₂(g)
Sodium	Na			
Lithium	Li			
Calcium	Ca			Reacts and gives off H₂(g)
Magnesium	Mg			
Aluminium	Al			
Carbon	C			
Zinc	Zn	Forms oxide when heated (metal powder burns)	No reaction	Reacts slowly and gives off H₂(g)
Iron	Fe			
Tin	Sn			
Lead	Pb			
Hydrogen	H			
Copper	Cu	Forms oxide when heated	No reaction	No reaction
Silver	Ag	No reaction		
Gold	Au			
Platinum	Pt			

Most reactive ↑ Least reactive

▲ **Figure 12.1** The reactivity series of some common metals.

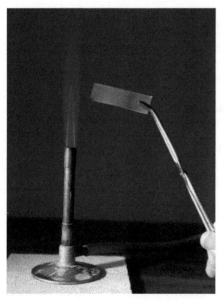

▲ **Figure 12.2** Copper reacts with oxygen but does not burn.

Reaction with oxygen

Most metals react with oxygen. Reactive metals burn when they are heated as they react with oxygen in the air. The metals in the middle of the reactivity series react with oxygen when heated and can burn if the metal is powdered. Copper, a low reactivity metal, reacts with oxygen forming a layer of copper oxide on the surface of the copper but does not burn (Figure 12.2). Metals with very low reactivity, such as gold, do not react with oxygen at all.

When metals react with oxygen they form a **metal oxide**:

metal + oxygen → metal oxide

For example:

sodium + oxygen → sodium oxide
$4Na + O_2 → 2Na_2O$

copper + oxygen → copper oxide
$2Cu + O_2 → 2CuO$

These are examples of **oxidation** reactions. An oxidation reaction can be defined as a reaction where a substance **gains** oxygen. A **reduction** reaction can be defined as a reaction where a substance **loses** oxygen.

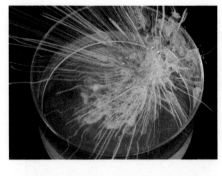

Reaction with water

Most metals do not react with cold water. However, metals with high reactivity react with **cold water** to form a **metal hydroxide** and **hydrogen gas** (Figure 12.3).

$$\text{metal} + \text{water} \rightarrow \text{metal hydroxide} + \text{hydrogen}$$

Table 12.1 shows how metals react with cold water.

◄ **Figure 12.3** Potassium reacts vigorously with water.

Table 12.1 How metals react with water.

Metal	Observations	Equation for reaction
Potassium (K)	Fizzes, melts, floats and moves on the surface of the water, lilac flame (Figure 12.3)	potassium + water → potassium hydroxide + hydrogen $2K + 2H_2O \rightarrow 2KOH + H_2$
Sodium (Na)	Fizzes, melts, floats and moves on the surface of the water (sometimes there is a yellow-orange flame)	sodium + water → sodium hydroxide + hydrogen $2Na + 2H_2O \rightarrow 2NaOH + H_2$
Lithium (Li)	Fizzes, floats and moves on the surface of the water	lithium + water → lithium hydroxide + hydrogen $2Li + 2H_2O \rightarrow 2LiOH + H_2$
Calcium (Ca)	fizzes, white solid forms	calcium + water → calcium hydroxide + hydrogen $Ca + 2H_2O \rightarrow Ca(OH)_2 + H_2$
Magnesium (Mg)	Very slow reaction	magnesium + water → magnesium hydroxide + hydrogen $Mg + 2H_2O \rightarrow Mg(OH)_2 + H_2$
Zinc (Zn)	No reaction	
Iron (Fe)	No reaction	
Copper (Cu)	No reaction	

▲ **Figure 12.4** Magnesium fizzing as it reacts with an acid giving off hydrogen gas.

Reaction with dilute acids

Metals that are more reactive than hydrogen react with dilute acids. When metals react with **dilute acids** they form a **salt** and **hydrogen gas**.

$$\text{metal} + \text{acid} \rightarrow \text{metal salt} + \text{hydrogen}$$

Hydrochloric acid makes **chloride salts**. Sulfuric acid makes **sulfate salts**. Nitric acid makes **nitrate salts**.

With high reactivity metals, the reaction with acids is explosive due to the hydrogen that is formed igniting. Metals that are less reactive than hydrogen do not react with dilute acids.

Table 12.2 shows how metals react with dilute hydrochloric acid.

Table 12.2 How metals react with dilute hydrochloric acid.

Metal	Observation	Equation for reaction
Potassium (K)	Explosive	potassium + hydrochloric acid → potassium chloride + hydrogen $2K + 2HCl \rightarrow 2KCl + H_2$
Sodium (Na)	Explosive	sodium + hydrochloric acid → sodium chloride + hydrogen $2Na + 2HCl \rightarrow 2NaCl + H_2$
Lithium (Li)	Explosive	lithium + hydrochloric acid → lithium chloride + hydrogen $2Li + 2HCl \rightarrow 2LiCl + H_2$
Calcium (Ca)	Fizzes	calcium + hydrochloric acid → calcium chloride + hydrogen $Ca + 2HCl \rightarrow CaCl_2 + H_2$
Magnesium (Mg)	Fizzes (Figure 12.4)	magnesium + hydrochloric acid → magnesium chloride + hydrogen $Mg + 2HCl \rightarrow MgCl_2 + H_2$
Zinc (Zn)	Fizzes slowly	zinc + hydrochloric acid → zinc chloride + hydrogen $Zn + 2HCl \rightarrow ZnCl_2 + H_2$
Iron (Fe)	Fizzes slowly	iron + hydrochloric acid → iron chloride + hydrogen $Fe + 2HCl \rightarrow FeCl_2 + H_2$
Copper (Cu)	No reaction	

▲ **Figure 12.5** A displacement reaction is used to weld railway lines together.

▲ **Figure 12.6** Silver forms on the copper wire as silver is displaced from the silver nitrate solution.

What happens to metal atoms when they react?

When metal atoms react, they **lose electrons** to form **positive ions**. For example:

- When sodium atoms react with oxygen, the sodium atoms lose electrons and form sodium ions (Na^+) in the product sodium oxide.
- When calcium atoms react with water, the calcium atoms lose electrons and form calcium ions (Ca^{2+}) in the product calcium hydroxide.
- When zinc atoms react with hydrochloric acid, the zinc atoms lose electrons and form zinc ions (Zn^{2+}) in the product zinc chloride.

The greater the tendency of a metal to lose electrons to form ions, the more reactive it is. **Reactive metals** like potassium and sodium **easily lose electrons** to **form ions,** but metals like gold and platinum do not tend to form ions and so are unreactive.

Displacement reactions

In a displacement reaction, a more reactive metal will take the place of a less reactive metal in a compound. For example, aluminium will displace iron from iron oxide because aluminium is more reactive than iron.

$$aluminium + iron\ oxide \rightarrow aluminium\ oxide + iron$$
$$2Al + Fe_2O_3 \rightarrow Al_2O_3 + 2Fe$$

This reaction is used to weld railway lines together (Figure 12.5). A mixture of aluminium and iron oxide is placed over the gap between the railway lines and the reaction started. The reaction gets very hot and produces molten iron which flows into a mould, cools and solidifies to weld the lines together.

Displacement reactions also take place in solution. For example, copper will displace silver from silver nitrate solution because copper is more reactive than silver. Copper nitrate is blue and so the solution turns blue as the copper nitrate is formed (Figure 12.6).

$$copper + silver\ nitrate \rightarrow copper\ nitrate + silver$$
$$Cu + 2AgNO_3 \rightarrow Cu(NO_3)_2 + 2Ag$$

Test yourself

1 Complete the following word equations, or write *no reaction* if no reaction would take place.

a) calcium + oxygen   **b)** gold + oxygen

c) copper + water   **d)** lithium + water

e) calcium + nitric acid   **f)** copper + sulfuric acid

g) zinc + hydrochloric acid   **h)** iron + sulfuric acid

i) tin + magnesium chloride   **j)** zinc + lead nitrate

k) magnesium + aluminium sulfate

2 Magnesium (Mg) reacts with oxygen (O_2) to form magnesium oxide (MgO).

a) Write a word equation for this reaction.

b) Write a balanced equation for this reaction.

c) What happens to the magnesium atoms in this reaction in terms of electrons?

3 Calcium (Ca) reacts with water to form calcium hydroxide ($Ca(OH)_2$) and hydrogen (H_2).

a) Describe what you would see in this reaction.

b) Write a word equation for this reaction.

c) Write a balanced equation for this reaction.

d) What happens to the calcium atoms in this reaction in terms of electrons?

4 Magnesium (Mg) and zinc (Zn) both react with sulfuric acid.

a) **i)** Which metal reacts more vigorously with sulfuric acid?

    ii) Explain, in terms of the tendency to form ions, why this metal reacts more vigorously.

b) For the reaction of magnesium with sulfuric acid:

    i) Write a word equation

    ii) Write a balanced equation.

5 The metal chromium can be made in a displacement reaction between aluminium and chromium oxide.

aluminium + chromium oxide → aluminium oxide + chromium

$2Al + Cr_2O_3 \rightarrow Al_2O_3 + 2Cr$

a) Why does aluminium displace chromium in this reaction?

b) Which substance is oxidised in this reaction?

c) Which substance is reduced in this reaction?

Show you can...

To determine the order of reactivity of the metals copper, magnesium, nickel and zinc each metal was heated with the oxides of other metals and the results obtained recorded in the table below.

Table 12.3

	Copper	Magnesium	Nickel	Zinc
Copper oxide		Reaction	Reaction	Reaction
Magnesium oxide	No reaction		No reaction	No reaction
Nickel oxide	No reaction	Reaction		Reaction
Zinc oxide	No reaction	Reaction	No reaction	

a) Determine the order of the four metals from the most reactive to the least reactive.

b) Write a balanced chemical equation for the reaction of nickel oxide (NiO) with magnesium.

c) From the list below, write word and balanced chemical equations for all reactions which occur (when nickel reacts, it forms Ni^{2+} ions).

    i) nickel + hydrochloric acid

    ii) zinc + water

    iii) nickel + water

    iv) zinc + sulfuric acid

    v) magnesium + zinc oxide

KEY TERMS

Oxidation A reaction in which a substance loses electrons (gains oxygen).

Reduction Reaction in which a substance gains electrons (loses oxygen).

Oxidation and reduction in terms of electrons

Oxidation can be defined as a reaction where a substance gains oxygen. However, a better definition of oxidation is a reaction where a substance loses electrons.

Reduction can be defined as a reaction where a substance loses oxygen. However, a better definition of reduction is a reaction where a substance gains electrons.

One way to remember this is the phrase OIL RIG (Figure 12.7).

When metals react with oxygen, water and acids, the metal atoms lose electrons and form metal ions (Table 12.4). This means that each reaction can be defined as oxidation in terms of the loss of electrons. However, the reaction with oxygen is the only one that can be defined as oxidation in terms of gaining oxygen.

▶ **Figure 12.7** OIL RIG – how to remember the role of electrons in oxidation and reduction.

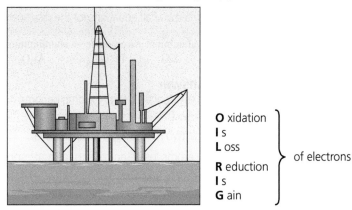

O xidation
I s
L oss
R eduction
I s
G ain
} of electrons

Table 12.4 Reactions of metals.

Reaction	General equation	Oxidation in terms of gaining oxygen	Oxidation in terms of losing electrons
+ oxygen	metal + oxygen → metal oxide	Metal gains oxygen	Metal atoms lose electrons to form metal ions in the metal oxide
+ water	metal + water → metal hydroxide + hydrogen		Metal atoms lose electrons to form metal ions in the metal hydroxide
+ acids	metal + acid → metal salt + hydrogen		Metal atoms lose electrons to form metal ions in the metal salt

Displacement reactions involve oxidation and reduction. In some reactions this can be explained in terms of oxygen and in terms of electrons. This is the case, for example, in the displacement of iron from iron oxide by aluminium (Figure 12.8).

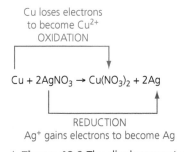

Cu loses electrons
to become Cu^{2+}
OXIDATION

$$Cu + 2AgNO_3 \rightarrow Cu(NO_3)_2 + 2Ag$$

REDUCTION
Ag^+ gains electrons to become Ag

▲ **Figure 12.9** The displacement of silver from silver nitrate by copper.

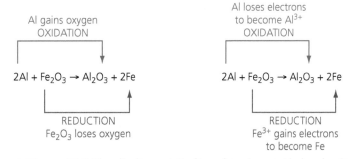

Al gains oxygen
OXIDATION

$$2Al + Fe_2O_3 \rightarrow Al_2O_3 + 2Fe$$

REDUCTION
Fe_2O_3 loses oxygen

Al loses electrons
to become Al^{3+}
OXIDATION

$$2Al + Fe_2O_3 \rightarrow Al_2O_3 + 2Fe$$

REDUCTION
Fe^{3+} gains electrons
to become Fe

▲ **Figure 12.8** The displacement of iron from iron oxide by aluminium.

In other displacement reactions, only the definitions in terms of electrons can be used to define the reaction as involving oxidation and reduction. This is the case, for example, in the displacement of silver from silver nitrate by copper (Figure 12.9).

In a reaction in which one substance loses electrons, another substance gains those electrons. This means that both reduction and oxidation take place and these are called redox reactions (**red**uction–**ox**idation).

KEY TERM

Redox reaction A reaction where both reduction and oxidation take place.

Writing ionic equations and/or half equations for displacement reactions

Ionic equations and/or half equations can be written for displacement reactions. (See Chapter 14 for more help on writing these equations.)

Example

Write two half equations for the displacement of iron from iron oxide by aluminium.

aluminium + iron oxide → aluminium oxide + iron
$2Al$ + Fe_2O_3 → Al_2O_3 + $2Fe$

Answer

In this reaction, the Al atoms become Al^{3+} ions in Al_2O_3 while the Fe^{3+} ions in Fe_2O_3 become Fe atoms.

The two half equations for this are:

Al atoms lose electrons to form Al^{3+} ions: $Al - 3e^- \rightarrow Al^{3+}$ (or $Al \rightarrow Al^{3+} + 3e^-$)

Fe^{3+} ions in Fe_2O_3 gain electrons to form Fe atoms: $Fe^{3+} + 3e^- \rightarrow Fe$

Example

Write an overall ionic equation and two half equations for the displacement of silver from silver nitrate by copper.

copper + silver nitrate → copper nitrate + silver
Cu + $2AgNO_3$ → $Cu(NO_3)_2$ + $2Ag$

Answer

In this reaction, the Cu atoms become Cu^{2+} ions in $Cu(NO_3)_2$ while the Ag^+ ions in $AgNO_3$ become Ag atoms. We can leave out the NO_3^- ions from the ionic equation as they do not change.

Therefore, the overall ionic equation is: $Cu + 2Ag^+ \rightarrow Cu^{2+} + 2Ag$

The two half equations for this are: Cu atoms lose electrons to form Cu^{2+} ions: $Cu - 2e^- \rightarrow Cu^{2+}$ (or $Cu \rightarrow Cu^{2+} + 2e^-$)

Ag^+ ions in $AgNO_3$ gain electrons to form Ag atoms: $Ag^+ + e^- \rightarrow Ag$

Example

Write an overall ionic equation and two half equations for the displacement of copper from copper sulfate by iron. Identify which half equation represents a reduction process and which represents an oxidation process.

iron + copper sulfate → iron sulfate + copper
Fe + $CuSO_4$ → $FeSO_4$ + Cu

Answer

In this reaction, the Fe atoms become Fe^{2+} ions in $FeSO_4$ while the Cu^{2+} ions in $CuSO_4$ become Cu atoms. We can leave out the SO_4^{2-} ions from the ionic equation as they do not change.

Therefore, the overall ionic equation is: $Fe + Cu^{2+} \rightarrow Fe^{2+} + Cu$

The two half equations for this are:

Fe atoms lose electrons to form Fe^{2+} ions: $Fe - 2e^- \rightarrow Fe^{2+}$ (or $Fe \rightarrow Fe^{2+} + 2e^-$)

Cu^{2+} ions in $CuSO_4$ gain electrons to form Cu atoms: $Cu^{2+} + 2e^- \rightarrow Cu$

The reduction half equation is: $Cu^{2+} + 2e^- \rightarrow Cu$

The oxidation half equation is: $Fe - 2e^- \rightarrow Fe^{2+}$ (or $Fe \rightarrow Fe^{2+} + 2e^-$)

Test yourself

6 Rubidium is a metal in Group 1 of the periodic table. It reacts vigorously with water.
   a) Write a word equation for this reaction.
   b) Write a balanced equation for this reaction.
   c) What happens to the rubidium atoms in this reaction in terms of electrons?
   d) Are the rubidium atoms oxidised or reduced in this reaction? Explain your answer.

7 Magnesium reacts with copper oxide in a displacement reaction to form copper.
   $Mg + CuO \rightarrow MgO + Cu$
   a) Explain, in terms of oxygen, why the magnesium is oxidised in this reaction.
   b) Write a half equation to show the oxidation of magnesium atoms to magnesium ions in this reaction.
   c) Explain, in terms of electrons, why the magnesium is oxidised in this reaction.
   d) Explain, in terms of oxygen, why the copper oxide is reduced in this reaction.
   e) Write a half equation to show the reduction of copper ions to copper atoms in this reaction.
   f) Explain, in terms of electrons, why the copper oxide is reduced in this reaction.

8 Write an overall ionic equation and two half equations for each of the following displacement reactions.
   a) Displacement of copper from copper sulfate by zinc.
      zinc + copper sulfate → zinc sulfate + copper
      $Zn + CuSO_4 \rightarrow ZnSO_4 + Cu$
   b) Displacement of silver from silver nitrate by zinc.
      zinc + silver nitrate → zinc nitrate + silver
      $Zn + 2AgNO_3 \rightarrow Zn(NO_3)_2 + 2Ag$
   c) Displacement of copper from copper sulfate by aluminium.
      aluminium + copper sulfate → aluminium sulfate + copper
      $2Al + 3CuSO_4 \rightarrow Al_2(SO_4)_3 + 3Cu$

9 Zinc displaces iron from a solution of iron(II) sulfate. For this reaction:
   a) Write a word equation for this reaction.
   b) Write a balanced equation for this reaction.
   c) Write an ionic equation for this reaction.
   d) Write the two half equations for this reaction.
   e) Identify which half equation is a reduction process.
   f) Identify which half equation is an oxidation process.
   g) Explain why this is a redox reaction.

10 Magnesium displaces silver from a solution of silver nitrate. For this reaction:
   a) Write a word equation for this reaction.
   b) Write a balanced equation for this reaction.
   c) Write an ionic equation for this reaction.
   d) Write the two half equations for this reaction.
   e) Identify which half equation is a reduction process.
   f) Identify which half equation is an oxidation process.
   g) Explain why this is a redox reaction.

Show you can...

Each of the following reactions can be classified as an oxidation or a reduction reaction.

Reaction 1: $CH_4 + 2O_2 \rightarrow CO_2 + 2H_2O$

Reaction 2: $CuO + Mg \rightarrow Cu + MgO$

Reaction 3: $ZnO + H_2 \rightarrow Zn + H_2O$

a) Write the formula of the substance which is oxidised in reaction 1.
b) Explain which substance in reaction 2 is reduced.
c) Explain which substance in reaction 3 is oxidised.
d) Write a balanced ionic equation for reaction 2 and state and use a half equation to explain which species is oxidised.

Extraction of metals

○ Where do metals come from?

A few metals, such as gold and platinum, occur naturally on Earth as elements. These are metals with very low reactivity.

Most metals are only found on Earth in compounds. For example, iron is often found in the compound iron oxide and aluminium in the compound aluminium oxide. In order to extract the metal from these compounds, a chemical reaction is required.

The metal compounds are found in rocks called ores. An ore is a rock from which a metal can be extracted for profit (Figure 12.10).

▲ **Figure 12.10** Aluminium metal is extracted from the aluminium oxide (Al_2O_3) in the ore bauxite.

> **KEY TERM** ⭐
>
> **Ore** A rock from which a metal can be extracted for profit.

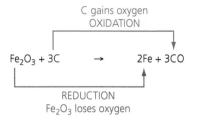

Most reactive ↑	Metal		Method of extraction
	Potassium	K	Electrolysis
	Sodium	Na	
	Lithium	Li	
	Calcium	Ca	
	Magnesium	Mg	
	Aluminium	Al	
	Carbon	C	
	Zinc	Zn	Heat with carbon
	Iron	Fe	
	Tin	Sn	
	Lead	Pb	
	Copper	Cu	
	Silver	Ag	Metals found as elements
	Gold	Au	
Least reactive	Platinum	Pt	

▲ **Figure 12.11** Extracting metals from their oxides.

○ Methods of extraction

Most of the compounds from which metals are extracted are oxides. In order to extract the metal, the oxygen is removed in a reduction reaction. The way in which this is done depends on the reactivity of the metal (Figure 12.11).

- Metals that are **less reactive** than **carbon** can be extracted by **heating the metal oxide with carbon**. For example, iron is extracted by heating iron oxide with carbon. The iron oxide is reduced in this reaction (Figure 12.12).
- Metals that are **more reactive** than **carbon** can be extracted by **electrolysis**. This is studied in detail in the section on electrolysis later in this chapter.

When metals are extracted from compounds, the metal ions in the compounds gain electrons. For example, when iron is extracted from iron oxide, Fe^{3+} ions in the iron oxide gain electrons to form Fe atoms. This is reduction because the iron oxide loses oxygen and also because the Fe^{3+} ions in the iron oxide gain electrons.

C gains oxygen
OXIDATION

$$Fe_2O_3 + 3C \rightarrow 2Fe + 3CO$$

REDUCTION
Fe_2O_3 loses oxygen

▲ **Figure 12.12** Extracting iron from iron oxide by heating with carbon.

Show you can...

From the list of elements below:

carbon	copper	gold
hydrogen	sodium	zinc

a) Choose an element which is most likely to be found in rocks as the metal itself.
b) Choose two elements which are likely to be found in ores.
c) Choose an element which can be heated with an ore to extract a metal.
d) Choose an element which is most likely to be extracted from its ore by electrolysis.

Test yourself

11 Most metals are extracted from compounds found in rocks called ores. A few metals are found as elements.
  a) Why are some metals, such as gold, found as elements and not in compounds?
  b) What is an ore?
12 Which method is used to extract the following metals?
  a) iron
  b) aluminium
  c) magnesium
  d) platinum
  e) zinc
13 Tin is extracted from the ore tin oxide by heating with carbon.
  a) Write a word equation for this reaction.
  b) Explain why this is a reduction reaction.

Reactions of acids

○ What are acids and alkalis?

KEY TERMS

Acid Solution with a pH less than 7; produces H^+ ions in water.

Aqueous Dissolved in water.

Alkali Solution with a pH more than 7; produces OH^- ions in water.

An acid is a substance that produces hydrogen ions, H^+, in aqueous solution. For example, solutions of:

- hydrochloric acid (HCl) contain hydrogen (H^+) ions and chloride (Cl^-) ions
- sulfuric acid (H_2SO_4) contain hydrogen (H^+) ions and sulfate (SO_4^{2-}) ions
- nitric acid (HNO_3) contain hydrogen (H^+) ions and nitrate (NO_3^-) ions.

An alkali is a substance that produces hydroxide (OH^-) ions, in aqueous solution. For example, solutions of:

- sodium hydroxide (NaOH) contain sodium (Na^+) ions and hydroxide (OH^-) ions
- potassium hydroxide (KOH) contain potassium (K^+) ions and hydroxide (OH^-) ions
- calcium hydroxide ($Ca(OH)_2$) contain calcium (Ca^{2+}) ions and hydroxide (OH^-) ions.

○ The pH scale

The pH scale is a measure of how acidic or alkaline a solution is. A solution with a pH of 7 is neutral, whereas a solution with a pH below 7 is acidic and one with a pH above 7 is alkaline. The further away from 7 the pH is, the more acidic or alkaline the solution is (Figure 12.13).

The scale is often shown as running from 0 to 14, but it does go further in both directions. For example, it is common for the solutions of acids

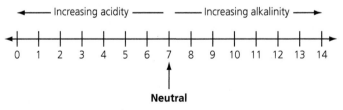

▲ **Figure 12.13** pH scale.

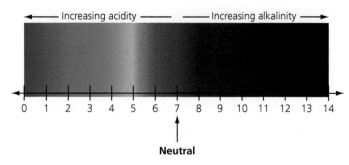

Increasing acidity ← → Increasing alkalinity

0 1 2 3 4 5 6 7 8 9 10 11 12 13 14

Neutral

▲ **Figure 12.14** The colours for universal indicator.

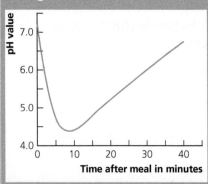

7.0

▲ **Figure 12.15** pH probe.

in school laboratories to have a pH that is less than 0 (typically about −0.3).

The approximate pH of a solution can be measured using universal indicator solution. A few drops of the indicator is added to the solution. The colour is compared to a colour chart to give the approximate pH of the solution (Figure 12.14).

A more accurate way of finding the pH of a solution is to use a pH probe (Figure 12.15). There are different types but the probe is dipped into the solution and the pH shown on the display, often to 1 or 2 decimal places.

The pH of a solution is based on the concentration of H^+ ions in the solution. The higher the concentration of H^+ ions the lower the pH.

As the pH decreases by one unit, the concentration of hydrogen ions increases by a factor of 10. For example, a solution with a pH of 2 has a concentration of H^+ ions that is 10 times greater than one with a pH of 3. A solution with a pH of 1 has a concentration of H^+ ions that is 100 times greater than one with a pH of 3.

In a neutral solution, the concentration of H^+ ions equals the concentration of OH^- ions.

Show you can...

In an experiment a sample of human saliva was removed from the mouth every five minutes after a meal and the pH value determined. The graph shows how the pH value of the saliva changed.

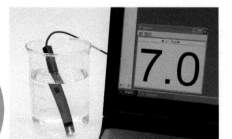

▲ **Figure 12.16**

a) How were the pH values of the saliva likely to have been determined in this experiment?

b) When the pH in the mouth is 5.0 or less tooth decay occurs. Use the graph to find the time after the meal at which teeth would start to decay.

c) At what time is the pH of the saliva most acidic?

Test yourself

14 Classify each of the following solutions as acidic, neutral or alkaline.
   a) A solution with pH 9.
   b) A solution with pH 3.
   c) A solution with pH 0.
   d) A solution with pH 7.

15 a) Three solutions had pH values of 8, 11 and 13. Which one was the most alkaline?
   b) Three solutions had pH values of 1, 2 and 5. Which one was the most acidic?

16 a) Give two ways in which the pH of a solution can be measured.
   b) Which method will give the most accurate value?

17 a) Which ion do aqueous solutions of acids all contain?
   b) Which ion do aqueous solutions of alkalis all contain?

18 The table gives some information about three solutions.

   Table 12.5

Solution	A	B	C
pH	4	3	1

   a) Which solution has the highest concentration of H^+ ions?
   b) By what factor is concentration of H^+ ions in solution **B** bigger or smaller than solution **A**?
   c) By what factor is concentration of H^+ ions in solution **C** bigger or smaller than solution **A**?

19 A solution has a pH of 7. Comment on the concentration of H^+ ions compared to the concentration of OH^- ions.

○ Strong and weak acids

Hydrogen chloride (HCl), hydrogen sulfate (H_2SO_4) and hydrogen nitrate (HNO_3) are molecules in their pure state. When they are added to water all of their molecules break down into ions forming hydrochloric acid (HCl), sulfuric acid (H_2SO_4) and nitric acid (HNO_3). They are strong acids because their molecules are **completely ionised** in water – this means that all their molecules break into ions in water (Figure 12.17).

(a)
$$H-Cl\,(g) \xrightarrow{H_2O} H^+(aq) + Cl^-(aq)$$
$$HCl(g) \xrightarrow{H_2O} H^+(aq) + Cl^-(aq)$$

(b)
$$H-O-\underset{\underset{O}{\|}}{\overset{\overset{O}{\|}}{S}}-O-H\,(l) \xrightarrow{H_2O} 2H^+(aq) + {}^-O-\underset{\underset{O}{\|}}{\overset{\overset{O}{\|}}{S}}-O^-\,(aq)$$
$$H_2SO_4\,(l) \xrightarrow{H_2O} 2H^+(aq) + SO_4{}^{2-}\,(aq)$$

▶ **Figure 12.17 (a)** Hydrogen chloride reacts with water and breaks up into ions. **(b)** Hydrogen sulfate reacts with water and breaks up into ions.

In weak acids the molecules are only **partially ionised** in water – this means that only a small fraction of the molecules break into ions when added to water. Figure 12.18 shows the difference between the strong acid HCl and a weak acid HX in water.

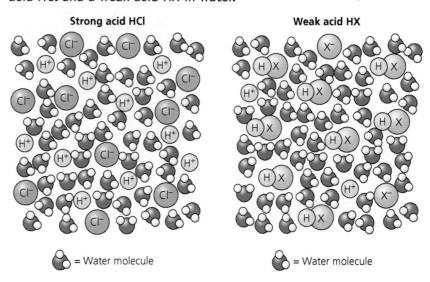

Strong acid HCl

Weak acid HX

😊 = Water molecule

😊 = Water molecule

Every HCl molecule breaks down to form H^+ and Cl^- ions

Only a small fraction of the HX molecules break down to form H^+ and X^- ions.

▶ **Figure 12.18** The difference between strong acid (HCl) and a weak acid (HX) in water.

If solutions of equal concentration of a strong acid and weak acid are compared, there will be more H^+ ions in the strong acid solution. This means that the solution of the strong acid will have a lower pH.

There are many weak acids in food and drink. They tend to have a sour taste and are not dangerous because there is only a low concentration of H^+ ions. **Ethanoic acid** in vinegar, **citric acid** in citrus fruits (Figure 12.19) and **carbonic acid** in fizzy drinks are examples of weak acids in food and drink.

▲ **Figure 12.19** Citrus fruits contain citric acid.

The terms strong and weak refer to the degree of ionisation in water of acids. The terms dilute and concentrated refer to the amount of acid dissolved in the solution. This is summarised in Figure 12.20.

Strong
(all of the molecules break down into ions in water)

Weak
(a small fraction of the molecules break down into ions in water)

Concentrated
(a lot of acid is dissolved)

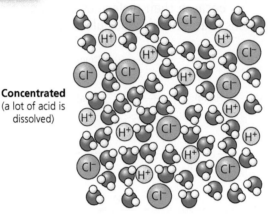
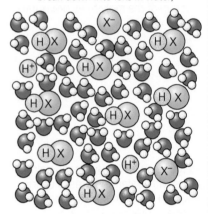

Dilute
(small amount of acid dissolved)

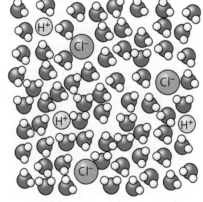
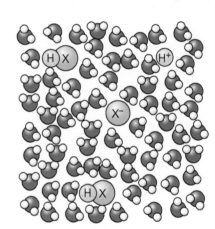

▲ **Figure 12.20** Differences between strong and weak acids, and concentrated and dilute acids.

Test yourself

20 a) Sulfuric acid is a strong acid. Explain what this means.

b) Citric acid is a weak acid. Explain what this means.

Show you can...

Two solutions A and B were tested using a pH meter, red litmus paper, blue litmus paper and universal indicator paper. The results are shown in the table.

a) Describe how the results with universal indicator may be converted into a pH value.

b) Are solutions A and B acidic, neutral or alkaline?

c) If the experiment was repeated using a more concentrated solution of A would the results be different? Explain your answer.

d) Explain why universal indicator gives more information than litmus.

Table 12.6

Test	Result for solution A	Result for solution B
pH meter	1.82	3.85
Red litmus	Red	Red
Blue litmus	Red	Red
Universal indicator paper	Red	Orange

○ Reaction of acids with metals

Metals that are more reactive than hydrogen react with dilute acids. When metals react with **dilute acids** they form a **salt** and **hydrogen gas**. The reaction fizzes as hydrogen gas is produced.

metal + acid → metal salt + hydrogen

Table 12.7 shows how different acids form different types of salts.

Table 12.8 shows how magnesium, zinc and iron react with dilute hydrochloric and sulfuric acids.

When a metal reacts with an acid, a redox reaction takes place (Figure 12.21). The metal atoms lose electrons and so are oxidised. The H^+ ions in the acid gain electrons and are reduced.

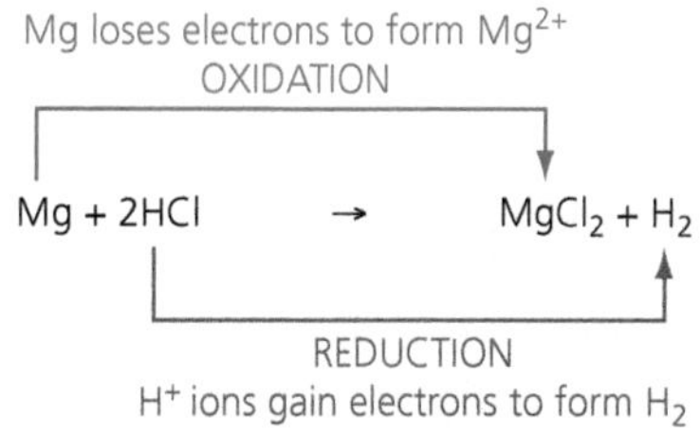

Mg loses electrons to form Mg^{2+}
OXIDATION

$Mg + 2HCl \rightarrow MgCl_2 + H_2$

REDUCTION
H^+ ions gain electrons to form H_2

▲ **Figure 12.21** A redox reaction takes place when a metal reacts with an acid.

Table 12.7 Salts formed when magnesium reacts with different acids.

Acid	Reaction	Type of salt formed
Hydrochloric acid HCl	magnesium + hydrochloric acid → magnesium chloride + hydrogen $Mg + 2HCl \rightarrow MgCl_2 + H_2$	**Chloride** (contains Cl^- ions)
Sulfuric acid H_2SO_4	magnesium + sulfuric acid → magnesium sulfate + hydrogen $Mg + H_2SO_4 \rightarrow MgSO_4 + H_2$	**Sulfate** (contains SO_4^{2-} ions)
Nitric acid HNO_3	magnesium + nitric acid → magnesium nitrate + hydrogen $Mg + 2HNO_3 \rightarrow Mg(NO_3)_2 + H_2$	**Nitrate** (contains NO_3^- ions)

Table 12.8 Reactions between some metals and acids.

Metal	Reaction with hydrochloric acid	Reaction with sulfuric acid
Magnesium	Fizzes vigorously magnesium + hydrochloric acid → magnesium chloride + hydrogen $Mg + 2HCl \rightarrow MgCl_2 + H_2$	Fizzes vigorously magnesium + sulfuric acid → magnesium sulfate + hydrogen $Mg + H_2SO_4 \rightarrow MgSO_4 + H_2$
Zinc	Fizzes gently zinc + hydrochloric acid → zinc chloride + hydrogen $Zn + 2HCl \rightarrow ZnCl_2 + H_2$	Fizzes gently zinc + sulfuric acid → zinc sulfate + hydrogen $Zn + H_2SO_4 \rightarrow ZnSO_4 + H_2$
Iron	Fizzes very slowly iron + hydrochloric acid → iron chloride + hydrogen $Fe + 2HCl \rightarrow FeCl_2 + H_2$	Fizzes very slowly iron + sulfuric acid → iron sulfate + hydrogen $Fe + H_2SO_4 \rightarrow FeSO_4 + H_2$

○ Reaction of acids with metal hydroxides

Acids react with **metal hydroxides** to form a **salt** and **water**. Some examples of this reaction are shown in the Table 12.9.

metal hydroxide + acid → metal salt + water

Table 12.9 Reactions between some metal hydroxides and acids.

Full equation	Ionic equation
sodium hydroxide + nitric acid → sodium nitrate + water $NaOH + HNO_3 \rightarrow NaNO_3 + H_2O$	$H^+(aq) + OH^-(aq) \rightarrow H_2O(l)$
potassium hydroxide + sulfuric acid → potassium sulfate + water $2KOH + H_2SO_4 \rightarrow K_2SO_4 + 2H_2O$	$H^+(aq) + OH^-(aq) \rightarrow H_2O(l)$
calcium hydroxide + hydrochloric acid → calcium chloride + water $Ca(OH)_2 + 2HCl \rightarrow CaCl_2 + 2H_2O$	$H^+(aq) + OH^-(aq) \rightarrow H_2O(l)$

The ionic equation for each of these reactions is the same. In each reaction, H^+ ions from the acid are reacting with OH^- ions from the metal hydroxide. This produces water in each case.

Metal hydroxides that dissolve in water are alkalis because they release OH^- ions into the water. Metal hydroxides that are insoluble in water do not release OH^- ions into the water and so are not alkalis. However, they still react with acids to produce a salt and water.

In these reactions, the H^+ ions from the acid are being used up and so they are examples of neutralisation reactions.

○ **Reaction of acids with metal oxides**

Acids react with **metal oxides** to form a **salt** and **water.** Most metal oxides are insoluble in water and the reactions usually need to be heated. Some examples of this reaction are shown in Table 12.10.

> metal oxide + acid → metal salt + water

Table 12.10 Reactions between some metal oxides and acids.

Full equation
calcium oxide + nitric acid → calcium nitrate + water $CaO \quad + \quad 2HNO_3 \quad \rightarrow \quad Ca(NO_3)_2 \quad + \quad H_2O$
copper oxide + sulfuric acid → copper sulfate + water $CuO \quad + \quad H_2SO_4 \quad \rightarrow \quad CuSO_4 \quad + \quad H_2O$
lithium oxide + hydrochloric acid → lithium chloride + water $Li_2O \quad + \quad 2HCl \quad \rightarrow \quad 2LiCl \quad + \quad H_2O$

In these reactions, the H^+ ions from the acid are being used up and so they are examples of neutralisation reactions.

○ **Reaction of acids with metal carbonates**

Acids react with **metal carbonates** to form a **salt**, **water** and **carbon dioxide**. These reactions usually take place readily. The reaction fizzes as carbon dioxide gas is produced. Some examples of this reaction are shown in the Table 12.11.

> metal carbonate + acid → metal salt + water + carbon dioxide

Table 12.11 Reactions between some metal carbonates and acids.

Full equation
sodium carbonate + nitric acid → sodium nitrate + water + carbon dioxide $Na_2CO_3 \quad + \quad 2HNO_3 \quad \rightarrow \quad 2NaNO_3 \quad + \quad H_2O \quad + \quad CO_2$
copper carbonate + sulfuric acid → copper sulfate + water + carbon dioxide $CuCO_3 \quad + \quad H_2SO_4 \quad \rightarrow \quad CuSO_4 \quad + \quad H_2O \quad + \quad CO_2$
calcium carbonate + hydrochloric acid → calcium chloride + water + carbon dioxide $CaCO_3 \quad + \quad 2HCl \quad \rightarrow \quad CaCl_2 \quad + \quad H_2O \quad + \quad CO_2$

In these reactions, the H^+ ions from the acid are being used up and so they are examples of neutralisation reactions.

Test yourself

21 Complete the following word equations.
   a) calcium + hydrochloric acid
   b) tin + sulfuric acid
   c) barium hydroxide + nitric acid
   d) lithium hydroxide + hydrochloric acid
   e) nickel oxide + nitric acid
   f) magnesium oxide + sulfuric acid
   g) potassium carbonate + nitric acid
   h) zinc carbonate + hydrochloric acid

22 a) There is fizzing when hydrochloric acid reacts with calcium. What gas causes this?
   b) There is fizzing when hydrochloric acid reacts with calcium carbonate. What gas causes this?

23 Hydrochloric acid reacts with sodium hydroxide as shown in this equation:
$HCl(aq) + NaOH(aq) \rightarrow NaCl(aq) + H_2O(l)$
   a) What does the symbol (aq) mean?
   b) Write an ionic equation for this reaction.
   c) Explain why sodium hydroxide is an alkali.

24 Write a balanced equation for each of the reactions in question 22(a), (d), (f) and (g).

Show you can...

Sodium sulfate, produced by the reaction between sulfuric acid and sodium hydroxide, is used in washing powders.

a) Write a word equation and a balanced symbol equation for the reaction of sulfuric acid with sodium hydroxide.

The composition of two washing powders, A and B are shown below.

Table 12.12

	Percentage composition in %				
	Sodium sulfate	Sodium carbonate	Sodium silicate	Soap	Detergent
A	29	20	20	0	15
B	35	0	26	6	13

Dilute nitric acid was added to each of the powders. Only one of the powders reacted.

b) Which powder reacted A or B? Explain your answer.
c) During the reaction, effervescence was noted. Name the gas produced in the reaction.
d) Describe a chemical test for this gas and state the result for a positive test.
e) The silicate ion is SiO_3^{2-}. Suggest the formula for sodium silicate.

Making salts

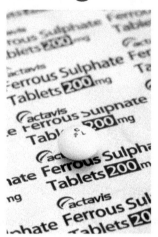

▲ **Figure 12.22** Iron tablets contain iron(ɪɪ) sulfate (sometimes called ferrous sulfate).

◯ What are salts?

Salts are substances made when acids react with metals, metal hydroxides, metal oxides and metal carbonates.

Salts are very useful substances. For example:

● many medicines are salts (Figure 12.22)
● fertilisers are salts
● toothpaste contains salts
● many food additives are salts.

Salts are made up of a metal ion combined with the ion left over from the acid when the H^+ ions react. The salt produced depends on which acid is used and what it reacts with. Table 12.13 below gives some examples of the salts formed when acids react.

Table 12.13 Examples of salts formed when acids react with metals, metal hydroxides, metal oxides and metal carbonates.

	Hydrochloric acid HCl	Sulfuric acid H_2SO_4	Nitric acid HNO_3
Magnesium Mg	Magnesium chloride ($+ H_2$) $MgCl_2$	Magnesium sulfate ($+ H_2$) $MgSO_4$	Magnesium nitrate ($+ H_2$) $Mg(NO_3)_2$
Sodium hydroxide NaOH	Sodium chloride ($+ H_2O$) NaCl	Sodium sulfate ($+ H_2O$) Na_2SO_4	Sodium nitrate ($+ H_2O$) $NaNO_3$
Copper oxide CuO	Copper chloride ($+ H_2O$) $CuCl_2$	Copper sulfate ($+ H_2O$) $CuSO_4$	Copper nitrate ($+ H_2O$) $Cu(NO_3)_2$
Zinc carbonate $ZnCO_3$	Zinc chloride ($+ H_2O + CO_2$) $ZnCl_2$	Zinc sulfate ($+ H_2O + CO_2$) $ZnSO_4$	Zinc nitrate ($+ H_2O + CO_2$) $Zn(NO_3)_2$

Stage 1   THE REACTION
● React the acid with an insoluble substance (e.g. a metal, metal oxide, metal carbonate, metal hydroxide) to produce the desired salt
● Add this substance until it no longer reacts
● This reaction may need to be heated

↓

Stage 2   FILTER OFF THE EXCESS
● Filter off the left over metal/metal oxide/metal carbonate/metal hydroxide

↓

Stage 3   CRYSTALLISE THE SALT
● Heat the solution to evaporate some water until crystals start to form
● Leave the solution to cool down – more crystals will form
● Filter off the crystals of the salt
● Allow the crystals to dry

▲ **Figure 12.23** Making salts that are soluble in water.

◯ Making soluble salts

It is important to be able to make pure samples of salts, especially if they are being used in medicines or food.

Some salts are soluble in water and some are not. Figure 12.23 shows the method for making salts that are soluble in water.

It is easier to make a pure salt by reacting an acid with a substance that is insoluble in water rather than one that is soluble. Suitable substances to use include some metals, metal oxides, metal hydroxides or metal carbonates. As it is insoluble, you can add an excess of that substance to ensure that all the acid is used up. The excess can then be filtered off. In this way, there is no left over acid or the substance it reacts with mixed in with the salt that is formed.

After filtration, you are left with a solution of the salt in water. If some of the water is boiled off, a hot saturated solution is formed. As this hot, saturated solution cools, crystals form as the salt is less soluble at lower temperatures and so cannot all stay dissolved.

TIP

Crystals of many salts contain water within the crystal structure, e.g. blue crystals of copper(II) sulfate have the formula, $CuSO_4.5H_2O$. These are hydrated salts.

TIP

If an aqueous solution of a soluble salt is heated to evaporate all the water, then it is not possible to form hydrated crystals.

TIP

Some salts only form anhydrous crystals (crystals containing no water). Sodium chloride (NaCl) is a good example.

Test yourself

25 Suggest two chemicals that could be reacted together to make the following salts.
  a) calcium chloride
  b) copper nitrate
  c) aluminium sulfate

26 The salt iron(II) sulfate is used in iron tablets. It can be made by reacting an excess of iron with sulfuric acid.
  a) Why is it important that the iron(II) sulfate used in iron tablets is pure?
  b) Why is an excess of iron used?
  c) How is the excess iron removed?
  d) Write a word equation for the reaction between iron and sulfuric acid.
  e) Write a balanced equation for the reaction between iron and sulfuric acid.

27 Crystals of the salt nickel nitrate form as a hot, saturated solution of nickel nitrate cools down.
  a) What is a saturated solution?
  b) Explain why crystals of nickel nitrate form as the hot, saturated solution cools down.

28 The salt copper chloride can be made by reacting hydrochloric acid with copper oxide, copper carbonate or copper hydroxide. It cannot be made by reacting copper with hydrochloric acid.
  a) Explain why copper chloride cannot be made by reacting copper with hydrochloric acid.
  b) For each of the reactions of hydrochloric acid with copper oxide, copper carbonate and copper hydroxide:
    i) Write a word equation.
    ii) Write a balanced equation.

Show you can...

A solution of the salt magnesium chloride can be prepared by any of reactions A to D in the diagram.

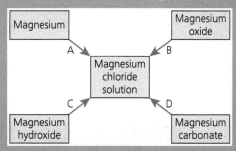

▲ Figure 12.24

a) Write word equations for each of the reactions A to D.
b) State two observations that you would make during reaction D.

Preparation of a pure dry sample of a soluble salt from an insoluble oxide or carbonate (magnesium sulfate from magnesium carbonate)

Bath crystals are a mixture of water soluble solids which are added to bathwater for health benefits. Bath crystals contain Epsom salts (Figure 12.25) (hydrated magnesium sulfate) which relax muscles, reduce inflammation and help muscle function. The name Epsom salts comes from the town of Epsom, which has mineral springs from which hydrated magnesium sulfate was extracted.

To produce pure dry crystals of hydrated magnesium sulfate ($MgSO_4.7H_2O$) in the laboratory by reacting magnesium carbonate with sulfuric acid, the following method was followed.

- Measure $25\,cm^3$ of dilute sulfuric acid and place in a conical flask.
- Warm the dilute sulfuric acid using a Bunsen burner and add magnesium carbonate, stirring until it is in excess.
- Filter the solution.
- Heat the filtered solution to make it more concentrated.
- Cool and crystallise.
- Filter the crystals from the solution.
- Dry the crystals.

▲ Figure 12.25 Epsom salts.

Questions

1 What piece of apparatus would you use to measure $25\,cm^3$ of sulfuric acid?
2 Draw a labelled diagram of the apparatus used for the second step of the method.
3 Explain why the magnesium carbonate is added until it is in excess.
4 State one way in which you would know that the magnesium carbonate is in excess.
5 What is the general name given to the solid trapped by the filter paper?
6 Why was the filtered solution evaporated using a water bath or electric heater?
7 Why is the solution not evaporated to dryness?
8 Why do crystals form as the solution is cooled?
9 State two methods of drying the crystals.
10 Write a word and balanced symbol equation for the reaction between magnesium carbonate and sulfuric acid.

Electrolysis

○ What is electrolysis?

Electrolysis is the decomposition of ionic compounds using electricity. Ionic compounds contain metals combined with non-metals. Examples include

- sodium chloride (NaCl) – a combination of the metal sodium with the non-metal chlorine
- copper sulfate ($CuSO_4$) – a combination of the metal copper with the non-metals sulfur and oxygen.

Ionic compounds are made up of positive and negative ions. As solids, ionic compounds cannot conduct electricity because the ions cannot

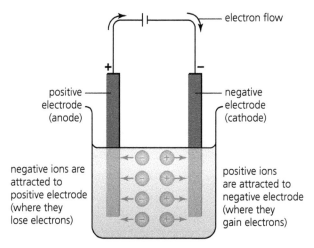

— electron flow

positive electrode (anode)

negative electrode (cathode)

negative ions are attracted to positive electrode (where they lose electrons)

positive ions are attracted to negative electrode (where they gain electrons)

▲ **Figure 12.26** Electrolysis.

TIP ✔

Remember that electrons have a negative charge.

TIP ✔

In the circuit as a whole, the current is carried by ions through the electrolyte and by electrons through the wires.

TIP ✔

In electrolysis the anode is the positive electrode and the cathode is the negative electrode.

move around. However, when ionic compounds are melted or dissolved, the ions are free to move and conduct electricity. These liquids or solutions are called electrolytes because they are able to conduct electricity.

If two electrodes connected to a supply of electricity are put into the electrolyte, the negative ions are attracted to the positive electrode (anode) and the positive ions are attracted to the negative electrode (cathode) (Figure 12.26). This happens because opposite charges attract each other.

When the ions reach the electrodes they are discharged. This means that they gain or lose electrons so that they lose their charge and become neutral. Positive ions gain electrons. Negative ions lose electrons.

The negative ions are discharged by **losing electrons** at the **positive electrode**. These electrons move around the circuit through the wires to the negative electrode. The **positive ions** are discharged by **gaining electrons** at the **negative electrode**.

Table 12.14 Electrodes and electrolysis.

Electrode	+ electrode (anode)	– electrode (cathode)
Which ions are attracted to the electrode	– ions	+ ions
What happens at the electrodes	– ions are discharged by losing electrons	+ ions are discharged by gaining electrons
Oxidation or reduction	Oxidation (loss of electrons) ≡Ⓗ	Reduction (gain of electrons) ≡Ⓗ

○ **Electrolysis of molten ionic compounds**

Binary ionic compounds are ones made from one metal combined with one non-metal. Examples include lead bromide ($PbBr_2$) (Figure 12.27), sodium chloride (NaCl) and aluminium oxide (Al_2O_3) (Table 12.15). The electrolysis of these compounds when molten produces the metal and non-metal.

The metal is produced at the cathode and the non-metal at the anode.

Table 12.15 Products from the electrolysis of aqueous solutions of ionic compounds.

Electrode	Ions	+ Electrode (anode)	– Electrode (cathode)
Lead bromide, $PbBr_2$	Lead ions, Pb^{2+} bromide ions, Br^-	Bromide ions (Br^-) lose electrons to form bromine (Br_2) $2Br^- - 2e^- \rightarrow Br_2$ ≡Ⓗ	Lead ions (Pb^{2+}) gain electrons to form lead (Pb) $Pb^{2+} + 2e^- \rightarrow Pb$ ≡Ⓗ
Sodium chloride, NaCl	Sodium ions, Na^+ chloride ions, Cl^-	Chloride ions (Cl^-) lose electrons to form chlorine (Cl_2) $2Cl^- - 2e^- \rightarrow Cl_2$ ≡Ⓗ	Sodium ions (Na^+) gain electrons to form sodium (Na) $Na^+ + e^- \rightarrow Na$ ≡Ⓗ
Aluminium oxide, Al_2O_3	Aluminium ions, Al^{3+} oxide ions, O^{2-}	Oxide ions (O^{2-}) lose electrons to form oxygen (O_2) $2O^{2-} - 4e^- \rightarrow O_2$ ≡Ⓗ	Aluminium ions (Al^{3+}) gain electrons to form aluminium (Al) $Al^{3+} + 3e^- \rightarrow Al$

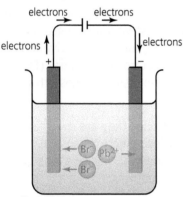

▲ **Figure 12.27** The electrolysis of molten lead bromide.

Bromide ions lose electrons to form bromine
$2\ Br^- - 2e^- \rightarrow Br_2$
(or $2\ Br^- \rightarrow Br_2 + 2e^-$)

Lead ions gain electrons to form lead
$Pb^{2+} + 2e^- \rightarrow Pb$

Show you can...

Copy and complete the table below for the electrolysis of molten lithium chloride.

Table 12.17

	Anode	Cathode
Product		
Observation		
Half equation		
Oxidation or reduction		

Test yourself

29 a) Why do ionic compounds conduct electricity when molten?
   b) Why do ionic compounds not conduct electricity when solid?
30 Copy and complete the table to show the products of electrolysis of some molten ionic compounds.

Table 12.16

Ionic compound (molten)	Product at the negative electrode (cathode)	Product at the positive electrode (anode)
Potassium iodide (KI)		
Zinc bromide ($ZnBr_2$)		
Magnesium oxide (MgO)		

31 a) What happens to negative ions at the positive electrode?
   b) What happens to positive ions at the negative electrode?
   c) What carries the electric charge through the electrolyte?
   d) What carries the electric charge through the wires?
32 a) What is an electrolyte?
   b) What is an anode?
   c) What is a cathode?
33 Balance each of the following half equations and state if each one is a reduction or oxidation process.
   a) $Cu^{2+} + e^- \rightarrow Cu$ b) $I^- - e^- \rightarrow I_2$
   c) $F^- \rightarrow F_2 + e^-$ d) $Fe^{3+} + e^- \rightarrow Fe$
34 Write a half equation for each of the following.
   a) The process at the positive electrode in the electrolysis of molten calcium oxide.
   b) The process at the negative electrode in the electrolysis of molten potassium bromide.

○ Metal extraction

Many metals are extracted from metal compounds in ores by heating with carbon in a reduction reaction. However, some metals cannot be extracted this way. This is because

● some metals are more reactive than carbon and/or
● some metals would react with carbon in the process.

Electrolysis is usually used to extract metals that cannot be extracted by heating with carbon. However, metals produced this way are expensive because of

● the high cost of thermal energy to melt the metal compounds and
● the high cost of the electricity for the process.

Extraction of aluminium

Aluminium is the second-most commonly used metal after iron/steel. It is too reactive to be extracted by heating with carbon and so is extracted by electrolysis.

The main ore of aluminium is bauxite which contains aluminium oxide. This has a very high melting point of 2072°C. The cost of the thermal

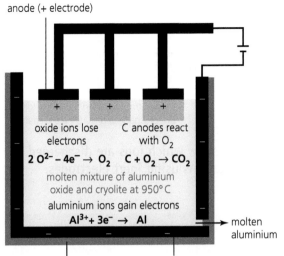

anode (+ electrode)

oxide ions lose electrons | C anodes react with O_2

$2\ O^{2-} - 4e^- \rightarrow O_2$ | $C + O_2 \rightarrow CO_2$

molten mixture of aluminium oxide and cryolite at 950°C

aluminium ions gain electrons

$Al^{3+} + 3e^- \rightarrow Al$

molten aluminium

outer casing of electrolysis cell | cathode (- electrode)

▲ **Figure 12.28** The extraction of aluminium by electrolysis.

energy to melt aluminium oxide for electrolysis is very high. However, if the aluminium oxide is mixed with a substance called cryolite, the mixture melts at about 950°C and so the cost of thermal energy to melt this mixture is lower.

The electrodes for the process are made of graphite, a form of carbon (Figure 12.28). Aluminium ions (Al^{3+}) are attracted to the negative electrode where they gain electrons and form aluminium metal. As it is so hot, this is produced as a liquid and is run off at the bottom. Oxide ions (O^{2-}) are attracted to the positive electrode where they lose electrons and form oxygen. This oxygen reacts with the carbon anode and so the anode burns to produce carbon dioxide. This means that the anode has to be replaced regularly.

> **TIP**
> Cryolite is added to the aluminium oxide to reduce the cost of thermal energy.

> **TIP** ✓
> Remember that aluminium forms at the cathode and oxygen at the anode.

Show you can...

Complete the table below for the electrolysis of molten aluminium oxide.

Table 12.18

	Anode	Cathode
Product		
Half equation		
Oxidation or reduction		
Conditions for the electrolysis		

Test yourself

35 **a)** Give two reasons why some metals cannot be extracted by heating with carbon.
   b) Name two metals other than aluminium that are extracted by electrolysis.
   c) Give two reasons why the energy cost of extracting metals by electrolysis is so high.

36 This question is about the extraction of aluminium by electrolysis.
   a) Identify the compound from which aluminium is extracted.
   b) Why is this compound mixed with cryolite?
   c) What happens at the positive electrode?
   d) What happens at the negative electrode?
   e) What are the electrodes made of?
   f) Why does the positive electrode have to be replaced regularly?

○ **Electrolysis of aqueous ionic compounds**

When an ionic compound is dissolved in water, the products of electrolysis are often different to those when the compound is molten. In water, a small fraction of the molecules break down into hydrogen ions (H^+) and hydroxide ions (OH^-). These ions can be discharged instead of the ions in the ionic compound. More water molecules can break down if these ions are used up.

At each electrode there are two ions that could discharge, one from the ionic compound and one from the water. The one that is easier to discharge is the one that is discharged. Table 12.19 shows which ions are discharged at each electrode when inert electrodes are used. Inert electrodes are electrodes that will allow the electrolysis to take place but do not react themselves. **Graphite electrodes** are the most common inert electrodes used.

> **KEY TERMS** ★
> **Inert electrodes** Electrodes that allow electrolysis to take place but do not react themselves.

Table 12.19 Products from the electrolysis of aqueous solutions of ionic compounds.

Electrode	Positive electrode	Negative electrode
Which ions are discharged	Negative ions discharged: **Oxygen** is produced from the discharge of hydroxide ions (unless the ionic compound contains halide ions when these are discharged producing a halogen)	Positive ions discharged: **Hydrogen** is produced from the discharge of hydrogen ions, unless the ionic compound contains metal ions from a metal that is less reactive than hydrogen when the metal ions are discharged producing the metal.

Table 12.20 gives some examples to show which ions are discharged in the electrolysis of some aqueous solutions using inert electrodes.

Table 12.20 Ion discharge in electrolysis.

		Sodium chloride NaCl(aq)	Copper(II) bromide $CuBr_2$(aq)	Silver nitrate $AgNO_3$(aq)	Potassium sulfate K_2SO_4(aq)
Negative electrode	Positive ions present	Na^+ and H^+	Cu^{2+} and H^+	Ag^+ and H^+	K^+ and H^+
	Positive ion discharged	H^+	Cu^{2+}	Ag^+	H^+
	Product	Hydrogen, H_2	Copper, Cu	Silver, Ag	Hydrogen, H_2
	Notes	Sodium is more reactive than hydrogen	Copper is less reactive than hydrogen	Silver is less reactive than hydrogen	Potassium is more reactive than hydrogen
	Half equation Ⓗ	$2H^+ + 2e^- \rightarrow H_2$ Ⓗ	$Cu^{2+} + 2e^- \rightarrow Cu$ Ⓗ	$Ag^+ + e^- \rightarrow Ag$ Ⓗ	$2H^+ + 2e^- \rightarrow H_2$ Ⓗ
	Process Ⓗ	Reduction	Reduction Ⓗ	Reduction Ⓗ	Reduction Ⓗ
Positive electrode	Negative ions present	Cl^- and OH^-	Br^- and OH^-	NO_3^- and OH^-	SO_4^{2-} and OH^-
	Negative ion discharged	Cl^-	Br^-	OH^-	OH^-
	Product	Chlorine , Cl_2	Bromine , Br_2	Oxygen, O_2	Oxygen, O_2
	Notes	Cl^- is a halide ion	Br^- is a halide ion	NO_3^- is not a halide ion	SO_4^{2-} is not a halide ion
	Half equation Ⓗ	$2Cl^- - 2e^- \rightarrow Cl_2$ Ⓗ	$2Br^- - 2e^- \rightarrow Br_2$ Ⓗ	$4OH^- - 4e^- \rightarrow 2H_2O + O_2$ Ⓗ	$4OH^- - 4e^- \rightarrow 2H_2O + O_2$ Ⓗ
	Process Ⓗ	Oxidation Ⓗ	Oxidation Ⓗ	Oxidation Ⓗ	Oxidation Ⓗ

Test yourself

37 Copy and complete the table to show the products of electrolysis of some aqueous solutions of ionic compounds with inert electrodes.

Table 12.21

Ionic compound (aqueous)	Product at the negative electrode (cathode)	Product at the positive electrode (anode)
Potassium iodide (KI)		
Copper(II) chloride ($CuCl_2$)		
Magnesium sulfate ($MgSO_4$)		
Copper(II) nitrate ($Cu(NO_3)_2$)		
Zinc bromide ($ZnBr_2$)		

38 For the electrolysis of aqueous sodium bromide solution using inert electrodes, write a half equation for the process at the electrode shown and state whether it is an oxidation or reduction process:
  a) at the positive electrode
  b) at the negative electrode.

39 For the electrolysis of aqueous copper(II) sulfate solution using inert electrodes, write a half equation for the process at the electrode shown and state whether it is an oxidation or reduction process:
  a) at the positive electrode
  b) at the negative electrode.

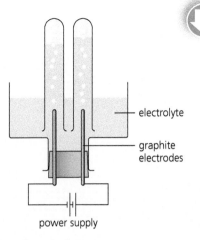

Required practical 9

Investigate what happens when aqueous solutions are electrolysed using inert electrodes

A student made a hypothesis which stated, 'Oxygen is produced at the positive electrode in the electrolysis of an aqueous solution, unless the compound contains halide ions.' To test this hypothesis the apparatus was set up as shown. The electricity supply was switched on and the observations at each electrode recorded. The experiment was repeated using different aqueous solutions as electrolyte and the results obtained recorded in the table.

> **KEY TERM**
>
> Hypothesis A proposal intended to explain certain facts or observations.

▲ Figure 12.29

Table 12.22

		Copper(II) chloride	Calcium nitrate	Silver sulfate	Potassium bromide	Sodium iodide	Sulfuric acid
Cathode (negative electrode)	Observations	Red brown solid	Colourless gas	Grey solid	Colourless gas	Colourless gas	Colourless gas
	Test used for product	Appearance	Squeaky pop when a burning splint is inserted	Appearance	Squeaky pop when a burning splint is inserted	Squeaky pop when a burning splint is inserted	Squeaky pop when a burning splint is inserted
	Identity of product						
Anode (positive electrode)	Observations	Colourless gas	Colourless gas	Colourless gas	Yellow-orange solution	Brown solution	Colourless gas
	Test used for product	Bleaches damp litmus paper	Relights a glowing splint	Relights a glowing splint	Bleaches damp litmus paper	Goes blue-black when starch is added	Relights a glowing splint
	Identity of product						

Questions

1 Why are graphite electrodes used?

2 Why are the aqueous solutions made up in distilled water?

3 What would you add to the circuit to show that a current is flowing?

4 Suggest why the experiment was only carried out long enough to make the necessary observations.

5 Write down the identity of the missing products at the cathode to complete the table.

6 Summarise the rules for deciding what is formed at the cathode when aqueous solutions of ionic compounds are electrolysed using inert electrodes.

7 Write down the identity of the missing products at the anode, to complete the table.

8 Using the results of this experiment, state if the hypothesis made by the student was correct.

Chapter review questions

1 The table shows the pH of some solutions.

Table 12.23

Solution	A	B	C	D	E
pH	12	9	1	6	7

a) Which solution is the most acidic?

b) Which solution is the most alkaline?

c) Which solution is neutral?

2 Complete the following word equations, or write *no reaction* if no reaction would take place.

a) iron + oxygen

b) zinc + sulfuric acid

c) magnesium oxide + hydrochloric acid

d) sodium carbonate + nitric acid

e) potassium hydroxide + sulfuric acid

f) gold + magnesium nitrate

g) iron + copper sulfate

h) zinc + iron nitrate

3 Metals are extracted from compounds found in ores. Which method is used for each of the following metals?

a) magnesium   b) aluminium

c) copper   d) zinc

e) gold

4 Balance the following equations.

a) $Ca + O_2 \rightarrow CaO$

b) $NaOH + H_2SO_4 \rightarrow Na_2SO_4 + H_2O$

c) $Fe_2O_3 + C \rightarrow Fe + CO$

5 Sodium hydroxide reacts with nitric acid to form the salt sodium nitrate.

a) Write a word equation for this reaction.

b) Write a balanced equation for this reaction.

c) Write an ionic equation for this reaction.

d) What type of reaction is this?

6 The salt magnesium sulfate is found in Epsom salts, which is used as a cure for constipation. It can be made by heating an excess of magnesium oxide with sulfuric acid. After the reaction is over, the left over magnesium oxide is filtered off. Crystallisation is used to produce the magnesium sulfate from the solution.

a) Why is an excess of magnesium oxide used?

b) What is the purpose of the filtration in this method?

c) Describe how the crystallisation of magnesium sulfate could be carried out.

d) Write a word equation for this reaction.

e) Write a balanced equation for this reaction.

7 Copy and complete the following table to show the products of electrolysis of some ionic compounds.

Table 12.24

Ionic compound	Product at the negative electrode (cathode)	Product at the positive electrode (anode)
Molten magnesium bromide ($MgBr_2$)		
Aqueous magnesium bromide ($MgBr_2$)		
Molten potassium oxide (K_2O)		
Molten sodium iodide (NaI)		
Aqueous calcium nitrate ($Ca(NO_3)_2$)		
Aqueous copper(II) chloride ($CuCl_2$)		

8 Zn metal is produced when zinc oxide reacts with aluminium in a displacement reaction.

$$3ZnO + 2Al \rightarrow 3Zn + Al_2O_3$$

a) Explain why aluminium displaces zinc.

b) Which substance is oxidised in this reaction? Give a reason for your answer.

c) Which substance is reduced in this reaction? Give a reason for your answer.

9 a) In the electrolysis of molten sodium chloride:

    i) Identify the products at the electrodes.

    ii) Write a half equation for the process at the negative electrode and state whether it is an oxidation or reduction process.

    iii) Write a half equation for the process at the positive electrode and state whether it is an oxidation or reduction process.

b) In the electrolysis of aqueous sodium chloride:

    i) Identify the products at the electrodes.

    ii) Write a half equation for the process at the negative electrode and state whether it is an oxidation or reduction process.

    iii) Write a half equation for the process at the positive electrode and state whether it is an oxidation or reduction process.

c) Explain why the electrolysis of molten and aqueous sodium chloride do not produce the same products.

10 Silver is produced when iron metal is added to a solution of silver nitrate.
$$Fe + 2AgNO_3 \rightarrow Fe(NO_3)_2 + 2Ag$$

a) Write an ionic equation for this reaction.

b) Write two half equations for this reaction.

c) Identify the substance that is oxidised in this reaction and explain your answer.

d) Explain why this is a redox reaction.

11 Write a balanced equation for each of the following reactions.

a) potassium + water

b) copper carbonate + hydrochloric acid

c) magnesium oxide + sulfuric acid

d) zinc + silver nitrate

e) aluminium + oxygen

12 $10\,cm^3$ of a solution of a strong acid had a pH 2. Some water was added to this to make the volume up to $1\,dm^3$. What is the pH of the new solution?

Explain your answer.

Practice questions

1 Which one of the following substances does not react with dilute hydrochloric acid at room temperature? [1 mark]

A calcium carbonate

B copper

C magnesium

D potassium hydroxide

2 The electrical conductivity of an electrolyte in electrolysis is due to the movement of particles through the substance when it is molten or in solution. Which of the following are the particles that move in the electrolyte? [1 mark]

A atoms

B electrons

C ions

D protons

3 Strips of four different metals were placed in solution of the nitrates of the same metals. Any reactions which occur are represented by a ✓ in the table. Which of the following is the correct order of reactivity with the most reactive first? [1 mark]

Table 12.25

Metal	P nitrate	Q nitrate	R nitrate	S nitrate
P	x	x	x	✓
Q	✓	x	✓	✓
R	✓	x	x	✓
S	x	x	x	x

A P Q R S

B Q R S P

C R S P Q

D Q R P S

4 Which one of the following could neutralise a solution of pH 10? [1 mark]

A ammonia solution

B hydrochloric acid

C sodium hydroxide solution

D water

5 Which one of the following pairs of substances could be used for the preparation of magnesium sulfate crystals? [1 mark]

A dilute hydrochloric acid and magnesium carbonate

B dilute nitric acid and magnesium oxide

C dilute sulfuric acid and magnesium chloride

D dilute sulfuric acid and magnesium carbonate.

6 What is the product formed at the anode in the electrolysis of aqueous sodium chloride solution? [1 mark]

A chlorine

B hydrogen

C oxygen

D sodium

7 Neutralisation occurs when an acid and an alkali react to form a salt and water.

a) i) Copy and complete the table below to give the names and formulae of the ions present in all acids and alkalis. [4 marks]

Table 12.26

	Ion present in all acids	Ion present in all alkalis
Name		
Formula		

ii) Write an ionic equation for neutralisation, including state symbols. [2 marks]

b) Sulfuric acid solution can be neutralised using an alkali such as sodium hydroxide or adding a solid oxide such as copper(II) oxide.

i) Write a balanced equation for the reaction between sodium hydroxide and sulfuric acid. [2 marks]

ii) Write a balanced equation for the reaction between copper(II) oxide and sulfuric acid. [2 marks]

c) In the preparation of hydrated copper sulfate(II) crystals, an excess of copper(II) oxide was added to warm dilute sulfuric acid. The excess copper(II) oxide was removed by filtration. Describe how you would obtain pure dry crystals of hydrated copper(II) sulfate from the filtrate collected. [4 marks]

8 The description that follows is about a metal, other than zinc, which belongs to the reactivity series.

At room temperature the metal is a silver coloured solid. It reacts rapidly with cold dilute sulfuric acid to produce hydrogen. It conducts electricity. On heating in air the metal burns with a very bright flame leaving a white powder.

a) State one property from the description above that is common to all metals. [1 mark]

b) Give two reasons why the metal is NOT copper. [2 marks]

c) What part of the air reacts with the metal? [1 mark]

d) Suggest one metal which fits the description. [1 mark]

e) Write a word equation for the reaction of your chosen metal with sulfuric acid. [1 mark]

9 Hydrochloric acid can react with calcium hydroxide and with calcium. **Compare and contrast** the reaction of hydrochloric acid with calcium hydroxide with the reaction of hydrochloric acid with calcium. In your answer you must include the names of all products for each reaction and the observations for each reaction. [6 marks]

10 Aluminium is the most abundant metal in the Earth's crust. Aluminium ore is first purified to give aluminium oxide and the metal is then extracted from the aluminium oxide by electrolysis.

a) What is meant by the term electrolysis?

b) Name the ore from which aluminium is extracted.

c) The electrolysis of the purified ore is carried out in the Hall-Héroult cell. The diagram below shows the cell used.

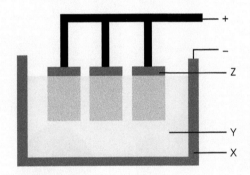

▲ Figure 12.30

i) Name X and Z. [2 marks]

ii) Y is the electrolyte. Name the substances in the electrolyte. [2 marks]

iii) Why is the electrolysis cell kept at about 950°C? [2 marks]

iv) Name the products formed at the positive and negative electrodes. [2 marks]

v) Write half equations for the reactions taking place at each electrode. [2 marks]

vi) Which electrode must be replaced regularly? Write a balanced symbol equation to explain your answer. [2 marks]

d) Give a reason in terms of electrons why the extraction of aluminium in this process is a reduction reaction. [1 mark]

11 Some substances, for example molten lead bromide and aqueous sodium chloride, are described as electrolytes. Other substances for example copper metal, are conductors.

An experiment, to investigate the electrolysis of the electrolyte molten lead bromide, was set up as shown in the diagram below.

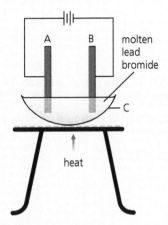

▲ Figure 12.31

a) Some pieces of apparatus in the diagram are labelled A–C. State the correct name for each piece of apparatus. [3 marks]

b) Name a piece of apparatus which could be connected in the circuit to show that an electric current is flowing through the molten lead bromide. [1 mark]

c) Copy and complete the table to state the names of the products, and the half equations for the electrodes. [4 marks]

Table 12.27

Electrode	Name of product	Half equation
A		
B		

d) Why does this electrolysis need to be carried out in a fume cupboard? [1 mark]

e) Describe the differences in conduction of electricity by copper metal and molten lead bromide. [2 marks]

f) Name the product at each electrode when aqueous sodium chloride is electrolysed. [2 marks]

Working scientifically:
Measurements and uncertainties

We often use apparatus to make measurements in chemistry. For example, we often measure the mass, volume or temperature of a substance.

Measurement instruments which you should be able to use correctly in chemistry include

▶ measuring cylinder (Figure 12.32)

▶ pipette (Figure 12.33)

▶ burette

▶ thermometer

▶ balance.

Measuring cylinders, pipettes and burettes are used to measure out volumes of liquids. Pipettes and burettes, which are used in titrations, measure volumes more accurately than measuring cylinders.

A meniscus is the curve seen at the top of a liquid in response to its container. When reading the volume of the liquid, the measurement at the bottom of the meniscus curve is read. This should be read at eye level (Figure 12.34).

▲ Figure 12.32 Measuring cylinders.

▲ Figure 12.33 Pipette.

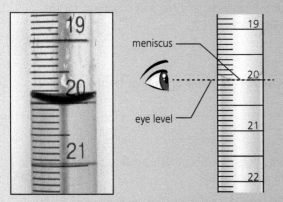

▲ Figure 12.34 The meniscus should be read at eye level.

The resolution of a piece of apparatus is the smallest change it can measure. For example, in Figure 12.34, the resolution of the burette is ±0.1 cm³.

When a measurement is made, there is always some doubt or uncertainty about its value. Uncertainty is often recorded after a measurement as a ± value.

The uncertainty can be estimated from the range of results that are obtained when an experiment is repeated several times.

KEY TERMS ★

Meniscus The curve at the surface of a liquid in a container.

Resolution The smallest change a piece of apparatus can measure.

Uncertainty The range of measurements within which the true value can be expected to lie.

For example, the volume of a gas produced in a reaction was measured five times. The results are 82, 77, 78, 96 and 80 cm^3.

The mean value is found after excluding any anomalous results (96 cm^3 is anomalous here as it is significantly different from all the others):

Mean volume = $\dfrac{82 + 77 + 78 + 80}{4}$ = 79 ± 3 cm^3

The mean is quoted to the nearest unit as all the values are measured to the nearest unit. The uncertainty is ±3 cm^3 as the highest and lowest values are within 3 cm^3 of the mean.

Questions

1 Record the volume of liquid in each measuring cylinder A-D. All scales are shown in cm^3.

▲ Figure 12.35

2 Record the volume of liquid in each of the burettes E-G shown below. All scales are shown in cm^3.

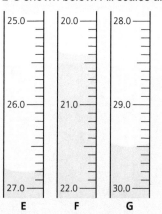

◄ Figure 12.36

3 Record the temperatures shown on each thermometer H-M. All scales are shown in °C.

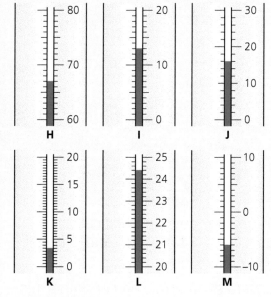

▲ Figure 12.37

4 In each pair of diagrams, find the difference in mass between the first reading and the second reading. All scales are shown in grams.

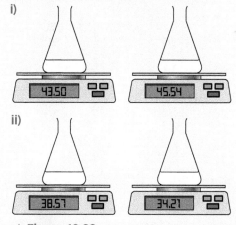

▲ Figure 12.38

5 Find the mean value and its uncertainty for the boiling point of a substance which was measured several times. Values measured were 124, 126, 123, 125 and 123°C.

6 Find the mean value and its uncertainty for the mass of gas produced in a reaction which was measured several times. Values measured were 0.36, 0.18, 0.33 and 0.40 g.

7 Find the mean value and its uncertainty for the volume of acid needed to neutralise an alkali which was measured several times. Values measured were 25, 27, 26, 20 and 25 cm^3.

13 Energy changes

Many people have a gas fire to keep warm at home. A chemical reaction takes place when the gas burns. This chemical reaction releases a lot of thermal energy that keeps us warm. This chapter looks at why some chemical reactions release thermal energy and increase the temperature while other reactions remove thermal energy and lower the temperature.

Specification coverage

This chapter covers specification points 5.5.1.1 to 5.5.1.3 and is called Energy changes.

It covers exothermic and endothermic reactions.

Prior knowledge

Previously you could have learnt:

> Energy cannot be made or destroyed – it can only be transferred from one form to another (this is the law of conservation of energy).

> An energy change is a sign that a chemical reaction has taken place.

> Some chemical reactions lead to an increase in temperature and some to a decrease in temperature.

Test yourself on prior knowledge

1 Chemical energy and thermal energy are two forms of energy. Write down three forms of energy besides these two.
2 What is the law of conservation of energy?

Exothermic and endothermic reactions

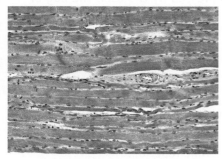

▲ Figure 13.1 Respiration takes place in all cells, including the cells in this muscle tissue.

KEY TERM

Exothermic reaction Reaction where thermal energy is transferred from the chemicals to the surroundings and so the temperature increases.

▲ Figure 13.2 An exothermic reaction takes place inside hand warmers.

○ Energy changes in reactions

When chemical reactions take place, thermal energy is transferred to or from the surroundings. Some reactions make their surroundings hotter and some make them colder.

Exothermic reactions

In exothermic reactions, **thermal energy** is transferred from the **chemicals to** the **surroundings**. As there is more thermal energy, the temperature increases and so it gets hotter. Most chemical reactions are exothermic.

In some exothermic reactions, only a small amount of thermal energy is transferred and the temperature may only rise by a few degrees. In some reactions a lot of thermal energy is transferred and the surroundings get very hot. Sometimes there is so much thermal energy transferred that the reactants catch fire (Table 13.1).

Applications of exothermic reactions

Hand warmers are used by people in cold places to keep their hands warm (Figure 13.2). There are many different types, but they all work by using an exothermic reaction that transfers thermal energy to the surroundings.

Table 13.1 Examples of exothermic reactions.

Example	Comments
Oxidation reactions e.g. respiration (Figure 13.1)	Oxidation reactions take place when substances react with oxygen. Many oxidation reactions are exothermic. Respiration takes place in the cells of living creatures and is the reaction of glucose with oxygen. Thermal energy is released in this reaction. For example: glucose + oxygen → carbon dioxide + water $C_6H_{12}O_6 + 6O_2 \rightarrow 6CO_2 + 6H_2O$
Combustion reactions e.g. burning fuels	Combustion reactions take place when substances react with oxygen and catch fire. They are a type of oxidation reaction. Fuels burning, such as methane (CH_4) in natural gas, are combustion reactions. These reactions are very exothermic, release a lot of thermal energy and catch fire. For example: methane + oxygen → carbon dioxide + water $CH_4 + 2O_2 \rightarrow CO_2 + 2H_2O$
Neutralisation reactions e.g. acids reacting with alkalis	Neutralisation reactions take place when acids react with bases. Many neutralisation reactions are exothermic. For example, when hydrochloric acid reacts with sodium hydroxide some thermal energy is transferred to the surroundings and the temperature rises by a few degrees. For example: hydrochloric acid + sodium hydroxide → sodium chloride + water $HCl + NaOH \rightarrow NaCl + H_2O$

▲ **Figure 13.3** Self-heating cans of food and drink.

Self-heating cans can also be very useful (Figure 13.3). They can be used to provide hot drinks, such as coffee and hot chocolate, or even foods. Inside these cans, the food or drink is separated from a layer containing the chemicals used for heating. This is often calcium oxide and a bag of water. When a button is pressed by the user, the bag of water is punctured and the water mixes with the calcium oxide. The calcium oxide reacts with the water in an exothermic reaction. The thermal energy released heats up the food or drink.

Endothermic reactions

In **endothermic reactions**, **thermal energy** is transferred from the **surroundings** to the **chemicals**. As there is less thermal energy, the temperature decreases and so it gets colder (Table 13.2). Endothermic reactions are less common than exothermic reactions.

Applications of endothermic reactions

Some sports injury packs act as a cold pack to put on an injury to prevent swelling (Figure 13.4). Inside the pack is a bag of water and a substance such as ammonium nitrate. When the pack is squeezed the water bag bursts and the ammonium nitrate dissolves in the water in an endothermic process.

Table 13.2 Examples of endothermic reactions.

Example		Comments
Decomposition, e.g. thermal decomposition of metal carbonates	Green copper carbonate decomposes into black copper oxide when heated	Decomposition reactions occur when substances break down into simpler substances. For example when metal carbonates are heated they break down into a metal oxide and carbon dioxide. Decomposition reactions are always endothermic. For example: copper carbonate → copper oxide + carbon dioxide $CuCO_3 \rightarrow CuO + CO_2$
Reaction of acids with metal hydrogencarbonates	Citric acid and sodium hydrogencarbonate are in sherbet sweets and react in your mouth in an endothermic reaction	When metal hydrogencarbonates (e.g. $NaHCO_3$) react with acids (e.g. citric acid), the temperature drops as this is an endothermic reaction. For example: citric acid + sodium hydrogencarbonate → sodium citrate + carbon dioxide + water

▲ **Figure 13.4** Sports injury packs use an endothermic reaction to keep the injury cold.

Test yourself

1 Is each of the following reactions endothermic or exothermic?
   a) the temperature started at 21°C and finished at 46°C
   b) the temperature started at 18°C and finished at 14°C
   c) the temperature started at 19°C and finished at 25°C

2 Is each of the following reactions endothermic or exothermic?
   a) burning alcohol
   b) thermal decomposition of iron carbonate
   c) reaction of vinegar (containing ethanoic acid) with baking powder (sodium hydrogencarbonate)
   d) reaction of vinegar (containing ethanoic acid) with sodium hydroxide

3 a) Why does the temperature increase when an exothermic reaction takes place in solution?
   b) Why does the temperature decrease when an endothermic reaction takes place in solution?

Show you can...

The reactions P and Q can be classified in different ways.

P calcium carbonate → calcium oxide + carbon dioxide
Q sodium hydroxide + nitric acid → sodium nitrate + water

Complete the table by placing one or more ticks (✔) in each row for reactions P and Q to indicate the terms which apply to each reaction. More than one term can apply to each reaction.

Table 13.3

	Combustion	Decomposition	Neutralisation	Oxidation	Respiration	Exothermic	Endothermic
P							
Q							

○ Reaction profiles

Chemical reactions can only occur when particles collide with each other with enough energy to react. The *minimum* energy particles must have to react is called the activation energy.

We can show the relative energy of reactants and products in a reaction profile (Figure 13.5). This can also show the activation energy.

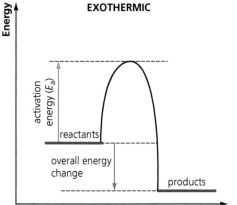

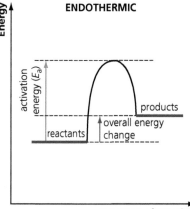

▲ **Figure 13.5** Reaction profiles for exothermic and endothermic reactions.

Investigate the variables that affect temperature change in reacting solutions – the temperature change in a neutralisation reaction

In an experiment the following method was followed:

1 25 cm³ of 40 g/dm³ sodium hydroxide was measured out and placed in a polystyrene cup.

2 A burette was filled with 36.5 g/dm³ hydrochloric acid.

3 The temperature of the sodium hydroxide was measured and recorded.

4 5 cm³ of hydrochloric acid was added from the burette to the plastic beaker, and the temperature recorded. Additional hydrochloric acid was added 5 cm³ at a time and the temperature recorded until the total volume of hydrochloric acid added was 40 cm³.

5 The experiment was repeated.

6 The results are shown in Table 13.4.

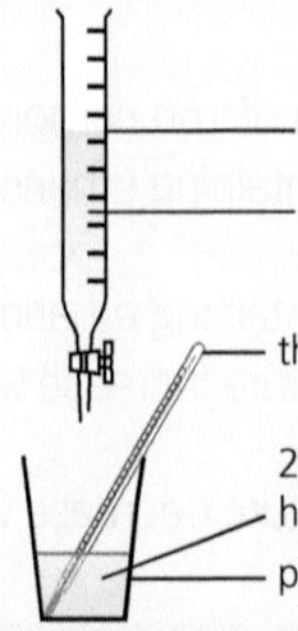

burette
hydrochloric acid
thermometer
25 cm³ of sodium hydroxide
polystyrene cup

adding hydrochloric acid to sodium hydroxide and measuring the temperature.

▲ Figure 13.6

Table 13.4 The temperature recorded when increasing volumes of 36.5 g/dm³ HCl was added to 25 cm³ of 40 g/dm³ NaOH.

Volume of acid added in cm³	Temperature in °C		
	Experiment 1	Experiment 2	Average
0	19.4	19.6	19.5
5	21.6	21.8	21.7
10	24.0	24.0	24.0
15	25.2	25.0	25.1
20	25.4	25.6	25.5
25	26.2	26.4	26.3
30	25.4	25.4	25.4
35	25.2	25.0	25.1
40	25.1	24.9	25.0

Questions

1 In this experiment identify the:
   a) independent variable
   b) dependent variable
   c) key control variables

2 Why was the experiment repeated?

3 Why was a polystyrene cup used rather than a glass beaker?

4 Why should the solution in the polystyrene cup be stirred after each addition of acid?

5 Plot a graph of average temperature (y-axis) against volume of acid added (x-axis).

6 Describe the trend shown by the results plotted.

7 It is thought that the highest temperature is reached when complete neutralisation has occurred. Suggest how you would experimentally confirm that the highest temperature reached is the point at which neutralisation has occurred.

A different experiment was carried out by adding 36.5 g/dm³ solutions of different acids to 25 cm³ of 40 g/dm³ sodium hydroxide in a plastic beaker. The highest temperature reached was recorded and presented in Table 13.5.

Table 13.5 The highest temperature recorded when 40 cm³ of different types of 1 mol/dm³ acid of was added to 25 cm³ of 1 mol/dm³ NaOH.

Type of acid	hydrochloric acid	sulfuric acid	ethanoic acid
Highest temp in °C	26.2	26.5	25.2
Repeat highest temperature in °C	26.4	26.3	25.4
Average highest temperature in °C	26.3	26.4	25.3

8 In this experiment identify the:
   a) independent variable
   b) dependent variable
   c) key control variables

9 State two conclusions you can draw from the results.

Example

Draw a reaction profile diagram for the reaction below which is exothermic.

$CH_4 + 2O_2 \rightarrow CO_2 + 2H_2O$

Answer

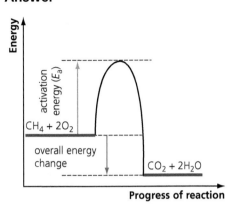

Show you can...

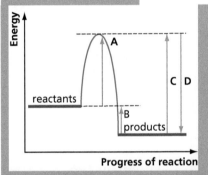

▲ Figure 13.7

The reaction profile for a reaction is shown.

a) Is it an exothermic reaction or an endothermic reaction?
b) Which letter represents the activation energy for the conversion of reactants to products?

Test yourself

4 Look at the following reaction profiles.

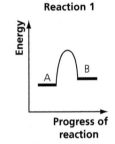
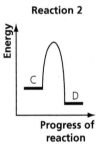
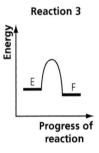

▲ Figure 13.8

a) Which reaction(s) is/are exothermic?
b) Which reaction(s) is/are endothermic?
5 a) Sketch the reaction profile for the following reaction which is endothermic:
$CuCO_3 \rightarrow CuO + CO_2$
b) Draw an arrow to show the overall energy change for the reaction and label it **O**.
c) Draw an arrow to show the activation energy for the reaction and label it **A**.

○ Bond energies

▲ Figure 13.9

Breaking a chemical bond takes energy. For example, 436 kJ of energy is needed to break one mole of H—H covalent bonds. Due to the law of conservation of energy, 436 kJ of energy must be *released* when making one mole of H—H covalent bonds (Figure 13.9).

During a chemical reaction:

- **energy** must be **supplied** to **break bonds** in the reactants
- **energy** is **released** when **bonds** in the products are made.

The overall energy change for a reaction equals the difference between the energy needed to break the bonds in the reactants and the energy released when bonds are made in the products (Table 13.6).

Energy change = energy needed breaking bonds in reactants
– energy released making bonds in products

Table 13.6

	Exothermic reaction	Endothermic reaction
Comparison of bond energies	More energy is released making new bonds than is needed to break bonds	More energy is needed to break bonds than is released making new bonds
Sign of energy change	–	+

Example

Find the energy change in the following reaction using the bond energies given.

Bond energies: C—H 412 kJ, O=O 496 kJ, C=O 743 kJ, O—H 463 kJ

H—C(H)(H)—H + 2 O=O ⟶ O=C=O + 2 H—O—H

Explain why the reaction is exothermic or endothermic using bond energies.

Answer

Bonds broken: Bonds made:

4C—H = 4(412) 2C=O = 2(743)

2O=O = 2(496) 4O—H = 4(463)

Total = 2640 kJ Total = 3338 kJ

Energy change = energy needed to break bonds
– energy released making bonds

= 2640 – 3338

= –698 kJ

This reaction is exothermic because more energy is released making bonds than is needed to break bonds.

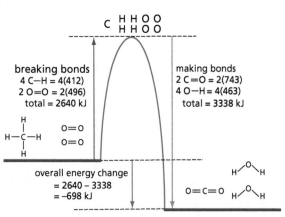

breaking bonds
4 C—H = 4(412)
2 O=O = 2(496)
total = 2640 kJ

making bonds
2 C=O = 2(743)
4 O—H = 4(463)
total = 3338 kJ

overall energy change
= 2640 – 3338
= –698 kJ

▲ **Figure 13.10**

Example

Find the energy change in the following reaction using the bond energies given.

Bond energies: H—H 436 kJ, O=O 496 kJ, O—H 463 kJ

2 H—H + O=O ⟶ 2 H—O—H

Explain why the reaction is exothermic or endothermic using bond energies.

Answer

Bonds broken: Bonds made:

2H—H = 2(436) 4O—H = 4(463)

1O=O = 496 Total = 1852 kJ

Total = 1368 kJ

Energy change = energy needed to break bonds
– energy released making bonds

= 1368 – 1852

= –484 kJ

This reaction is exothermic because more energy is released making bonds than is needed to break bonds.

Example

Find the energy change in the following reaction using the bond energies given.

Bond energies: C—C 348 kJ, C—H 412 kJ, C—O 360 kJ, O—H 463 kJ, C=C 612 kJ

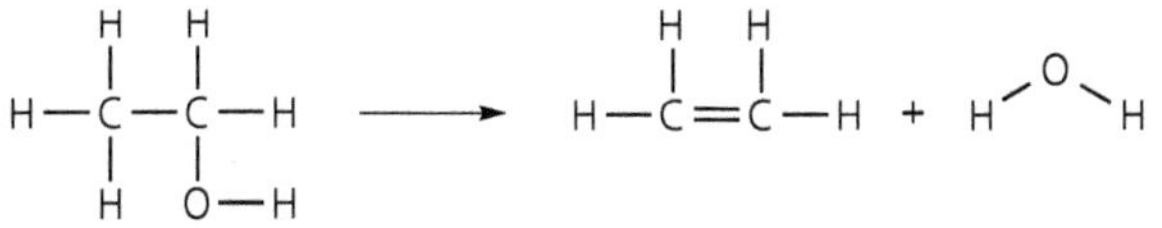

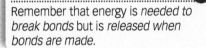

Explain why the reaction is exothermic or endothermic using bond energies.

Answer

Bonds broken:	Bonds made:
5C—H = 5(412)	4C—H = 4(412)
C—C = 348	C=C = 612
C—O = 360	2O—H = 2(463)
O—H = 463	Total = 3186 kJ
Total = 3231 kJ	

Energy change = energy needed to break bonds
  – energy released making bonds

= 3231 – 3186

= +45 kJ

This reaction is endothermic because more energy is needed breaking bonds than is released making bonds.

TIP

Remember that energy is *needed to break bonds* but is *released when bonds are made*.

Test yourself

6 For each of the following reactions, calculate the energy change and explain why the reaction is endothermic or exothermic by discussing bond energies.

Table 13.7

Bond	C—C	C=C	C—H	C—Br	C=O	O=O	O—H	H—H	H—Br	N≡N	N—H	Br—Br
Bond energy in kJ	348	612	412	276	743	496	463	436	366	944	388	193

a) H—H + Br—Br ⟶ 2 H—Br

b)
N≡N + 3 H—H ⟶ 2 H—N—H (with H above)

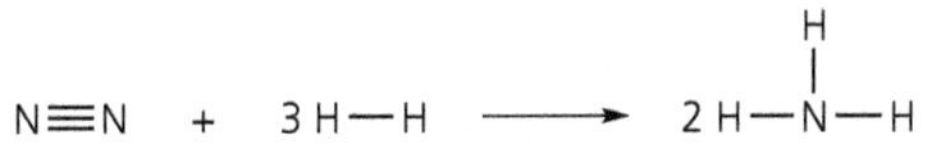

c)

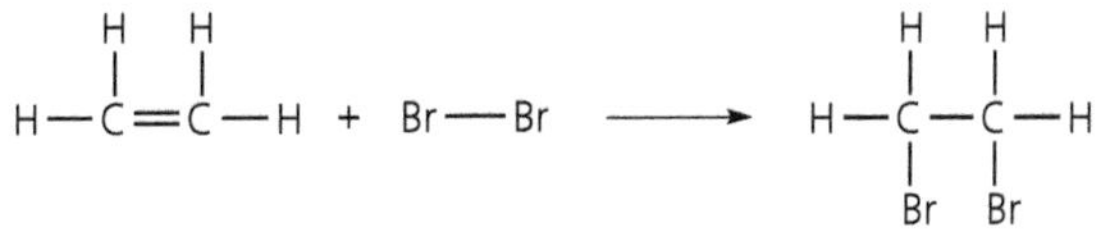

d)
H—C—C—C—H + 5 O=O ⟶ 3 O=C=O + 4 H—O—H (propane combustion structure)

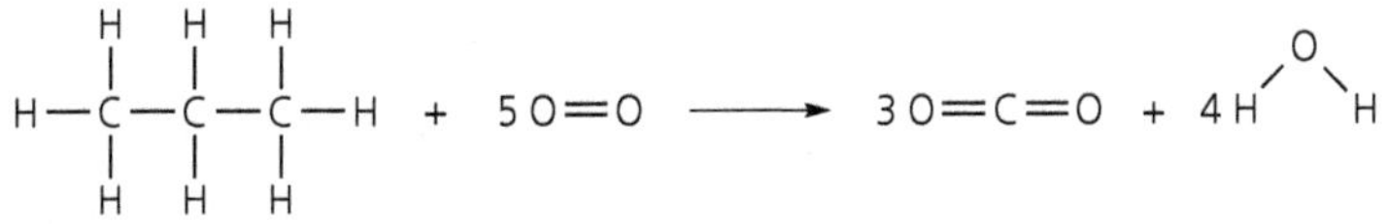

Show you can...

The energy change is −30 kJ in the following reaction:

I—I + Cl—Cl ⟶ 2 I—Cl

Bond energies: I—I = 150 kJ,
Cl—Cl = 242 kJ

Calculate the bond energy of I—Cl.

Chapter review questions

1 The following reactions all took place in solution in a beaker. The temperature before and after the chemicals were mixed was recorded in each case. Decide whether each reaction is exothermic or endothermic.

Table 13.8

	Start temperature in °C	End temperature in °C
Reaction 1	21	15
Reaction 2	20	27
Reaction 3	22	67

2 Copy and complete the spaces in the following sentences.

In an exothermic reaction, thermal energy is transferred from the chemicals to their surroundings and so the temperature _____. In an _____ reaction, thermal energy is transferred away from the surroundings to the chemicals and so the temperature _____.

3 Copy and complete the spaces in the following sentences.

Chemical reactions can only take place when particles _____ with each other and have enough energy. The minimum energy particles need to react is called the _____ energy.

4 The reaction profile for a reaction is shown.

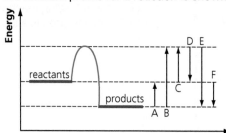

▲ Figure 13.11

a) Give the letter of the arrow that shows the activation energy for the reaction.

b) Give the letter of the arrow that shows the overall energy change for the reaction.

c) Is this reaction endothermic or exothermic?

5 Decide whether each of the following reactions is likely to be endothermic or exothermic.

a) burning magnesium

b) decomposition of silver oxide

c) reaction inside a sports injury cold pack

d) reaction inside a self-heating food can

e) neutralisation of sulfuric acid by sodium hydroxide

f) neutralisation of sulfuric acid by sodium hydrogencarbonate

6 a) Calculate the energy change for the following reaction using these bond energies.

H—H = 436 kJ, Cl—Cl = 242 kJ, H—Cl = 431 kJ

H—H + Cl—Cl ⟶ 2 H—Cl

b) Is this reaction endothermic or exothermic? Explain your answer by discussing bond energies.

7 Calculate the energy change for the following reaction using these bond energies.

N—H = 388 kJ, N—N = 158 kJ, N≡N = 944 kJ, O—H = 463 kJ, O=O = 496 kJ

$$\underset{\substack{H \\ }}{\overset{\substack{H \\ }}{N}}-\underset{\substack{H \\ }}{\overset{\substack{H \\ }}{N}} \quad + \quad O=O \quad \longrightarrow \quad N\equiv N \quad + \quad 2\,H\underset{\substack{}}{\overset{O}{\diagup}}H$$

Practice questions

1 Which one of the following is an endothermic reaction? [1 mark]

A   the combustion of hydrogen

B   the reaction of citric acid and sodium hydrogencarbonate

C   the reaction of magnesium and hydrochloric acid

D   the reaction of sodium hydroxide and hydrochloric acid

2 When potassium hydroxide dissolves in water the temperature of the solution rises. Which of the following is this an example of?

A   an endothermic change

B   an exothermic change

C   a neutralisation reaction

D   a thermal decomposition reaction   [1 mark]

3 a) For each of the reactions A to E, choose the appropriate word from the list below to describe the type of reaction. Each word may be used once, more than once or not at all.   [5 marks]

combustion   decomposition   neutralisation
oxidation   reduction

A   copper carbonate → copper oxide + carbon dioxide

B   ethanoic acid + sodium hydroxide → sodium ethanoate + water

C   magnesium + oxygen → magnesium oxide

D   methane + oxygen → carbon dioxide + water

E   sodium hydroxide + hydrochloric acid → sodium chloride + water

b) Explain the difference between an exothermic reaction and an endothermic reaction.   [2 marks]

c) For each of the reactions A to E above decide if it is an exothermic or an endothermic reaction.   [5 marks]

d) Describe how you would experimentally prove that the reaction between magnesium and hydrochloric acid is an exothermic reaction.   [4 marks]

4 Photosynthesis is an endothermic process used by plants to produce carbohydrates.

$6CO_2 + 6H_2O → C_6H_{12}O_6 + 6O_2$

a) What is meant by the term endothermic?   [1 mark]

b) Describe what is meant by the term activation energy.   [1 mark]

c) Draw a labelled reaction profile for this reaction. You must show: the position of the reactants and products; the activation energy; and the energy change of the reaction.   [4 marks]

d) Explain in terms of bond making and breaking why this reaction is endothermic.   [2 marks]

5 The diagram shows the profile for a reaction with a catalyst and without a catalyst.

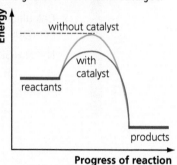

▲ Figure 13.12

a) Is this an exothermic or an endothermic reaction?   [1 mark]

b) Does a catalyst have an effect on the overall energy change of the reaction?   [1 mark]

c) On a copy of the diagram label the activation energy for the catalysed reaction as A and the activation energy for the uncatalysed reaction as B.   [2 marks]

d) From the information shown in the graph, state the effect of a catalyst on the activation energy of a reaction.   [1 mark]

6 Hydrogen reacts with fluorine to form hydrogen fluoride.

$H_2 + F_2 → 2HF$

Use the bond energies in the table to calculate the energy change for this reaction. Explain if the reaction is exothermic or endothermic.   [4 marks]

Table 13.9

Bond	Bond energy in kJ
H—H	436
F—F	158
H—F	568

7 Ethanol burns in oxygen.

$$H-\overset{\overset{\displaystyle H}{|}}{\underset{\underset{\displaystyle H}{|}}{C}}-\overset{\overset{\displaystyle H}{|}}{\underset{\underset{\displaystyle O-H}{|}}{C}}-H \quad + \quad 3\,O=O \quad \longrightarrow \quad 2\,O=C=O \quad + \quad 3\,H^{O}\!\diagdown^{H}$$

Use the bond energies below to calculate the energy change for this reaction.   [3 marks]

Bond energies:
C—H 412 kJ
O=O 496 kJ
C=O 743 kJ
O—H 463 kJ
C—C 348 kJ
C—O 360 kJ

Working scientifically:
Identifying variables when planning experiments

When carrying out an experiment different variables are used. Variables are the things that can change. When we plan experiments, we choose to change some variables while keeping others the same.

A variable is a physical, chemical or biological quantity or characteristic that can have different values. It may be for example, temperature, mass, volume, pH or even the type of chemical used in an experiment.

There are different types of variables which you need to be familiar with.

A continuous variable has values that are numbers. Mass, temperature and volume are examples of continuous variables. The values of these variables can either be found by counting (e.g. the number of drops) or by measurement (e.g. the temperature).

A categoric variable is one which is best described by words. Variables such as the type of acid or the type of metal are categoric variables.

▲ **Figure 13.13** The volume of carbon dioxide gas produced over time, from the reaction between calcium carbonate and $36.5\,g/dm^3$ HCl was recorded using this apparatus. The experiment was repeated using different concentrations of HCl. Can you identify the independent, dependent and controlled variables?

Questions

1 Decide if the following are categoric or continuous variables.
   a) temperature
   b) type of acid
   c) volume
   d) name of gas produced
   e) concentration of solution
   f) volume of gas produced
   g) mass
   h) name of metal
   i) colour of solution
   j) drops of acid

Scientists often plan experiments to investigate if there is a relationship between two variables, the independent and the dependent variable.

The independent variable is the variable for which values are changed or selected by the investigator (i.e. it is the one which you deliberately change during an experiment).

The dependent variable is one which may change as a result of changing the independent variable. It is the one which is measured for each and every change in the independent variable.

A control variable is one which may, in addition to the independent variable, affect the outcome of the investigation. Control variables must be kept constant during an experiment to make it a fair test.

As an example, the variables are shown below for an experiment to find the effect of temperature on the rate of the reaction between magnesium and excess hydrochloric acid. The rate was measured by timing how long it took for all the magnesium to react.

Table 13.10

Independent variable	Dependent variable	Control variables
Temperature	Time taken for all the magnesium to react	The mass of magnesium The surface area of magnesium The volume of hydrochloric acid The concentration of hydrochloric acid

Questions

2 For the following experiments identify the
   (i) independent variable
   (ii) dependent variable
   (iii) control variables.

a) Some magnesium was added to hydrochloric acid and the temperature recorded. The experiment was repeated several times using different volumes of hydrochloric acid.

b) In the reaction between copper carbonate and hydrochloric acid the time taken for a mass of copper carbonate to all be used up was recorded. The experiment was repeated using different masses of copper carbonate.

c) 2g of magnesium was added to copper sulfate solution and the highest temperature reached was recorded. The experiment was repeated using five different concentrations of copper sulfate.

d) In an experiment to find the effect of stirring on speed of dissolving, the time taken to dissolve some copper sulfate in water was measured. This was repeated stirring the solution.

e) The volume of carbon dioxide gas produced when calcium carbonate reacts with hydrochloric acid was measured, and the experiment repeated using different masses of calcium carbonate.

f) The temperature of nitric acid was recorded before and after some sodium hydroxide was added. The experiment was repeated using sulfuric acid, ethanoic acid and methanoic acid.

14 Formulae and equations

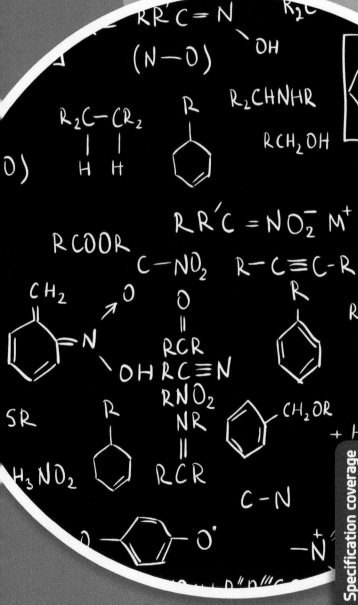

Chemists use formulae and equations as a quick way of identifying substances and showing what happens in chemical reactions. Being formula literate is vital for any chemist. This chapter looks at writing formulae and understanding what they mean, as well as how to write chemical, ionic and half equations.

Specification coverage

This chapter is called Formulae and Equations and brings together points from throughout the specification.

Writing formulae

Chemists use formulae a lot and it is important that you are formula literate meaning that you can write and recognise formulae.

○ Elements

The formula for most elements is just its symbol. For example, the formula of argon is Ar and that of magnesium is Mg (Table 14.1).

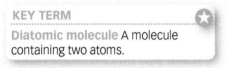

KEY TERM

Diatomic molecule A molecule containing two atoms.

However, this is not the case for elements made of molecules. Many of these molecules contain two atoms (called diatomic molecules) such as hydrogen (H_2) and oxygen (O_2). Some elements that are made of molecules contain more than two atoms in their molecules, such as phosphorus molecules which contain four atoms (P_4).

Table 14.1 Formulae of some common elements.

Common elements whose formula is the symbol				Common elements whose formula is not the symbol			
Al	aluminium	Pb	lead	Br_2	bromine	I_2	iodine
Ar	argon	Li	lithium	C_{60}	carbon (buckminsterfullerene)	N_2	nitrogen
Be	beryllium	Mg	magnesium	Cl_2	chlorine	O_2	oxygen
B	boron	Ne	neon	F_2	fluorine	P_4	phosphorus
Ca	calcium	Ni	nickel	H_2	hydrogen	S_8	sulfur
C	carbon (diamond)	K	potassium				
C	carbon (graphite)	Si	silicon				
Cu	copper	Ag	silver				
Au	gold	Na	sodium				
He	helium	Sn	tin				
Fe	iron	Zn	zinc				

○ Compounds

Some common compounds

It is very useful to know the formula of some common compounds. Some are listed in Table 14.2.

Table 14.2 Formulae of some common compounds.

Common compounds	
NH_3	ammonia
CO_2	carbon dioxide
CO	carbon monoxide
CH_4	methane
NO	nitrogen monoxide
NO_2	nitrogen dioxide
SO_2	sulfur dioxide
SO_3	sulfur trioxide
H_2O	water

Ionic compounds

Compounds made from metals combined with non-metals have an ionic structure. The formula of each of these compounds can be worked out using ion charges. The charges of common ions are shown in Tables 14.3 and 14.4.

TIP

You should be able to work out the charge of ions of elements in Groups 1, 2, 6 and 7.

Table 14.3 Positive ions.

Group 1 ions (form 1+ ions)		Group 2 ions (form 2+ ions)		Group 3 ions (form 3+ ions)		Others	
Li^+	lithium	Mg^{2+}	magnesium	Al^{3+}	aluminium	NH_4^+	ammonium
Na^+	sodium	Ca^{2+}	calcium			Cu^{2+}	copper(II)
K^+	potassium	Ba^{2+}	barium			H^+	hydrogen
						Fe^{2+}	iron(II)
						Fe^{3+}	iron(III)
						Pb^{2+}	lead
						Ag^+	silver
						Zn^{2+}	zinc

Table 14.4 Negative ions.

Group 6 ions (form 2– ions)		Group 7 ions (form 1– ions)		Others	
O^{2-}	oxide	F^-	fluoride	CO_3^{2-}	carbonate
S^{2-}	sulfide	Cl^-	chloride	OH^-	hydroxide
		Br^-	bromide	NO_3^-	nitrate
		I^-	iodide	SO_4^{2-}	sulfate

In an ionic substance the total number of positive charges must equal the total number of negative charges. This allows us to work out the formula of ionic substances.

Examples

Sodium oxide: contains sodium ions (Na^+) and oxide ions (O^{2-})

There must be the same number of positive and negative charges, so we need two Na^+ ions (total of two positive charges) for every one O^{2-} ion (two negative charges)	Na^+ Na^+	O^{2-}	Formula = Na_2O

Iron(III) sulfide: contains iron(III) ions (Fe^{3+}) and sulfide ions (S^{2-})

There must be the same number of positive and negative charges, so we need two Fe^{3+} ions (total of six positive charges) for every three S^{2-} ions (six negative charges)	Fe^{3+} Fe^{3+}	S^{2-} S^{2-} S^{2-}	Formula = Fe_2S_3

TIP

The charge on the nitrate (NO_3^-) ion is 1–, not 3–. The charge on the ammonium (NH_4^+) ion is 1+ not 4+.

Some ions contain atoms of different elements. Examples include sulfate (SO_4^{2-}), hydroxide (OH^-) and nitrate (NO_3^-). These are sometimes called compound ions or molecular ions. If you need to write more than one of these in a formula, then these ions should be placed in a bracket.

Example

Magnesium hydroxide: contains magnesium ions (Mg^{2+}) and hydroxide ions (OH^-)

There must be the same number of positive and negative charges, so we need one Mg^{2+} ion (total of two positive charges) for every two OH^- ions (two negative charges)	Mg^{2+}	OH^- OH^-	Formula = $Mg(OH)_2$

Table 14.5 Ions of some common compounds.

Name	+ Ions		− Ions		Formula
Sodium chloride	Na^+	(1+ charge)	Cl^-	(1− charge)	$NaCl$
Magnesium chloride	Mg^{2+}	(2+ charges)	Cl^- Cl^-	(2− charges)	$MgCl_2$
Magnesium sulfide	Mg^{2+}	(2+ charges)	S^{2-}	(2− charges)	MgS
Copper(ii) sulfate	Cu^{2+}	(2+ charges)	SO_4^{2-}	(2− charges)	$CuSO_4$
Sodium carbonate	Na^+ Na^+	(2+ charges)	CO_3^{2-}	(2− charges)	Na_2CO_3
Ammonium sulfate	NH_4^+ NH_4^+	(2+ charges)	SO_4^{2-}	(2− charges)	$(NH_4)_2SO_4$
Calcium nitrate	Ca^{2+}	(2+ charges)	NO_3^- NO_3^-	(2− charges)	$Ca(NO_3)_2$
Aluminium oxide	Al^{3+} Al^{3+}	(6+ charges)	O^{2-} O^{2-} O^{2-}	(6− charges)	Al_2O_3
Iron(iii) hydroxide	Fe^{3+}	(3+ charges)	OH^- OH^- OH^-	(3− charges)	$Fe(OH)_3$

Test yourself

1 Write the formula of each of the following elements and compounds.
- a) copper
- b) hydrogen
- c) carbon dioxide
- d) argon
- e) silver
- f) oxygen
- g) ammonia
- h) chlorine
- i) carbon (diamond)
- j) carbon (buckminsterfullerene)
- k) sulfur dioxide
- l) methane

2 Write the formula of each of the following ionic compounds.
- a) potassium oxide
- b) sodium sulfate
- c) aluminium fluoride
- d) iron(iii) sulfide
- e) copper(ii) nitrate
- f) lithium carbonate
- g) ammonium bromide
- h) barium hydroxide
- i) silver nitrate
- j) aluminium sulfate
- k) strontium oxide
- l) potassium selenide

3 Name each of the following substances.
- a) Br_2
- b) Na
- c) Cu
- d) CO
- e) SO_3
- f) CaO
- g) AlF_3
- h) CuS
- i) KNO_3
- j) $(NH_4)_2CO_3$
- k) FeO
- l) Fe_2O_3

Classifying substances

○ Structure types

It is very useful to be able to identify what type of structure a substance has from its name or formula. Table 14.6 gives some general guidance on this.

Table 14.6 Types of structures.

Structure type	Description of structure	Which substances have this structure
Monatomic	Made of individual atoms	Group 0 elements
Molecular	Made of individual molecules	Some non-metal elements (e.g. H_2, C_{60}, N_2, O_2, F_2, P_4, Cl_2, Br_2, I_2) Compounds made from non-metals (e.g. CH_4, CO_2, H_2O, NH_3, $C_6H_{12}O_6$)
Giant covalent	Lattice of atoms joined by covalent bonds	Diamond (C), graphite (C), graphene (C), silicon (Si), silicon dioxide (SiO_2)
Ionic	Lattice of positive and negative ions	Compounds made from metals combined with non-metals (e.g. NaCl, Fe_2O_3, $CuSO_4$)
Metallic	Lattice of metal atoms in a cloud of delocalised outer shell electrons	Metals (e.g. Cu, Fe, Al, Na, Ca, Mg, Au, Ag, Pt)

○ Acids, bases, alkalis and salts

Some compounds act as acids, bases, alkalis or salts. It is very useful if you can identify an acid, base, alkali or salt although not all substances are one of these (Table 14.7).

Table 14.7 Acids, bases, alkalis and salts.

Acids	Bases	Salts
Substances that react with water to release H^+ ions	Substances that react with acids to form a salt and water (and sometimes carbon dioxide as well)	Ionic substances formed when acids react with bases
Common acids: H_2SO_4 sulfuric acid HCl hydrochloric acid HNO_3 nitric acid H_3PO_4 phosphoric acid CH_3COOH ethanoic acid	*Common bases:* Metal oxides e.g. CaO, Na_2O Metal hydroxides e.g. $Ca(OH)_2$, NaOH Metal carbonates e.g. $CaCO_3$, Na_2CO_3 **Alkalis** Substances that react with water to release OH^- ions (they are a special type of water-soluble base) *Common alkalis:* NH_3 ammonia *plus water-soluble metal hydroxides:* NaOH sodium hydroxide KOH potassium hydroxide $Ca(OH)_2$ calcium hydroxide	*Common salts:* Sulfates from sulfuric acid Chlorides from hydrochloric acid Nitrates from nitric acid Phosphates from phosphoric acid Ethanoates from ethanoic acid Citrates from citric acid

Alkalis are a special type of base and so any substance that is an alkali is also a base.

○ Acid–base character of oxides

Most metal oxides are basic (Table 14.8). For example, calcium oxide (CaO) is used as a base to neutralise acidic soil on farms.

Most non-metal oxides are acidic. For example, carbon dioxide (CO_2) dissolves in rain water to make rain naturally slightly acidic.

Table 14.8 Oxides.

Type of oxide	Metal oxides	Non-metal oxides
Acidic or basic	Basic (react with acids)	Acidic (react with bases)
Examples	Calcium oxide (CaO) Sodium oxide (Na_2O) Copper oxide (CuO)	Carbon dioxide (CO_2) Sulfur dioxide (SO_2) Phosphorus oxide (P_4O_{10})

Test yourself

4 What type of structure does each of the following substances have?
  a) lead (Pb)
  b) argon (Ar)
  c) potassium iodide (KI)
  d) oxygen (O_2)
  e) diamond (C)
  f) methane (CH_4)
  g) ethanol (C_2H_5OH)
  h) aluminium oxide (Al_2O_3)
  i) chromium (Cr)
  j) silicon dioxide (SiO_2)
  k) sulfur dioxide (SO_2)
  l) potassium nitrate (KNO_3)

5 Classify each of these substances as an acid, base, alkali or salt.
  a) Fe_2O_3
  b) Na_2SO_4
  c) KOH
  d) $ZnCO_3$
  e) HNO_3
  f) $Ca(NO_3)_2$
  g) NH_3
  h) K_2O
  i) HCl
  j) $MgCl_2$
  k) NaBr
  l) H_2SO_4

6 Classify each of the following oxides as acidic or basic.
  a) NO_2
  b) K_2O
  c) MgO
  d) SiO_2

Common reactions

There are some general reactions that are useful to know. Many of these involve acids and/or metals. Remember that hydrochloric acid produces chloride salts, sulfuric acid produces sulfate salts and nitric acid produces nitrate salts.

Examples

element + oxygen → oxide of element
  e.g. calcium + oxygen → calcium oxide

compound + oxygen → oxides of each element in compound
  e.g. methane (CH_4) + oxygen → carbon dioxide + water

water + metal → metal hydroxide + hydrogen (for metals that react with water)
  e.g. water + sodium → sodium hydroxide + hydrogen

acid + metal → salt + hydrogen (for metals that react with dilute acids)
  e.g. hydrochloric acid + magnesium → magnesium chloride + hydrogen

acid + metal oxide → salt + water
  e.g. sulfuric acid + copper oxide → copper sulfate + water

acid + metal hydroxide → salt + water
  e.g. nitric acid + potassium hydroxide → potassium nitrate + water

acid + metal carbonate → salt + water + carbon dioxide
  e.g. hydrochloric acid + calcium carbonate → calcium chloride + water + carbon dioxide

acid + ammonia → ammonium salt
  e.g. nitric acid + ammonia → ammonium nitrate

Chemical reactions

Chemical reactions take place in one of the three ways shown in Table 14.9.

Table 14.9 Chemical reactions.

Way in which the reaction takes place	Examples
1 Transfer of electrons	**Metals reacting with non-metals** e.g. sodium + chlorine → sodium chloride $2Na + Cl_2 \rightarrow 2NaCl$ Sodium atoms lose electrons to form sodium ions. These electrons are transferred to chlorine atoms which form chloride ions. This forms the ionic compound sodium chloride **Displacement reactions** e.g. zinc + copper sulfate → zinc sulfate + copper $Zn + CuSO_4 \rightarrow ZnSO_4 + Cu$ Zinc atoms lose electrons to form zinc ions. These electrons are transferred to copper ions in copper sulfate forming copper atoms.
2 Sharing of electrons	**Non-metals reacting with non-metals** e.g. hydrogen + oxygen → water $H_2 + O_2 \rightarrow 2H_2O$ Hydrogen atoms share electrons with oxygen atoms to form covalent bonds in water
3 Transfer of protons	All acids contain H^+ ions. An H^+ ion is simply a proton. When acids react, they transfer H^+ ions to the substance they react with **Acids reacting with alkalis** e.g. hydrochloric acid + sodium hydroxide → sodium chloride + water $HCl + NaOH \rightarrow NaCl + H_2O$ H^+ ions (protons) are transferred from the hydrochloric acid to the OH^- ions in sodium hydroxide to form water **Acids reacting with metal oxides** e.g. sulfuric acid + copper oxide → copper sulfate + water $H_2SO_4 + CuO \rightarrow CuSO_4 + H_2O$ H^+ ions (protons) are transferred from the sulfuric acid to the O^{2-} ions in copper oxide to form water

Test yourself

7 Write a word equation for each of the following reactions with oxygen.
  a) magnesium (Mg) + oxygen
  b) hydrogen sulfide (H_2S) + oxygen
  c) phosphorus (P_4) + oxygen
  d) silane (SiH_4) + oxygen
  e) propane (C_3H_8) + oxygen
  f) methanol (CH_3OH) + oxygen

8 Write a word equation for each of the following reactions.
  a) potassium + water
  b) nitric acid + zinc
  c) sulfuric acid + nickel oxide
  d) hydrochloric acid + potassium hydroxide
  e) nitric acid + sodium carbonate
  f) hydrochloric acid + ammonia
  g) magnesium hydroxide + sulfuric acid
  h) calcium + water
  i) copper carbonate + nitric acid
  j) ammonia + sulfuric acid
  k) magnesium oxide + nitric acid
  l) cobalt + hydrochloric acid

9 Which of the following reactions involves the
  • transfer of electrons?
  • sharing of electrons?
  • transfer of protons?
  a) nitric acid + sodium oxide → sodium nitrate + water
  b) aluminium + iron oxide → aluminium oxide + iron
  c) hydrogen + sulfur → hydrogen sulfide
  d) aluminium + bromine → aluminium bromide
  e) hydrochloric acid + sodium carbonate → sodium chloride + water + carbon dioxide

Balancing equations

Word equations show the names of the reactants and products in a reaction. A balanced symbol (formula) equation shows the formula of each substance and how many particles of each are involved in the reaction. An example of this is shown in Table 14.10.

Table 14.10 Word equations and balanced symbol (formula) equations.

Type of equation	Word equation	Balanced equation
Equation	nitrogen + hydrogen $\rightarrow$ ammonia	$N_2 + 3H_2 \rightarrow 2NH_3$
What it tells us	Nitrogen reacts with hydrogen to form ammonia	One molecule of nitrogen (N_2) reacts with three molecules of hydrogen (H_2) to form two molecules of ammonia (NH_3)

In a balanced equation, the total number of atoms of each element on both sides of the equation must be the same. This is because atoms cannot be created or destroyed. In the equation for the reaction between nitrogen and hydrogen above, there are two nitrogen atoms and six hydrogen atoms in both the reactants and products.

You are often required to write a balanced equation. Here are some steps to follow plus two examples.

Step 1 Write the word equation.

Step 2 Rewrite the equation with formulae (be very careful to ensure the formulae are correct).

Step 3 Count the number of atoms of each element on each side of the equation. If they are the same then the equation is already balanced and nothing more needs to be done.

Step 4 If the equation is not balanced, then add in extra molecules to try and balance it. You must never change the formulae themselves. For example, you could not change the formula of water from H_2O to H_4O in example 1 to balance the H atoms because it is water that is formed and that has the formula H_2O and not H_4O.

Step 5 Write out the final balanced equation.

> **TIP**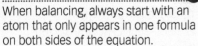
>
> When balancing, always start with an atom that only appears in one formula on both sides of the equation.

Examples

Table 14.11

	Example 1	Example 2
Step 1	methane + oxygen → carbon dioxide + water	aluminium hydroxide + nitric acid → aluminium nitrate + water
Step 2	$CH_4 + O_2 → CO_2 + H_2O$	$Al(OH)_3 + HNO_3 → Al(NO_3)_3 + H_2O$
Step 3	reactants products C = 1 C = 1 H = 4 H = 2 O = 2 O = 3 The equation is not balanced	reactants products Al = 1 Al = 1 O = 6 O = 10 H = 4 H = 2 N = 1 N = 3 The equation is not balanced
Step 4	Add another H_2O to the products (so there are now $2H_2O$) to balance the H atoms: $CH_4 + O_2 → CO_2 + 2H_2O$ reactants products C = 1 C = 1 H = 4 H = 4 O = 2 O = 4 Then add another O_2 to the reactants (so there are now $2O_2$) to balance the O atoms: $CH_4 + 2O_2 → CO_2 + 2H_2O$ reactants products C = 1 C = 1 H = 4 H = 4 O = 4 O = 4 The equation is now balanced	It is easiest to start with N as it is the only unbalanced atom that is in just one substance on both sides of the equation Add two more HNO_3 to the reactants (so there are now $3HNO_3$) to balance the N atoms $Al(OH)_3 + 3HNO_3 → Al(NO_3)_3 + H_2O$ reactants products Al = 1 Al = 1 O = 12 O = 10 H = 6 H = 2 N = 3 N = 3 Then add two more H_2O to the products (so there are now $3H_2O$) to balance the O and H atoms $Al(OH)_3 + 3HNO_3 → Al(NO_3)_3 + 3H_2O$ reactants products Al = 1 Al = 1 O = 12 O = 12 H = 6 H = 6 N = 3 N = 3 The equation is now balanced
Step 5	$CH_4 + 2O_2 → CO_2 + 2H_2O$	$Al(OH)_3 + 3HNO_3 → Al(NO_3)_3 + 3H_2O$

Balanced equations sometimes include state symbols to show the state of each substance:

- (s) = solid
- (l) = liquid
- (g) = gas
- (aq) = aqueous (dissolved in water).

For example, the equation:

$$CaCO_3(s) + 2HCl(aq) → CaCl_2(aq) + H_2O(l) + CO_2(g)$$

means that calcium carbonate solid reacts with an aqueous solution of hydrochloric acid to form an aqueous solution of calcium chloride, water liquid and carbon dioxide gas.

Test yourself

10 Magnesium reacts with sulfuric acid as shown:

$Mg(s) + H_2SO_4(aq) \rightarrow MgSO_4(aq) + H_2(g)$

    a) What does the (s) mean?   b) What does the (aq) mean?

    c) What does the (g) mean?

11 Balance the following equations.

    a) $K + I_2 \rightarrow KI$ b) $Na + H_2O \rightarrow NaOH + H_2$

    c) $CuCO_3 \rightarrow CuO + CO_2$ d) $Al + O_2 \rightarrow Al_2O_3$

    e) $Ca + HCl \rightarrow CaCl_2 + H_2$ f) $KOH + H_2SO_4 \rightarrow K_2SO_4 + H_2O$

    g) $C_5H_{12} + O_2 \rightarrow CO_2 + H_2O$ h) $H_3PO_4 + NaOH \rightarrow Na_3PO_4 + H_2O$

    i) $NH_3 + H_2SO_4 \rightarrow (NH_4)_2SO_4$ j) $NO + H_2O + O_2 \rightarrow HNO_3$

12 Write a balanced equation for each of the following reactions.

    a) sodium + oxygen → sodium oxide

    b) propane (C_3H_8) + oxygen → carbon dioxide + water

    c) calcium + water → calcium hydroxide + hydrogen

    d) chlorine + sodium bromide → sodium chloride + bromine

    e) magnesium oxide + nitric acid → magnesium nitrate + water

Ionic equations

In a solid ionic compound, the positive and negative ions are bonded to each other strongly in a lattice. When it dissolves in water, the ions separate and become surrounded by water molecules (Figure 14.1).

When ionic compounds dissolved in water react, it is usual for some of the ions not to react and remain unchanged in the water. These are often called spectator ions as they are present but do not take part in the reaction.

We can write ionic equations for reactions involving ions. These ionic equations only show what happens to the ions that react. We do not include the spectator ions in these ionic equations.

> ### KEY TERMS
>
> **Spectator ions** Ions that do not take part in a reaction and do not appear in the ionic equation for the reaction.
>
> **Ionic equation** Balanced equation for reaction that omits any spectator ions.

The overall electric charge of the ions in the reactants must equal the overall electric charge of the ions in the products. This can sometimes be useful to help you check that the ionic equation is balanced or actually to help you balance the ionic equation.

Reaction of acids with alkalis

When sulfuric acid reacts with sodium hydroxide solution, it is only the hydrogen ions from the sulfuric acid and the hydroxide ions from the sodium hydroxide that react. These hydrogen ions and hydroxide ions

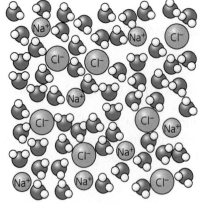

NaCl(s)   **NaCl(aq)**

▲ **Figure 14.1** Sodium chloride (NaCl) as a solid and when dissolved in water.

react to form water (Figure 14.2). The sulfate ions and sodium ions remain unchanged as they do not react and are left out of the ionic equation as they are spectator ions.

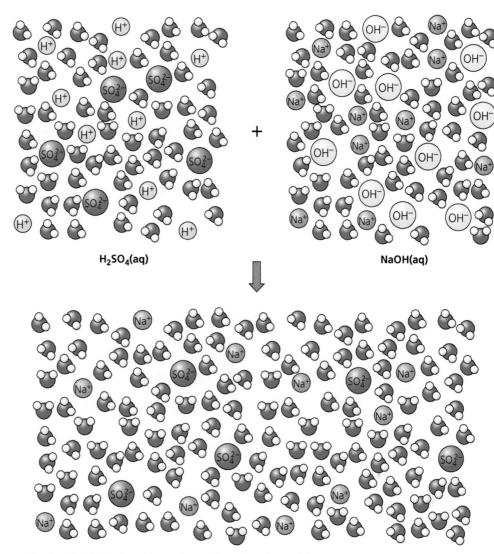

▲ **Figure 14.2** Sulfuric acid reacting with sodium hydroxide.

The ionic equation for this reaction can be written as:

$$H^+(aq) \quad + \quad OH^-(aq) \quad \rightarrow H_2O(l)$$
from the acid   from the alkali

When any acid reacts with any alkali, the ionic equation is the same. Some examples are shown in Table 14.12.

Table 14.12 Reactions between some acids and alkalis.

Examples	What reacts	Ions that do not react	Ionic equation
sulfuric acid (aq) + sodium hydroxide (aq)	H^+ ions from H_2SO_4 OH^- ions from NaOH	SO_4^{2-} ions from H_2SO_4 Na^+ ions from NaOH	$H^+(aq) + OH^-(aq) \rightarrow H_2O(l)$
hydrochloric acid (aq) + potassium hydroxide (aq)	H^+ ions from HCl OH^- ions from KOH	Cl^- ions from HCl K^+ ions from KOH	$H^+(aq) + OH^-(aq) \rightarrow H_2O(l)$
nitric acid (aq) + calcium hydroxide (aq)	H^+ ions from HNO_3 OH^- ions from $Ca(OH)_2$	NO_3^- ions from HNO_3 Ca^{2+} ions from $Ca(OH)_2$	$H^+(aq) + OH^-(aq) \rightarrow H_2O(l)$

Displacement reactions

Ionic equations can also be written for displacement reactions that take place when a more reactive metal displaces a less reactive metal from a metal compound.

For example, when zinc reacts with copper sulfate solution, the zinc atoms in the zinc metal react with the copper ions in the copper sulfate (Figure 14.3). The sulfate ions are spectator ions and so do not appear in the ionic equation.

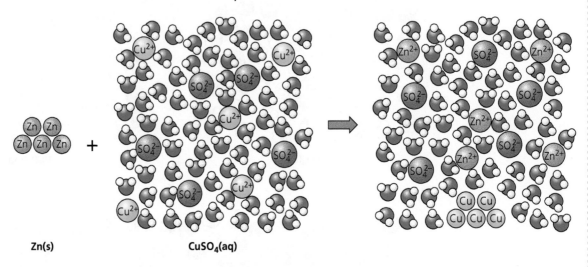

Zn(s)   CuSO$_4$(aq)

▲ **Figure 14.3** Zinc reacting with copper sulfate solution.

The ionic equation for this reaction can be written as:

$$Zn(s) \quad + \quad Cu^{2+}(aq) \quad \rightarrow \quad Zn^{2+}(aq) + Cu(s)$$
from the
copper sulfate

This and some other examples of displacement reactions are shown in Table 14.13. In these reactions it helps to use the electric charges of the ions to balance the equation. For example, in the second example in the table, two Ag$^+$ ions are needed giving an overall 2+ charge on the left side of the equation to balance with the 2+ charge on the Fe^{2+} ion on the right side of the equation.

Table 14.13 Some displacement reactions.

Examples	What reacts	Ions that do not react	Ionic equation
zinc (s) + copper sulfate (aq)	Zn atoms in Zn metal Cu^{2+} ions from CuSO$_4$	SO$_4^{2-}$ ions from CuSO$_4$	Zn(s) + Cu^{2+}(aq) → Zn^{2+}(aq) + Cu(s)
iron (s) + silver nitrate (aq)	Fe atoms in Fe metal Ag$^+$ ions from AgNO$_3$	NO$_3^-$ ions from AgNO$_3$	Fe(s) + 2Ag$^+$(aq) → Fe^{2+}(aq) + 2Ag(s)
aluminium(s) + copper chloride (aq)	Al atoms in Al metal Cu^{2+} ions from CuCl$_2$	Cl$^-$ ions from CuCl$_2$	2Al(s) + 3Cu^{2+}(aq) → 2Al^{3+}(aq) + 3Cu(s)

Test yourself

13 Write an ionic equation (including state symbols) for each of the following reactions between acids and alkalis.
- **a)** hydrochloric acid (aq) + sodium hydroxide (aq)
- **b)** nitric acid (aq) + potassium hydroxide (aq)
- **c)** sulfuric acid (aq) + calcium hydroxide (aq)
- **d)** phosphoric acid (aq) + sodium hydroxide (aq)

14 Write an ionic equation (including state symbols) for each of the following displacement reactions.
- **a)** Displacement of copper from copper(ii) sulfate (aq) by magnesium.
- **b)** Displacement of silver from silver nitrate (aq) by magnesium.
- **c)** Displacement of zinc from zinc sulfate (aq) by aluminium.

15 Write an ionic equation (including state symbols) for each of the following reactions.
- **a)** Reaction of barium hydroxide (aq) with hydrochloric acid (aq).
- **b)** Displacement of silver from silver nitrate (aq) by zinc (s).
- **c)** Reaction of sulfuric acid (aq) with potassium hydroxide (aq).
- **d)** Displacement of nickel from nickel(ii) sulfate (aq) by zinc (s).

Half equations

Many chemical reactions involve the transfer of electrons and half equations can be written for these reactions. These equations show the number of electrons that are gained or lost.

In these half equations:

- positive ions gain electrons
  (e.g. a 3+ ion gains 3 electrons, e.g. $Al^{3+} + 3e^- \rightarrow Al$)
- negative ions lose electrons
  (e.g. a 2– ion loses 2 electrons, e.g. $S^{2-} - 2e^- \rightarrow S$).

Half equations for the loss of electrons can be written in two ways. The electrons can be shown being taken away from the left-hand side, or shown on the right-hand side with the products.

For example, the half equation for the loss of electrons from S^{2-} ions to form S can be written as:

$$S^{2-} - 2e^- \rightarrow S \qquad \text{or} \qquad S^{2-} \rightarrow S + 2e^-$$

Some elements contain diatomic molecules (e.g. H_2, O_2, Cl_2, Br_2, I_2). When balancing half equations that produce these elements, two ions are needed to make one diatomic molecule.

For example, when H_2 is formed from H^+ ions, two H^+ ions are needed which both gain one electron and so two electrons are gained altogether:

$$2H^+ + 2e^- \rightarrow H_2$$

When O_2 is formed from O^{2-} ions, two O^{2-} ions are needed which both lose two electrons and so four electrons are lost altogether:

$$2O^{2-} - 4e^- \rightarrow O_2 \quad \text{or} \quad 2O^{2-} \rightarrow O_2 + 4e^-$$

In half equations, the total electric charge on the left-hand side must equal the total electric charge on the right-hand side of the equation. This can be used to check that the half equation is balanced.

In this example, both the left and right-hand sides of the equation add up to the same overall charge (which is 0 in this case):

$$Al^{3+} + 3e^- \quad \rightarrow \quad Al$$

total charge: $\quad \underbrace{3+ \quad 3-}_{0} \quad\quad\quad 0$

left-hand side $\quad\quad$ right-hand side

Oxidation takes place when a substance loses electrons. Reduction takes place when a substance gains electrons. See the section on oxidation and reduction in Chapter 12. One way to remember whether electrons are lost or gained in oxidation and reduction is the phrase OIL RIG (see Figure 12.7).

Electrolysis

Half equations can be written for the reactions that take place at each electrode in electrolysis. Some examples are shown in Table 14.14.

Table 14.14

Substance (in molten state)	Negative electrode reaction		Positive electrode reaction	
	What happens	Half equation	What happens	Half equation
Copper sulfide	Cu^{2+} ions gain electrons to form Cu	$Cu^{2+} + 2e^- \rightarrow Cu$	S^{2-} ions lose electrons to form S	$S^{2-} - 2e^- \rightarrow S$ (or $S^{2-} \rightarrow S + 2e^-$)
Sodium chloride	Na^+ ions gain electrons to form Na	$Na^+ + e^- \rightarrow Na$	Cl^- ions lose electrons to form Cl_2	$2Cl^- - 2e^- \rightarrow Cl_2$ (or $2Cl^- \rightarrow Cl_2 + 2e^-$)
Aluminium oxide	Al^{3+} ions gain electrons to form Al	$Al^{3+} + 3e^- \rightarrow Al$	O^{2-} ions lose electrons to form O_2	$2O^{2-} - 4e^- \rightarrow O_2$ (or $2O^{2-} \rightarrow O_2 + 4e^-$)

Displacement reactions

Two half equations can be written for displacement reactions.

For example, in the reaction where zinc displaces copper from copper sulfate solution the overall ionic equation is:

$$Zn + Cu^{2+} \rightarrow Zn^{2+} + Cu$$

The two half equations for this are:

Zn atoms lose electrons to form Zn^{2+} ions:

$$Zn - 2e^- \rightarrow Zn^{2+} \text{ (or } Zn \rightarrow Zn^{2+} + 2e^-)$$

Cu^{2+} ions in $CuSO_4$ gain electrons to form Cu atoms:

$$Cu^{2+} + 2e^- \rightarrow Cu$$

For example, in the reaction when copper displaces silver from silver nitrate solution the overall ionic equation is:

$$Cu + 2Ag^+ \rightarrow Cu^{2+} + 2Ag$$

The two half equations for this are:

Cu atoms lose electrons to form Cu^{2+} ions:

$$Cu - 2e^- \rightarrow Cu^{2+} \text{ (or } Cu \rightarrow Cu^{2+} + 2e^-)$$

Ag^+ ions in $AgNO_3$ gain electrons to form Ag atoms:

$$Ag^+ + e^- \rightarrow Ag$$

Test yourself

16 Write a balanced half equation for each of the following conversions.
   a) $Mg^{2+} \rightarrow Mg$
   b) $Se^{2-} \rightarrow Se$
   c) $K^+ \rightarrow K$
   d) $Br^- \rightarrow Br_2$
   e) $O^{2-} \rightarrow O_2$
   f) $H^+ \rightarrow H_2$

17 Write two half equations to show what happens in the following displacement reactions.
   a) displacement of copper from copper(II) sulfate (aq) by magnesium
   b) displacement of silver from silver nitrate (aq) by magnesium
   c) displacement of zinc from zinc sulfate (aq) by aluminium

15 Energy

We are concerned that energy reserves are running out. What does the future hold? How will we generate electricity for the projected world population of 10 billion people in 2050?

Specification coverage

This chapter covers specification points 6.1.1.1 to 6.1.3 and is called Energy.

It covers energy changes in a system, the ways energy is stored before and after such changes, conservation and dissipation of energy and national and global energy resources.

Prior knowledge

Previously you could have learnt:

> Our primary source of energy is the Sun.
> Food provides us with energy to live.
> Energy reaches us from the Sun in the form of electromagnetic radiation.
> Fossil fuels are a source of energy.
> We generate electricity using fossil fuels and other sources of energy.
> Energy can be transferred from a hot object to colder objects.
> Metals are good thermal conductors. They allow energy to be transferred quickly.
> Fluids (liquids and gases) transfer energy by convection.
> Energy is measured in joules, J.

Test yourself on prior knowledge

1 State three examples of how you use energy every day.
2 Give an example of a fossil fuel.
3 Why are metals good thermal conductors?

Energy changes in a system, and the ways energy is stored before and after such changes

○ Energy stores and systems

We can begin to understand energy by studying changes in the way energy is stored when a **system** changes. A 'system' is an object or a group of objects that interact. Here are some situations with which you should be familiar.

● **Throwing an object upwards**
When you throw a ball upwards, just after the ball leaves your hand it has a store of kinetic energy. When the ball reaches its highest point, it has a store of gravitational potential energy. Just before you catch it again, it has a store of kinetic energy.

● **Boiling water in a kettle**
When you turn on your electric kettle, the water in the kettle gets hotter. There is now more internal (or thermal) energy stored in the hot water than there was in the cold water.

● **Burning coal**
When we burn coal there is a chemical reaction. Coal has a store of chemical energy which is transferred to thermal energy as it burns. A coal fire can warm up a room.

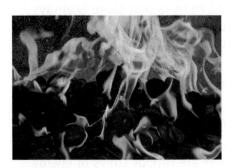

▲ Figure 15.1 One store of energy is transferred to another store.

- **A car using its brakes to slow down**
  A moving car has a store of kinetic energy. When the car slows to a halt, it has lost this store of kinetic energy. The brakes exert a frictional force on the wheels, and the brakes get hot. The store of kinetic energy in the car has been transferred to a store of thermal energy in the brakes. This energy is then transferred to the surroundings.

- **Dropping an object which does not bounce**
  Just before the object hits the ground, it has a store of kinetic energy. After the object has stopped moving, the kinetic energy has been transferred to a store of internal energy in the object and the surroundings. So the object and the surroundings warm up a little. (You might hear a noise, but the energy carried by the sound is also transferred to the internal energy of the surroundings.)

- **Accelerating a ball with a constant force**
  We have a store of chemical potential energy in our muscles. When we throw a ball, our store of chemical potential energy decreases, and the ball's store of kinetic energy increases. The hand applies a force to the ball and does work to accelerate it.

Energy stores

In the simple everyday events and processes that were described above, we identified objects that had gained or lost energy. For example, objects slow down or get hotter. We saw that the way energy is stored changes.

We use the following labels to describe the stores of energy you will meet:

- kinetic
- chemical
- internal (or thermal)
- gravitational potential

- magnetic
- elastic potential
- nuclear.

	gravitational energy store	kinetic energy store
(b) ball stationary above the ground	100 J	0
(a) ball moving upwards	0	100 J

▲ **Figure 15.2** A ball is thrown upwards from the ground. In (a) the ball has 100 J of kinetic energy and zero gravitational potential energy. In (b) the ball has 100 J of gravitational potential energy and zero kinetic energy.

Counting the energy

Energy is a quantity that is measured in joules, J. Large quantities of energy are measured in kilojoules, kJ, and megajoules, MJ.

$$1\,kJ = 1000\,J\ (10^3\,J) \qquad 1\,MJ = 1\,000\,000\,J\ (10^6\,J)$$

The reason that energy is so important to us is that there is always the **same energy at the end** of a process **as there was at the beginning**. If we add up the total energy in all the stores, that number stays the same.

Figures 15.2 and 15.3 show some examples of counting the energy.

The principle of conservation of energy

The **principle of conservation of energy** states that the **amount of energy always remains the same**. There are various stores of energy. In any process energy can be transferred from one store to another, but energy cannot be destroyed or created.

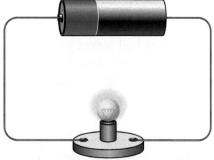

chemical energy store reduces by 200 J

internal energy store of the lamp and surroundings increases by 200 J

▲ **Figure 15.3** A cell passes a charge through a lamp. The charge flows for a few minutes. After this time the store of chemical energy in the cell has decreased by 200 J. The store of internal energy in the lamp and the surroundings has increased by 200 J.

Transferring energy from one store to another

Light, sound and electricity are useful, but they are not stores of energy. They are ways of transferring energy from one store to a different energy store. You cannot go into a shop to buy a box of 'electrical energy', but you can buy a cell or battery. In a circuit, the chemical energy stored in a cell or battery causes electric charge to flow.

In a torch, the chemical energy stored in the battery causes an electric current (a flow of charge). The electric current causes the temperature of the bulb to increase so much that the bulb lights up. The light cannot be stored but it is useful. When the light strikes an object and is absorbed, the internal energy of the object increases.

If we drop a bunch of keys onto a table, the collision will make the air vibrate and we hear a sound. The sound wave transfers energy; it is not an energy store. The energy will transfer to the air and surrounding objects causing an increase in their store of internal energy.

Test yourself

1 Describe the energy stored in each of the following:
   a) a moving bicycle
   b) a compressed spring
   c) a bowl of breakfast cereal
   d) a rock lifted off the ground.
2 Explain what is wrong with this statement:
   'A car battery stores electrical energy for the lights, horn and starter motor.'
3 Describe how the stores of energy change from the beginning to the end of the following processes.
   a) A catapult launches a marble.
   b) A ball rolls along the ground and comes to rest.
   c) A butane gas camping cooker heats up a pan of water.
   d) A lump of soft putty falls to the ground.
4 Figure 15.4 shows a ball falling. Copy the diagram and fill in the values for the ball's kinetic energy and gravitational potential energy at each height.

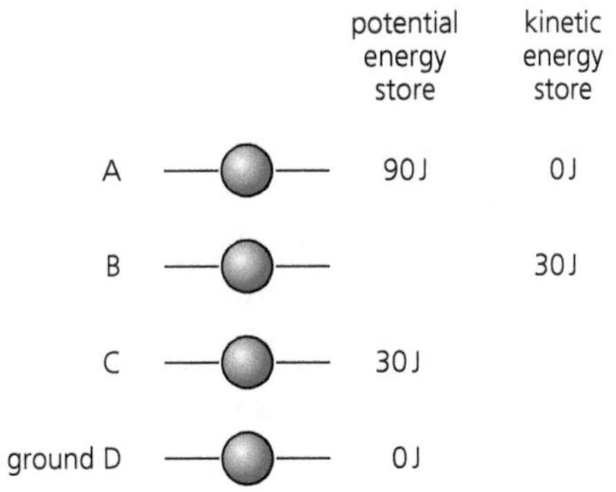

	potential energy store	kinetic energy store
A	90 J	0 J
B		30 J
C	30 J	
ground D	0 J	

▲ Figure 15.4

Show you can...

Complete this task to show you understand the different stores of energy.

Name four stores of energy, and describe three examples of how energy can be transferred from one store to another.

○ **Calculating the energy**

In this section you will learn how to calculate the amount of energy associated with a moving object, a stretched spring and an object raised above the ground. These calculations are useful to us. For example, we can show how the energy in a system is redistributed when a change happens to the system.

Kinetic energy

The kinetic energy stored by a moving object can be calculated using the equation:

TIP
You need to be able to **recall** and/or **apply** the equation:
$$E_k = \frac{1}{2}mv^2$$

> $$\textbf{kinetic energy} = \frac{1}{2} \times \textbf{mass} \times \textbf{(speed)}^2$$
> Where energy is in joules, J
>   mass is in kilograms, kg
>   speed is in metres per second, m/s.

Elastic potential energy

The amount of elastic potential energy stored in a stretched spring can be calculated using the equation:

TIP
- You need to be able to apply the equation:
$$E_e = \frac{1}{2}ke^2$$
It is given on the physics equation sheet.
- You need to be aware of which equations to learn.

> $$\textbf{elastic potential energy} = \frac{1}{2} \times \textbf{spring constant} \times \textbf{(extension)}^2$$
> Where energy is in joules, J
>   spring constant is in newtons per metre, N/m
>   extension is in metres, m.

The spring constant, k, is a measure of the spring's stiffness; k is equal to the force needed to stretch the spring one metre.

The extension of the spring is the increase in its length from its original unstretched length.

Gravitational potential energy

The amount of gravitational potential energy gained by an object raised above ground level can be calculated using the equation:

TIP
You are not expected to know the gravitational field strength (g) when carrying out calculations.

> $$E_p = m\,g\,h$$
> **gravitational** = **mass × gravitational field strength × height**
> **potential energy**
> Where energy is in joules, J
>   mass is in kilograms, kg
>   gravitational field strength is in newtons per kilogram, N/kg
>   height is in metres, m.

Example 1

A crate has a mass of 80 kg. A crane lifts the crate from a height of 3 m above ground to a height 18 m above the ground. Calculate the increase in gravitational potential energy of the crate g = 10 N/kg.

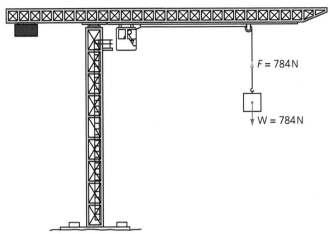

$F = 784\,N$

$W = 784\,N$

▲ **Figure 15.5** A crane does work to increase the gravitational potential energy of the crate.

Answer
Increase in height = 15 m

$$\text{Increase in } E_p = mgh$$
$$= 80 \times 9.8 \times 15$$
$$= 11\,760\,J \text{ or } 12\,kJ \text{ (to 2 significant figures)}$$

Example 2

A car has a kinetic energy store of 64 800 J. It is travelling at a speed of 12 m/s. Calculate the mass of the car.

Answer

$$E_k = \frac{1}{2}\,mv^2$$
$$64\,800 = \frac{1}{2} \times m \times (12)^2$$
$$m = \frac{2 \times 64\,800}{(12)^2}$$
$$m = 900\,kg$$

Example 3

A spring has a force constant of 60 N/m. The spring is extended by 5 cm. Calculate the elastic potential energy stored in the spring.

Answer

$$E_e = \frac{1}{2}\,ke^2$$
$$= \frac{1}{2} \times 60 \times (0.05)^2$$
$$= 0.075\,J$$

Remember to change the extension of 5 cm into metres.

Example 4

Calculate the change in kinetic energy stored when a car of mass 1200 kg slows down from 30 m/s to 20 m/s.

Answer

$$\begin{aligned}\text{Change in kinetic energy} &= \frac{1}{2}\,mv_1^2 - \frac{1}{2}\,mv_2^2 \\ &= \frac{1}{2} \times 1200 \times 30^2 \\ &\quad - \frac{1}{2} \times 1200 \times 20^2 \\ &= 540\,000 - 240\,000 \\ &= 300\,000\,J \text{ or } 300\,kJ\end{aligned}$$

Test yourself

5 Calculate the kinetic energy of a bullet of mass 0.015 kg, travelling at a speed of 240 m/s.

6 Calculate the increase in the gravitational potential energy store of a boy of mass 50 kg after he has climbed the Taipei 101 Tower, which is 440 m high.

7 A car has a mass of 1500 kg. It accelerates from a speed of 15 m/s to a speed of 20 m/s. Calculate the increase in kinetic energy of the car.

8 A car suspension spring has a spring constant of 2000 N/m. Calculate the elastic potential energy stored in the spring when it is compressed by 8 cm.

9 A meteor has a mass of 0.05 kg and it is travelling towards Earth at a speed of 30 km/s. Calculate the kinetic energy of the meteor.

○ Changes in energy

In this section you will learn how to use the energy equations to make predictions about changes to a system: which energy is transferred from one type of store to another.

Example 1

A ball of mass 100 g is thrown vertically upwards with a speed of 15 m/s. What is the maximum height the ball reaches?

Answer

As the ball rises, work is done by gravity to slow the ball down. Energy is transferred from the kinetic energy store of the ball to the gravitational potential energy store of the ball.

The principle of the conservation of energy tells us that the kinetic energy stored when the ball is at the bottom of its path is equal to the potential energy stored when the ball is at the top of its path, (if we assume that none of the energy is transferred to the surroundings).

$$\frac{1}{2}\,mv^2 = mgh$$
$$\frac{1}{2} \times 0.1 \times 15^2 = 0.1 \times 9.8 \times h$$
$$\text{So} \quad h = \frac{11.25}{0.98}$$
$$= 11.5\,\text{m}$$

Example 2

A stretched bow stores 64 J of elastic potential energy. The bow fires an arrow of mass 20 g. Calculate the speed of the arrow as it leaves the bow.

Answer

As the bow does work to speed up the arrow, energy is transferred from the bow's store of elastic potential energy to the arrow's store of kinetic energy.

$$\text{So}\ \frac{1}{2}\,mv^2 = 64$$
$$\frac{1}{2} \times 0.02 \times v^2 = 64$$
$$v^2 = \frac{64}{0.01}$$
$$v^2 = 6400$$
$$v = 80\,\text{m/s}$$

Example 3

A car of mass 1200 kg is parked on a 1 in 5 slope. The handbrake is released and the car rolls down the slope. When the car has travelled 20 m down the slope, how fast is it travelling?

Answer

A 1 in 5 slope means that the car goes up (or down) 1 m in height, when it travels 5 m along the slope. So, when the car has travelled 20 m along the slope it has gone down by a height of 4 m. The gravitational potential energy transferred can be calculated from this height.

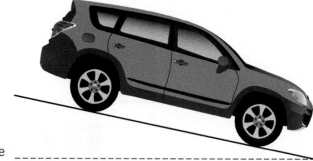

▲ **Figure 15.6** Car on a 1 in 5 slope.

$$mgh = \frac{1}{2}\,mv^2$$
$$\text{gravitational field strength}, g = 9.8\,\text{N/kg}$$
$$1200 \times 9.8 \times 4 = \frac{1}{2} \times 1200 \times v^2$$
$$\frac{1}{2}v^2 = 39.2$$
$$v^2 = 78.4$$
$$v = 8.9\,\text{m/s (to 2 significant figures)}$$

Test yourself

You can do these questions for practice, but you can also set up some experiments like this in the laboratory.

10 A spring with a spring constant of 200 N/m is fixed, so that it points vertically on to a table by using some Blu-Tack. The spring is compressed by 1 cm.

   a) Calculate the elastic potential energy stored in the spring.
     A polystyrene ball of mass 0.5 g is held on the spring and launched vertically upwards.

   b) i) How much gravitational potential energy does the ball have when it reaches its highest point?

     ii) Calculate the height the ball reaches when it is released. Assume the ball travels vertically upwards.

11 In Figure 15.8 a trolley is attached to a spring as shown. The trolley is pulled back to extend the spring by 15 cm. The spring has a spring constant of 80 N/cm.

   a) Calculate the elastic potential energy stored in the spring.

   The trolley is now released and it is accelerated by the spring.

   b) i) State the kinetic energy stored in the trolley just after the spring reaches its unstretched length.

     ii) Calculate the maximum speed of the trolley.

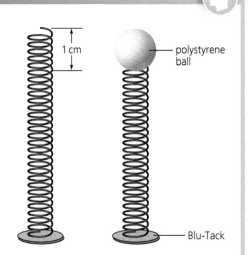

▲ **Figure 15.7** Using a spring to propel a ball.

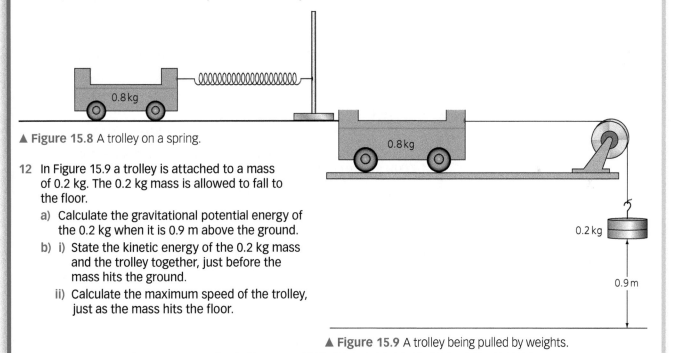

▲ **Figure 15.8** A trolley on a spring.

12 In Figure 15.9 a trolley is attached to a mass of 0.2 kg. The 0.2 kg mass is allowed to fall to the floor.

   a) Calculate the gravitational potential energy of the 0.2 kg when it is 0.9 m above the ground.

   b) i) State the kinetic energy of the 0.2 kg mass and the trolley together, just before the mass hits the ground.

     ii) Calculate the maximum speed of the trolley, just as the mass hits the floor.

▲ **Figure 15.9** A trolley being pulled by weights.

Show you can...

State the principle of the conservation of energy. Describe two demonstrations you have seen that help to explain this principle.

KEY TERM

Work When a force causes an object to move.
work = force × distance

○ Work

In this section you are introduced to the definition of **work**, because by doing work we can transfer energy from one store to another.

A force does work on an object when the force causes the object to move, in the direction of the force. Work can be calculated using the equation:

$$W = F s$$

work = force × distance moved in the direction of the force

Where work is in joules, J
force is in newtons, N
distance is in metres, m.

One joule of work is done when a force of 1 newton causes a displacement of 1 metre.

1 joule = 1 newton – metre

TIP
The unit of work is the joule (J).

When we do work, by applying a force to move an object, we change the energy store of that object.

- When 200 J of work is done to lift a box upwards, the gravitational potential energy store of the box increases by 200 J.
- When 3000 J of work is done to accelerate a car, the kinetic energy store of the car increases by 3000 J.
- When 2 J of work is done to stretch a spring, the spring stores 2 J of elastic potential energy.

TIP
The total amount of energy always remains the same. It may, however, change from one form to another.

Electrical work is done by a battery when the battery makes a charge flow. (Current and charge is covered further in Chapter 16.)

Example

We can use the idea of work to help us calculate the braking distance of a car.

A car of mass 1500 kg is travelling at a speed of 20 m/s. The brakes apply a force of 5000 N to slow down and stop the car.

Calculate the braking distance of the car.

Answer

The decrease in the kinetic energy store of the car (transferred to the internal energy store in the brakes) = work done by the brakes.

$$\frac{1}{2} mv^2 = Fs$$
$$\frac{1}{2} \times 1500 \times 20^2 = 5000 \times s$$
$$s = \frac{300\,000}{5000}$$
$$= 60\,m$$

○ Power

Often when we want to do a job of work, we want to do it quickly. We say that a crane that lifts a crate more quickly than another crane lifting the same crate is more **powerful.**

KEY TERM

power is the rate at which energy is transferred.

$$\text{power} = \frac{\text{energy transferred}}{\text{time}}$$

▲ **Figure 15.10** When this train travels at 60 m/s, its engines run at a power of 2 MW.

Power is defined as the rate at which energy is transferred or the rate at which work is done. Power can be calculated by using these equations.

$$P = \frac{E}{t}$$

$$\text{power} = \frac{\text{energy transferred}}{\text{time}}$$

$$P = \frac{W}{t}$$

$$\text{power} = \frac{\text{work done}}{\text{time}}$$

where power is in watts, W

energy transferred is in joules, J

time is in seconds, s

work done is in joules, J

An energy transfer of 1 joule per second is equal to a power of 1 watt.

Large powers are also measured in kilowatts, kW, and megawatts, MW.

$$1\,\text{kW} = 1000\,\text{W}\,(10^3\,\text{W}) \qquad 1\,\text{MW} = 1\,000\,000\,\text{W}\,(10^6\,\text{W})$$

TIP

Power is measured in watts (W).

Example

A weight lifter lifts a mass of 140 kg a height of 1.2 m in 0.6 s. Calculate the power developed by the weight lifter.

Answer

The potential energy transferred = mgh

$E_p = 140 \times 9.8 \times 1.2 = 1646.4\,\text{J}$

$\text{power} = \dfrac{\text{energy transferred}}{\text{time}}$

$= \dfrac{1646.4}{0.6} = 2700\,\text{W}$ or 2.7 kW (to 2 significant figures)

Practical

Measuring your own power

Work out your personal power by running up a flight of steps. You need to know your mass, the time it takes you to run up the stairs and the vertical height of the stairs. Remember $g = 9.8\,\text{N/kg}$.

1 Record your time and calculate the increase in your gravitational potential energy store.

2 Now calculate your power.

3 Explain why you need the vertical height of the staircase and not the length along the staircase.

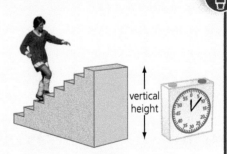

▲ **15.11** When you go up a flight of stairs, you are lifting your body weight and doing work against the force of gravity. The faster you run up the stairs, the greater your power.

Test yourself

13 What is the unit of power?

14 What is the connection between the energy transferred and power?

15 A crane lifts a weight of 12000N through a height of 30m in 90s. Calculate the power output of the crane in kW.

16 Two students have an argument about who is more powerful. Peter says he is more powerful because he is bigger. Hannah says she is more powerful because she is quicker.
To settle the argument, they run up stairs of height 4.5m. Use the information about their weights and times to settle the argument.

Table 15.1

	Weight in N	Fastest time in s
Peter	760	3.80
Hannah	608	3.04

17 When an express train travels at a speed of 80m/s, the resistive forces acting against it add up to 150kN.
   a) Calculate the work done against the resistive forces in 1s.
   b) Calculate the power output of the train, travelling at 80m/s.

Show you can...

Design an experiment to measure the power of your arm as you lift a weight.

Energy changes in systems

You will find that some of this section – specifically, specific heat capacity – is also covered in Chapter 17.

The amount of energy stored or released from a system, as its temperature changes, can be calculated using the equation:

$$\Delta E = mc\,\Delta\theta$$

change in thermal = mass × specific heat × temperature
energy   capacity   change

Where change in thermal energy is in joules, J
   mass is in kilograms, kg
   specific heat capacity is in joules per kilogram per degree Celsius, J/kg°C
   temperature change is in degrees Celsius, °C.

KEY TERM

Specific heat capacity The energy needed to raise the temperature of 1 kg of substance by 1 °C.

The specific heat capacity of a substance is the amount of energy required to raise the temperature of one kilogram of the substance by one degree Celsius.

The specific heat capacity varies from substance to substance.

An investigation to measure the specific heat capacity of a material

There are several different ways to obtain the data needed to calculate the specific heat capacity of a material. All of the methods involve the same idea; the decrease in one energy store (or work done) leads to an increase in the temperature of the material.

In this method energy is transferred from an electrical immersion heater to a metal block. The increase in the temperature of the metal block depends on the mass of the block and the specific heat capacity of the block.

Method

1 Measure the mass of the metal block (*m*) in kilograms.

2 Put the thermometer and immersion heater into the holes in the block.

3 Connect the immersion heater, joulemeter and power supply together as shown in Figure 15.12.

4 Measure the temperature of the metal block (θ_1) and then switch on the power supply.

5 Wait until the temperature of the block has gone up by about 10 °C then switch off the power supply. Write down the reading on the joulemeter (*E*). This gives you the amount of energy transferred to the immersion heater.

6 Do not take the immersion heater out of the block. Keep looking at the temperature and write down the highest temperature shown by the thermometer (θ_2).

If a joulemeter is not available set up the circuit shown in Figure 15.13. Use the following method to measure the energy transfer.

Switch the power supply on and again wait for the temperature of the block to increase by about 10 °C.

Watch the voltmeter and ammeter and write down the readings (*V* and *I*). The readings may change a little as the block gets warmer. Switch the power supply off and write down how many seconds the power supply was on for (*t*).

The energy transferred to the block can be calculated using the equation:

$$E = VIt$$

Analysing the results

1 Calculate the increase in temperature of the block $\Delta\theta = (\theta_2 - \theta_1)$.

2 Use the following equation to calculate the specific heat capacity (*c*) of the metal block:

$$c = \frac{E}{m\Delta\theta}$$

3 It is likely that the value you calculate for the specific heat capacity of the metal will not be accurate.

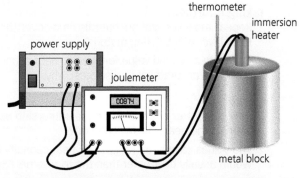

▲ Figure 15.12 Calculating the specific heat capacity of a metal block using a joulemeter.

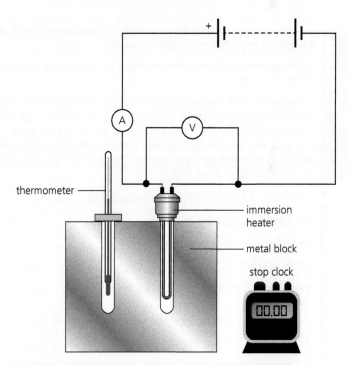

▲ Figure 15.13 Calculating the specific heat capacity of a metal block using an electrical circuit with a voltmeter and ammeter.

> **KEY TERM**
>
> **Accurate** A measurement or calculated value that is close to the true value.

Look up the true value. How close is the true value to your experimental value? Calculate the difference between the two values. Do you think your experimental value is accurate?

4 If other people in the class have used different types of metal, then compare the different specific heat capacity values with the temperature rise of the metal. If you do this, it is important that the blocks used by everyone have the same mass and that the same amount of energy is transferred to each immersion heater.

You should find that the higher the temperature rise, the smaller the specific heat capacity of the metal.

Taking it further

1 Repeat the investigation, but this time cover the side of the block in a thick layer of insulating material.

2 Calculate a second value for specific heat capacity. Is this second value any more accurate? If it is, suggest why.

Questions

1 Why is it better not to remove the immersion heater from the block as soon as the heater is switched off?

2 The values calculated for specific heat capacity from this investigation would usually be greater than the true value. Explain why.

Demonstration experiment

(a)

Your teacher might demonstrate how you can work out the temperature of a hot object.

In Figure 15.14a) a small piece of steel is heated until it is red hot in a Bunsen flame. The steel is then quickly transferred to an insulated beaker, which contains 0.1 kg (100 ml) of water as shown in Figure 15.14b). *[Caution, the steel is red hot, and will burn the bench if dropped. The water will 'spit' as the steel is put into the beaker. Do not use a piece of steel of mass more than about 20 g.]*

Table 15.2 **Example data and calculation**

Specific heat capacity of water	4200 J/kg °C
Specific heat capacity of steel	450 J/kg °C
Mass of steel	20 g
Temperature of water at the start of the experiment	19 °C
Temperature of water after the hot steel has been placed in it	34 °C

The thermal energy transferred by the hot steel equals the thermal energy gained by the water.

The thermal energy gained by the water

$$= (mc\Delta\theta)_{water}$$

$$= 0.1 \times 4200 \times 15$$

$$= 6300\,J$$

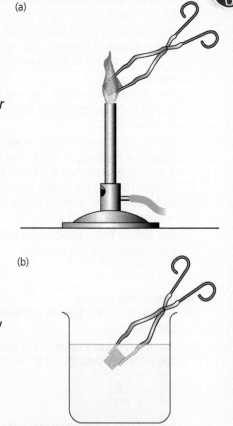

▲ Figure 15.14 Calculating the temperature of a very hot solid.

Practical

Note: the temperature change is $34\,°C - 19\,°C = 15\,°C$

The thermal energy transferred by the hot steel is $6300\,J$.

So $6300 = (mc\Delta\theta)_{steel}$

$6300 = 0.02 \times 450 \times \Delta\theta$

$\Delta\theta = \dfrac{6300}{0.02 \times 450}$

$\Delta\theta = 700\,°C$

So the initial temperature of the steel was:

$700\,°C + 34\,°C = 734\,°C$

Note: the final temperature was $34\,°C$, which is added to $700\,°C$.

Give two reasons why the steel might have been even hotter than this.

Test yourself

18 State the units of specific heat capacity.

19 Air has a specific heat capacity of $1000\,J/kg\,°C$. Calculate the energy transferred by a heater when it warms the $80\,kg$ of air in a room from $10\,°C$ to $22\,°C$.

20 a) A night storage heater contains $60\,kg$ of concrete, which has a specific heat capacity of $800\,J/kg\,°C$. How much energy must be supplied to the concrete to warm it from $15\,°C$ to $45\,°C$?

   b) A $200\,W$ heater is used to heat the concrete. How long does the heater take to supply the energy calculated in part (a)?

21 Milk has a specific heat capacity of $3800\,J/kg\,°C$. A cup of milk, of mass $0.3\,kg$, is placed into a microwave oven in a plastic insulating cup. When switched on the oven heats the milk at a rate of $700\,W$.

   a) Calculate how much energy has been transferred to the milk after 1 minute.

   b) When the milk was put in the oven its temperature was $6\,°C$. Calculate the temperature of the milk after 1 minute of heating.

Show you can...

Show you understand what is meant by specific heat capacity by completing this task.

Describe an experiment to measure the specific heat capacity of water. Explain what measurements you would take and the calculations you would do.

Conservation and dissipation of energy

KEY TERM

Dissipate To scatter in all directions or to use wastefully. When energy has been dissipated, it means we cannot get it back. The energy has spread out and heats up the surroundings.

You have already learnt that energy is conserved: energy cannot be created or destroyed. However, when energy is transferred from a source, it is not all transferred usefully. Often when energy is transferred some of the energy is dissipated or 'wasted'.

When petrol is used to power a car, much of the energy is wasted. Only about a quarter of the chemical energy stored in the petrol is used to drive the car forwards (by doing work) and about three quarters of the chemical energy is wasted by heating up the engine.

◯ Reducing energy dissipation

Every day in our lives we use energy from fuels for transport or for heating our homes. In all cases the chemical energy stored in the fuels is eventually transferred to the thermal energy store of the surrounding area. This is a less useful way of storing energy; the energy is being 'wasted'. However, we try to ensure that as much energy is transferred usefully as possible. We try to minimise the amount of wasted energy.

Power stations

The purpose of a power station is to generate electricity, which can do useful work to light and heat our homes. Engineers design the generators in the power station to reduce the amount of waste energy in the power station. Generators are large machines which can dissipate energy by heating or by unwanted mechanical vibrations.

Car design

When we drive a car we want to make sure that as much of the chemical energy stored in the fuel as possible does useful work for us.

- Engineers design fuel efficient cars, which dissipate less energy.
- The car is made streamlined to reduce air resistance on the car.
- Moving parts of the car are lubricated with oil to reduce friction.

Keeping warm at home

When we heat our homes, the energy stored inside the house is dissipated through the roof, walls, windows or doors of the house to warm up the air outside. We want to make sure the energy escapes as slowly as possible.

Here are some ways in which we reduce unwanted energy dissipation at home.

Chimneys

Figure 15.15 shows a coal fire burning in a sitting room. Some of the energy from the burning coal is transferred to the air outside the house. This is wasted energy. By having the chimney inside the house, thermal energy can be transferred into the bedrooms upstairs. This is useful energy.

Walls

The rate at which energy is transferred through the walls of a house depends on four factors.

- The temperature difference between inside and outside. (Our heating bills are larger in winter than in summer.)
- The area of the walls. (Large houses cost more to heat than small houses.)
- The thermal conductivity of the walls. Some materials conduct heat well, metals, for example. These materials have a high thermal conductivity. Brick and glass are not good thermal conductors. They have relatively low thermal conductivities, but energy still flows out of a warm house through the walls and windows. The higher the thermal conductivity of a material, the higher the **rate of energy transfer by conduction** across the material.
- The thickness of the walls (or windows) is important. The thicker the walls, the slower the rate of energy loss.

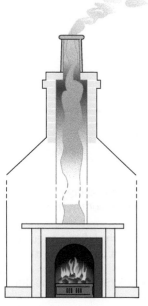

▲ **Figure 15.15** In this house, some of the thermal energy is transferred upstairs.

▲ **Figure 15.16** This house was built in 1820. The walls are 70 cm thick. This keeps the house warm.

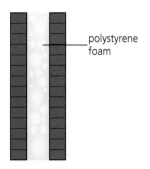

▲ **Figure 15.17** Cavity wall insulation reduces heat loss.

Modern houses are built with two layers of brick as shown in Figure 15.17. Then the house is insulated with **cavity wall insulation**, between the two layers of brick. The foam which insulates the walls is full of trapped air. The air is a good insulator; it has a much lower thermal conductivity than brick or glass.

Loft insulation and carpets

The most efficient way to reduce energy loss from our house is to **insulate the loft**. A thick layer of loft insulation reduces energy loss through the roof (Figure 15.18). We also use insulating carpets to reduce energy loss through the floor.

The tiles on the kitchen floor feel cold when you walk on them in bare feet. These tiles are much better thermal conductors than carpets.

Double glazing

A thin pane of glass in a window transfers energy out of the house. We use **double glazing** to reduce energy loss through the windows. A layer of gas trapped between two panes of glass provides good insulation (Figure 15.19).

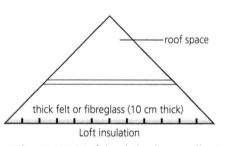

▲ **Figure 15.18** Loft insulation is very effective.

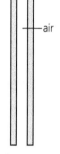

▲ **Figure 15.19** Double glazing.

Test yourself

22 What is meant by the term 'energy dissipation'?

23 Explain how engineers can design cars to be more efficient.

24 Give two examples of how we reduce unwanted energy transfers. Try to think of two not mentioned earlier in the text.

25 List four ways that unwanted energy transfer is reduced in your home.

Show you can...

Show you understand the principle of conservation of energy by planning an experiment to demonstrate and prove this principle. State the apparatus you will use, the measurements you will take and the calculations you will do.

▲ **Figure 15.20** Both of these have done some useful work, but they have also wasted energy; they are inefficient.

○ Efficiency

Efficiency is a way of expressing the proportion of energy that is **usefully transferred** in a process as a number. The most efficient machines transfer the highest proportions of input energy to useful output energy.

To calculate efficiency we use the equation:

$$\text{efficiency} = \frac{\textbf{useful output energy transfer}}{\textbf{total input energy transfer}}$$

Efficiency may also be calculated using the equation:

$$\text{efficiency} = \frac{\textbf{useful power output}}{\textbf{total power input}}$$

Efficiency is a ratio of energies or powers. So efficiency has no unit. We write efficiencies as a decimal or a **percentage.**

> **TIP**
> - Efficiency can be expressed as a decimal or a percentage.
> - Efficiency cannot be more than 1, or 100%.
> - Values closer to 1, or 100%, represent greater efficiency.

Example 1

A steam engine uses coal as its source of energy. When the chemical energy store of the coal in the engine's furnace goes down by 150 kJ, the engine does 18 kJ of useful work against resistive forces.

Calculate the efficiency of the engine.

Answer

$$\text{efficiency} = \frac{\text{useful output energy transfer}}{\text{total input energy transfer}}$$
$$= \frac{18}{150}$$
$$= 0.12 \text{ or } 12\%$$

Note: here both energies were expressed in kJ. Make sure you use the same unit on the top and bottom of the fraction.

Example 2

A joulemeter records that 18.2 J of electrical work is done when an electric motor lifts a 0.3 kg load through a distance of 0.90 m. Calculate the efficiency of the motor.

Answer

$$\text{efficiency} = \frac{\text{useful output energy transfer}}{\text{total input energy transfer}}$$
$$= \frac{\text{gain in } E_p}{18.2} = \frac{mgh}{18.2}$$
$$= \frac{0.3 \times 9.8 \times 0.90}{18.2}$$
$$= 0.15 \text{ or } 15\% \text{ (2 significant figures)}$$

Increasing the efficiency of an intended energy transfer

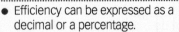

Whenever we do a job of work, we want to ensure that as much energy as possible is transferred usefully, and that little energy is dissipated wastefully.

When we move an object we transfer energy by doing work, and work is calculated using the equation:

$$W = Fs$$

where *F* is the force applied and *s* is the distance moved in the direction of the force. We do less work if the forces of friction or air resistance that act against us are small. When frictional forces act, energy is transferred to the thermal energy store of the surroundings. This is wasteful.

We reduce **friction** by:

● using **wheels**
● applying **lubrication**.

We reduce **air resistance** by:

● **travelling slowly**
● **streamlining**.

When we streamline a car (for example), we are shaping its surface so that air flows past the car and offers as little resistance to the motion of the car as possible.

Increasing efficiency using machines

We can also increase the efficiency of a job by using a machine.
Figure 15.21 shows two men lifting a load of bricks on a building site. The man on the ground pulls with a force of 250 N. Because there are four ropes in their system of pulleys, the ropes apply a force of 4 × 250 N (1000 N), which is enough to lift the bricks and the pulley, and to overcome friction.

Without the machine the men would have to carry the bricks up the ladder in smaller loads. This means that they would have to work to carry the bricks and to lift their own weight. The machine allows them to apply a smaller force to lift the bricks, but they have to pull the rope 4 times as far. Using the machine is far more efficient than climbing up and down the ladder. It saves time and is much safer.

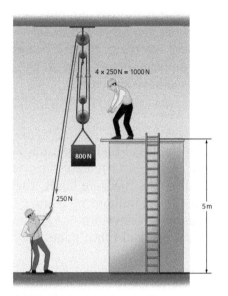

▲ Figure 15.21

Test yourself

26 Define efficiency.

27 A student compares two machines, A and B. Both machines transfer the same input energy. The student discovers that machine A wastes less energy than machine B. Which machine is more efficient?

28 Phil is in the gym doing pull-ups (Figure 15.22). Each time he does a pull-up his store of chemical energy decreases by 1500 J. Phil's mass is 72 kg and he lifts himself up 0.5 m in one pull-up.

  a) Calculate the gravitational potential energy stored after one pull-up.

  b) Calculate the efficiency of Phil's body during this exercise.

29 A car's engine is supplied with one kilogram of fuel, which stores 45 MJ of chemical energy. The efficiency of the car is 36%.

Calculate the amount of energy available for useful work against resistive forces.

30 Why does a streamlined car use fuel more efficiently than another similar car which experiences larger air resistance forces?

31 a) Why does the machine in Figure 15.21 allow the men to work more efficiently?

  b) Suggest another machine that increases the efficiency of a job. How does the machine ensure that more energy is usefully transferred?

▲ Figure 15.22

National and global energy resources

Every day we depend on various energy resources to make our lives comfortable. A hundred and fifty years ago our ancestors walked to school, lived in cold houses and did not have electricity in their homes. Now we travel by car, train or bus, we live in warm homes, and all of us use electricity to run the many appliances we have at home.

○ Fossil fuels

Much of our energy in the UK comes from the fossil fuels: coal, oil and gas. Like all fuels, they store energy. However, to release the energy, the fossil fuels must be burned. Once the fuels are burnt, they are gone forever, because fossil fuels have taken millions of years to be formed. Fossil fuels are described as non-renewable energy resources, because there is a **finite** supply of them (once gone, they cannot be replaced).

By contrast renewable energy resources will never run out. We can obtain renewable energy from the Sun, tides, waves and rivers, from the wind and from the thermal energy of the Earth itself.

Finite resources

Figure 15.23 shows that we only have relatively small supplies of fossil fuels left. Unless we find substantial new resources, the world's supply of oil and gas will run out by about 2070 and coal will run out by about 2130. Our descendants will have to use different energy resources.

○ Using fuels

In the UK the three main fossil fuels provide most of the energy for our needs. These needs include:

- **Transport.** Fuels such as petrol, diesel and kerosene are produced from oil. These fuels drive our cars, trains and planes. Electricity is also used to run our trains, and cars are being developed to run from electricity supplies. In 50 years' time, we may not be able to fly on holiday.
- **Heating.** Most of our home heating is provided by gas and electricity. Gas pipes run into our houses to provide energy to our boilers. Some homes are warmed by oil-fired boilers or by burning solid fuels such as coal and wood.

KEY TERMS

Non-renewable energy resources Energy resources which will run out, because they are finite reserves, and which cannot be replenished.

Renewable energy resources Energy resources which will never run out and (or can be) replenished as they are used.

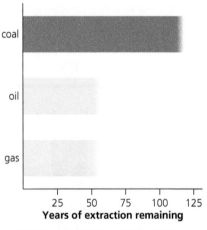

▲ **Figure 15.23** The world's supply of fossil fuels will not last forever. The information here shows current estimates based on known reserves; many more may be discovered.

- **Electricity**. In the UK, electricity is generated using different energy resources, but most of our electricity is generated by burning fossil fuels. Table 15.3 shows the percentage of electricity generated using different energy resources. Gas and coal provide a convenient and relatively cheap way to generate our electricity. However, the burning of gas and coal produces carbon dioxide, which most scientists think is responsible for global warming. In November 2015, the UK government announced that coal fuelled power stations will be phased out by 2025. More electricity will be generated by gas, which produces less carbon dioxide than coal.

Table 15.3 Percentage of UK electricity generated by different energy resources. (Source: Department of Energy and Climate Change, September 2015.) We can expect a greater amount of electricity to be generated by renewable sources in future.

Energy resource used to generate electricity	Percentage of UK electricity generated from each energy resource 2015
Gas	30
Coal	20
Nuclear	22
Biomass	9
Wind	11
Hydroelectric	2
Oil	2
Solar	4

Fossil fuels and acid rain

One of the products of burning coal is sulfur dioxide. When sulfur dioxide combines with water, acid rain is produced. Acid rain damages buildings and kills plants.

Sulfur dioxide can be removed from the waste gases of burning coal, but this is expensive.

▲ **Figure 15.24** Trees killed by acid rain.

KEY TERM

1 GW, 1 gigawatt = 10^9 W

Power stations

The average consumption of electrical power in the UK is about 36 GW and our peak consumption (in the evening) reaches 57 GW. (1 GW is 1 gigawatt, which is 1000 million watts or 10^9 watts.)

Figure 15.25 shows the layout in a coal-fired power station. You are not expected to remember this for your GCSE but it is helpful to understand the principle behind generating electricity: an energy resource such as gas or water drives a turbine; the turbine then drives the generator, which makes electricity.

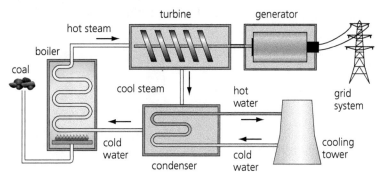

▶ **Figure 15.25** A coal-fired power station.

◯ **Other energy resources**

Due to concerns over global warming, which is caused (in part) by the burning of fossil fuels, governments must find alternative energy resources. Climate change meetings attended by governments usually involve agreements to reduce fossil fuel use.

Nuclear power

Nuclear power generates about 22% of the electricity in the UK. There are currently plans for a new nuclear power station at Moorside in West Cumbria. The Moorside power station will begin generating electricity in 2024. The peak generating capacity of the power station will be 3.4 GW.

The nuclear fuels used are mainly uranium and plutonium. These are also non-renewable energy resources. However, nuclear fuel contains a huge amount of nuclear energy, and it is estimated that there is enough uranium to last thousands of years.

Nuclear power has the advantage of producing **no pollutant gases**. However, we need to be very careful how we store **nuclear waste**.

Biomass

Waste products can provide fuel for some small electrical generators. Much of the waste is wood, so this is a renewable energy resource. Many of the biomass generators are also used to heat factories or houses directly.

Biofuels emit carbon dioxide when they are burnt. However, the plants that become biofuels used carbon dioxide as they grew. So, overall biofuels do not add to the amount of carbon dioxide in the atmosphere. They are '**carbon neutral**'.

Using tides

Every day tides rise and fall. Massive amounts of water move in and out of river estuaries. It is estimated that the energy of the tides could generate about 20% of Britain's electricity.

At present there is little electricity generated by tidal energy. **Expensive barriers** must be constructed. The largest tidal barriers in the world generate about 250 MW of electricity.

A barrage is like a dam built across a river estuary (Figure 15.26). The barrage has underwater gates that open as the tide comes in and then close to keep the water behind the barrage (Figure 15.27). When the tide goes out, a second set of gates is opened. Water flows out of these gates and drives turbines that are connected to generators as it does so.

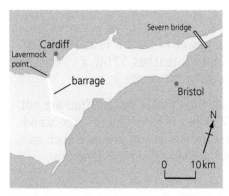

▲ **Figure 15.26** A tidal barrage across the River Severn could generate about 6% of Britain's electricity. The peak power would be about 7 GW.

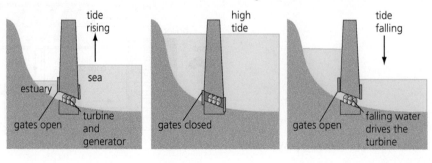

▶ **Figure 15.27** How a tidal barrage works.

TIP ✓

Environmental issues are different from **ethical issues**. This section lists many environmental issues, but it also raises ethical issues. For example, is it right to build a dam and destroy animal habitats?

Hydroelectric power

▲ **Figure 15.28** Hydroelectric power stations can be huge and generate vast amounts of power.

Hydroelectric power stations generate about 2% of Britain's electricity, but they generate about 10% of the world's electricity.

Many hydroelectric power schemes have had **environmental impacts**, as a new lake is formed.

- Forests have been cut down.
- Farmland is lost.
- Wildlife habitats have been destroyed.
- Many people have had to move homes.

Wind power

When the wind blows with sufficient force, the blades of a wind turbine rotate. The blades turn a generator which produces electricity. Wind power has the advantage of being **environmentally clean** and is a **renewable** energy source. There are no waste products.

However, wind power also has some **disadvantages**.

- Wind power is **unreliable**. If the wind is too light, little power is produced. If the wind is too strong, generators can overheat, so the blades have to be stopped from moving.
- Some people think that wind farms are **unsightly** and spoil the look of the countryside.
- Wind generators make a low frequency sound, which creates **noise pollution**. Some people find it most unpleasant living close to a wind farm.

At present 11% of Britain's electricity comes from wind power. The government's target is to increase this figure to about 20%. Although wind power is unreliable, when the wind does blow, we can turn off gas generators. So wind power will save fossil fuels and reduce greenhouse gas emissions.

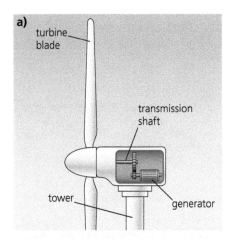

a)
turbine blade

transmission shaft

tower

generator

b)

▲ **Figures 15.29a) and b)** In 2010 the UK government approved the building of 6000 new wind turbines to be erected off the coast of Britain. This will give a total of 12000 turbines with a maximum power capacity of 32 GW.

Solar power

Solar cells use energy directly from the Sun to generate electricity. Solar cells generate electricity on a small scale.

In some countries solar electricity generation is **unreliable**, due to the weather, but it can still make a useful contribution to a country's overall power supply and does not contribute to **global warming** and is **renewable**.

Geothermal energy

A small number of countries are able to make good use of geothermal energy to generate electricity. Iceland generates 30% of its electricity by taking advantage of the volcanic activity on the island. In some cases hot water is used directly to warm houses. Hot water or steam is also used to generate electricity. This form of electricity generation has the advantage of being **environmentally clean** and renewable.

▲ **Figure 15.30** This traffic slowing sign is powered by solar cells.

▲ **Figure 15.31** Solar panels on a house.

▲ **Figure 15.32** Generating energy from geothermal sources.

Test yourself

32 a) What is a non-renewable energy resource? Give one example.

  b) What is a renewable energy resource? Give one example.

33 Name a common fuel used in a nuclear power station. Is this fuel renewable or non-renewable?

34 In Britain 20% of our electricity is generated in coal-fired power stations.

  a) What are the advantages of using coal to generate electricity?

  b) State two environmental problems caused by using coal to generate electricity.

35 a) What environmental problems are caused by building a hydroelectric power station?

  b) Give one advantage of hydroelectric power.

36 Why are tides a more reliable way of generating electricity than wind power?

37 Britain plans to have 12000 wind turbines spread from the south to the north of the country. Why does spreading out the wind farms increase the reliability of wind power?

▶

38 A wind turbine is designed to produce a maximum power of 4 MW. However, due to variations of wind speed, the generator only produces this power for 10% of the time.
How many such wind turbines are required to replace a coal-fired power station which generates 2000 MW of power all the time?

39 Figure 15.33 shows the layout of a pumped storage power station. Water from the high level lake generates electricity by flowing through the turbines which are connected to generators. These are placed above the low level lake. When there is a low demand for electricity, the generators are driven in reverse to pump water back into the high level lake. This ensures there is enough water to generate electricity next time the demand is high.

a) Why is this sort of power station useful to electricity companies?

b) Does this power station produce any pollution or greenhouse gases
   i) when generating electricity, ii) when pumping water back up the hill?

c) Use the information in Figure 15.33 to calculate the gravitational potential energy transferred per second when the generators are working.

d) The generators are 80% efficient. Calculate the power output of the power station in MW.

Figure 15.34 shows the typical power use in the UK on a spring day.

e) At what time of the day does the pumped storage station i) generate electricity, ii) pump water back up the hill? Give reasons for your answer.

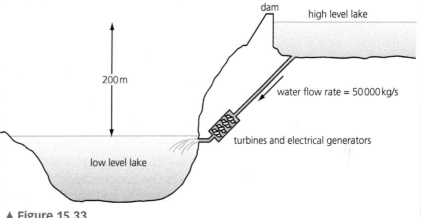

▲ Figure 15.33

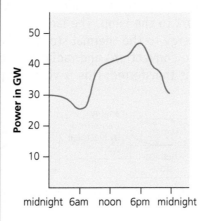

◀ **Figure 15.34** A typical day's use of electricity on a spring day in the UK.

Energy: a summary

In this book we provide clear definitions using 'key terms' where possible. No such easy definition exists for energy. Instead we have included this brief summary to pull the ideas about energy together.

Energy is an idea that cannot be described by a single process, nor is energy something we can hold or measure directly. However, we pay an enormous amount of attention to energy because it is conserved.

There are many different stores of energy. In any process, energy can be transferred from one store to another store, but energy is never destroyed or created.

Energy stores

Stores of energy include:

- kinetic
- chemical
- internal (or thermal)
- gravitational potential
- magnetic
- elastic potential
- nuclear.

Energy transfers

There are various ways that energy can be transferred:

- by mechanical work
- by electrical work
- by heating
- by radiation.

The word 'radiation' includes light and all electromagnetic waves. Radiation also includes 'mechanical radiation' such as sound and shock waves. Sometimes people refer to light, sound and electrical energy. However, light, sound and electricity are not energy stores, they transfer energy from one store to another. Here are some examples.

A hot cup of tea has a store of thermal energy. As the tea cools, it heats the surroundings. Energy is transferred to the thermal store of the surroundings and the temperature of the surroundings goes up (but is too small to measure).

A battery stores chemical energy. Energy is transferred by electrical work to the lamp. The lamp does not store energy. The lamp transfers energy to the thermal store of the surroundings by heating (conduction and convection) and radiation. Some of this radiation is useful to us as it is transferred; this is visible light.

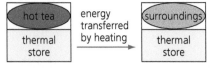

▲ Figure 15.35 Thermal energy transfer.

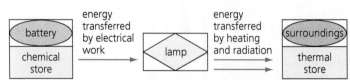

▲ Figure 15.36 Energy in a battery is eventually transferred to its surroundings.

A motor is used to lift a mass. A battery stores chemical energy. Energy is transferred by electrical work from a battery to a motor. The motor does not store energy, but it has a temporary store of kinetic energy when it is turning. The motor does mechanical work and transfers energy to the gravitational potential store of the raised mass. The motor also transfers energy to the thermal store of the surroundings by heating and by making a noise.

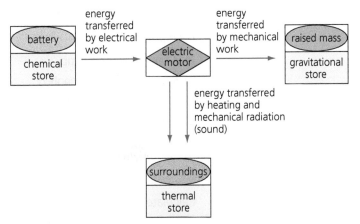

▲ **Figure 15.37** Energy transfer to and from an electric meter.

And finally...

Energy can be transferred from one energy store to a different store, and there is as much energy stored at the end of the process as there was at the beginning. The reason energy is useful to us is that it allows us to do various jobs of work and to keep warm.

Chapter review questions

1. Describe the energy store of each of the following:

   a) an electrical battery

   b) a moving car

   c) a stretched rubber band

   d) a lake of water behind a hydroelectric dam.

2. In the following processes energy is transferred from one energy store to another store or stores. State the stores of energy at the beginning and end of each process.

   a) A battery lights a lamp.

   b) A bowl of hot soup cools.

   c) A battery is connected to a motor which is lifting a load.

   d) A firework rocket has been launched into the air and is travelling upwards.

3. Your feet feel warmer on a carpet than they do a tiled kitchen floor. Give one reason why.

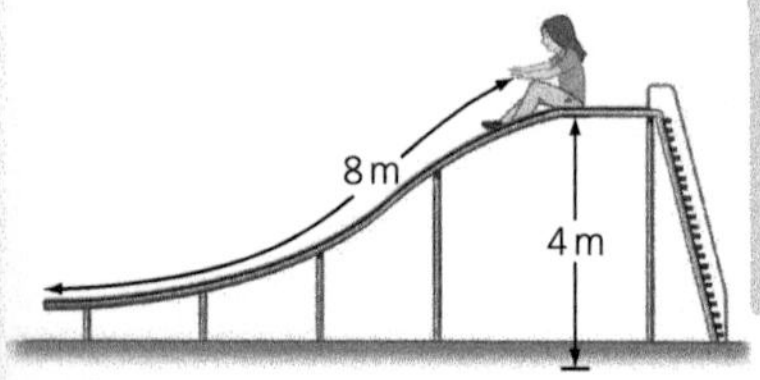

▲ Figure 15.38

4. Calculate the energy stored in each of these examples.

   a) A car of mass 1400 kg travels at a speed of 25 m/s.

   b) A suspension spring for a truck with a spring constant of 40 000 N/m is compressed by 5 cm.

   c) A suitcase of mass 18 kg is placed in a luggage rack 2.5 m above the floor of a train.

5. Figure 15.38 shows a girl on a slide. Her mass is 45 kg.

   a) Calculate the girl's speed at the bottom of the slide. Ignore the effects of friction.

   b) Explain why the girl's speed is likely to be less than the answer calculated in part (a).

6. A gymnast of mass 55 kg lands from a height of 5 m onto a trampoline (see Figure 15.39). Calculate how far the trampoline stretches before the gymnast comes to rest.
   The trampoline has a spring constant of 35 000 N/m.

▲ Figure 15.39

7. Some lead shot with a mass of 50 g is placed into a cardboard tube as shown in Figure 15.40. The ends of the tube are sealed with rubber bungs to keep the lead shot in place. The tube is rotated so that the lead shot falls and hits the bung at the bottom.

   a) Why does the temperature of the lead shot increase after it has fallen and hit the lower bung?

   b) A student rotates the tube 50 times. Calculate the total decrease in the gravitational potential energy store of the lead shot in this process.

   c) The specific heat capacity of lead is 160 J/kg °C. Calculate the temperature rise of the lead shot after the student has rotated the tube 50 times.

   d) In practice, the temperature rise of the lead shot is likely to be less than your answer in part (c). Give a reason why.

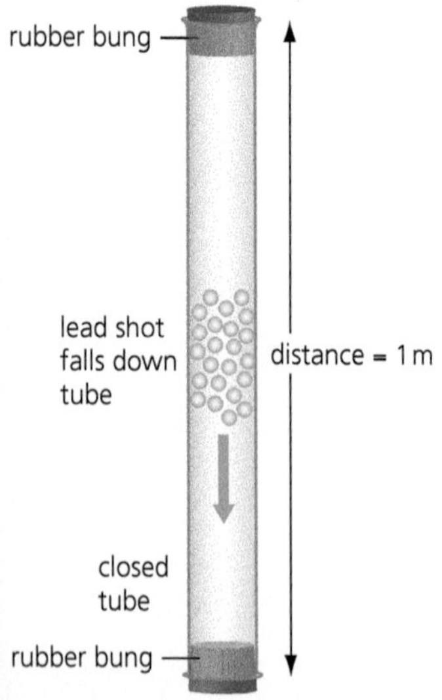

▲ Figure 15.40

8. A girl kicks a football with a force of 300 N. The girl's foot is in contact with the ball for a distance of 0.2 m. The ball has a mass of 450 g. Calculate the speed of the ball just after it has been kicked.

9 An electricity supply provides electrical power to a motor at a rate of 800W. The motor lifts a crate of mass 80kg through a height of 3m in 12 seconds. Calculate the efficiency of the motor.

10 A student wants to calculate the specific heat capacity of a liquid. First he determines that the liquid has a density of 900 kg/m³.

a) The student places a volume of 200 cm³ of liquid into an insulated beaker. Calculate the mass of the liquid in kg. [1 cm³ = 10⁻⁶ m³]

The student measures the temperature of the liquid. It is 22 °C. The student then heats the liquid with a heater that has a power rating of 24 W. He heats the liquid for 10 minutes.

b) Calculate the energy transferred to the liquid in 10 minutes.

c) After 10 minutes the student finds the liquid has risen to a temperature of 72 °C. Calculate the specific heat capacity of the liquid.

11 A scientist observes a grasshopper as it jumps. The grasshopper takes 25 milliseconds to take off, and reaches a speed of 3.0 m/s. The grasshopper has a mass of 1.5 g.

a) Calculate the kinetic energy of the grasshopper after its jump.

b) Calculate the mechanical power developed by the grasshopper's legs.

Practice questions

1 Which of the following is the correct unit for power?

   newtons   joules   watts   [1 mark]

2 Which of the following is the correct unit for specific heat capacity?

   J/kg °C   J kg/ °C   J kg °C   [1 mark]

3 The energy input to a machine is 2000 J. The machine transfers 600 J of useful energy. Calculate the efficiency of the machine. [2 marks]

4 The British government has planned to build up to 12 000 wind generators. Give one advantage of wind power and one disadvantage. [2 marks]

5 Many of the world's electricity power stations burn fossil fuels.

  a) Burning fossil fuels produces carbon dioxide.

    What effect can an increase in carbon dioxide levels have on the Earth's atmosphere? [1 mark]

  b) Figure 15.41 shows how much carbon dioxide is produced for each unit of electricity generated in coal-, gas- and oil-burning power stations.

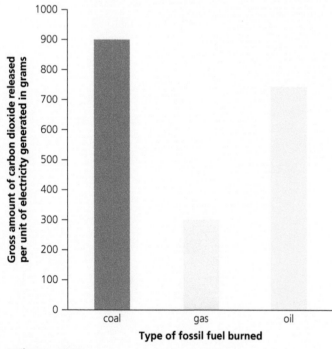

▲ Figure 15.41

    i) Which type of fossil fuel produces the most carbon dioxide for each unit of electricity generated? [1 mark]

    ii) Why is a bar chart drawn to show the data and not a line graph? [1 mark]

  c) Biofuels are renewable energy resources that can be used to generate electricity.

    i) Name one other type of renewable energy resource. [1 mark]

    ii) Most biofuels are derived from plants. When biofuels are burned carbon dioxide is produced. Why does burning a biofuel have less effect on the atmosphere than burning coal? [1 mark]

6 An advertisement for solid fuel firelighters claims:

Be certain of a fast fire …

… use H&S, the firelighters that give out more energy than any others

▲ Figure 15.42

  a) To test this claim a student plans the following investigation.

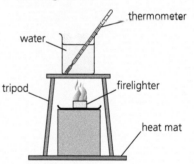

▲ Figure 15.43

- Place 1 g of H&S firelighter on a tin lid.
- Put 80 cm³ of water into a beaker.
- Measure the temperature of the water.
- Set fire to the firelighter and then use it to heat the water.
- When all of the firelighter has burned, measure the new water temperature.
- Repeat with two different brands of firelighter.

    i) What type of variable is the brand of firelighter? [1 mark]

    ii) Name two control variables in this investigation. [2 marks]

    iii) Give one experimental hazard in this investigation. [1 mark]

    iv) Suggest one change that the student could have made to improve the resolution of the temperature readings. [1 mark]

b) To compare the data the student drew the bar chart shown in Figure 15.44.

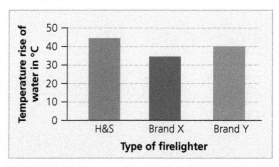

▲ Figure 15.44

   i) Was the data collected by the student sufficient to confirm the claim made by the maker of H&S firelighters? Give a reason for your answer. [2 marks]

   ii) Give **two** reasons why the decrease in the chemical energy store of the firelighter is greater than the increase in the thermal energy store of the water. [2 marks]

7 The roof of the Tokyo Skytree is 495 m high and can be climbed using its steps. A tourist decided to climb the tower. He took 35 minutes to do it and he had a mass of 60 kg.

a) Calculate the increase in the gravitational potential energy store of the tourist in climbing the steps. [3 marks]

b) Calculate his average power output during the climb. [3 marks]

c) The energy to make the climb is transferred from the chemical energy store of the tourist. Eating one slice of bread provides 400 kJ of chemical energy. Calculate the number of slices of bread he should eat for breakfast to provide the energy for the climb.

(Assume his body is 20% efficient at transferring energy from his chemical energy store to gravitational potential energy.) [3 marks]

d) Where is most of the energy from the tourist's food transferred to? [1 mark]

8 In Figure 15.45 a conveyor belt is used to lift bags of cement on a building site.

▲ Figure 15.45

a) A 40 kg bag of cement is lifted from the ground to the top of the building. Calculate its gain in gravitational potential energy. [3 marks]

b) The machine lifts five bags per minute to the top of the building. Calculate the useful energy delivered by the machine each second. [2 marks]

c) The machine is 35% efficient. Calculate the input power to the machine while it is lifting the bags. [3 marks]

9 A car and its passengers have a combined mass of 1500 kg. The car is travelling at a speed of 15 m/s. It then increases speed to 25 m/s.

Calculate the increase in kinetic energy of the car. [3 marks]

10 Figure 15.46 shows a pirate boat theme park ride which swings from A to B to C and back.

a) As the boat swings from A to B a child increases her kinetic energy store by 10 830 J. The child has a mass of 60 kg and sits in the centre of the boat. Calculate the speed of the child as the boat passes through B. [3 marks]

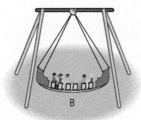

▲ Figure 15.46

b) Sketch a graph to show how the child's gravitational potential energy changes as the boat swings from A to B to C. [3 marks]

c) Calculate the change in height of the ride.

(Assume the decrease in the gravitational potential energy store as the child falls is transferred to the kinetic energy store of the child.) [3 marks]

11 An electric winch is used to pull a truck up a slope, as shown in Figure 15.47.

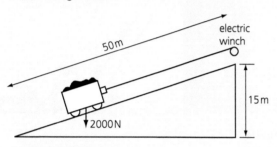

▲ Figure 15.47

a) How much work is done in lifting the truck 15 m? [3 marks]

The winch uses a 6 kW electrical supply, and pulls the truck up the slope at a rate of 5 m/s.

b) How long does it take to pull the truck up the slope? [2 marks]

c) How much work is done by the winch? [2 marks]

d) Calculate the efficiency of the winch. [2 marks]

12 A barrage could be built across the estuary of the River Severn. This would make a lake with a surface area of about 200 km² (200 million m²).

Figure 15.48 shows that the sea level could change by 9 m between low and high tide; but the level in the lake would only change by 5 m.

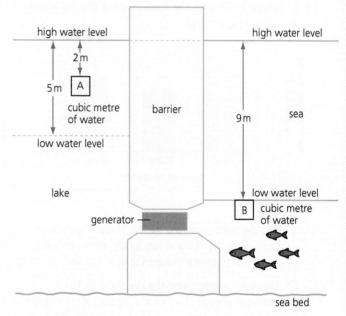

▲ Figure 15.48

a) Calculate the gravitational potential energy transferred when a mass of 1 kg falls from A to B. [3 marks]

b) Calculate the number of cubic metres of water that flow out of the lake between high and low tides. [2 marks]

c) Calculate the mass of water that flows out between high and low tides. A cubic metre of water has a mass of 1 kg. [2 marks]

d) Use your answers to parts (a) and (c) to calculate how much energy can be transferred from the tide. Assume that position A is the average position of a cubic metre of water between high and low tide. [3 marks]

e) The time between high and low tide is approximately 6 hours. Use this figure to calculate the average power available from the dam. Give your answer in megawatts. [3 marks]

f) What are the advantages and disadvantages of the Severn barrage as a possible source of power? [4 marks]

Working scientifically:
Uncertainty, errors and precision

When an electric motor is used to lift a weight, the power supply does electrical work to make the motor turn.

The motor transfers some energy usefully. This increases the gravitational potential energy stored by the weight. The rest of the energy is dissipated to the surroundings.

Susan decided to find out if the efficiency of an electric motor depends on the size of the weight being lifted. To do this she set up the apparatus shown in Figure 15.49.

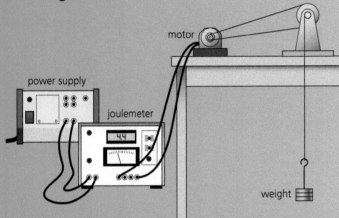

▲ **Figure 15.49** Calculating the efficiency of an electric motor.

Susan started with a 2N weight. She switched on the power supply and increased the potential difference (p.d.) until the motor just lifted the weight from the floor to the bench top; a distance of 0.8 m. The joulemeter recorded the energy transfer to the motor. Susan repeated this step twice more. Her results are recorded in the table.

Table 15.4

Trial	Joulemeter reading
1	14.8
2	15.3
3	14.9

1 Show that the mean (average) joulemeter reading was 15.0.

The three readings taken from the joulemeter are all close to the mean. The values for the energy transferred are precise.

When the same quantity is measured several times, the bigger the spread of the measurements about the mean, the less precise the measurements are. The precision of a set of measurements depends on the extent of the random errors.

The uncertainty in these energy values is ±0.3. The uncertainty is worked out by calculating the difference between the mean value and the value furthest away from the mean (in this case 15.3 − 15.0).

An uncertainty in a set of data can be caused by random errors or a systematic error. In this investigation, judging when the weight reaches the bench top and then switching off the power supply is a random error. If whenever the joulemeter is reset it does not go back to zero, this is a systematic error. This type of systematic error is also called a zero error.

Susan repeated the procedure with a range of different weights. For each weight, she obtained three energy values and recorded the mean. These values are shown in the table.

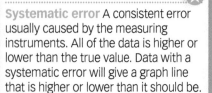
Table 15.5

Mean energy input to the motor in J	Weight lifted in N	Percentage efficiency of the motor
15.0	2	10.7
16.9	3	14.2
21.2	4	
25.5	5	15.7
33.5	6	14.3
46.0	7	12.2

2 In Susan's investigation:

   a) What range of weights was used?

   b) What was the independent variable?

   c) Which variable was controlled during the investigation?

3 How could Susan tell that the joulemeter she used did not have a zero error?

4 Copy and complete the table above by calculating the missing efficiency value.

5 Draw a graph of percentage efficiency against weight lifted.

6 What can you conclude from this investigation?

16 Electricity

At the beginning of the 20th century, very few people had electricity supplied to their homes. Now, a century later, the supply of electricity to our homes, offices and streets is an essential part of life. However, electricity comes at a price and we must be careful how much we use it. We must plan ahead to make sure we are able to provide electricity far into the future.

Specification coverage

This chapter covers specification points 6.2.1.1 to 6.2.4.3 and is called Electricity.

It covers current, potential difference and resistance, series and parallel circuits, domestic uses and safety and energy transfers.

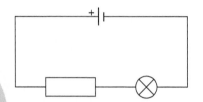

▲ **Figure 16.1** The resistor and lamp are in series.

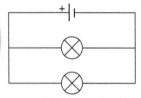

▲ **Figure 16.2** The two lamps are in parallel.

Previously you could have learnt

› When two objects rub against each other, electrons can transfer from one object to another. When electrons transfer to an object it becomes negatively charged. When electrons leave an object it becomes positively charged.

› Two like charges repel each other.

› Two unlike charges attract each other.

› An electrical current is a flow of charge.

› In metals, current is a flow of electrons.

› Some materials are good conductors of electricity.

› Some materials do not conduct electricity. These are called **insulators**.

› A cell or battery has a store of chemical energy. The energy stored decreases when a current flows.

› When the same current passes through a number of components they are said to be in **series**.

› In a **parallel** circuit, the current divides into different branches.

Test yourself on prior knowledge

1 Are the lights in your home in series or in parallel? How can you tell?

2 a) Name three good conductors of electricity.

   b) Name three good insulators of electricity.

Current, potential difference and resistance

○ Circuit symbols

Figure 16.3 shows the circuit symbols for the electrical components that you will meet in this section. A brief explanation of their function is given here and you will learn more about them later on.

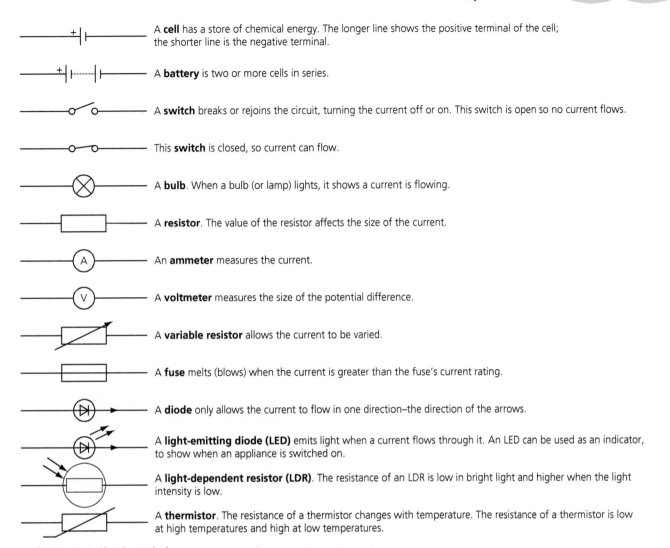

A **cell** has a store of chemical energy. The longer line shows the positive terminal of the cell; the shorter line is the negative terminal.

A **battery** is two or more cells in series.

A **switch** breaks or rejoins the circuit, turning the current off or on. This switch is open so no current flows.

This **switch** is closed, so current can flow.

A **bulb**. When a bulb (or lamp) lights, it shows a current is flowing.

A **resistor**. The value of the resistor affects the size of the current.

An **ammeter** measures the current.

A **voltmeter** measures the size of the potential difference.

A **variable resistor** allows the current to be varied.

A **fuse** melts (blows) when the current is greater than the fuse's current rating.

A **diode** only allows the current to flow in one direction–the direction of the arrows.

A **light-emitting diode (LED)** emits light when a current flows through it. An LED can be used as an indicator, to show when an appliance is switched on.

A **light-dependent resistor (LDR)**. The resistance of an LDR is low in bright light and higher when the light intensity is low.

A **thermistor**. The resistance of a thermistor changes with temperature. The resistance of a thermistor is low at high temperatures and high at low temperatures.

▲ **Figure 16.3** Circuit symbols.

Test yourself

1 Which one of the following is the correct symbol for an LDR?

(a) (b) (c)

▲ **Figure 16.4**

2 Use words from the list below to label each of the components in the circuit in Figure 16.5.
   cell   resistor   fuse   lamp
   switch   diode

3 Draw a circuit diagram to show a cell in series with an ammeter, variable resistor and bulb.

▲ **Figure 16.5**

Show you can...

Complete this task to show that you understand how to draw and design electrical circuits. Draw a circuit diagram to show how two lamps can be connected to a battery, with components that allow the two lamps to be dimmed independently.

○ Current and charge

Figure 16.6 shows a circuit diagram. In this circuit a cell provides a potential difference of 1.5 V, giving a current of 0.1 A in the circuit.

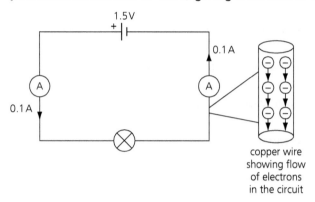

▶ **Figure 16.6** A simple circuit showing the conventional (traditional) direction of current flow.

KEY TERMS ★

Potential difference (p.d.) A measure of the electrical work done by a cell (or other power supply) as charge flows round the circuit. Potential difference is measured in volts (V).

Electric current A flow of electrical charge. The size of the electric current is the rate at which electrical charge flows round the circuit.

The **potential difference** (or p.d.) is a measure of the electrical work done by a cell (or other power supply) as charge flows round the circuit. The potential difference is measured in volts (V). Here the cell provides a potential difference of 1.5 V. (Remember the positive terminal of the cell is shown with the long line and the negative terminal with a shorter line.)

It is quite common to call the potential difference 'voltage'. However, you will find that the examination papers (and this book) will use the term 'potential difference'.

In a metal the current is carried by **electrons** which are free to move. The electrons are repelled from the negative terminal of the cell and attracted towards the positive terminal.

In a circuit the direction of the electric current is always shown as the direction in which positive charge would flow – from the positive terminal of the battery to the negative terminal. Current was defined in this way before the electron was discovered, at a time when people did not understand how a wire carried a current. So in Figure 16.6 the direction of current is shown from positive to negative.

The amount of charge flowing round in the circuit is measured in **coulombs**, C. One coulomb of charge is equivalent to the charge on 6 billion billion electrons.

The unit of current is the **ampere**, A. This unit is often abbreviated to amp. Small currents can be measured in milliamps (mA).

$1\,mA = 0.001\,A\ (10^{-3}\,A)$

The current at all points of the circuit shown in Figure 16.6 is the same. So the two ammeters on either side of the lamp read the same current – in this case 0.1 A.

Current and the flow of charge are linked by the equation.

$$Q = I\,t$$
charge flow = current × time
where charge is in coulombs, C
current is in amps, A
time is in seconds, s.

TIP ✓
Potential difference (voltage) is measured in **volts**, and current in **amperes** (amps).

Example

In Figure 16.6 the current of 0.1 A flows for 30 minutes. How much charge flows round the circuit?

Answer

$Q = It$

$Q = 0.1 \times 1800$

$\quad = 180\,C$

Test yourself

4 State the unit of each of the following quantities:
  a) potential difference
  b) current
  c) charge.

5 In Figure 16.7 the current is 0.08 A at point X. What is the current at points Y and Z?

6 a) A charge of 3 C flows round a circuit in 2 seconds. Calculate the current.
  b) A torch battery delivers a current of 0.3 A for 20 minutes. Calculate the charge which flows round the circuit.
  c) A thunder cloud discharges 5 C of charge in 0.2 ms. Calculate the current.
  d) A mobile phone battery delivers a current of 0.1 mA for 30 minutes. Calculate the charge which flows through the battery in this time.

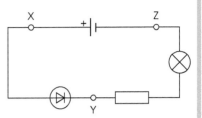

▲ **Figure 16.7**

Show you can...

Show you understand the nature of an electrical current, by explaining the relationship between a current and charge.

◯ Controlling the current

You can change the size of the current in a circuit by changing the potential difference of the cell or battery, or by changing the components in the circuit.

In Figure 16.8a) a current of 1 A flows. In Figure 16.8b) an extra cell has been added. Now the current is larger.

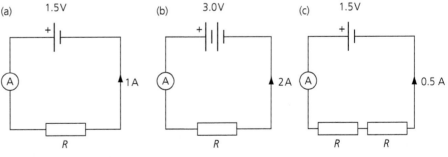

▲ **Figure 16.8** Changing the current in an electrical circuit.

In Figure 16.8c), the potential difference (voltage) of the cell is 1.5 V but now the resistance has been increased by adding a second resistor to the circuit. This makes the current smaller.

TIPS

- In a series circuit, increasing the potential difference increases the current.
- In a series circuit, increasing the resistance makes the current smaller. Resistance is a measure of a component's opposition to the current.

KEY TERM

Resistor A component that acts to limit the current in a circuit. When a resistor has a high resistance, the current is low.

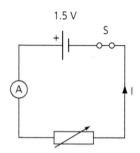

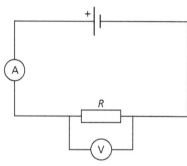

▲ **Figure 16.9** A variable resistor controls the size of the current.

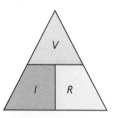

▲ **Figure 16.10** An electrical circuit with an ammeter in series and a voltmeter in parallel.

In Figure 16.9 a variable resistor has replaced the fixed resistor in the circuit. You can control the size of the current by adjusting the resistor.

○ Ammeters and voltmeters

Figure 16.10 shows you how to set up a circuit using an ammeter and a voltmeter.

- The **ammeter** is set up in **series** with the resistor. The same current flows through the ammeter and the resistor.
- The **voltmeter** is placed in **parallel** (on a separate branch) with the resistor. The voltmeter measures the potential difference across (between the ends of) the resistor.
- The voltmeter only allows a very small current to flow through it, so it does not affect the current flowing around the circuit.

○ Resistance

The **current** through a component depends on two things:

- the **resistance** of the component
- the **potential difference** across the component.

The circuit in Figure 16.10 can be used to determine the value of the resistor in the circuit.

The current, potential difference or resistance can be calculated using the equation:

$$V = IR$$

potential difference = current × resistance

where potential difference is in volts, V
current is in amps, A
resistance is in ohms, Ω

Resistances are also measured in kilohms (kΩ) and megohms (MΩ).

$$1\,k\Omega = 1000\,\Omega\ (10^3\,\Omega)\quad 1\,M\Omega = 1\,000\,000\,\Omega\ (10^6\,\Omega)$$

It is very important that you can use this equation in each of its three forms. The triangle in Figure 16.11 is a useful way to remember it.

Example

A potential difference of 12V is applied across a resistor of 240Ω. Calculate the current through the resistor.

Answer

If you are in doubt about rearranging the formula on the examination paper, try to remember the triangle. Put your finger over the symbol you want to work out, in this case *I*, and we get:

$$I = \frac{V}{R}$$

$$= \frac{12}{240} = 0.05\,A$$

▲ **Figure 16.11**

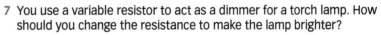

KEY TERMS

Directly proportional When two quantities are directly proportional, doubling one quantity will cause the other quantity to double. When a graph is plotted, the graph line will be straight and pass through the origin (0, 0).

Inversely proportional When two quantities are inversely proportional, doubling one quantity will cause the other quantity to halve.

Test yourself

7 You use a variable resistor to act as a dimmer for a torch lamp. How should you change the resistance to make the lamp brighter?

8 Copy and complete this table.

Table 16.1

Electrical device	Potential difference across device in V	Current through device in A	Resistance of device in Ω
Resistor	1.5		20
Lamp	230	0.05	
Heater	230		23
LED		0.04	75
Electric car motor	72	48	

Required practical 15

Investigate how the resistance of a wire depends on the length of the wire

Method

1 Use electrical insulating tape to attach a 100 cm length of wire to a wooden metre rule. Very thin constantan wire is suitable with a diameter of 0.1 mm or less.

2 Connect the wire into the circuit as shown in Figure 16.12. Start by connecting the crocodile clips across 20 cm of the wire. Do not allow the current to rise higher than 1 A.

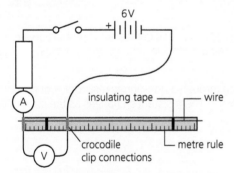

▲ **Figure 16.12** Investigating resistance in different lengths of wire.

3 Draw a suitable table to record the length of the wire, the current through the wire, the potential difference across the wire and the calculated value for the resistance of the wire.

4 Close the switch then measure the current in the wire (*I*) and the potential difference (*V*) across the wire. Write these values and the length of the wire into your table.

5 Open the switch.

6 Use the equation $R = V/I$ to calculate the resistance of the wire.

7 Connect different lengths of wire into the circuit to obtain the data needed to plot a graph of resistance against length of wire.

If you have an ohmmeter you can measure the resistance of the wire directly; you do not need to calculate it.

Analysing the results

1 Plot a graph of resistance against the length of the wire.

2 Your graph should give a straight line going through the origin (0, 0). If it does, then you have shown that the resistance is directly proportional to the length of the wire.

Taking it further

1 Use the same circuit and a range of wires of different cross-sectional area to show that the resistance of a wire is inversely proportional to the area of the wire.

2 Set up the circuit shown in Figure 16.10. Use the circuit to investigate the resistance of combinations of resistors in series and in parallel. (See the section on series and parallel circuits later in this chapter.) First take measurements to calculate the value of a single fixed value resistor. Add a second resistor in series with the first resistor. Measure the p.d. across both resistors and the current through the resistors. Use these values to calculate the resistance of the two resistors in series. Repeat the experiment but with the two resistors connected in parallel. What do your results show happens to the resistance of a circuit when a) resistors are connected in series b) resistors are connected in parallel?

Questions

1 What was the dependent variable in this investigation?

2 What aspects of the investigation were important in trying to stop the wire from getting hot?

3 The width of the crocodile clips makes it difficult to measure the exact length of wire connected into the circuit. What type of error will this cause?

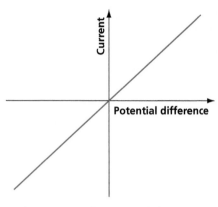

▲ **Figure 16.13** The *I–V* graph for a resistor or metal wire at constant temperature.

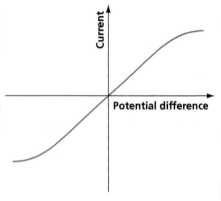

▲ **Figure 16.14** The current in a filament lamp does not increase in proportion to the potential difference.

Current in A

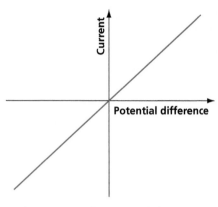

▲ **Figure 16.15** The resistance of a diode depends on the direction of the potential difference; the current only flows for a positive or forwards potential difference.

○ Current–potential difference characteristic graphs

An ohmic conductor

For some resistors, at constant temperature, the current through the resistor is proportional to the potential difference across it. A graph of current against potential difference gives a straight line. If the direction of the p.d. is reversed, the graph has the same shape. The resistance is the same when the current is reversed.

The resistor in this case is said to be ohmic.

A filament lamp

The current–potential difference graph for a filament lamp does not give a straight line; the line curves away from the current axis (*y*-axis).

The current is not proportional to the applied potential difference. The lamp is a non-ohmic resistor.

As the current increases, the resistance gets larger. The temperature of the filament increases when the current increases. So we can conclude that the resistance of the filament increases as the temperature increases. Reversing the p.d. makes no difference to the way the resistance of the lamp changes. The resistance always increases when the temperature of the filament increases.

> **KEY TERMS** ⭐
>
> **Ohmic** The current flowing through an ohmic conductor is proportional to the potential difference across it. If the p.d. doubles, the current doubles. The resista.nce stays the same.
>
> **Non-ohmic** The current flowing through a non-ohmic resistor is not proportional to the potential difference across it. The resistance changes as the current flowing through it changes.

A diode

A **diode** is a component that allows current to go in only **one direction**. For a forward potential difference, current starts to flow when the potential difference reaches about 0.7 V. When the potential difference is 'reversed', there is no current at all. The diode has a very high resistance in the reverse direction.

A **light-emitting diode** (LED) is a special type of diode that lights up when a current flows through it. This is useful because it allows an LED to be used as an indicator to show us that a small current is flowing.

Changing resistance

Some resistors change their resistance as they react to their surroundings. The resistance of a **thermistor** decreases as the temperature increases. You can control its temperature by putting it into a beaker of warm or cold water, as in Figure 16.16.

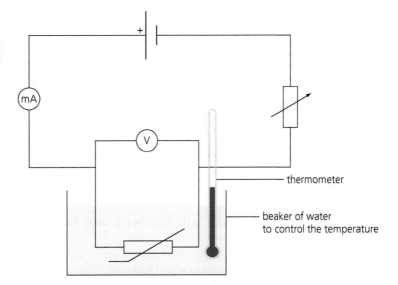

▶ **Figure 16.16** Investigating how the resistance of a thermistor changes with temperature.

By gently heating the water, the resistance of the thermistor can be found at different temperatures (Figure 16.17). You could use an ohmmeter to measure the resistance directly.

A **thermistor** can be used as the sensor in a temperature-operated circuit, such as a **fire alarm**, so when the temperature and current increase to a certain level, the alarm sounds. Some electronic thermometers use a thermistor to detect **changes in temperature**. The change in the resistance of a thermistor can be used to switch on (or off) other electrical circuits automatically.

The resistance of a **light-dependent resistor** (LDR) changes as the light intensity changes. In the dark the resistance is high but in bright light the resistance of an LDR is low. This is shown in Figure 16.18. A higher current flows through the resistor in bright light because the resistance is lower.

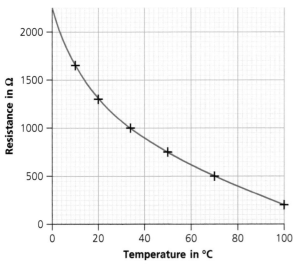

▲ **Figure 16.17** A graph to show how the resistance of a thermistor changes with temperature.

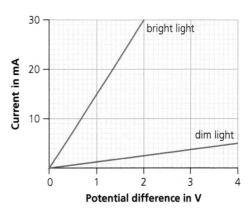

▲ **Figure 16.18** A current–potential difference graph for a light-dependent resistor in bright and dim light.

LDRs can be used as sensors in light-operated circuits, such as security lighting. The change in resistance of LDRs is used in digital cameras to control the total amount of light that enters the camera.

Investigating the *I–V* characteristic of a circuit component

1 Set up the circuit shown in Figure 16.19. A suitable power supply to use is four 1.5V cells joined in series.

2 Connect a 6V filament lamp into the circuit where it says component.

3 Adjust the variable resistor to give a potential difference (p.d.) of 1V across the lamp.

4 Write the readings on the voltmeter (p.d.) and the ammeter (current) in a suitable table.

5 Adjust the variable resistor so that you can obtain a set of p.d. and current values. Write the new values in your table.

6 Reverse the connections to the power supply. The readings on the voltmeter and ammeter should now be negative. Obtain a new set of data with the p.d. increasing negatively.

You could leave the variable resistor out of the circuit and change the p.d. and current by simply connecting across one cell, then two cells, then three cells and lastly all four cells.

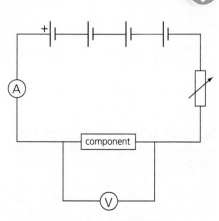

▲ Figure 16.19 Investigating the relationships between potential difference and current in a range of different circuits.

Analysing the results

1 Plot a graph of current against potential difference (*I–V* characteristic graph). Draw the axes so that you can show all of the data, the positive values and the negative values.

2 You should notice that plotting the negative values for p.d. and current gives the same shape graph line as plotting the positive values.

3 If you used a variable resistor you would have been able to increase the p.d. using a smaller interval than if you simply connected across 1, 2, 3 then 4 cells. The advantage of using smaller intervals is that you can be more confident that the shape you draw for your graph line is correct.

> **KEY TERM**
> **Interval** The difference between one value in a set of data and the next.

Taking it further
Replace the filament lamp with a low value resistor and then a diode. Obtain a set of p.d./current data for each component. Plot an *I–V* characteristic graph for each component. Remember to obtain the data needed to plot the negative part of the graph.

Questions

1 The *I–V* graph for a resistor (at constant temperature) is a straight line. Reversing the power supply does not change the shape of the graph line. What does this tell you about the resistance of the resistor and the direction of the current through the resistor?

2 How is the *I–V* graph for a diode different from the *I–V* graph for a filament lamp?

3 Why does having a small interval between values allow you to be more confident that the shape you draw for your graph line is correct?

4 If the p.d. were increased from 0V to 6V in 13 equal intervals, what would be the interval between p.d. values?

Test yourself

9 Use the graph in Figure 16.17 to find the resistance of the thermistor at temperatures of:
   a) 0°C
   b) 40°C
   c) 90°C.
10 What is meant by an ohmic resistor?
11 Figure 16.20 shows the current–potential difference graph for a filament bulb.
   a) Calculate the resistance of the filament when the applied voltage is:
      i) 1V
      ii) 3V.
   b) What causes the resistance to change?
12 Figure 16.18 shows a current–potential difference graph for an LDR in dim and bright light. Calculate the resistance of the LDR in:
   a) bright light
   b) dim light.
13 a) Draw a circuit diagram to show a cell, a 1kΩ resistor and an LED used to show that there is a current flowing through the resistor.
   b) Draw a circuit diagram to show how you would investigate the effect of light intensity on the resistance of an LDR.

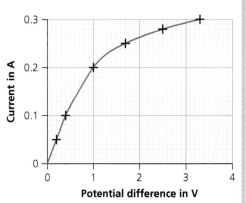

▲ **Figure 16.20** A current–potential difference graph for a filament bulb.

Show you can...

Show that you understand about various types of resistors, by explaining how the following behave in a circuit:
a) an LDR
b) an LED
c) a thermistor.

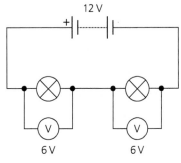

▲ **Figure 16.21** Two identical bulbs connected in series.

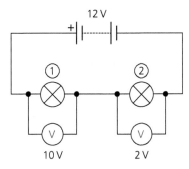

▲ **Figure 16.22** Bulbs with different resistances connected in series.

◯ Series and parallel circuits

Series circuits

In Figure 16.21 two identical bulbs are connected in series with a 12 V battery. The p.d. of 12 V from the battery is shared equally so that each bulb has 6 V across it.

Having two bulbs in the circuit rather than one increases the resistance, so the current decreases. The two bulbs in series will not be as bright as a single bulb.

The potential differences do not always split equally. In Figure 16.22 bulb 1 has a larger resistance than bulb 2. There is a larger potential difference across bulb 1 than across bulb 2. As the current flows, bulb 1 transfers more energy to the surroundings than bulb 2. Therefore bulb 1 is brighter than bulb 2.

Series circuit rules

The rules for series circuits are as follows:

- there is the **same current** through each component
- the **total potential difference** of the power supply is **shared** between the components. So if there are just two components then;

$$V_{supply} = V_1 + V_2$$

- the **total resistance** of two components is the **sum** of the resistance of each component.

$$R_{total} = R_1 + R_2$$

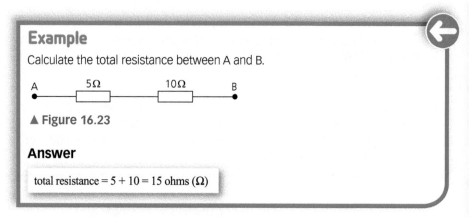

Example

Calculate the total resistance between A and B.

▲ **Figure 16.23**

Answer

total resistance = $5 + 10 = 15$ ohms (Ω)

Parallel circuits

In Figure 16.24 two identical bulbs have been connected in parallel with the 12 V battery. Now there is a 12 V potential difference across each bulb, and the same current flows through each bulb.

When two bulbs are joined in parallel, the total current in the circuit increases. So the combined resistance of the two bulbs is less than either bulb by itself. The total current in the circuit is the sum of the currents through the two bulbs.

Parallel circuit rules

The rules for parallel circuits are as follows:

- the **potential difference** across each component is the **same**
- the **total current** through the whole circuit is the **sum** of the currents through the separate components
- the **total resistance** of two resistors in parallel is less than the resistance of the smaller individual resistor.

Cells and batteries

A battery consists of two or more electrical cells. When cells are joined in series, the total potential difference of the battery is worked out by adding the separate potential differences together. This only works if the cells are joined facing the same way, positive (+) to negative (−) (see Figure 16.25).

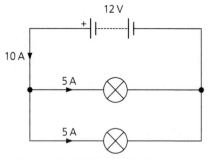

▲ **Figure 16.24** Identical bulbs connected in parallel.

(a)

1.5V 1.5V 1.5V

▲ **Figure 16.25** a) The cells are joined correctly – the separate p.d.s add to give 4.5V.

(b)

1.5V 1.5V 1.5V

▲ **Figure 16.25** b) One cell is the wrong way round. The p.d. of this cell cancels out the p.d. of one of the other cells – the total p.d. is only 1.5V.

Test yourself

14 Figure 16.26 shows a circuit with a cell, two ammeters and a resistor. What reading does the ammeter on the right–hand side give?

15 a) In Figure 16.27, what is the resistance between

   i) AB

   ii) CD?

b) Which of the following correctly states the resistance between E and F?

30 Ω	more than 20 Ω
less than 10 Ω	between 20 Ω and 10 Ω

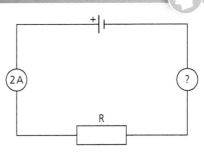

▲ Figure 16.26

16 a) What is the potential difference of the cell in Figure 16.28a)?

b) What is the potential difference across R₂ in Figure 16.28b)?

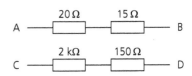

▲ Figure 16.27a

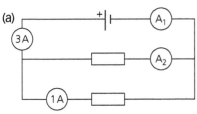

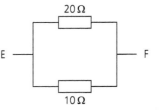

▲ Figure 16.27b

(a)

(b)

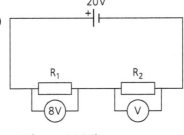

▲ Figure 16.28a   ▲ Figure 16.28b

17 State the values of A₁, A₂, A₃ and A₄ in Figures 16.29a) and b).

(a)

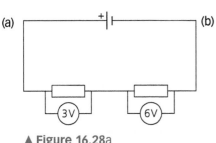

(b)

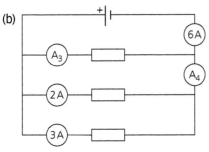

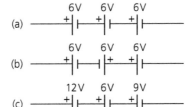

▲ Figure 16.29   ▲ Figure 16.30

18 Work out the potential difference of each cell combination shown in Figure 16.30.

○ Circuit calculations

You can use the series circuit rules to solve circuit problems. Two worked examples are given here.

Example

Use the information in Figure 16.31 to work out the resistance of the bulb.

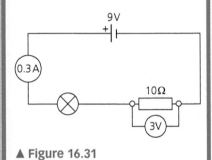

▲ **Figure 16.31**

Answer

The potential difference across the bulb is 9V – 3V = 6V

$$\text{So} \quad R = \frac{V}{I}$$
$$= \frac{6V}{0.3A}$$
$$= 20\,\Omega$$

Example

a) The light dependent resistor in Figure 16.32 is in bright light. Use the information in the diagram to work out its resistance.

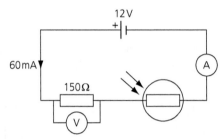

▲ **Figure 16.32**

b) Explain what happens to the voltmeter reading when the light intensity drops.

Answer

a) The total resistance of the series circuit can be calculated using:

$$R = \frac{V}{I}$$
$$= \frac{12}{0.06}$$
$$= 200\,\Omega$$

Therefore the LDR's resistance is:

$$200\,\Omega - 150\,\Omega = 50\,\Omega$$

OR you could work out the potential difference across the 150 Ω resistance:

$$V = I R$$
$$= 0.06 \times 150$$
$$= 9\,V$$

The potential difference across the LDR is:

$$12\,V - 9\,V = 3\,V$$

$$\text{Then } R = \frac{V}{I}$$
$$= \frac{3}{0.06}$$
$$= 50\,\Omega$$

b) When it gets dark the resistance of the LDR increases. So the current decreases. Therefore, the potential difference across the 150 Ω resistor drops, and the LDR gets a larger share of the potential difference.

Test yourself

19 a) Calculate the reading on the ammeter in Figure 16.33.

   b) Now work out the potential difference, *V*, of the battery.

20 a) Calculate the reading on the ammeter in Figure 16.34.

   b) Now work out the resistance *R*.

21 Work out the value of the resistance *R* in Figure 16.35.

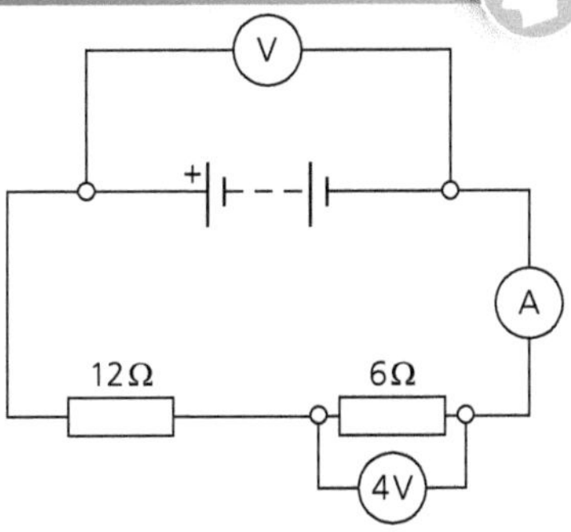

▲ Figure 16.33

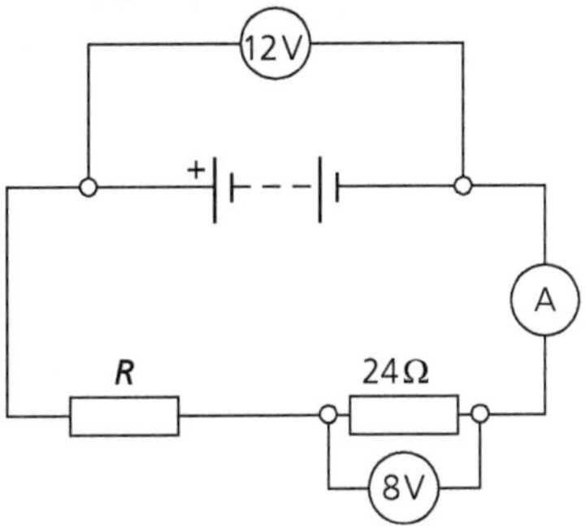

▲ Figure 16.34

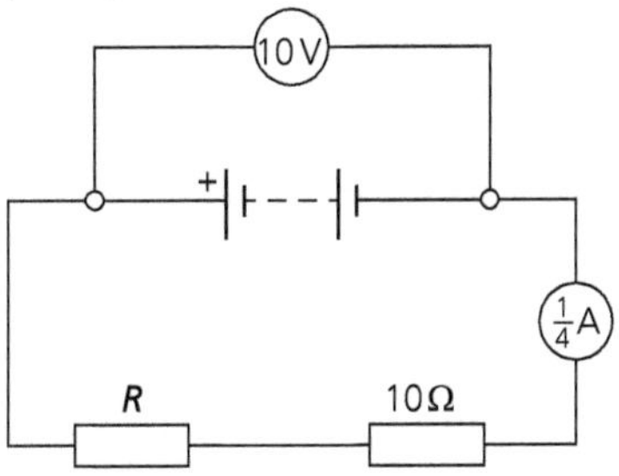

▲ Figure 16.35

○ Domestic use and safety

In the home we rely on electricity for heating, lighting, cooking, washing and powering devices which we use for work or leisure.

Direct and alternating potential difference

A cell or battery provides potential difference.
A **direct potential difference** remains always in the same direction, and causes a current to flow in the same direction. This is a **direct current (d.c.)**.

If an alternating power supply is used in a circuit, the potential difference switches direction many times each second. This is an **alternating potential difference**, which causes the current to switch direction. So an **alternating current (a.c.)** is one that constantly changes direction, passing one way around a circuit and then the other.

Figure 16.37 shows graphs of how a.c. and d.c. power supplies change with time. The d.c. supply remains constant at 6 V; the a.c. supply changes from positive to negative.

Note that the peak a.c. potential difference is a little higher than 6 V. This is to make up for the time when the potential difference is close to zero. A 6 V a.c. supply and a 6 V d.c. supply will light the same bulb equally brightly.

Mains supply

The mains electricity (domestic electricity supply) is supplied by **alternating current**. In the UK it has a potential difference of about **230 V** and a frequency of about **50 Hz**.

▲ **Figure 16.36** Appliances in this kitchen are reliant on electricity to run

KEY TERM

Frequency The number of waves produced or passing a point each second. The unit of frequency is the hertz, Hz; 1 hertz means there is 1 cycle per second.

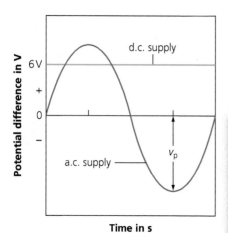

▲ **Figure 16.37** Direct and alternating current supplies. V_p is the peak potential difference for the a.c. supply. You can use a cathode ray oscilloscope (CRO) to plot potential differences.

A **frequency** of 50 Hz means that the cycle shown in Figure 16.37 repeats itself 50 times per second – or one cycle takes 1/50 of a second.

In Figure 16.38, a 230 V a.c. supply provides current to a cooker. In diagram (a) the current goes one way round the circuit, then in diagram (b) the current is reversed. In each case energy is transferred to the cooker.

Test yourself

22 a) What is meant by the terms a.c. and d.c.?
  b) Explain the difference between direct and alternating potential difference.
23 In the USA the mains electricity supply is 115 V 60 Hz. Explain the difference between the mains electricity supply in the USA and the mains electricity supply in the UK.

Cables

We use many electrical appliances at home, which we connect to a wall socket using a three-core cable and plug. The wires inside the cables connect the appliance to the plug and have a cross-sectional area of $2.5\,mm^2$. These cables should carry no more than a 13 A current. Appliances such as showers and cookers need larger currents. These appliances are connected to the mains supply using thicker cables.

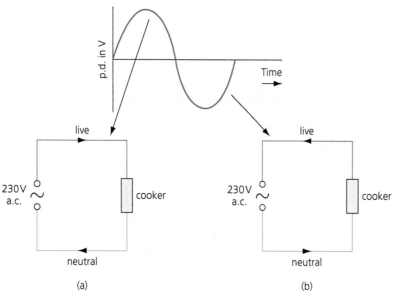

▲ **Figure 16.38** Alternative current being supplied to a cooker.

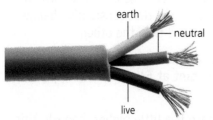

▲ **Figure 16.39** A cable has three wires: neutral (blue), earth (green/yellow), live (brown).

Live, neutral and earth wires

The insulation covering the three wires inside a cable is colour coded so we know which is which.

- **Live** wire – **brown**
- **Neutral** wire – **blue**
- **Earth** wire – green and yellow stripes

The **live wire** carries the **alternating potential difference** from the mains supply. The neutral wire completes the circuit. So the live and neutral wires carry the current to and from an electrical appliance. The earth wire is there to stop an appliance becoming live; it only carries a current if there is a fault.

However, the three wires have different potentials.

- The earth wire is at 0 V. It only carries a current if there is a fault.
- The potential difference between the live wire and earth (0 V) is about 230 V. A bare live wire is a hazard even if it is not delivering a current to an appliance. If a person touches a bare live wire, the current will pass through the person's body, giving an electric shock. In effect we would act as an earth.
- The neutral wire is close to earth potential (0 V). So touching the neutral wire would not give us an electric shock.

TIP

A live wire may be dangerous even when a switch in the mains circuit is open.

Power

○ Energy and charge

When charge passes through a resistor, the resistor gets hot. This is because electrons collide with the atoms in the resistor as they pass through it. The atoms increase their thermal energy store and vibrate faster, making the resistor hotter. The energy in the chemical store of the battery decreases as work is done to move the electrons round the circuit. The temperature of the resistor goes up and the resistor also heats up the surroundings.

The amount of energy transferred by electrical work depends on two factors:

- the potential difference, V, across the resistor
- how much charge, Q, flows through the resistor.

The energy transferred by electrical work can be calculated using this equation:

$$E = V Q$$

energy transferred = potential difference × charge

where energy is in joules, J
potential difference is in volts, V
charge is in coulombs, C

Example

A 6V torch battery passes 250 C of charge through a bulb. How much energy has been transferred to the bulb?

Answer

$E = V Q$
$= 6 \times 250$
$= 1500\,J$

TIP

You need to be able to recall and apply both of the equations for energy transferred.

○ Power and energy

When an electricity company sends a bill to a household, the company is charging for the energy transferred. So knowing how much energy is transferred by an electrical appliance is important.

The energy transferred by an electrical appliance depends on two things:

- the **power** of the appliance
- the **time** the appliance is switched on.

So, the energy transferred by electrical work can also be calculated using this equation:

$$E = P t$$

energy transferred = power × time

where energy is in joules, J
power is in watts, W
time is in seconds, s

By rearranging this equation, the power of an appliance can be calculated.

Example

A kettle has 0.5 kg of water in it. To heat the water from room temperature to boiling, 180 kJ of energy must be transferred. It takes 2 minutes for the kettle to boil. Calculate the power of the kettle.

Answer

$$P = \frac{E}{t}$$

$$= \frac{180\,000}{120}$$

$$= 1500\,\text{W or } 1.5\,\text{kW}$$

Remember: $180\,\text{kJ} = 180\,000\,\text{J}$

$2 \text{ minutes} = 120\,\text{s}$

In fact, the power of the kettle will be a little more than this, because some energy is wasted as the hot kettle will transfer some energy to heat up the surroundings.

Example 1

The information plate on a convection heater is marked as follows:

230 V 50 Hz 1800 W

Calculate the current the heater draws from the mains supply.

Answer

$$P = VI$$

$$1800 = 230 \times I$$

$$I = \frac{1800}{230}$$

$$= 7.8\,\text{A}$$

Example 2

A current of 4.7 A passes through a 30 Ω resistor. Calculate the power transferred to the resistor.

Answer

$$P = I^2 R$$

$$= 4.7^2 \times 30$$

$$= 663\,\text{W}$$

We can combine the two equations as follows.

$$P = \frac{VQ}{t}$$

but since

$$I = \frac{Q}{t}$$

$$P = VI$$

power = potential difference × current

where power is in watts, W
potential difference is in volts, V
current is in amps, A.

Since potential difference, $V = IR$, this equation can also be written as:

$$P = I^2 R$$

power = current squared × resistance

where power is in watts, W
current is in amps, A
resistance is in ohms, Ω.

Most electrical appliances have an information plate, which tells us the power of the appliance and the potential difference of the supply. From this you can work out the current the appliance will draw.

Test yourself

24 Calculate the power rating in watts of:
   a) a car starter motor that draws a current of 90 A from a 12 V supply
   b) a toaster that draws a current of 2.5 A from a 230 V supply
   c) a phone that draws a current of 0.0003 A from a 3 V battery.
25 Calculate the electrical work done by the supply in each of the following.
   a) A 12 V battery supplies 200 C of charge to a circuit.
   b) A current of 0.2 A is drawn by a bulb from the mains 230 V supply for 30 minutes.
   c) A current of 2 mA is supplied by a 6 V battery for 2 hours.

○ National Grid

The electricity we use in our homes is generated in power stations. Our homes are often a long way from a power station and electricity is transmitted to us along overhead transmission cables (Figure 16.41). It is important to keep the current low, so that energy is not wasted in heating up the transmission cables.

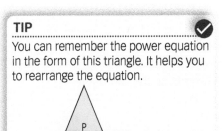

▲ **Figure 16.41** Electricity is transmitted through the National Grid using overhead transmission cables.

KEY TERMS ⭐

Step-up transformer A transformer that increases potential difference and decreases current.

Step-down transformer A transformer that decreases potential difference and increases current.

TIP ✓

Transformers make the transfer of energy from power stations to customers **more efficient** due to **reduced energy losses**.

Figure 16.42 shows how we can reduce the current in the transmission cables.

- The generator in the power station sends a current of 1000 A at a potential difference of 25 000 V into the National Grid.
- A device called a step-up transformer steps the potential difference up to 400 000 V, but reduces the current to 62.5 A.
- With a current of 62.5 A, less energy is transferred into heating up the transmission cables than with a current of 1000 A.
- Near our homes, step-down transformers step the potential difference down to a safe level of 230 V, so a larger current is available to use in the home.
- This makes the National Grid system an efficient way to transfer energy.

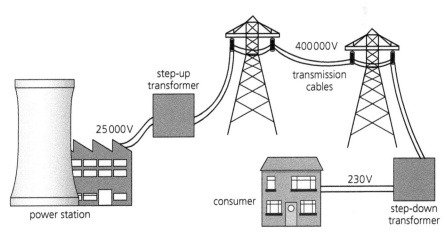

▲ **Figure 16.42** A step-up transformer increases the potential difference across the transmission cables. This allows the same power to be transmitted with a much lower current, reducing the energy wasted by heating up the cables.

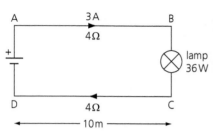

▲ **Figure 16.43** Calculating power.

Test yourself

26 In Figure 16.42 the resistance of the power line is 200 Ω.
   a) Calculate the power transferred in heating up the line when
      i) a current of 100 A passes through it
      ii) a current of 1000 A passes through it.
   b) Explain why transformers are used to reduce the current passing through the transmission cables of the National Grid.
27 A man decides to use a large cell in his house to supply current to light a 36 W lamp in his garden shed 10 m away (Figure 16.43). The resistances of the two wires to the shed are each 4 Ω.
   a) Calculate the power transferred in each wire.
   b) Calculate the fraction of the power delivered by the cell which is used to light the bulb.

Show you can...

Show you understand about the transmission of electricity by completing this task.

a) Why is electricity transmitted across the country at very high potential differences?
b) Why must we use only low potential differences in our homes?

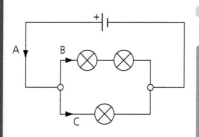

▲ Figure 16.44

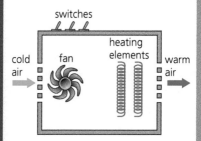

▲ Figure 16.45

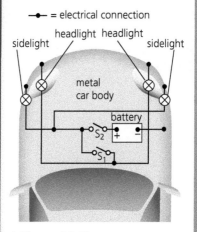

▲ Figure 16.46

Chapter review questions

1 Which current in Figure 16.44 is the largest, A, B or C? Which is the smallest?

2 Figure 16.45 shows a simplified picture of the inside of a small fan heater. The electrical wiring is not shown.

Draw, using circuit symbols, a diagram to show how the heating element, fan and switches would be connected together so that:

- when the mains is switched on, the fan comes on
- both heating elements can be switched on independently
- on hot days the heater can be used as a cooling fan.

3 Figure 16.46 shows a simplified circuit diagram for the front lights of a car. The metal body acts as a wire for the circuit.

a) Which switch operates the headlights?

b) Can the sidelights be switched on without the headlights? Give a reason for your answer.

c) If one headlight breaks, will the other still work? Give a reason for your answer.

d) What change would you have to make to the circuit if the car's body was made of plastic?

4 A $10\,\Omega$ resistor is placed in series with a bulb, a switch and a 9V battery.

a) Draw the circuit diagram.

b) When the switch is closed a current of 0.3A flows. Calculate:

i) the power supplied by the battery

ii) the power transferred to the resistor

iii) the power transferred by the bulb.

5 In Figure 16.47 each cell has a potential difference of 1.5V.

a) What potential difference does the battery produce?

b) State the reading on the voltmeter.

c) Calculate the current through the ammeter.

d) Calculate the resistance of i) the resistor, ii) the bulb.

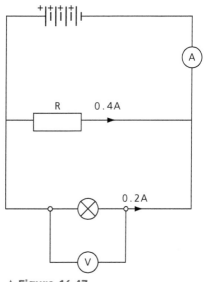

▲ Figure 16.47

6 In Figure 16.48 a 24 Ω resistor is placed in series with component X. Calculate the resistance of X.

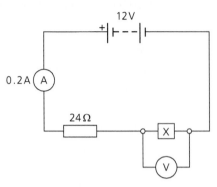

▲ Figure 16.48

7 A set of decorative lights has 115 identical bulbs connected in series. Each bulb is designed to take a current of 0.05 A. The set of bulbs is connected directly into the 230 V mains electricity supply.

a) What is the potential difference across one of the bulbs?

b) Calculate the resistance of one of the bulbs.

c) Calculate the resistance of all of the bulbs in series.

d) Calculate the power of the set of bulbs.

8 Explain why the resistance of a filament bulb increases as the current flowing through the bulb increases.

9 When the switch is closed in Figure 16.49, the bulb lights dimly at first. However, the bulb gets brighter slowly. Explain why.

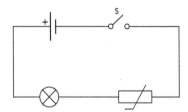

▲ Figure 16.49

Practice questions

1 Figure 16.50 shows a simple circuit.

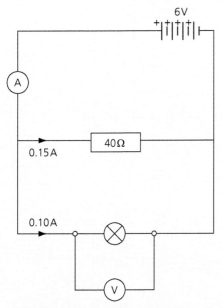

▲ Figure 16.50

a) The four cells are identical.

What is the potential difference of one cell?   [1 mark]

b) State the reading on the voltmeter.   [1 mark]

c) The current through the 40 Ω resistor is 0.15 A.

The current through the bulb is 0.10 A.

What is the reading on the ammeter?   [1 mark]

d) Use the correct answer from the box to copy and complete the sentence.

| less than   equal to   greater than |

The bulb has a resistance _____ 40 Ω

Give a reason for your answer.   [2 marks]

2 Figure 16.51 shows a simple circuit. The circuit includes an LDR.

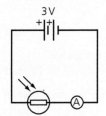

▲ Figure 16.51

a) How does the resistance of an LDR change with changing light intensity?   [1 mark]

b) Figure 16.52 shows how the reading on the ammeter changes with light intensity.

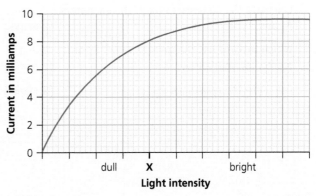

▲ Figure 16.52

i) What is the current in the circuit when the light intensity is equal to the value marked 'X'?   [1 mark]

ii) Calculate the resistance of the LDR when the light intensity is equal to the value 'X'. Give the unit.   [2 marks]

iii) Suggest a practical use for this circuit.   [1 mark]

c) Figure 16.53 shows the current–potential difference graph for an LDR in a dark room.

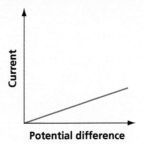

▲ Figure 16.53

Copy the graph and draw a second line to show how the current would change with potential difference if the LDR were in bright sunlight.   [1 mark]

3 Mains electricity provides an alternating current (a.c.); a cell provides a direct current (d.c.).

Describe the difference between a.c. and d.c.   [2 marks]

4 An electric iron has been wired without an earth connection. After years of use the live wire becomes loose and touches the metal part of the iron.

a) A man touches the iron and receives an electric shock. Sketch a diagram to show the path taken by the current.   [1 mark]

b) The mains potential difference is 230 V. The man's resistance is 46 kΩ.

Calculate the current that passes through the man.   [3 marks]

5 Figure 16.54 shows three resistors connected to a 12 V battery.

a) Calculate the currents through the ammeters A$_1$ and A$_2$. [2 marks]

b) Which resistor has the greater value, R$_1$ or R$_2$. Give a reason for your answer. [2 marks]

c) Calculate the resistance R$_1$. [3 marks]

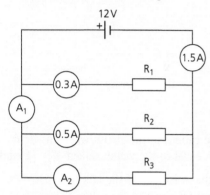

▲ Figure 16.54

6 A filament lamp is connected to a 12 V battery. The current through the lamp is recorded by a data logger when the lamp is switched on. Figure 16.55 shows how the current changes just after the lamp is switched on.

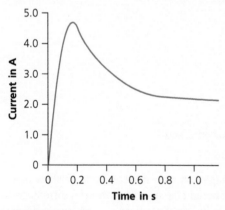

▲ Figure 16.55

a) Describe how the current changes just after the bulb is switched on. [2 marks]

b) Use the graph to determine:

i) the maximum current

ii) the current after 1 second. [2 marks]

c) The resistance of the filament increases as it gets hotter. Use this information to explain the shape of the graph. [3 marks]

d) Use the graph to calculate the power of the bulb when it is working at its steady temperature. [3 marks]

7 Figure 16.56 shows a circuit diagram which includes a diode. Figure 16.57 shows how the current through the diode varies with the potential difference across the diode.

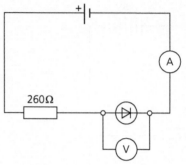

▲ Figure 16.56

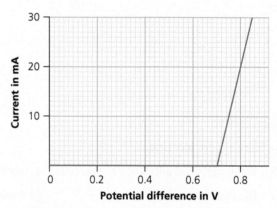

▲ Figure 16.57

a) A student sets up the circuit and measures the current through the diode as 20 mA. Use Figure 16.56 to determine the potential difference across the diode for this current. [1 mark]

b) Calculate the potential difference across the 260 Ω resistor when the current through the resistor is 20 mA. [3 marks]

c) Calculate the potential difference of the cell. [1 mark]

8 Figure 16.58 shows a circuit which includes a fixed resistor of 750 Ω and a component, X. The resistance of X changes with temperature. Figure 16.59 shows how the resistance of X changes with temperature.

a) What is component X? [1 mark]

b) At what temperature does X have a resistance of 250 Ω? [1 mark]

c) Calculate the current flowing through the circuit when X has a resistance of 250 Ω. [3 marks]

The component is now placed in a beaker of water and warmed from 20°C to 100°C.

d) Describe how the reading on the voltmeter changes as the water warms from 20°C to 100°C. Give reasons for your answer. [3 marks]

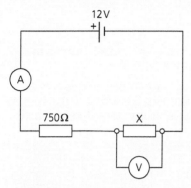

▲ Figure 16.58

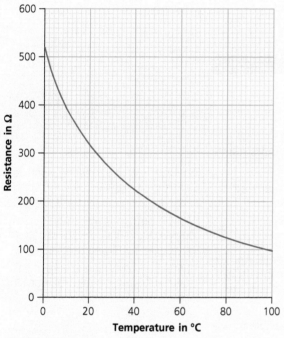

▲ Figure 16.59

9 a) The table shows the current in three different electrical appliances when connected to a 230 V a.c. supply.

Table 16.2

Appliance	Current in A
Kettle	11.5
Bulb	0.05
Toaster	4.2

i) Which appliance has the greatest resistance? How does the data show this? [2 marks]

ii) The bulb is connected to the mains supply using a thin, twin-cored cable, consisting of live and neutral connections. State two reasons why this cable should not be used to connect the kettle to the mains supply. [2 marks]

b) Calculate the power rating of the kettle when it is operated from the 230 V a.c. mains supply. [3 marks]

The kettle is taken to the USA where the mains supply has a potential difference of 115 V.

c) i) Calculate the current flowing through the kettle when it is connected to a 115 V mains supply. [3 marks]

ii) The kettle is filled with water. The water takes 90 s to boil when working from the 230 V supply. Calculate how the time it takes to boil changes when the kettle operates on the 115 V supply. [3 marks]

10 a) In which circuit does the smallest current flow from the battery? [1 mark]

b) In which circuit does the largest current flow from the battery? [1 mark]

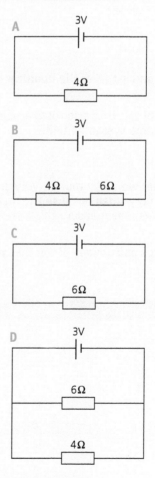

▲ Figure 16.60

Working scientifically:
Units and calibration

International System of Units (SI)

The three thermometers in Figure 16.61 are all measuring the same temperature but each one gives a different reading. This is because each thermometer has been calibrated using a different scale of units.

Scientists around the world have an agreed set of units that are used to measure quantities such as mass, time, current and temperature. These are known as the SI units. The units used in this book are SI units.

1 The table lists three quantities and the SI unit for that quantity.

Table 16.3

Quantity	Unit	Symbol
Current	ampere	A
Resistance	ohm	Ω
Energy	joule	J

Draw your own table and list all of the quantities and their SI units found in this chapter.

2 Suggest why it is important that scientists around the world measure quantities using the same system of units.

Powers of ten

Some measurements that we make may be very small or very large. An electric current may be very small because the resistance of the circuit is very large. In this case, we may measure the current in milliamps and the resistance in kilohms.

The table lists the prefixes and powers of ten that you need to be able to use.

Table 16.4

Prefix	Symbol	Power of ten
tera	T	10^{12}
giga	G	10^{9}
mega	M	10^{6}
kilo	k	10^{3}
centi	c	10^{-2}
milli	m	10^{-3}
micro	μ	10^{-6}
nano	n	10^{-9}

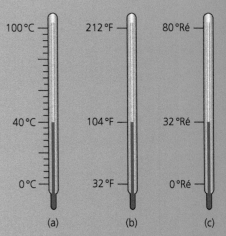

▲ Figure 16.61 a) This thermometer is calibrated in degrees Celsius (°C). This scale is the one you use in science.

b) This thermometer is calibrated in degrees Fahrenheit (°F). This scale is often used in UK weather forecasts.

c) This thermometer is calibrated in degrees Réaumur (°Ré). This is an old scale that is not used any more.

Calibrating a voltmeter to measure temperature

Zach has designed the circuit shown in Figure 16.62 to measure temperature. The circuit includes a thermistor. We know that the resistance of a thermistor changes with temperature, so the thermistor is being used as the temperature sensor.

The reading on the voltmeter changes as the resistance of the thermistor changes. This means that the voltmeter can be used to measure temperature, but first it must be calibrated.

To calibrate the voltmeter to measure temperature Zach put the thermistor into a beaker of ice cold water at exactly 0 °C. Zach then heated the water. Using an accurate thermometer, Zach measured and recorded different water temperatures and the reading of the voltmeter at those temperatures. Zach then drew the calibration graph shown in Figure 16.63.

3 Explain how Zach can use the calibration graph to convert a voltmeter reading into a temperature value.

4 Explain why Zach could not have drawn an accurate calibration graph if he had only put the thermistor into melting ice and boiling water.

5 Estimate the reading on the voltmeter if the thermistor were to be placed inside an oven at 120 °C. (The rest of the circuit is outside the oven). To do this you must assume the pattern shown in Figure 16.63 continues – this is called **extrapolating** the results. Extrapolating results is easiest if the pattern is a straight line.

6 Explain how the resolution of the voltmeter as a thermometer changes as the temperature of the thermistor increases above 60 °C.

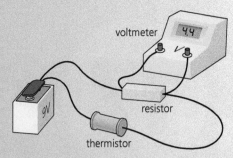

▲ Figure 16.62

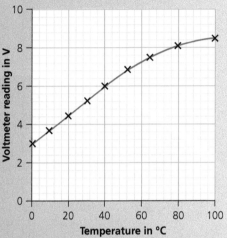

▲ Figure 16.63

> **KEY TERM**
>
> **Extrapolation** To make an estimate (or prediction) assuming that an existing trend or pattern continues to apply in an unknown situation.

17 Particle model of matter

In the 19th century scientists proposed the existence of atoms to explain some of their observations – for example the random motion of small particles of pollen, seen floating on water. By the early 20th century, we had worked out the size of atoms, but in the 21st century we can use a scanning tunnelling microscope to form an image of atoms in solids. In the photograph, each green sphere is a silicon atom. The image is generated by a computer, based on the tiny electron current detected by a probe near the surface of the silicon sample.

This chapter covers specification points 6.3.1.1 to 6.3.3.1 and is called Particle model of matter.

It covers changes of state and the particle model, internal energy and energy transfers, and particle model and pressure.

Density

A tree with a mass of $1000\,kg$ obviously has a greater mass than a steel nail with a mass of $0.01\,kg$. However, sometimes we hear people say: 'steel is heavier than wood.' What they mean is that a piece of steel has a greater mass than a piece of wood with the same volume. Steel has a greater **density** than wood.

The density of a material is defined by the equation:

$$\rho = \frac{m}{V}$$

$$\text{density} = \frac{\text{mass}}{\text{volume}}$$

where density, ρ, is in kilograms per metre cubed, kg/m^3

mass, m, is in kilograms, kg

volume, V, is in metres cubed, m^3.

Example

Aluminium has a density of $2700\,kg/m^3$. Calculate the volume of $135\,g$ of aluminium.

Answer

$$\rho = \frac{m}{V}$$

$$2700 = \frac{0.135}{V}$$

$$V = \frac{0.135}{2700}$$

$$= 0.00005\,m^3$$

$$= 5 \times 10^{-5}\,m^3$$

Remember: always make sure to work in kg and m^3.

An investigation to calculate the density of liquids and solids.

To calculate the density of a substance, the mass and volume of the substance must be measured. Density can then be calculated using the equation:

$$density = \frac{mass}{volume}$$

The density of a liquid

Method

1 Draw a table to write your results in. Make sure each column in the table has a heading that includes the quantity and the unit.

2 Use an electronic balance to measure the mass of an empty 100 millilitre (ml) measuring cylinder.

3 Take the measuring cylinder off the balance then carefully pour 20 ml of water into the measuring cylinder.

4 Measure the new mass of the measuring cylinder and water, then calculate the mass of water in the measuring cylinder.

5 Add another 20 ml of water to the measuring cylinder then measure the new mass. Repeat this by adding another 20 ml of water to the measuring cylinder.

6 You now have three sets of results for different masses and volume of water.

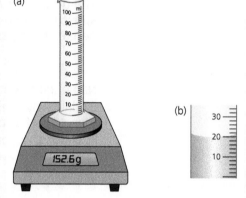

▲ Figure 17.1 a) Use an electronic balance with a resolution of at least 0.1 g. b) The surface of the water is curved (the meniscus). Measure the volume at the lowest point of the curve.

Analysing the results

1 Use each set of results to calculate a value for the density of water. Calculate the mean (average) value.

Note – a volume of 1 ml is the same as 1 centimetre cubed (cm^3).

2 If you have mass in grams (g) and volume in cm^3 then density will be in g/cm^3. Change the density to kg/m^3 by multiplying by 1000.

Taking it further

Olive oil and vinegar are often used to make a salad dressing. When left in a bottle they eventually separate out. The one with the highest density will sink to the bottom. Which one would you predict would go to the bottom of the bottle, the oil or the vinegar?

Measure the densities of olive oil and of vinegar and see if your prediction was right.

Questions

1 Why is it better to measure a large volume of liquid rather than a small volume?

2 What is meant by a resolution of 0.1 g?

The density of a regular solid

(b)

(a)

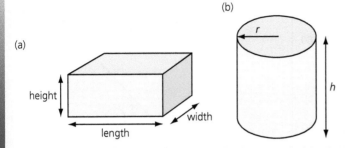

▲ **Figure 17.2** a) The volume of a cuboid = length × width × height.
b) The volume of a cylinder = $\pi r^2 h$.

Method

1 Measure the dimensions of the solid. If it is a cuboid this will be the length, width and height. Measure each dimension several times in different places. If the measurements for, say, the length are different, calculate a mean value. Then calculate the volume of the solid.

2 Measure the mass of the solid.

Analysing the results

1 Calculate the density of the solid.

2 If you know the type of material your solid is, look up its true density. Calculate the difference between your value and the true value. Do you think your value is accurate?

Taking it further

Describe how to measure the density of a sheet of aluminium cooking foil.

Questions

1 Measuring the length of a cuboid three times may give three slightly different values. Suggest why.

2 Describe how you can measure the thickness of paper using an ordinary 30 cm ruler.

The density of an irregularly shaped solid

Method

1 Make sure the object to be used fits easily into the measuring cylinder.

2 Measure the mass of the object.

3 Put enough water into the measuring cylinder to submerge the object. Measure the volume of water in the measuring cylinder.

4 Tilt the measuring cylinder and slide the object in.

5 Measure the new position of the water surface in the measuring cylinder. Then calculate the volume of the object.

Analysing the results

1 Calculate the density of the object.

Taking it further

Use several different shaped objects all of the same material. This could simply be five or six stones. Measure the mass and volume of each object. Plot a graph of mass against volume. Your graph may look like Figure 17.4.

} volume of water displaced

▲ **Figure 17.3** The volume of the object = volume of water displaced.

Use the graph to calculate the density of the objects.

Your graph may include an anomalous data point. This could be due to a big measurement error but more likely one of the objects has a different density. If so then the object must be a different material.

Questions

1 Why should a graph of mass against volume go through the origin?

2 How does plotting a graph allow you to identify anomalous data?

> **KEY TERM**
>
> **Anomalous** A result that does not fit the expected pattern.

▶ **Figure 17.4** If the line misses the origin, there may have been an error in the measurements.

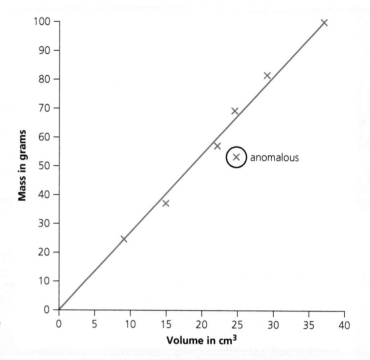

Test yourself

1 When answering an exam question a student wrote the following.

'A cork floats on water because it is lighter than water. A stone sinks because it is too heavy to float on water.'

Correct the mistakes in the student's answer.

2 A student wants to determine the density of a cuboid of material. He takes these measurements.

- Mass of the cuboid = 173.2 g
- Length = 10.1 cm; Width = 4.8 cm; Height = 1.3 cm

Calculate the density of the material in kg/m³.

3 A geologist needs to determine the density of a rock. First she weighs the rock and calculates that its mass is 90 g. Next she measures the volume of the rock by immersing it in water.

a) i) Use the information in Figure 17.5 to determine the volume of the rock in ml.

   ii) Express this volume in m³. [1 ml = 1 cm³ = 10^{-6} m³]

b) Calculate the density of the rock in kg/m³.

4 Copy the table and fill in the gaps.

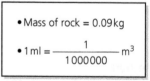

- Mass of rock = 0.09 kg
- 1 ml = $\dfrac{1}{1\,000\,000}$ m³

▲ **Figure 17.5** Measuring the volume of an irregular solid.

Table 17.1

Material	Volume in m³	Mass in kg	Density in kg/m³
Water	3	3000	
Alcohol		3200	800
Titanium	0.5		4500
Cork		0.2	200
Gold	0.02	390	

Show you can...

Show that you understand the definition of density by describing an experiment to calculate the density of a liquid. Explain what measurements you will take and the calculations you will do.

Solid, liquids and gases

Ice, water and steam are three different **states** of the same substance. We call these three states solid, liquid and gas.

TIP ✓

The forces between atoms are strong in a solid, less strong in a liquid, and weak in a gas.

- **Solid:** In a solid the atoms (or molecules) are packed in a regular structure. The atoms cannot move out of their fixed position, but they can vibrate. The atoms are held close together by strong forces. So it is difficult to change the shape of a solid.
- **Liquid:** The atoms (or molecules) in a liquid are also close together. The forces between the atoms keep them in contact, but atoms can move from one place to another. A liquid can flow and change shape to fit any container. Because the atoms are close together, it is very difficult to compress a liquid.
- **Gas:** In a gas the atoms (or molecules) are separated by relatively large distances. The forces between the atoms are very small. The atoms in a gas are in a constant random motion. A gas can expand to fill any volume.

(a) 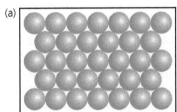 (b) 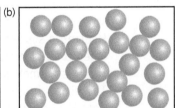 (c)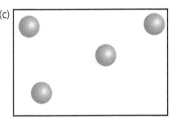

▲ **Figure 17.6** The particle arrangement in a) a solid, b) a liquid and c) a gas.

○ Density of solids, liquids and gases

The densities of solids and liquids are usually much higher than the density of gases. In solids and liquids, the atoms are closely packed together, so there is a lot of mass in a small volume. In gases the atoms are much further apart, so there is less mass in the same small volume.

Table 17.2 The densities of some solids, liquids and gases. (The gases are at room temperature and pressure.)

Material	Density in kg/m³
Lead (solid)	11 400
Glass (solid)	2 500
Water (liquid)	1 000
Lithium (solid)	500
Cork (solid)	200
Air (gas)	1.3
Hydrogen (gas)	0.09

Test yourself

5 Draw diagrams to show the arrangement of the particles in each of the three states of matter.

6 Why are gases less dense than liquids and solids?

- Lead has a much higher density than lithium because the atoms of lead have a much greater mass than lithium atoms. (The lead atoms are only slightly larger than lithium atoms.)
- The density of air is greater than the density of hydrogen, because the nitrogen and oxygen molecules in the air have a greater mass than hydrogen molecules.

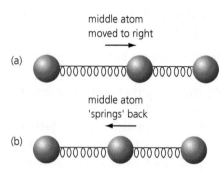

▲ **Figure 17.7** Model representing potential and kinetic energy in atoms and bonds.

○ Internal energy

Energy is stored inside a system by the particles (atoms or molecules) that make up that system. This is called **internal energy**. Internal energy is the **total kinetic and potential energy** of all the particles that make up the system.

We can use a model of several balls and springs to help us understand the nature of internal energy in a solid. The balls represent the atoms and the springs represent the forces or 'bonds' that keep the atom in place.

In Figure 17.7a) the middle atom has been displaced to the right. Now potential energy is stored in the stretched bond. In Figure 17.7b) the atom has kinetic energy as it moves to the left.

Heating

Heating changes the energy stored within a system by increasing the energy of the particles that make up the system.

● Heating can **increase the temperature** of a system. For example, when a gas is heated, the atoms (or molecules) move faster and the kinetic energy of the atoms increases.
● Heating a system can also cause a **change of state**; for example, when a solid melts to become a liquid. Usually when a solid melts, there is a small increase in volume, as the solid turns to liquid. The atoms increase their separation and there is an increase in the potential energy stored. So the internal energy increases.

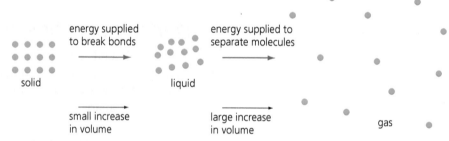

▲ **Figure 17.8** Heating can cause a change of state.

Changes of state

We use these terms to describe changes of state.

● **Melting** occurs when a solid turns to a liquid. The internal energy of the system increases.
● **Freezing** occurs when a liquid turns to a solid. The internal energy of the system decreases.
● **Boiling** or **evaporation** occurs when a liquid turns to a gas. The internal energy of the system increases.
● **Condensation** occurs when a gas turns to a liquid. The internal energy of the system decreases.
● **Sublimation** occurs when a solid turns directly into a gas. The internal energy of the system increases. Sublimation is rare. An example is carbon dioxide (CO_2): solid CO_2 (dry ice) turns directly into the gas CO_2 missing out the liquid state at normal atmospheric pressure.

A change of state of a substance is a **physical change**. The change does not produce a new substance and the process can be reversed. For example, a cube of ice from the freezer can be allowed to melt into water. The water can be put back into its container and then into the freezer. The water will freeze back into ice. No matter what its state, water or ice, the mass is the same. So, when a substance changes state, the **mass is conserved**. This is because the **total number of particles** (atoms or molecules) **stays the same**.

Test yourself

7 a) What is meant by a change of state of a substance?
  b) Give two examples of changes of state.
8 Which of the following changes are physical changes?
  ● Melting snow
  ● Burning a matchstick
  ● Breaking a matchstick
  ● Boiling an egg
  ● Mixing salt and sugar together.
9 a) What happens to the internal energy of a system when the system is heated?
  b) How is it possible to heat a system without the temperature of the system increasing?

Specific heat capacity

When the temperature of a system is increased by supplying energy to it, the increase in temperature depends on:

● the **mass** of the substance heated
● the **type of material** (what the substance is made of)
● the **energy** put into the system.

Water needs much more energy to increase its temperature by 1 °C than the same mass of concrete. This also means that when water cools by 1 °C, it gives out more energy than the same mass of concrete cooling by 1 °C.

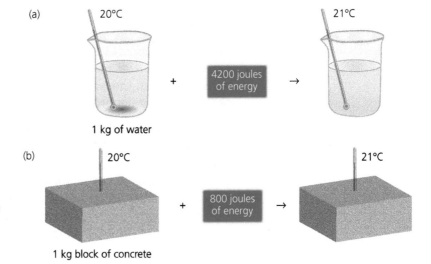

▶ **Figure 17.9** a) 4200 joules of energy are needed to increase the temperature of 1 kg of water by 1 °C. b) 800 joules of energy are needed to increase the temperature of 1 kg of concrete by 1 °C.

Example

A domestic hot water tank contains 200 kg of water. How much energy is required to warm the tank from 15 °C to 45 °C?

Answer

> Temperature rise
> $= 45\,°C - 15\,°C = 30\,°C$

Energy supplied = increase in thermal energy of the water

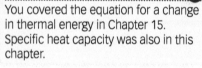

> $\Delta E = m\, c\, \Delta\theta$
> $= 200 \times 4200 \times 30$
> $= 25\,200\,000\,J$
> $= 25.2\,MJ$

TIP ✓

You covered the equation for a change in thermal energy in Chapter 15. Specific heat capacity was also in this chapter.

Based on Figure 17.9, we say that the specific heat capacity of water is 4200 joules per kilogram per degree Celsius (J/kg °C).

The **specific heat capacity** of a substance is the amount of energy required to raise the temperature of one kilogram of the substance by one degree Celsius.

To calculate the change in **thermal energy** in a substance we use the equation:

> $$\Delta E = mc\,\Delta\theta$$
>
> | change in thermal energy | = mass × specific heat capacity × temperature change |
>
> Where change in thermal energy is in joules, J
>
> mass is in kilograms, kg
>
> specific heat capacity is in joules per kilogram per degree Celsius, J/kg °C
>
> temperature change is in degrees Celsius, °C.

Table 17.3 gives some examples of specific heat capacities for various substances at 20 °C.

Table 17.3 Examples of specific heat capacities (c).

Substance	Specific heat capacity in J/kg°C
Water	4200
Alcohol	2400
Ice	2100
Dry air	1000
Aluminium	880
Concrete	800
Glass	630
Steel	450
Copper	380
Lead	160

○ The specific heat capacity of water

Water has a very high specific heat capacity. This means that 1 kg of water requires a lot of energy to heat it up and a lot of energy must be transferred from the water when it cools down. This high specific heat capacity is very important.

- We are made mostly of water. A high specific heat capacity of water means that our body temperature does not increase too much when we take exercise or cool too quickly when we go outside on a cold day.
- Water is used for keeping many homes warm. A house central heating system pumps hot water around the house. The hot water transfers a lot of energy as it flows through radiators. If water had a low specific heat capacity, water would cool down before it got to some of the radiators in your house.

Test yourself

In these questions you will need to refer to the information in Table 17.1.

10 a) A night storage heater contains 60 kg of concrete. A heater embedded in the concrete heats the concrete up from 12 °C to 37 °C. How much energy is transferred to the concrete?

   b) A heater supplies 4180 J to a block of copper of mass 0.5 kg. Calculate the temperature rise of the block.

   c) A heater supplies 21 120 J to a block of aluminium. The temperature of the block rises from 18 °C to 34 °C. Calculate the mass of the block.

11 a) In Figure 17.10 a block of tin is heated from a temperature of 20 °C to 65 °C. The mass of the block is 1.2 kg. Use the reading on the joulemeter to calculate the specific heat capacity of tin.

   b) Give two reasons why the specific heat capacity calculated in part a) is likely to be inaccurate.

12 An electric kettle has a power rating of 2.0 kW. The kettle is filled with 0.75 kg of water.

   a) Calculate the energy required to warm the water from 20 °C to 100 °C.

   b) Calculate how long it takes the kettle to bring the water to the boil at 100 °C from 20 °C.

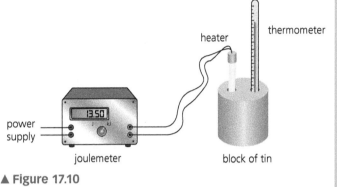

▲ Figure 17.10

○ Latent heat

When you heat a pan of water on the cooker, energy is transferred to the water and the temperature of the water increases. After a while the water begins to boil, and the temperature of the water stays constant at 100 °C. Yet, the cooker is still supplying energy to the water at the same rate. So where is the energy transferred to now? The answer is that energy is transferred into the internal energy of the steam. The molecules in steam at 100 °C have more internal energy than the same molecules of water at 100 °C.

The energy needed for 1 kg of a substance to change state is called specific latent heat.

The **specific latent heat** of a substance is the amount of energy required to change the state of one kilogram of the substance with no change in temperature.

The energy required to change the state of a substance can be calculated using this equation:

$$E = mL$$

energy required for a change of state = mass × specific latent heat

where energy is in joules, J
   mass is in kilograms, kg
   specific latent heat is in joules per kilogram, J/kg.

TIPS ✓

- Specific heat capacity has units of J/kg°C.
- Do not confuse **specific heat capacity** and **specific latent heat**.

TIP ✓

When a change of state happens, the energy supplied changes the energy stored (the internal energy) but not the temperature.

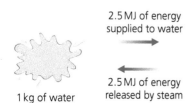

2.5 MJ of energy supplied to water

2.5 MJ of energy released by steam

1 kg of water at 100 °C

1 kg of steam at 100 °C

▲ **Figure 17.11** Specific latent heat of vaporisation of water.

Matter has three states: solid, liquid and gas. So a substance has two specific latent heats.

- The **specific latent heat of fusion** is the energy required to change 1 kg of a solid into 1 kg of a liquid at the same temperature.
- The **specific latent heat of vaporisation** is the energy required to change 1 kg of a liquid into 1 kg of a gas (vapour) at the same temperature.

Melting and freezing

When a substance melts, energy must be supplied to the substance.

When a substance freezes (or solidifies), energy is transferred from the substance to the surroundings.

Vaporising and condensing

When a substance vaporises, energy is supplied to the substance to turn it from a liquid into a gas.

When a substance condenses, energy is transferred from the substance as it changes from a gas into a liquid.

○ **Measuring latent heat**

Figure 17.12 shows how you can calculate the latent heat of vaporisation of water.

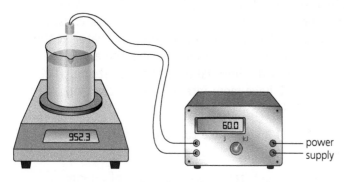

▲ **Figure 17.12** Measuring the specific latent heat of vaporisation of water.

- A beaker of water is placed on top of a balance. The beaker is on a heatproof mat. The water is then brought to boiling point with a heater. At the moment the water boils, the joulemeter is reset to zero.
- The heater is then allowed to boil the water for 5 minutes.
- The following measurements are taken:
  i) joulemeter reading after 5 minutes – 60 kJ
  ii) mass of beaker and contents at the start – 968 g
  iii) mass of beaker and contents after 5 minutes – 944 g.

Calculation

Mass of water turned to steam = 968 g − 944 g

$$= 24\,g = 0.024\,kg$$

$$E = m\,L$$

So $60\,000 = 0.024 \times L$

$$L = \frac{60\,000}{0.024}$$

$$= 2\,500\,000\,J/kg$$

$$= 2.5\,MJ/kg$$

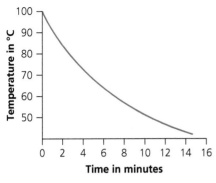

▲ **Figure 17.13** The cooling curve for water.

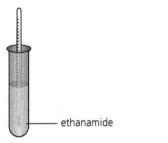

— ethanamide

▲ **Figure 17.14** Cooling ethanamide in a boiling tube.

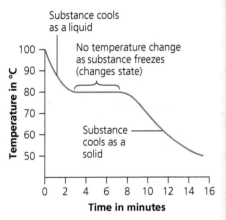

Substance cools as a liquid

No temperature change as substance freezes (changes state)

Substance cools as a solid

▲ **Figure 17.15** The cooling curve for ethanamide.

▲ **Figure 17.16** We cool when water evaporates from our bodies.

▲ **Figure 17.17**

Cooling graphs

When a boiling tube containing water is heated and then left to cool down, the temperature of the water drops gradually (see Figure 17.13).

When the temperature of the water is high, the temperature drops quickly. When water is closer to room temperature, the temperature drops more slowly.

If a substance changes state as it cools, the cooling curve takes a different shape. Ethanamide is a substance that melts at 80 °C. When a boiling tube containing ethanamide (Figure 17.14) is allowed to cool from 100 °C, it cools quickly from 100 °C to 80 °C (Figure 17.15). Then the temperature remains constant for a few minutes as the ethanamide solidifies (or freezes). Although the boiling tube continues to transfer energy to the surroundings, the temperature of the ethanamide remains constant at 80 °C. This is possible because the ethanamide releases energy as the internal energy of its molecules decreases. When all the ethanamide has solidified, its temperature begins to fall again.

Test yourself

13 a) When we take exercise we sweat. The sweat evaporates from our skin (Figure 17.16). Why does sweating help us stay cool?

   b) Even on a warm day having wet skin can soon make you feel cold (Figure 17.17). Explain why.

14 A solid is heated at a constant rate until it becomes a gas. Figure 17.18 shows how the temperature of the substance increases with time.

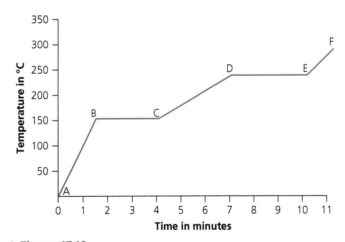

▲ **Figure 17.18**

a) Explain why the temperature of the substance remains constant over the periods:

   i) B to C

   ii) D to E.

▶

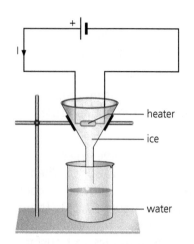

▲ **Figure 17.19** Melting ice to calculate its specific latent heat of fusion.

b) At what temperature does the substance melt?

c) Which specific latent heat is higher, fusion or vaporisation? Give a reason for your answer.

15 A student designs an experiment to calculate the specific latent heat of fusion of ice. He uses a heater to melt ice as shown in Figure 17.19.

a) Explain why the ice should be allowed to reach 0°C rather than being taken straight out of the freezer at a temperature of –18°C. The heater is turned on for 1 minute and 8g of ice melts. The heater has a power of 50W.

b) Calculate the energy supplied by the heater in 1 minute.

c) Calculate the specific latent heat of fusion of ice. Give your answer in joules per kilogram.

d) Give two reasons why the value obtained from this experiment is likely to be inaccurate.

Show you can...

Show you understand about latent heat by completing this task. Explain to a friend why it is much more painful to be burnt by 1 gram of steam at 100°C than to be burnt by 1 gram of water at 100°C.

Particle model and pressure

○ The particle model of gases

As a result of studying the behaviour of gases, we have built up a model (or theory) to help us understand, explain and predict the properties of gases. This is called the **particle model** or **kinetic theory** of gases. The main points of the model are listed here.

- The particles in a gas (molecules or atoms) are in a constant state of random motion.
- The particles in a gas collide with each other and the walls of their container without losing any of their kinetic energy.
- The temperature of the gas is related to the average kinetic energy of the molecules.
  - As the kinetic energy of the molecules increases the temperature of the gas increases.

Gas pressure

When the particles of a gas collide with a wall of their container, the particles exert a force on the wall. Figure 17.20 shows three particles bouncing off a container wall. Each particle exerts a force on the wall at right angles to the wall.

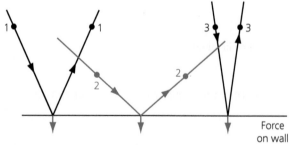

▲ **Figure 17.20** Gas particles bouncing off a wall exert a force on the wall.

The pressure inside a container of gas, with a fixed volume, is increased when the temperature of the gas is increased. When the temperature of a gas increases, the average speed of the particles in the gas increases. This means that the particles hit the walls of the container with a greater force and the particles hit the walls more frequently. So the pressure increases.

Demonstrating gas pressure

Your teacher might demonstrate the effect of gas pressure as follows:

- Take a tin with a close fitting lid and put a small amount of water in it. Press the lid firmly in place.
- Then put the tin on a tripod and heat with a Bunsen burner. After a while the lid flies off.

So why does the lid fly off?

- In Figure 17.21a) the molecules inside the tin move at the same speed as the molecules outside the tin. There is no force on the tin lid.
- In Figure 17.21b) two things have happened. As the tin warms up, some water evaporates so the number of molecules of gas inside the tin increases. Then the molecules travel faster as the temperature rises (shown with longer arrows on the molecules). The molecules inside the tin exert a force on the lid large enough to blow it off.

(a)

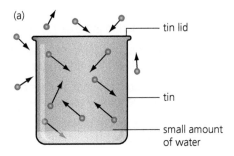

tin lid

tin

small amount of water

(b)

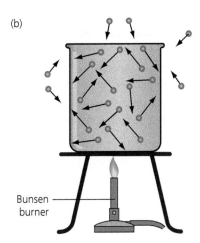

Bunsen burner

▲ **Figure 17.21** Safety note: This experiment should be done behind a safety screen and everyone should wear safety glasses.

Chapter review questions

1 The sides of a block of wood measure 4.0 cm, 3.0 cm and 5.0 cm. The block of wood has a mass of 30.0 grams.

   a) Calculate the volume of the wood in m^3.

   b) Calculate the density of the wood in kg/m^3.

2 A man buys a 'gold' ornament from an antiques shop. He decides to check if the ornament is made of solid gold.

   The results of his measurements are shown below.
   * mass of ornament = 320 g
   * volume of ornament = $26 \times 10^{-6} m^3$

   a) Explain how the man might have measured the mass and volume of the ornament.

   b) Calculate the density of the ornament.

   c) Use the data below to suggest what the man might find if he cuts his ornament in half.
   * density of gold = 19 300 kg/m^3
   * density of lead = 11 600 kg/m^3.

3 When a drop of ether is placed on the skin, the skin feels cold. Explain why.

4 a) State one difference between the arrangement of the molecules in water and the molecules in ice. Draw a diagram to illustrate your answer.

   b) An ice cube at a temperature of 0 °C is more effective in cooling a drink that the same mass of water at 0 °C. Give the reason why.

   c) Give the reason why ice floats on water.

5 Use the information in Table 17.1 to help you answer these questions.

   a) A heater supplies 200 kJ of energy to 40 kg of dry air at a temperature of 15 °C. Calculate the temperature of the air after it has been heated.

   b) A block of concrete has a mass of 60 kg. It cools down from 48 °C to 13 °C. Calculate the energy transferred from the block.

6 a) Explain in terms of molecular motion why a gas exerts a pressure on the walls of its container.

   b) The pressure in a container of gas increases when the temperature increases. Explain why.

7 A pan of water is placed on top of a cooker. The cooker transfers energy to the water at a rate of 500 W.

   a) When the water is boiling, the pan is left on the cooker for 5 minutes. Calculate the energy transferred to the cooker in this time.

   b) Calculate the mass of water that turns into steam in 5 minutes. (The specific latent heat of vaporisation of water is 2.5 MJ/kg.)

Practice questions

1 Which of the following is the correct unit for density?

kg/m²   kg/m   kg/m³   m³/kg   [1 mark]

2 Figure 17.22 represents four measuring cylinders each containing a liquid. The mass and volume of the liquid in each cylinder are given.

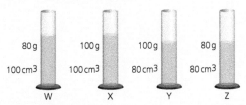

80 g	100 g	100 g	80 g
100 cm3	100 cm3	80 cm3	80 cm3
W	X	Y	Z

▲ Figure 17.22

Which two measuring cylinders could contain the same liquid?

a) W and X

b) W and Y

c) X and Y

d) X and Z   [1 mark]

3 Each of the following statements describes either a solid, a liquid or a gas.

Copy each statement and write the correct words, solid, liquid or gas, next to each one.

a) It takes the shape of its container, but does not always fill the container.   [1 mark]

b) The particles are in a regular pattern.   [1 mark]

c) The particles are free to move over each other. [1 mark]

d) The particles move in random directions.   [1 mark]

e) The particles always fill the whole of their container.   [1 mark]

f) The particles vibrate about a fixed position.   [1mark]

4 Energy is supplied to a substance and its temperature remains the same. Explain how this is possible.   [2 marks]

5 Describe the differences between the arrangement of the atoms in a solid and in a gas.   [4 marks]

6 Explain how you would use a measuring cylinder, electronic balance and some glass marbles to calculate the density of glass.   [4 marks]

7 The apparatus in Figure 17.23 is used to heat up a block of metal of mass 2 kg. When the heater is turned on, the temperature of the block of metal increases as shown in Figure 17.24.

a) Use the graph to determine the temperature rise of the metal in the first 10 minutes of heating.   [1 mark]

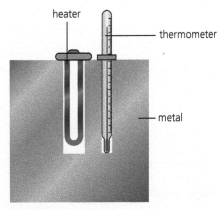

▲ Figure 17.23

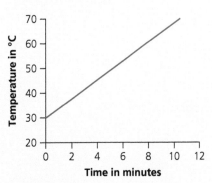

▲ Figure 17.24

b) During 10 minutes of heating, 48 000 J of energy is supplied to the block. Calculate the specific heat capacity of the block.   [4 marks]

c) Use the information in part (b) and information from the graph to show that the power of the heater is 80 W.   [3 marks]

8 Figure 17.25 shows a heater at the bottom of a boiling tube of solid wax. The heater is then connected to a power supply. A joulemeter measures the energy supplied to the heater as it melts the wax. The graph in Figure 17.26 shows how the temperature of the wax changes with the energy supplied.

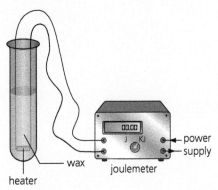

▲ Figure 17.25

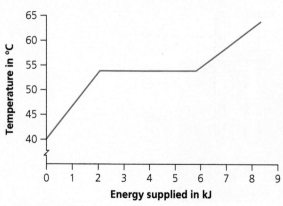

▲ Figure 17.26

a) State the melting temperature of wax. [1 mark]

b) The temperature of the wax remains constant as the wax melts. Explain why. [2 marks]

c) The mass of the wax is 50 g. Use information from the graph to calculate the specific latent heat of fusion for the wax. Give the unit. [4 marks]

Working scientifically:
Physical models

We all know what a scale model is and you have probably seen a model aircraft. It may not have been an exact replica but you would be able to recognise it as an aircraft. If you can't, then the model probably needs replacing.

In science we often use a physical model to help visualise objects or systems that are too big, too small or impossible to see. A scale model of the Solar System helps us to understand the position and distances between the planets and the Sun.

In this chapter you have read about the particle model of a gas. Atoms and molecules are too small for us to see directly so a physical model can help us to visualise what is happening and understand the real thing.

1 Describe another physical model given in this chapter.

▲ Figure 17.27 This model aircraft looks and flies just like the real thing.

Scientific models

A scientific model is an idea or related group of ideas used to explain something in the real world. A model is sometimes called a theory. Some people think that the word 'theory' means a guess; it does not. A theory or model is the idea used to explain observations and patterns in data.

A scientific model should be able to:

▶ explain observations

▶ be used to predict outcomes

▶ fit with other ideas in science.

Scientists use graphs, diagrams, equations, computer graphics and physical structures to represent and communicate models to others.

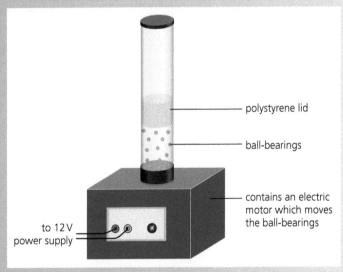

▲ Figure 17.28 The ball-bearings represent the atoms or molecules of a gas. The ball-bearings whizz around hitting the sides of the container. Turning the motor up is like increasing the temperature. The ball-bearings move faster and push the polystyrene lid upwards. So the volume expands but the pressure remains the same, just like a real gas.

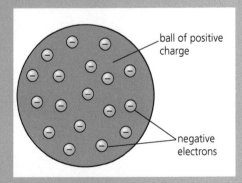

▲ Figure 17.29 A diagram used to represent the plum pudding model of the atom (see Chapter 18.)

The particle model of a gas can be represented by mathematical equations. The equations can be used to explain and predict how changing one variable, for example the volume, affects the pressure and temperature of the gas.

2 The equation $F = ma$ is a mathematical model. What would you predict happens to the acceleration of an object when the resultant force on the object is increased? How would you be able to test this model?

Why do models change?

A scientific model is only as good as the evidence that supports it. A model is supported if any predictions made using the model turn out to be correct. However, a test that gives data which the model cannot predict provides evidence that the model may be wrong.

So models change to give a better fit to any new experimental results or observations.

In Chapter 18 you can read how new experimental evidence led to the plum pudding model of the atom being replaced by the nuclear model. This model itself has been changed several times, each time to explain the most up-to-date observations.

Alternative models

Sometimes scientists have more than one model to explain the same thing. Light is sometimes described as a wave. This is one model. However, sometimes light is better explained using the idea that it is a stream of particles. So is one model right and the other wrong? The answer is no. Each of the models is appropriate but in different situations.

Scientists know that they cannot explain everything. Sometimes there are alternative models but insufficient evidence to support or reject any of them. For example, how will the Universe end? Will it expand for ever, will it eventually shrink or will it stop expanding and stay at that size? A group of scientists using a new mathematical model believe the Universe will rip itself apart. However, there is no need to panic. The model predicts it will not happen for another 22 billion years.

3 What would be needed for one of the models explaining the end of the Universe to be accepted by scientists and the other models rejected?

18 Atomic structure

Henry Becquerel discovered radioactivity in 1896. Over the last century, we have learnt how to use radioactive materials safely and to put them to good use. Radioactive sources are used in industry, agriculture and in medicine. The image shows the concentration of radioactive sugar 2 hours after tracer molecules were fed into a plant. The red colour shows a high concentration of sugar in the young leaves of the plant, which are growing. The fast growing young leaves take the sugar from the older leaves, which appear blue.

Specification coverage

This chapter covers specification points 6.4.1.1 to 6.4.2.4 and is called Atomic structure.

It covers atoms and isotopes as well as atoms and nuclear radiation.

Previously you could have learnt:

> All materials are made up of tiny particles called atoms.
> Elements are made up of only one type of atom.
> An atom has a very small positively charged nucleus.
> The nucleus contains protons and neutrons.
> Negatively charged electrons orbit the nucleus.
> The proton carries a positive charge and the electron carries a negative charge of the same size as the proton.
> An atom is neutral in charge, because the positive charge on the nucleus is balanced by the negative charge of the electrons.

Test yourself on prior knowledge

1 Which two particles are in the nucleus of an atom?
2 Which particle is more massive, the proton or the electron?
3 Why are the number of protons and electrons the same for a particular atom?

Atoms and isotopes

(a) hydrogen atom

KEY
⊕ proton
⊖ electron
● neutron

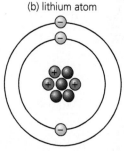
(b) lithium atom

▲ **Figure 18.1** These figures show the numbers of protons, neutrons and electrons in a) a hydrogen and b) a lithium atom.

KEY TERMS

Proton A positively charged particle found in the nucleus of an atom.

Neutron A neutral particle found in the nucleus of an atom.

Electron A negatively charged particle that orbits the nucleus of an atom.

○ Neutrons, protons and electrons

We cannot see atoms directly because they are so small. However, indirect measurements show us that the **radius** of an **atom** is about 10^{-10} m (0.000 000 000 1 m). The radius of the nucleus is less than 1/10 000 of the radius of the entire atom.

Inside the nucleus there are two types of particle, protons and neutrons. The protons and neutrons have approximately the same mass. A proton has a positive charge and the neutron has no charge. Outside the nucleus there are electrons which orbit the nucleus at distances of about 10^{-10} m. Electrons are able to move in different orbits around the nucleus, and change their orbit by absorbing or emitting electromagnetic radiation.

Electrons have very little mass in comparison with protons and neutrons. Electrons carry a negative charge which is the same size as the positive charge on the proton. Because protons and neutrons are much more massive than electrons, most of the mass of an atom is in its nucleus.

TIP

The **radius of a nucleus** is less than 1/10000 of the **radius of an atom**.

Example

A gold atom has a radius of 1.34×10^{-10} m and a gold nucleus has a radius of 7.3×10^{-15} m. How many times larger is the radius of a gold atom than the radius of a gold nucleus?

Answer

$$\frac{\text{radius of gold atom}}{\text{radius of gold nucleus}} = \frac{1.4 \times 10^{-10}\,\text{m}}{7 \times 10^{-15}\,\text{m}}$$

$$= 20\,000$$

This answer gives an approximate ratio of the atomic and nuclear radii for gold. This ratio is different for different elements.

TIPS

- Atoms are electrically neutral, because there are always the same number of electrons as protons.
- Ions are **not** atoms, because they are not electrically neutral.

○ Atoms and ions

A hydrogen atom has one proton and one electron; it is **electrically neutral** because the charges of the electron and proton cancel each other out. A helium atom has two protons and two neutrons in its nucleus, and two electrons outside that. The helium atom is also neutral because it has the same number of electrons as it has protons.

Ions

Atoms are electrically neutral because the number of protons balances exactly the number of electrons. However, it is possible either to add extra electrons to an atom or to take them away. When an electron is added to an atom a **negative ion** is formed; when an electron is removed a **positive ion** is formed. Ions are made in pairs because an electron that is removed from one atom attaches itself to another atom, so a positive and negative ion pair is formed.

○ Atomic and mass numbers

The number of protons in the nucleus of an atom determines what element it is. Hydrogen atoms have one proton, helium atoms two protons, uranium atoms 92 protons. The number of protons in the nucleus is the same a the number of electrons surrounding the nucleus. The number of protons in the nucleus is called the atomic number of the atom (symbol Z). So the proton number of hydrogen is 1; Z = 1.

KEY TERMS

Atomic number Number of protons in an atom.
Mass number Number of neutrons plus protons in an atom.

The mass of an atom is determined by the number of neutrons and protons added together. Scientists call this number the mass number of an atom.

atomic number = number of protons

mass number = number of protons plus neutrons

mass number ^{23}
atomic number $_{11}$Na

For example, an atom of sodium has eleven protons and twelve neutrons. So its atomic number is 11 and its mass number is 23. To save time in describing we can write it as $^{23}_{11}\text{Na}$; the mass number appears on the left and above the symbol Na, for sodium, and the atomic number on the left and below.

○ Isotopes

Not all the atoms of a particular element have the same mass. For example, two sodium atoms might have mass numbers of 23 and 24. The nucleus of each atom has the same number of protons, 11, but one atom has 12 neutrons and the other 13 neutrons. Atoms of the same element (sodium in this case) that have different masses are called **isotopes**. These two isotopes of sodium can be written as sodium-23, $^{23}_{11}$Na, and sodium-24, $^{24}_{11}$Na.

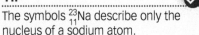

KEY TERM

Isotopes Different forms of a particular element. Isotopes have the same number of protons but different numbers of neutrons.

TIP

The symbols $^{23}_{11}$Na describe only the nucleus of a sodium atom.

Test yourself

1 a) What is the approximate radius of an atom?

10^{-3}m 10^{-6}m 10^{-10}m

  b) Use an answer from the box to complete the sentence below.

1000   10000   100000

   The radius of an atom is about _____ times larger than the radius of a nucleus.

2 The diagram shows the nuclei of three atoms. Which two atoms are isotopes of the same element? Give a reason for your answer.

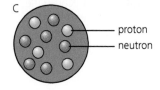

A   B   C — proton, neutron

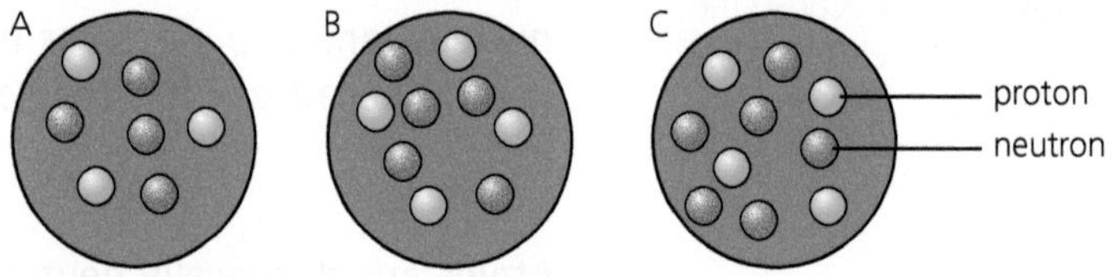

▲ Figure 18.2

3 a) A nitrogen atom has 7 protons, 7 neutrons and 7 electrons.
   i) What is the atomic number of nitrogen?
   7   14   21
   Give a reason for your answer.
   ii) What is the mass number of nitrogen?
   7   14   21
   Give a reason for your answer.
  b) Explain why a nitrogen atom is neutral.

4 Calculate the number of protons and neutrons in each of the following nuclei.
  a) $^{17}_{8}$O   b) $^{200}_{80}$Hg   c) $^{238}_{92}$U   d) $^{3}_{1}$H

5 Gadolinium-156 and gadolinium-158 are two isotopes of gadolinium, which has an atomic number of 64.
  a) Explain what the numbers 64, 156 and 158 mean.
  b) i) What do these two isotopes have in common?
     ii) How are the two isotopes different?

6 Explain the term isotope.

7 The radius of a magnesium nucleus is 3.0×10^{-15}m and the radius of a magnesium atom is 1.5×10^{-10}m. How many times larger is the radius of a magnesium atom than the radius of a magnesium nucleus?

Show you can...

Complete this task to show that you understand the model of the atom.

Write a paragraph to explain and describe the structure of an atom. In your answer include these words: nucleus, proton, neutron, electron, mass number and atomic number.

○ Discovery of the nucleus

The discovery of the nucleus and the development of our understanding of the atom provide a clear example of how new experimental evidence can lead to a scientific model being changed or replaced.

Democritus, a Greek philosopher, lived from 460–370 BC. He was the first person to suggest the idea of atoms as small particles that cannot be cut or divided.

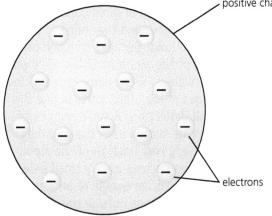

▲ **Figure 18.3** The 'plum pudding' atomic model.

The discovery of the electron

In 1897, **J. J. Thompson** discovered that electrons were emitted from the surface of hot metals. Thompson showed that electrons are negatively charged and that they are much less massive than atoms. This discovery led to a change in the accepted atomic theory.

In 1904, Thompson proposed a new model for the atom. His idea was that atoms were made up of a **ball of positive charge** with **electrons dotted around inside it** (Figure 18.3). This idea is known as the **'plum pudding'** model of the atom as it looks rather like a solid pudding with plums in it.

The nuclear model of the atom

In 1909, **Geiger and Marsden** discovered a way of exploring the insides of atoms. They directed a beam of **alpha particles** at a thin sheet of gold foil. Alpha particles were known to be positively charged helium ions, He^{2+}, which travel very quickly. They had expected all of these energetic particles to pass straight through the thin foil because they thought the atom was like Professor Thompson's soft plum pudding model. Much to their surprise they discovered that although most of them travelled through the foil without any noticeable change of direction, a very small number of the alpha particles bounced back.

> **TIP** ✅
> The symbol He^{2+} shows that the helium ion has two fewer electrons than it has protons.

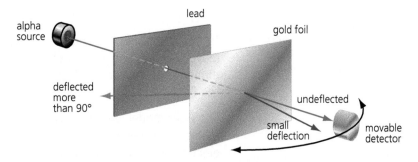

▲ **Figure 18.4** Geiger and Marsden's experiment.

In 1911, Ernest Rutherford produced a theory to explain these results. This is illustrated in Figure 18.5.

Rutherford suggested that the deflection of an alpha particle was due to an electrostatic interaction between it and a very small charged nucleus. He also suggested that the nucleus must be massive, because

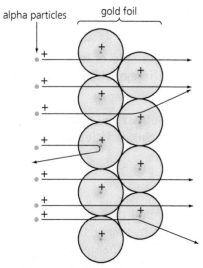

▲ **Figure 18.5** Today we know that the nucleus is positively charged. Alpha particles which are deflected by a large amount are repelled by the strong electric field of the nucleus.

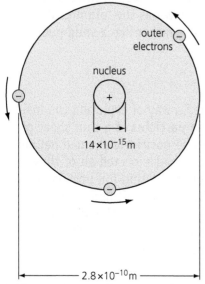

▲ **Figure 18.6** A gold atom (not drawn to scale). The diameter of the atom is about 20000 times larger than the diameter of the nucleus (though this ratio is not the same for all elements).

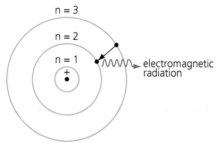

▲ **Figure 18.7** Bohr's nuclear model of the atom showing electrons orbiting the nucleus at specific distances.

it did seem to be moved by the energetic alpha particle. In Rutherford's original paper he suggested that the nucleus might be either positively or negatively charged. A positive nucleus would repel the positively charged alpha particle backwards, and a negative nucleus might pull the alpha particle round it, in the same way that a comet falling towards the Sun has its direction changed by the pull of gravity.

Alpha scattering explained

Now we know that the gap between the nucleus and electrons is large; the diameter of the atom is about 20 000 times larger than the diameter of the nucleus itself (Figure 18.6). After Rutherford's original paper, scientists confirmed that the nucleus is positively charged, and that allows us to explain the scattering of the particles as follows. Because so much of the atom is empty space (Figure 18.6), most of the alpha particles could pass through it without getting close to the nucleus. Some particles pass close to the nucleus and so the positive charges of the alpha particle and the nucleus repel each other, causing a small deflection. A small number of particles met the nucleus head on; these are turned back the way they came (Figure 18.5). The fact that only a very tiny fraction of the alpha particles bounce backwards tells us that the nucleus is very small indeed. Rutherford proposed that all the mass and positive charge of an atom are contained in the nucleus and that the electrons outside the nucleus balance the charge of the protons.

Later experiments led to the idea that the positive charge of any nucleus could be subdivided into a whole number of smaller particles. These particles were called protons, and the positive charge on a proton was discovered to be the same size as the negative charge on an electron. Rutherford did not know that there are **neutrons** in the nucleus; these were discovered by **James Chadwick** in 1932. Chadwick's discovery allowed scientists to account for the mass of the atom.

> **TIP** ✓
> The results from alpha scattering showed that the mass of an atom is concentrated at the centre (nucleus) and that the nucleus is positively charged.

The Bohr model of the atom

In 1913, **Niels Bohr** suggested a model of the atom in which electrons move round the nucleus in **circular orbits at specific distances from the nucleus**. In this model electrons can change their orbits. Figure 18.7 shows the Bohr model of a hydrogen atom, with the first three energy levels.

The Bohr model was successful in explaining why hydrogen emits particular wavelengths of electromagnetic radiation. In Figure 18.7 an electron emits electromagnetic radiation and so loses some energy. It moves closer to the nucleus as it falls from level 3 to level 2. The reverse process is possible too: if an electron absorbs electromagnetic radiation it can move into a higher energy level further away from the nucleus.

Although the Bohr model was successful up to a point, it does not allow a full explanation of the behaviour of electrons in larger atoms. However, Bohr adapted his model for the hydrogen atom by suggesting that electrons in larger atoms are also confined to specific orbits. The most recent theories suggest that electrons move in very complex orbits, which have a variety of different shapes. However, once an electron is in an orbit, it has a fixed amount of energy.

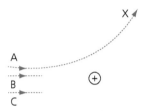

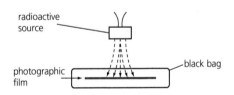

▲ **Figure 18.8** Alpha particles being deflected by the positive charge of the nucleus.

Test yourself

8 a) How is the mass of the atom distributed in the plum pudding model?
   b) Where is most of the mass of the atom in the nuclear model?
9 Explain how you know that the polarity of the charge on an alpha particle is the same as that of the nucleus.
10 Why did most of the alpha particles pass through the foil without being deflected?
11 Describe the plum pudding model of the atom.
12 a) Describe Bohr's model for the atom. Draw a diagram to help your explanation.
   b) Explain what happens to an electron when
      i)   it absorbs electromagnetic radiation.
      ii) it emits electromagnetic radiation.
13 Figure 18.8 shows the path of an alpha particle being deflected by the charge of a gold nucleus.
   Make a copy of the diagram to show the paths of two more alpha particles B and C.

Show you can...

Complete this task to show you understand how new evidence may lead to a scientific model being changed or replaced.

In the 19th century, scientists thought that atoms were the smallest particles. Explain what evidence led scientists to change this model of the atom.

Atoms and nuclear radiation

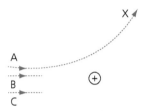

▲ **Figure 18.9** How Becquerel discovered radioactivity. Radioactive particles pass through a lightproof bag to expose a photographic film.

Henri **Becquerel** discovered radioactivity in 1896. He placed some uranium salts next to a photographic plate which had been sealed in a thick black bag to prevent light exposing the plate. When the plate was later developed it had been affected as if it had been exposed to light. Becquerel realised that new particles were emitted from uranium salts, which passed through the bag.

○ **Nuclear decay**

The nuclei of most atoms are very stable. The atoms that we are made of have been around for thousands of millions of years. Atoms may lose or gain a few electrons during chemical reactions, but the nucleus does not change during such processes.

However, there are some atoms that have **unstable nuclei** which throw out particles to make the nucleus more stable. This is a random process that depends only on the nature of the nucleus. The rate at which particles are emitted from a nucleus is not affected by other factors such as temperature or chemical reactions. One element discovered that emits these particles is radium, and the name **radioactivity** is given to this process.

There are four types of radioactive emission.

Alpha particles (α) are identical to the **nuclei of helium atoms**. The **alpha** particle is formed from two protons and two neutrons, so it has a mass number of four and an atomic number of two. When an alpha particle is emitted from a nucleus it causes the nucleus to change into another nucleus with a mass number four less and an atomic number two less than the original one, for example:

$$\underset{92}{\overset{238}{}}U \quad \rightarrow \quad \underset{90}{\overset{234}{}}Th \quad + \quad \underset{2}{\overset{4}{}}He$$

uranium   thorium   alpha particle
nucleus   nucleus   (helium nucleus)

This is called **alpha decay**.

Beta particles (β) are **fast moving electrons**. In a nucleus there are only protons and neutrons, but a beta particle is made and ejected from a nucleus when a neutron turns into a proton and an electron. Since an electron has a very small mass, when it leaves a nucleus it does not alter the mass number of that nucleus. However, the electron carries away a negative charge so the removal of an electron increases the atomic number of a nucleus by 1. For example, carbon-14 decays into nitrogen by emitting a beta particle:

$$\underset{6}{\overset{14}{}}C \quad \rightarrow \quad \underset{7}{\overset{14}{}}N \quad + \quad \underset{-1}{\overset{0}{}}e$$

carbon   nitrogen   beta particle
nucleus   nucleus   (electron)

TIP

When a beta particle is emitted from a nucleus, a neutron turns into a proton. The mass number of the nucleus remains the same, but the atomic number increases by 1.

When some nuclei decay by sending out an alpha or beta particle, they also give out a **gamma ray**. Gamma (γ) rays are **electromagnetic waves**, like radio waves or visible light. They carry away from the nucleus a lot of energy, so that the nucleus is left in a more stable state. Gamma rays have no mass or charge, so when one is emitted there is no change to the mass or atomic number of a nucleus.

Neutrons are emitted from some highly unstable nuclei. The effect of this is to reduce the mass number by 1, but the atomic number does not change. Neutron emission is rare, but neutrons are a dangerous radiation. An example of neutron emission from helium-5 is given below:

$$\underset{2}{\overset{5}{}}He \quad \rightarrow \quad \underset{2}{\overset{4}{}}He \quad + \quad \underset{0}{\overset{1}{}}n$$

Table 18.1 A summary of the four types of radioactive emission.

	Radiation emitted from nucleus	Change in mass number	Change in atomic number
alpha (α) decay	Helium nucleus $\underset{2}{\overset{4}{}}He$	−4	−2
beta (β) decay	Electron $\underset{-1}{\overset{0}{}}e$	0	+1
gamma (γ) decay	Electromagnetic waves	0	0
neutron (n) decay	Neutron	−1	0

TIP

Make sure you know and understand the nature of alpha, beta, gamma and neutron radiations.

○ Ionisation

All types of radiation cause **ionisation**. This is why we must be careful when we handle radioactive materials. The radiation makes ions in our bodies and these ions can then damage our body tissues.

Your teacher can show the ionising effect of radium by holding some close to a charged gold leaf electroscope (Figure 18.11). The electroscope is charged positively at first so that the gold leaf is repelled from the metal stem. When a radium source is brought close to the electroscope, the leaf falls, showing that the electroscope has been discharged. The reason for this is that the alpha particles from the radium create ions in the air above the electroscope. This is because the charges on the alpha particles pull some electrons out of air molecules (Figure 18.10). Both negative and positive ions are made; the positive ones are repelled from the electroscope, but the negative ones are attracted so that the charge on the electroscope is neutralised. It is important that you understand that it is not the charge of the alpha particles that discharges the electroscope, but the ions that they produce.

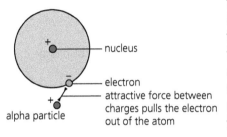

▲ **Figure 18.10** Alpha particles cause heavy ionisation.

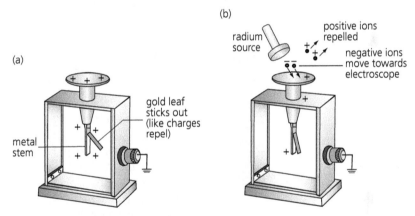

▲ **Figure 18.11** a) Positively charged electroscope. b) Negative ions neutralise the electroscope. **Safety note**: we always handle a radioactive source using tongs, to keep the source away from our body.

Test yourself

14 What is the nature of each of the following:
   a) an alpha particle
   b) a beta particle
   c) a gamma ray?
15 Which row in the table, A, B or C, shows what happens to the mass number and atomic number of a nucleus when a gamma ray is emitted from it?

Table 18.2

	Mass number	Atomic number
A	Increases	Decreases
B	Does not change	Does not change
C	Decreases	Increases

16 a) What is meant by the term ionisation?

b) Why are ions always produced in pairs?

17 Fill in the gaps in the following radioactive decay equations:

a) $^{3}_{?}H \rightarrow ^{?}_{2}He + ^{0}_{?}e$

b) $^{229}_{90}Th \rightarrow ^{?}_{?}Ra + ^{4}_{2}He$

c) $^{14}_{6}C \rightarrow ? + ^{0}_{-1}e$

d) $^{209}_{82}Pb \rightarrow ^{?}_{83}Bi + ?$

e) $^{225}_{89}Ac \rightarrow ^{?}_{87}Fr + ?$

18 $^{238}_{92}U$ decays by emitting an alpha particle then two beta particles. Which element is produced after these three decays?

19 Explain how a radioactive source emitting alpha particles can discharge a negatively charged electroscope.

Show you can...

Complete this task to show you understand the nature of radioactivity.

Explain the nature of alpha, beta, gamma and neutron radiation.

State examples of balanced nuclear equations for the emission of alpha and beta particles from a nucleus.

Detecting particles

We make use of the ionising properties of alpha, beta and gamma radiations to detect them. Those three radiations are emitted by radioactive sources permitted in schools.

Practical

The range and penetration of radiation

Safety note: Only your teacher can do this experiment as you are not allowed to handle radioactive sources until you are over 16.

Figure 18.12 shows how your teacher can use a Geiger-Müller (GM) tube and a radioactive source to investigate the range of different radiations.

1 A source emitting alpha particles, for example, can then be placed in front of the GM tube. By varying the separation (x) of the tube and the source, the range of the radiation may be calculated.

2 You can also check which materials stop a type of radiation. Now you keep the distance, x, constant. Then you can place various absorbers such as paper or metal foils in between the source and the GM tube.

KEY TERM

Geiger-Müller (GM) tube A device which detects ionising radiation. An electronic counter can record the number of particles entering the tube.

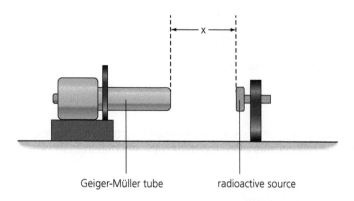

Geiger-Müller tube   radioactive source

▲ **Figure 18.12** Investigating the range of different types of radiation.

Properties of radiation

Alpha particles travel about 5 cm through air but can be stopped by a sheet of paper (Figure 18.13). They ionise air very strongly.

Beta particles can travel several metres through air. They are stopped by a sheet of aluminium that is a few millimetres thick (Figure 18.13). They do not ionise air as strongly as alpha particles.

Gamma rays can only effectively be stopped by a very thick piece of lead (Figure 18.13). Gamma rays only ionise air very weakly and travel great distances through air.

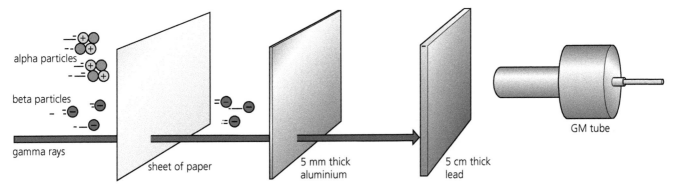

alpha particles

beta particles

gamma rays

sheet of paper

5 mm thick aluminium

5 cm thick lead

GM tube

▲ **Figure 18.13 The** penetrating power of alpha, beta and gamma radiation.

Table 18.3 The range and penetration of radiation.

Radiation	Nature	Range in air	Ionising power	Penetrating power
Alpha α	Helium nucleus	A few centimetres	Very strong	Stopped by paper
Beta β	Electron	A few metres	Medium	Stopped by aluminium
Gamma γ	Electromagnetic waves	Great distances	Weak	Stopped by thick lead

Show you can...

Complete this task to show you understand how to investigate the properties of radiations.

Design an experiment to investigate the range of alpha particles in air. How would you adapt this experiment to investigate the penetrating power of alpha particles?

Test yourself

20 Which one of the following, A, B or C, is a property of beta radiation?
   A It is the most strongly ionising radiation.
   B It will travel through several metres of air.
   C It can be easily stopped by paper.
21 a) Which type of radiation has a range of a few centimetres in air?
   b) Which types of radiation are stopped by thin metal sheets?
22 Explain why a teacher uses long tongs to handle a source of radiation.

○ Radioactive decay

The atoms of some radioactive materials decay by emitting alpha, beta or gamma radiations from their nuclei. However, it is not possible to predict when the nucleus of one particular atom will decay. It could be the next second or sometime next week or not for a million years. Radioactive decay is a **random process**.

Random process

The radioactive decay of an atom is rather like rolling dice or tossing a coin. You cannot say with certainty that the next time you toss a coin it will fall heads up. However, if you throw a lot of coins you can start to predict how many of them will fall heads up. You can use this idea to help you understand how radioactive decay happens. Imagine you start off with a thousand coins. When any coin falls heads up then it has 'decayed' and you take it out of the game. Table 18.2 shows the likely result (on average). Every time you throw a lot of coins, about half of them will turn up heads.

▲ **Figure 18.14** On average how many sixes will turn up when you roll 100 dice? Why can you not be certain what will happen on each occasion?

Table 18.4 Modelling radioactive decay using coins.

Throw	Number of coins left
0	1000
1	500
2	250
3	125
4	62
5	31
6	16
7	8
8	4
9	2
10	1

Radioactive decay

Radioactive materials decay in a similar way. If we start off with a million atoms then after a period of time (for example 1 hour), half of them will have decayed. In the next hour we find that half of the remaining atoms have decayed, leaving us with a quarter of the original number. The period of time taken for half the number of nuclei to decay in a radioactive sample is called the half-life and it is given the symbol $t_{\frac{1}{2}}$. It is important to understand that we have chosen a half-life here of 1 hour to explain the idea. Different radioisotopes have different half-lives.

Count rate and activity

The activity of a source is equal to the number of particles emitted per second. We can express this as counts per second, but in honour of Henry Becquerel, this unit is called the becquerel (Bq).

The count rate is the term we use when a GM tube is measuring the radiation emitted from a radioactive source. This is different from the activity of a source, because not all the radiation emitted from the source goes into the GM tube. Count rates may be in counts per second or sometimes counts per minute.

Measurement of half-life

If you look at Table 18.5, you can see that the number of nuclei that decayed in the first hour was 500 000, then in the next hour 250 000 and in the third hour 125 000. So as time passes not only does the number of nuclei left get smaller but so does the rate at which the nuclei decay. So by measuring the activity of a radioactive sample we can determine its half-life.

Figure 18.15 shows the result of a laboratory experiment to determine the half-life of a radioactive material. You can see that the count rate detected halves every half-life. At the start of the experiment the count rate was 40 per second, after 50 seconds (one half-life) it has reduced to 20 per second, and after a further 50 seconds the count rate has halved again to 10 per second.

> **KEY TERM**
>
> **Half-life** The time taken for the number of nuclei in a radioactive isotope to halve. In one half-life the activity or count rate of a radioactive sample also halves.
>
> **1 becquerel** (1 Bq) = An emission of 1 particle/second.

> **TIP**
>
> **Activity** is the rate at which a source of unstable nuclei decays.

Table 18.5 Radioactive decay and half life.

Time (hour)	Number of nuclei left
0	1000000
1	500000
2	250000
3	125000
4	62500
5	31250
6	15620
7	7810
8	3900
9	1950
10	975

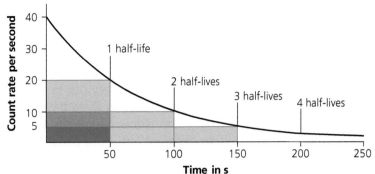

▲ **Figure 18.15** A typical curve showing radioactive decay over time.

Table 18.6 Half-lives of some radioisotopes.

Isotope	Half-life
Potassium-40	1.3 billion years
Carbon-14	5700 years
Caesium-137	30 years
Iodine-131	8 days
Lawrencium-260	3 minutes
Nobelium-252	2.3 seconds

Test yourself

23 Use a word from the box to complete the sentence.

count rate   half-life   reaction

The _____ is the number of alpha, beta or gamma emissions detected from a radioactive source in one second.

24 The graphs in Figure 18.16 show the decay of three different radioactive isotopes.
Which isotope has:
a) the longest half-life?
b) the shortest half-life?

25 Explain what the word random means.

26 A radioactive material has a half-life of 15 minutes. What does this mean? How much of the original material will be left after 60 minutes?

27 The results in the table opposite for the count rate of a radioactive source were recorded every minute. Plot a graph of the count rate (*y*-axis) against time (*x*-axis) and use the graph to work out the half-life of the source.

28 A radioactive isotope has a half-life of 8 hours. At 12 noon on 2 March a GM tube measures a count rate of 2400 per second.
a) What will be the count rate at 4.00 am on 3 March?
b) At what time will a count rate of approximately 75 per second be measured?

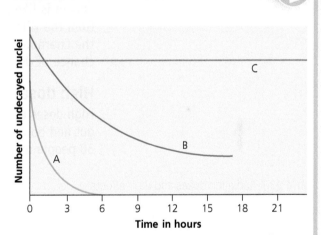

▲ **Figure 18.16** Three radioactive isotopes with different half-lives.

Table 18.7

Count rate in Bq	Time in minutes
1000	0
590	1
340	2
200	3
120	4
70	5

○ Radiation damage

If radiation gets into our body, damage can occur to cells and tissues. The ions which are produced by the radiation produce chemicals which destroy cells they come into contact with. Alpha particles cause the most damage if they get inside the body, as they are strongly ionising; this could happen if we breathed in a radioactive gas. An alpha source in school is less dangerous as the radiation only travels short distances, so does not enter our body.

Although gamma rays are less ionising than alpha particles, a gamma ray source in a laboratory is a hazard, as the rays are so penetrating. A gamma ray can pass into our body from a source several metres away from us.

Table 18.8 gives some examples of when we are exposed to radiation and the risk.

Low doses

Low doses of radiation, are unlikely to cause us harm. However, it is possible that any exposure to radiation will increase our chances of cancer.

Moderate doses

Moderate doses of radiation are unlikely to kill someone, but the person will be very unwell. Damage will be done to cells in the body, but not enough to be fatal. The body will be able to replace dead cells and the person is likely to recover completely. However, studies of the survivors from the Hiroshima and Nagasaki bombs in 1945, and of survivors from the Chernobyl reactor disaster of 1986, show that there is an increased chance of dying from cancer some years after the dose of radiation.

High doses

High doses of radiation are likely to be fatal. A high dose damages the gut and bone marrow so much that the body cannot work normally. About 30 people died of acute radiation syndrome in the Chernobyl disaster.

Table 18.8 Radiation doses and their effects.

Source of dose	Effect of dose
Airport security scan; Eating a banana	Low risk
Dental X-ray	Low risk
Transatlantic flight	Low risk
Chest X-ray	Low risk
Release limit from a nuclear plant per person per year	Low risk
Yearly dose from food	Low risk
Mammogram	Low risk
Average computer tomography (CT) scan	Low risk
Maximum yearly dose permitted for radiation workers	Medium risk
Lowest annual dose where increased risk of cancer is evident	Medium risk
Highly targeted dose used in radiotherapy (single dose)	High risk, but balanced by a likely cure of cancer
Extremely severe dose, received in a nuclear accident	Death probable within 6 weeks
Maximum radiation dose per day found at the Fukushima plant in 2011	Fatal dose; death within 2 weeks
10 minutes exposure to the Chernobyl reaction meltdown in 1986	Death within hours

Test yourself

29 Name four nuclear radiations which are dangerous to us.

30 The table below shows the results of research into the number of deaths caused by various types of radiation.

Table 18.9

Source of radiation	Type of radiation	Number of people studied	Extra number of cancer deaths caused by radiation
Uranium miners	Alpha	3400	60
Radium luminisers	Alpha	800	50
Medical treatment	Alpha	4500	60
Hiroshima bomb	Gamma rays and neutrons	15000	100
Nagasaki bomb	Gamma rays	7000	20

a) Discuss whether the table supports the suggestion that alpha particles are more dangerous than gamma rays.

b) What conclusion can you draw about the relative dangers of neutrons?

c) A student comments that the table does not provide a fair test for comparison of the radiations. Comment on this.

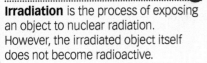

Show you can...

Complete this task to show that you understand the hazards of radiation.

Explain how radiation can damage our bodies. Explain how measuring radiation doses can help reduce risk to people.

TIP ✓

Irradiation is the process of exposing an object to nuclear radiation. However, the irradiated object itself does not become radioactive.

◯ Irradiation and contamination

When a teacher brings a radioactive source into a laboratory to demonstrate to her pupils, she **irradiates** the surroundings (with a very small safe dose) but does not cause any contamination. When the source is near a GM tube, the tube is irradiated. This means that radiation from the source is entering the tube. However, as soon as the source is removed and put away, the GM tube will not be radioactive, because there are no radioactive nuclei inside it to emit radiation.

Radioactive isotopes cause **contamination** when they get into places where we do not want them. For example, iodine-131 and caesium-137 were emitted in the Fukushima and Chernobyl disasters. Iodine-131 has a half-life of 8 days, which means that it is highly active for a short time. Iodine disperses in the atmosphere and gets into the food chain. This poses a very high risk, but for a short period. Caesium-137, however, has a half-life of 30 years. Consequently, the ground and water in the region close to the two nuclear reactors will be contaminated for many years to come.

Both irradiation and contamination are potentially hazardous to us. However, when a patient is exposed to a dose of radiation, for example, in hospital, the dose will be carefully calculated and controlled. The problem with contamination is that a person is exposed to radiation in a way which is unknown, and it is possible that a large and very dangerous dose of radiation is consumed by mistake.

Test yourself

31 Injecting a radioactive substance into a person to diagnose a medical problem always involves some risk to a person's health. Use an answer from the box to complete the sentence.

| greater than   the same as   less than |

A radioactive substance may be used to diagnose a medical problem if the potential benefit of the diagnosis is _____ than the risk to the person's health.

Chapter review questions

1 Lithium atoms have three protons and four neutrons.

   a) What is the atomic number of lithium?

   b) What is the mass number of lithium?

   c) How many electrons are there in a lithium atom?

2 There are two stable isotopes of carbon: carbon-12 and carbon-13.

   a) Explain what is meant by the words

      i)   stable

      ii) isotope.

   b) The atomic number of carbon is 6. How many protons and neutrons are there in each of the isotopes mentioned above?

3 How do the atoms of one particular element differ in atomic structure from the atoms of all other elements?

4 a) Write nuclear equations for the alpha decay of:

      i) $^{241}_{94}$Pu to uranium (U)

      ii) $^{229}_{90}$Th to radium (Ra)

      iii) $^{213}_{84}$Po to lead (Pb).

   b) Write nuclear equations for the beta decay of:

      i) $^{237}_{92}$U to neptunium (Np)

      ii) $^{59}_{26}$Fe to cobalt (Co)

      iii) $^{32}_{14}$Si to phosphorus (P).

5 Why did the work done by Geiger and Marsden convince scientists that the 'plum pudding' model of the atom needed to be replaced?

6 In the alpha scattering experiment about 1 in 10 000 alpha particles bounced back from the gold foil.

   Explain how you think the number of alpha particles bouncing back will change when:

   a) thicker gold foil is used

   b) aluminium foil of the same thickness is used.

7 Bismuth-213 emits alpha particles.

   a) What is an alpha particle?

   b) Explain why bismuth-213 would be highly dangerous if put inside the body.

8 Zak's teacher carried out an experiment to measure the half-life of
   protactinium-234. His results are shown in the table.

Table 18.10

Time in seconds	Count rate in counts per second
0	66
40	44
80	30
120	20
160	13
200	9
240	6
280	4

a) In the experiment, what was:

   i)   the independent variable

   ii) the dependent variable?

b) What instrument was used to measure the count rate?

9 The table gives information about some radioactive sources that emit ionising
   radiation.

Table 18.11

Source	Radiation emitted	Half-life
Bismuth-213	Alpha	45 minutes
Cobalt-60	Gamma	5 years
Uranium-233	Alpha	150000 years
Radon-226	Beta	6 minutes
Technetium-99	Gamma	6 hours

a) What is 'half-life'?

b) i) Explain what is meant by the term 'ionising radiation'.

   ii) Which of the radiations shown in the table is the most ionising?

Practice questions

1 The diagram represents an atom of beryllium-9.

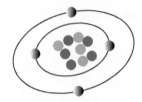

▲ Figure 18.17

a) i) Copy and complete the following table of information for an atom of beryllium-9. [3 marks]

Table 18.12

Number of electrons	
Number of protons	
Number of neutrons	

ii) What is the atomic number of a beryllium-9 atom?

Choose the correct answer from the box.

| 4 5 9 13 |

[2 marks]

Give the reason for your answer.

b) Beryllium-10 is a radioactive isotope of beryllium.

i) Choose the correct answer from the box to complete the sentence. [1 mark]

| electron   neutron   proton |

The nucleus of a beryllium-10 atom has one more _____ than the nucleus of a beryllium-9 atom.

ii) Beryllium-10 decays by emitting beta particles.

Which statement, A, B or C, describes a beta particle?

A the same as a helium nucleus

B an electromagnetic wave

C an electron from the nucleus [1 mark]

c) The graph shows how the count rate from a sample of beryllium-10 changes with time.

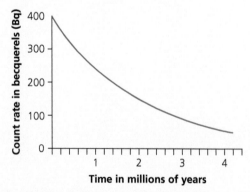

▲ Figure 18.18

i) How many millions of years does it take for the count rate to fall from 400 Bq to 40 Bq? [1 mark]

ii) What is the half-life of beryllium-10? [1 mark]

2 a) The statements A, B and C give three properties of nuclear radiation.

A It will pass through cardboard but not thin aluminium.

B It is weakly ionising.

C It can travel only a few centimetres through the air.

i) Which **one** of the statements, A, B or C, gives a property of alpha radiation? [1 mark]

ii) Which **one** of the statements, A, B or C, gives a property of gamma radiation? [1 mark]

b) Fresh strawberries grown abroad are sometimes irradiated before being sent to the UK.

The irradiation process kills bacteria on the strawberries.

i) Which **one** of the statements, X, Y or Z, is correct? [1 mark]

X The irradiated strawberries do not become radioactive.

Y Particles containing radioactive atoms settle on the strawberries.

Z The strawberries cannot be eaten for a few days after irradiation.

ii) Suggest **one** reason why the farmers growing the strawberries want the strawberries to be irradiated. [1 mark]

3 a) Phosphorus is an element with an atomic number of 15. Its most common isotope is phosphorus-31, $^{31}_{15}P$. Another isotope, phosphorus-32, is radioactive.

i) State the number of protons, neutrons and electrons in phosphorus-31. [2 marks]

ii) Why does phosphorus-31 have a different mass number from phosphorus-32? [1 mark]

iii) Atoms of phosphorus-32 change into atoms of sulfur by beta decay. Copy and complete the equation to show the atomic and mass numbers of this isotope of sulfur.

$^{32}_{15}P \rightarrow\ ^{?}_{?}S + beta$ [2 marks]

b) i) Name a suitable detector that could be used to show that phosphorus-32 gives out radiation. [1 mark]

ii) Name a disease that can be caused by too much exposure to a radioactive substance such as phosphorus-32. [1 mark]

4 A radioactive source emits alpha (α), beta (β) and gamma (γ) radiation.

   a) Which **two** types of radiation will pass through a sheet of card? [1 mark]

   b) Which type of radiation has the greatest range in air? [1 mark]

5 The table gives information about some of the radioactive substances released into the air by the explosion at the Fukushima nuclear plant in 2011.

   Table 18.13

Radioactive substance	Half-life	Type of radiation emitted
Iodine-131	8 days	Beta and gamma
Caesium-134	2 years	Beta
Caesium-137	30 years	Beta

   a) How is the structure of a caesium-134 atom different from the structure of a caesium-137 atom? [1 mark]

   b) What are beta and gamma radiations? [2 marks]

   c) A sample of soil is contaminated with some iodine-131. Its activity is 10 000 Bq. Calculate how long it will take for the activity to drop to 2 500 Bq. [2 marks]

   d) Which of the three isotopes will be the most dangerous 50 years after the accident? Explain your answer. [2 marks]

6 The table gives information about five radioactive isotopes.

   Table 18.14

Isotope	Radiation emitted	Half-life
A	Alpha	4 minutes
B	Gamma	5 years
C	Beta	12 years
D	Beta	28 years
E	Gamma	6 hours

   a) What is a 'beta particle'? [1 mark]

   b) What does the term 'half-life' mean? [1 mark]

   c) Radioactive waste needs to be stored. One suggestion is to seal it in steel drums and bury these in deep underground caverns. Suggest why people may be worried about having such a storage site close to where they live. [3 marks]

7 Some patients who suffer from cancer are given an injection of boron, which is absorbed by cancer cells. The cancerous tissue is then irradiated with neutrons. After this, the reaction that occurs is:

$$^{11}_{5}B \rightarrow \,^{7}_{x}Li + \,^{y}_{z}He$$

   a) Copy the equation and write numbers in place of x and y. [3 marks]

   b) Why do the lithium and helium nuclei repel each other? [1 mark]

   c) Why is this process dangerous for healthy patients? [1 mark]

8 In the early 20th century scientists thought that atoms were made up of electrons embedded inside a ball of positive charge. They called this the 'plum pudding' model of the atom.

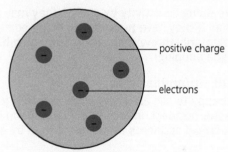

plum pudding model

▲ Figure 18.19

Geiger and Marsden fired a beam of alpha particles at a thin gold foil. Explain how the results of their experiment led to a new model of the atom. Illustrate your answer with suitable diagrams. [6 marks]

Working scientifically:
Risk and perception of risk

Before starting an investigation you complete a risk assessment. You identify the hazards, the risks and the controls needed to reduce the identified risks. For example, if you use boiling water there is always a risk you may scald yourself.

1 Look at the information on radiation and on radiation damage earlier in this chapter. What controls should be used to reduce the risk to the teacher and students of using a radioactive source?

Perception of risk

The perception of risk is often different from the actual risk. This happens for different reasons.

Voluntary versus imposed risk

When people are doing an activity by choice, they may perceive it to have a smaller risk than if they have no choice in doing something. For example, some people voluntarily throw themselves from high bridges and cranes. They call it bungee jumping!

Familiar versus unfamiliar risk

People may be happy to use electrical appliances but worry about the risk of a nuclear reactor accident. In fact, many more people have been killed or injured by electrical accidents in the home than by accidents at nuclear reactors.

Visible versus invisible hazard

Every day we accept the risks involved in crossing the road. The hazards are real but our perception of the risk is likely to be reduced because generally we can see the hazards. Nuclear radiation, however, is invisible. We can't see it, we can't feel it, we can't smell it. The perceived risk is usually high.

2 Which of the following involves a voluntary risk?

   a) Flying to a holiday destination.

   b) Riding a bicycle to school.

   c) Exposure to background radiation.

   d) Learning to drive a car.

Electrical safety in the home

A recent survey found that many people perceive the risk of injury from using electrical appliances to be much lower than the accident figures suggest. On average, fires caused by the misuse of electrical appliances kill one person each week and seriously injure about 2500 people each year. In addition, thousands of people are injured each year (with about 30 deaths) by electric shocks. Perhaps it's because electrical appliances are so familiar to us that we simply forget or underestimate the risks.

▲ Figure 18.20 In a fall, a horse rider reduces the risk of serious injury by wearing a hard hat and body protector.

▲ Figure 18.21 For some people, having a suntan outweighs the risk of getting skin cancer from over exposure to ultraviolet radiation.

▲ Figure 18.22 An overloaded socket increases the risk of a fire.

19 Homeostasis and the human nervous system

In order that you can read this sentence, light must reflect into your eyes from the page. Your eyes really are amazing. They contain millions of receptor cells, which detect different wavelengths of light and start electrical signals. These travel along your nerves in your central nervous system to your brain. Your nervous system is your brain, spinal cord and the network of nerves that spread throughout your body. You have a staggering 45 miles of them!

Specification coverage

This chapter covers specification points 4.5.1 to 4.5.2 and is called Homeostasis and the human nervous system.

It covers homeostasis and the structure and function of the nervous system.

Prior knowledge

Previously you could have learnt:

> **The body consists of a range of different types of cells and systems.**

Test yourself on prior knowledge

1 What is a neurone (nerve cell)?
2 What does the nervous system do?

Homeostasis

There are certain conditions that your body needs to keep stable in order to survive. You need to have enough glucose in your blood for your cells to respire. Not enough would leave you without sufficient energy, and too much would send you into a coma. You need to be warm. Too hot or too cold and your enzymes wouldn't be able to control your cellular reactions. You need to have sufficient water. Too little or too much water in your body would kill you. So your body maintains at particular levels or concentrations your:

- **blood glucose concentration**
- **body temperature**
- **water balance.**

KEY TERM

Homeostasis The maintenance of a constant internal environment.

The maintenance of these three key conditions (and many others) is called homeostasis. This is the detection of changes to these conditions and responses to return the body to normal. The definition of homeostasis is the regulation of the internal environment of a cell or organism to maintain optimal conditions for function in response to internal and external changes. These changes are **automatic**. You do not know that they are occurring. They are described as involuntary.

TIP ✓

It is important that you can explain the importance of homeostasis in maintaining **optimal conditions** for **enzyme action** and cell function.

Many of your body's systems are involved in maintaining internal conditions. Your **nervous system** coordinates your voluntary and involuntary actions. It does this by transmitting electrical impulses along your nerve cells. These electrical impulses move very quickly along your neurones. As a consequence, homeostatic responses that involve your nervous system happen **very quickly**.

TIP ✓

You should be aware that homeostatic systems can adjust levels up or down. For example, if blood glucose is too low, the level can be increased, and if blood glucose is too high, the level can be reduced.

Other parts of your body produce chemicals called **hormones**. These are proteins, and they are released by glands into your bloodstream. Hormones travel around the blood until they reach their **target organ**, where they act. Because hormones travel in the blood, their effects are much **slower** than those of nerve impulses.

The electrical impulses that travel along your nerve cells from receptors reach parts of your body called **coordination control centres**. These include your brain, spinal cord and pancreas, and they process the information and respond accordingly. These centres can respond by releasing a hormone in the case of the glands in your brain or your pancreas. Coordination centres can also send electrical impulses back along

KEY TERM

Effector A muscle or a gland.

your nerve cells. These usually end in glands or muscles, which are called **effectors** because they can bring about a response. For example, the sweat glands in your skin might produce more sweat in response to a high temperature.

Receptors in the body detect stimuli (changes in the environment).	→	**Coordination centres** such as the brain, spinal cord and pancreas receive and process information from receptors.	→	**Effectors** such as muscles or glands bring about responses that result in optimum levels being restored.

▲ **Figure 19.1** Control systems in homeostasis.

Show you can...

Explain why maintenance of blood glucose is an example of homeostasis.

Test yourself

1 Give an example of homeostasis.
2 Give an example of a control centre.
3 Describe why responses along nerve cells are faster than hormonal ones.

The human nervous system – structure and function

An animal's nervous system controls its voluntary and involuntary actions and is responsible for transmitting and receiving impulses in different parts of its body. In most animals, including humans, it comprises the **central nervous system (CNS)** and the network of nerves that spreads from it throughout your body. The CNS is composed of your **brain** and **spinal cord**. Nerves are made from bundles of individual neurones (nerve cells).

KEY TERM

Central nervous system (CNS) The brain and spinal cord.

TIP

Neurones link receptors and effectors to the coordinator.

Receptors in your body detect a change inside or outside your body.	→	Your **central nervous system** coordinates your body's response.	→	**Effectors** cause a response by moving part of your body or secreting a hormone.

▲ **Figure 19.2** This flow diagram shows how the nervous system enables the body to respond to changes.

Generating electrical impulses

All messages sent along nerve cells are electrical. This means they move quickly along them. In longer nerve cells that have a myelin sheath to insulate the impulse, these speeds can reach 120 metres per second. This is over 250 miles per hour! Think about how fast you react if you put your hand on a hot surface.

These electrical impulses are generated by special cells called **receptors** and travel to your brain and/or spinal cord. There are many different types of receptor that measure and respond to different stimuli. These are often associated with your sense organs. Table 19.1 shows your senses and the associated stimuli.

KEY TERM

Receptor A cell or group of cells at the beginning of a pathway of neurones that detects a change and generates an electrical impulse.

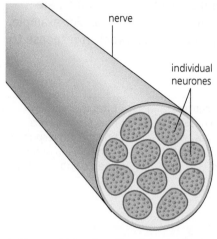

▲ **Figure 19.3** A bundle of neurones forms a nerve.

KEY TERMS ★

Sensory neurone A neurone that carries an electrical impulse from a receptor towards the central nervous system.

Relay neurone A neurone that carries an electrical impulse within the central nervous system (brain and spinal cord).

Motor neurone A neurone that carries an electrical impulse away from the central nervous system to an effector (muscle or gland).

Table 19.1 Your senses, the organs and the stimuli involved.

Sense	Organ	Stimuli
Sight	Eyes	Light
Hearing	Ears	Sound
Taste	Tongue	Chemicals in food
Smell	Nose	Chemicals in air
Touch	Skin	Touch, pressure, temperature, pain and itching

Your skin is covered with a large number of different receptors. They are not always found in the same number in the same place in your skin. Which parts of your skin are the most sensitive? This is where the most receptors are found. Usually your fingertips and lips have the most receptors, and the soles of your feet and elbows have the least.

◯ Sensory, relay and motor neurones

Once a receptor has started an electrical impulse, the impulse moves along a neurone towards your brain or spinal cord. Any neurone that takes an impulse in this direction is called a sensory neurone. Once inside your CNS the electrical impulse is passed along relay neurones. Finally the impulse passes from your brain or spinal cord to your muscles or glands along a motor neurone. Your muscles can then contract and relax to help you move in response. Your glands can produce and secrete hormones into your bloodstream. Because muscles and glands both effect a response, we call them **effectors**. This can be shown in a pathway:

stimulus → receptor → coordinator → effector → response

This can also be shown in more detail:

stimulus → receptor → sensory neurones → relay neurones → motor neurones → effector → response

Remember that the only cells in this pathway that are completely within your CNS are the relay neurones. The differences in the structures of sensory, relay and motor neurones are shown in Figure 19.4.

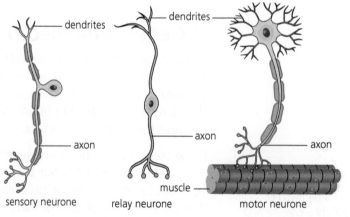

▲ **Figure 19.4** The three types of neurone.

⃝ Synapses

Neurones in this network do not join directly to each other. There is a small gap between each of them called a synapse. When an electrical impulse reaches the end of the axon it spreads out into the ends of the cell, which look a little like the roots of a plant. At the tips of these 'roots' are special areas that convert the electrical impulse into a chemical signal. These are called neurotransmitters and quickly diffuse across the synapse. On the other side the neurotransmitters meet the dendrites of the next nerve cell. Here they bind to receptors and trigger the start of the electrical impulse, which travels along that neurone until it reaches the next synapse.

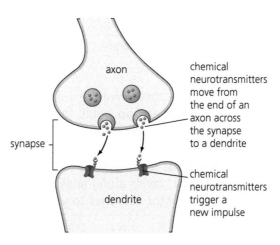

axon

chemical neurotransmitters move from the end of an axon across the synapse to a dendrite

synapse

chemical neurotransmitters trigger a new impulse

dendrite

▲ **Figure 19.5** How a synapse functions.

⃝ **The reflex arc**

Some of your responses are automatic. They happen quickly. An example of this might be moving your hand away from a hot radiator you have just touched by mistake. Other automatic responses include regulating your heart rate and controlling the amount of light that enters your eyes. These responses do not require your brain to make a decision. They are called reflex responses.

Because reflex responses happen quickly they often occur to stop you damaging your body by mistake. Remember how quickly you move your hand from a hot radiator! Just as described before, the electrical impulse is first generated by a receptor. In this case, this is a pain receptor in the skin of your hand. The impulse travels quickly through sensory neurones to your spinal cord. Here relay neurones take over the impulse. However, this impulse does not travel immediately to your brain. Your spinal cord is able to send the impulse back along motor neurones to your muscles. These then quickly contract, moving your hand from the radiator. This nervous pathway is called a reflex arc.

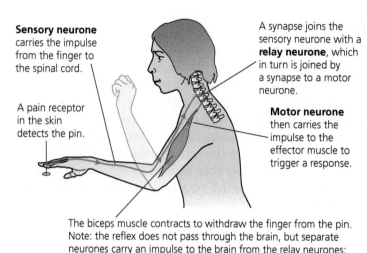

Sensory neurone carries the impulse from the finger to the spinal cord.

A pain receptor in the skin detects the pin.

A synapse joins the sensory neurone with a **relay neurone**, which in turn is joined by a synapse to a motor neurone.

Motor neurone then carries the impulse to the effector muscle to trigger a response.

The biceps muscle contracts to withdraw the finger from the pin. Note: the reflex does not pass through the brain, but separate neurones carry an impulse to the brain from the relay neurones; therefore, you are aware of the reaction just after it has happened.

▲ **Figure 19.6** A reflex arc. Note the route.

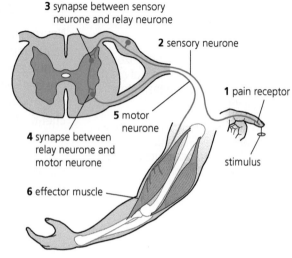

3 synapse between sensory neurone and relay neurone

2 sensory neurone

1 pain receptor

5 motor neurone

4 synapse between relay neurone and motor neurone

stimulus

6 effector muscle

▲ **Figure 19.7** The neurones involved in a reflex arc.

TIPS ✓
- Reflex actions are **rapid** and **automatic**. They do not involve the conscious part of the brain.
- Reflex actions are **protective**, because they are **fast** and do not require thinking time.

Other responses require a decision to be made. Think about how much longer a decision would take to answer the question 'How comfortable is your new school uniform?' compared with moving your hand away from a hot radiator. Decisions like this are not automatic. We call them conscious decisions. When you make conscious decisions the electrical impulse travels along relay neurones to your brain before moving back along motor neurones to your muscles (and glands).

Required practical 6

Plan and carry out an investigation into the effect of a factor on human reaction time

In this experiment you test the reaction time of yourself and a partner using visual and auditory cues.

Method

1 In pairs decide who will have their reaction time tested and who will be the tester.

2 The tester should hold two 30 cm rulers vertically at the end near the 30 cm mark, one in each hand.

3 The other person places their index finger and thumb of each hand either side of the 0 cm marks, holding them as wide as possible ready to catch a ruler when one falls.

TIP ✓
Be careful when you talk about changing reaction times. Reducing your reaction time means that your reaction is faster.

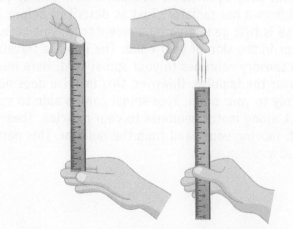
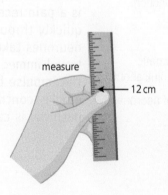

measure

12 cm

▲ **Figure 19.8** How to test your partner's reaction time.

4 Without warning, the tester lets go of one of the rulers and the other student tries catch the ruler as soon as possible.

5 Record the distance travelled by measuring where the ruler was caught just above the student's first finger.

6 Repeat this two more times, ensuring that the ruler is always dropped at random times, and then calculate a mean (charts are available to convert centimetres to seconds).

7 Repeat the process, but this time the student being tested should close their eyes and the tester should say 'left' or 'right' as they drop the corresponding ruler. Again do three repeats so that a mean can be determined.

8 Swap over and repeat the whole experiment.

Questions

1 Who had the quickest reaction time?

2 How repeatable and reproducible were your data?

3 Were your reaction times quicker with visual or auditory clues?

4 What is the stimulus in both experiments?

5 What is the response?

Test yourself

4 Name the two parts of the central nervous system.

5 Name the two types of effector.

6 Describe what happens when an electrical impulse reaches the end of a neurone.

Show you can...

Explain how the reflex arc protects you from damage when you touch a hot radiator.

Chapter review questions

1 Name the two parts of the nervous system.

2 Describe what nerve cells (neurones) are.

3 Name five senses.

4 Explain how reflex responses are different from voluntary responses.

5 Give an example of a reflex response.

6 Describe an experiment in which you investigate the effects of caffeine on reaction time.

7 In which direction do sensory neurones send electrical impulses?

8 Where in your body would you find relay neurones?

9 Name one type of structure that is found at the end of motor neurones.

10 Define the term 'synapse'.

11 Name the chemicals that move across a synapse.

12 Explain how chemical neurotransmitters work in a synapse.

13 Explain how a reflex arc works.

Practice questions

1 a) What is homeostasis? [1 mark]

b) Complete the following sentence.

If blood glucose levels get too high, homeostatic systems _____ the amount of glucose in the blood, so that they return to their _____ level. [2 marks]

2 The flow diagram (Figure 19.9) shows a response to a stimulus. Of stages 1–5:

a) Which stage involves the CNS? [1 mark]
b) Which stage involves a receptor? [1 mark]

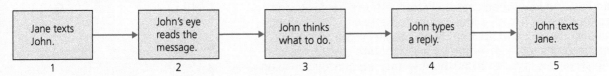

| Jane texts John. | John's eye reads the message. | John thinks what to do. | John types a reply. | John texts Jane. |
| 1 | 2 | 3 | 4 | 5 |

▲ **Figure 19.9**

3 a) Using the words below, put the stages of a reflex action into the correct sequence. [4 marks]

effector stimulus response coordinator receptor

b) During a reflex action, as shown in Figure 19.10, the impulse travels along different types of neurones.

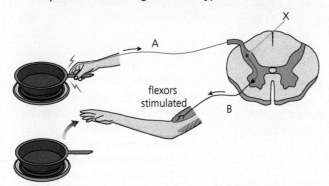

▲ **Figure 19.10**

Name the type of neurone represented by:

i) Neurone A [1 mark]

ii) Neurone B [1 mark]

c) i) On Figure 19.10, X shows a gap between two neurones. What is the term for this gap? [1 mark]

A Synapse   B Axon
C Dendrite   D Myelin

ii) Explain how a message is transmitted across such gaps. [3 marks]

d) During a reflex action the impulse travels approximately 1.5 m at a speed of 120 metres per second. How long would it take in seconds to respond? Show your working. [2 marks]

e) Name a substance that can slow your reaction speed. [1 mark]

Working scientifically:
Experimental skills

Accuracy and precision

Most people think of the human body temperature as being 37 °C. However, the normal range is between 36.5 and 37.5 °C. Temperatures outside this range can lead to medical problems. Hypothermia is when core body temperature falls too low. When core body temperature begins to fall, people become drowsy; if body temperature continues to fall, a person can become unconscious and die from hypothermia. If the core body temperature becomes too high, enzymes will denature causing cellular processes to stop. This is called hyperthermia and can quickly cause death.

Thermometers are used to measure core body temperature. Body temperature can be measured in many locations such as the mouth, ear, armpit, rectum and forehead. It is important that the thermometer takes both precise and accurate readings in order to give correct information about a person's body temperature.

The terms 'precise' and 'accurate' are often used incorrectly or to mean the same thing. However, there are important differences. An accurate reading is one that is close to the true value (the actual value), whereas precise readings are measurements that are close to each other and show little spread about the mean. These two terms are not interchangeable. You can have precise readings that are not accurate, and accurate readings that are not precise.

The data in Table 19.2 show the readings taken from three thermometers measuring the temperature of a thermostatically controlled water bath. The performance of the thermometers is being measured against a clinical (medical) thermometer, which is giving a reading considered to be the true value.

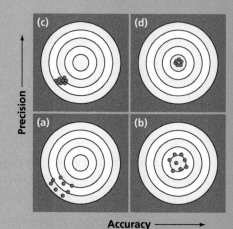

▲ Figure 19.11 The difference between precision and accuracy.

Table 19.2 Temperature of a thermostatically controlled water bath.

Thermometer	Temperature reading in °C				
	1	2	3	4	5
Clinical	51.0	51.0	51.0	51.0	51.0
1	51.2	51.0	47.6	48.9	50.0
2	48.0	47.9	48.0	48.0	48.1
3	51.4	51.2	51.6	51.4	51.7

Questions

1 What are the most accurate and least accurate readings for each thermometer?

2 Work out the mean temperature readings for each thermometer and use this to decide which thermometer is the most accurate.

3 Why does it matter if a thermometer is inaccurate?

4 Which location in the body do you think will give the most accurate reading for core body temperature?

5 Which thermometer produces the most precise readings? Explain your choice.

6 Which of the three thermometers in Table 19.2 do you think should be used? Explain your choice.

7 Why do we say that the clinical thermometer 'gives a value considered to be the true value', rather than it 'takes the true value'?

20 Hormonal coordination in humans

Puberty involves a lot of changes in your body. It can also affect your emotions and behaviour. These drastic changes in your body mean you have reached sexual maturity and are now able to have children. But why does this happen and what causes it? The answer is the release of hormones from your glands. These are like chemical messages which travel around your bloodstream telling bits of your body what to do at certain times.

Specification coverage

This chapter covers specification points 4.5.3.1 to 4.5.3.6 and is called Hormonal coordination in humans.

It covers homeostasis, the human endocrine system, hormones in human reproduction, contraception, hormones in infertility treatment and negative feedback.

Prior knowledge

Previously you could have learnt:

> An unbalanced diet can have consequences, including obesity.
> The reproductive system includes the parts of the body concerned with reproduction in humans.

Test yourself on prior knowledge

1 Name a medical condition that can arise from having too much sugar in your diet.
2 Describe how obesity can be prevented.
3 Explain why the lining of the uterus thickens during the menstrual cycle.

Human endocrine system

KEY TERM

Gland A structure in the body that produces hormones.

TIP

Compared with the effects of the nervous system, the effects of hormones are usually slower but last for longer.

TIP

Copy out the table headings in the first row and column and test yourself by filling in the rest of the table from memory. This will help you remember the detail.

The **endocrine system** is a group of glands that secrete hormones directly into the blood. **Hormones** are proteins, which are large chemical molecules. Hormones travel in the **blood** until they reach their **target organ**. Here they effect a change. Examples of hormones, their target organs and their effects are given in Table 20.1, and Figure 20.1 on the next page shows the locations of these glands. Because hormones travel in the blood, their effects are usually slower than those of impulses that travel along your nerves as a part of your nervous system. However, the effects of hormones are usually longer lasting than the effects of the nervous system.

Table 20.1 Common examples of hormones and their functions.

Hormone	Produced	Target organ	Function
TSH (thyroid-stimulating hormone)	Pituitary gland	Thyroid	Controls the release of hormones from your thyroid gland
Adrenaline	Adrenal gland	Heart (and other vital organs)	Prepares the body to fight or run away (flight)
Insulin (and glucagon)	Pancreas	Liver	Insulin increases (and glucagon decreases) the conversion of blood glucose to glycogen
Thyroid hormones (e.g. thyroxine)	Thyroid	Various	Control how quickly you use energy, and your overall metabolic rate
Oestrogen	Ovaries	Reproductive organs	Controls puberty and the menstrual cycle
Testosterone	Testes	Reproductive organs	Controls puberty

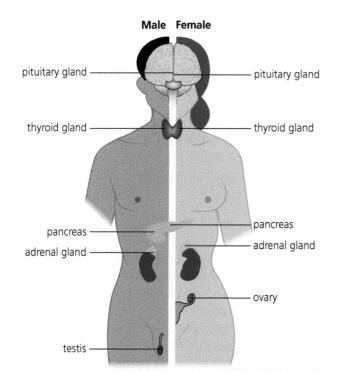

Male   Female

pituitary gland —————— —————— pituitary gland

thyroid gland —————— —————— thyroid gland

pancreas —————— —————— pancreas

adrenal gland —————— —————— adrenal gland

—————— ovary

testis ——————

▲ **Figure 20.1** The positions of key glands in your endocrine system.

The pituitary gland in the brain is your **'master gland'**. Your pituitary gland is about the size of a pea. It secretes hormones that help control growth and blood pressure, as well as partly controlling functions of the ovaries or testes, pregnancy, childbirth and the kidneys. Your pituitary gland releases hormones that have other glands as their target organs. Thus your pituitary gland releases hormones that 'turn on' other glands. An example of this is the thyroid gland. This controls how quickly your body uses energy and makes proteins and how sensitive it is to other hormones.

> **KEY TERM**
>
> Pituitary gland The master gland in your brain that produces a number of hormones, including TSH, FSH (in women) and LH (again in women).

Test yourself

1 Name two hormones produced in the pituitary gland.
2 What does adrenaline prepare the body for?
3 Describe where your pituitary gland is located.
4 Describe what effect insulin and glucagon have on the body.

Show you can...

Explain why your pituitary gland is called the 'master gland'.

Control of blood glucose concentration

The purpose of your digestive system is to break down large lumps of food into molecules small enough to be absorbed into your blood. Carbohydrase enzymes break down carbohydrates into sugars, which are absorbed through villi into your bloodstream. Your pancreas monitors and controls the amount of blood sugar (glucose) that remains in your blood. Too little and your cells cannot respire, and too much and you become comatose. Because it involves your body automatically returning to a normal state, control of blood glucose is an example of homeostasis.

Many people eat three meals a day. So there will be three times when lots of glucose from their food rushes into their bloodstream. Just before meals there are times when your blood glucose is likely to be low because your cells have been respiring. So your body, and the body of every other mammal, has evolved a way to store excess glucose after a meal and release it when your body needs it.

TIPS

- Your body produces **IN**sulin when there is too much glucose **IN** your blood.
- Your body produces gluca**GON** when the glucose has **GON**e from your blood.

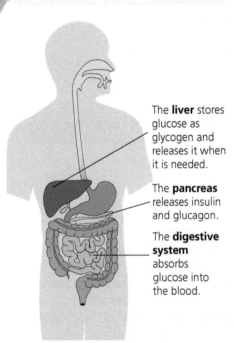

The **liver** stores glucose as glycogen and releases it when it is needed.

The **pancreas** releases insulin and glucagon.

The **digestive system** absorbs glucose into the blood.

▲ Figure 20.2 Absorption of glucose, control of blood glucose concentration and storage of glucose in the body.

○ Insulin

A hormone called insulin helps to control the concentration of blood glucose. Two organs, the pancreas and liver, are also involved. When your blood glucose is too high, the **pancreas** detects this and produces insulin, which is released into the bloodstream and reaches its target organ, the liver. The **liver cells** then absorb the glucose from the bloodstream and start to turn the excess glucose into an insoluble larger molecule called glycogen. The formation of glycogen also occurs in your **muscles**. Its formation reduces your blood glucose concentrations back to normal.

○ Glucagon

If your blood glucose is too low, your **pancreas** produces a second hormone, called glucagon. This travels in your blood to your **liver** and **muscles**, where it converts the insoluble glycogen back into glucose. This is released into your blood to return your blood glucose concentrations to normal. Like all examples of homeostasis, this two-way process is constantly happening all the time in your body.

The control of blood glucose is an example of negative feedback control. Here your body has detected a change (high or low blood glucose) and made an adjustment to return it back to normal. All examples of homeostasis involve negative feedback.

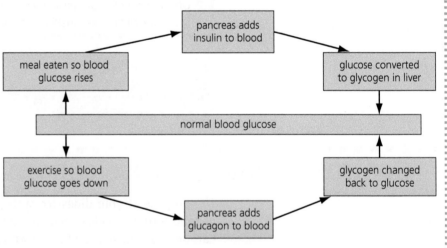

▲ Figure 20.3 Two hormones work to control blood glucose level.

○ Diabetes

Diabetes is a non-communicable disease. People with diabetes cannot easily control their blood glucose levels. There are approximately three million people with diabetes in the UK, and this figure is rising. Some people develop the disorder early in life, while others develop it later, usually as a result of an unhealthy lifestyle. Regardless of the way a person develops diabetes, the symptoms are similar. Because there are two different ways in which diabetes develops there are two different names for it.

Type 1 diabetes

The causes of type 1 diabetes are unknown. About 10% of the total number of diabetic people in the UK have type 1 diabetes. It usually develops in children or young adults when the insulin-producing cells in the pancreas are destroyed. This happens because the sufferer's immune system mistakenly makes antibodies to attack and destroy these cells. Without these cells the blood glucose concentration of diabetic people can quickly and easily rise to harmful levels, which may become fatal.

People with type 1 diabetes usually **inject insulin** to help reduce their blood glucose. They can also help keep blood glucose levels low by **reducing the sugar in their diet** and **exercising** regularly. Reducing sugar intake means they need to inject less insulin, and exercise increases the amount of glucose that is used by their muscle cells for respiration. Often diabetic people test their blood glucose level after a meal. If they have a high blood glucose concentration they will need to inject a greater volume of insulin.

It is important that type 1 diabetes is diagnosed as early as possible. There is currently no cure, so injecting insulin, eating carefully and exercising manage the symptoms of the disorder without treating them.

Type 2 diabetes

Type 2 diabetes usually develops later in life. Its cause is different from that of type 1 diabetes. People with type 2 diabetes cannot produce enough insulin or, if they can, their liver and muscle cells won't respond to it. This prevents the conversion of glucose into glycogen and blood glucose concentrations remain high.

The number of cases of type 2 diabetes is rising. This is partly because the number of older people is increasing as people are generally living longer. People who do not exercise regularly, have an unhealthy diet (often high in sugar) and are obese are more likely to develop type 2 diabetes. **Obesity** is an important risk factor for this disorder.

People with type 2 diabetes control their blood glucose concentrations by eating a balanced, healthy diet and by exercising regularly. Because the liver and muscle cells of many people with type 2 diabetes do not respond to insulin, injecting insulin is not usually a treatment.

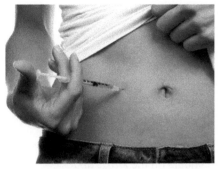

▲ **Figure 20.4** A person with diabetes needs to inject more insulin after eating sugary foods.

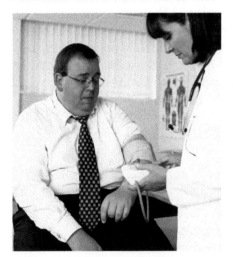

▲ **Figure 20.5** Obesity can lead to serious health issues, including type 2 diabetes.

Controlling blood sugar concentration

Data were collected on the concentration of blood glucose for two patients over 6 hours following a high glucose meal (at time 0) (see Table 20.2).

Table 20.2 Blood glucose concentrations for patient A and patient B.

Time in h	Concentration of blood glucose in mg/dL	
	Patient A	Patient B
0	82	120
1	144	223
2	99	190
3	95	172
4	92	163
5	87	145
6	82	138

Questions

1 Plot the data as line graphs with time in hours on the *x*-axis and concentration of blood glucose in mg/dL on the *y*-axis.

2 After looking at the graph, which patient would you expect to have diabetes? Quote data to support your answer.

3 Why did it take an hour for the blood glucose concentration of both patients to rise?

4 Use your knowledge of blood sugar regulation to explain the changes in patient A's blood sugar concentration over the 6 hours.

TIPS

- You should be able to compare type 1 and type 2 diabetes and explain how they can be treated.
- You should be able to extract information and interpret data from graphs that show blood glucose concentrations for diabetes and non-diabetes, such as is given in the activity in this section.

▲ **Figure 20.6** Sir Steven Redgrave won five Olympic medals for rowing, despite having type 1 diabetes.

Managing diabetes

Diabetes is a common medical condition and the number of people that have the disorder is increasing. It does have serious medical effects for some people, but many are able to lead relatively normal lives. In fact, some very successful people are diabetic. These include actors Tom Hanks and Halle Berry, as well as Olympic rower Sir Steven Redgrave. He has had type 1 diabetes since 1997 but won five gold medals in five different and successive Olympic Games (from 1984 to 2000). This makes him one of the most successful athletes of all time.

Test yourself

5 Which type of diabetes is treated using insulin?
6 Name the organ in which insulin is produced.
7 Describe how blood sugar concentrations are increased.

Show you can...

Explain the differences between type 1 and type 2 diabetes, including their causes and treatments.

Hormones in human reproduction

The sex hormones in women include oestrogen and progesterone. Testosterone is the male sex hormone. These hormones are involved in the development of secondary sex characteristics in puberty. These are shown in Table 20.3.

Table 20.3 The secondary sex characteristics caused by sex hormones during puberty.

Female secondary sex characteristics	Male secondary sex characteristics
Growth of breasts	Increased growth of testes and penis
Hips widen for childbirth	Increased muscle mass and broadening of shoulders
Growth of facial and underarm hair	Growth of facial and underarm hair
Growth of pubic hair	Growth of pubic hair
Growth spurt	Growth spurt
	Deepening of voice

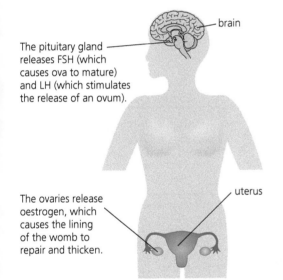

The pituitary gland releases FSH (which causes ova to mature) and LH (which stimulates the release of an ovum).

The ovaries release oestrogen, which causes the lining of the womb to repair and thicken.

▲ **Figure 20.7** Hormones produced by the pituitary and ovaries control the uterus. Examine carefully the sites of production of the hormones and the target organs.

brain

uterus

Testosterone in males is secreted by the **testes. Oestrogen** and **progesterone** in females are produced by the **ovaries**. The testes and ovaries are therefore glands. As well as causing the secondary sex characteristics shown in Table 20.3, testosterone stimulates the production of **sperm**. Oestrogen and progesterone are two key hormones in the menstrual cycle.

○ The menstrual cycle

After puberty, women undergo a reproductive cycle of around 28 days until they reach the menopause. At this point a woman's menstrual cycle stops and she is no longer able to have children. This occurs between approximately 45 and 55 years of age. There is no menopause for men, and they are often able to produce sperm for much longer than women can release eggs. But older men produce less testosterone than younger ones and their sperm become less able to fertilise an ovum (i.e. become less fertile).

Oestrogen is responsible for the lining of a woman's uterus. After menstruation, the lining thickens in preparation for a fertilised ovum to settle after travelling down an oviduct (fallopian tube) to the uterus. The release of a mature ovum from an ovary into the oviduct happens on about day 14 of the menstrual cycle. This is called ovulation. The ovum takes several days to travel down an oviduct and be fertilised by a sperm cell. If a fertilised ovum does embed in the lining of the uterus, the hormone **progesterone** continues to be produced to continue the build-up of the uterus lining and to prevent menstruation. (Menstruation would abort a developing fetus.) If an ovum is not fertilised, less progesterone is produced and a woman menstruates as normal. This lasts several days and can be painful, causing cramps. Its beginning marks the start of the next 28-day menstrual cycle. The hormonal control of the menstrual cycle is very delicate. It is often altered by stress or exercise.

There are four main hormones involved in the menstrual cycle. They are shown in Table 20.4, together with the effects they have. Figure 20.8 shows these effects in a flow diagram.

TIP

It is important that you can explain the interactions of hormones in the menstrual cycle. Draw a flow diagram of the process to help you remember.

Table 20.4 The main hormones involved in the control of the menstrual cycle.

Hormone	Released by	Target organ and effect
Follicle-stimulating hormone (FSH)	Pituitary gland	Ovary • Causes an ovum to mature in the ovary • Stimulates ovaries to produce oestrogen
Oestrogen	Ovaries	Uterus • Causes lining to thicken in first half of the cycle Pituitary • High oestrogen concentration switches off release of FSH and switches on release of LH
Luteinising hormone (LH)	Pituitary gland	Ovary • Stimulates ovulation (release of the ovum from the ovary)
Progesterone (produced if fertilised ovum implants in uterus)	Ovaries	Uterus • Maintains thick uterus lining if fertilised ovum implants • High concentrations of progesterone in pregnancy stop the cycle

TIP

You should be able to extract and interpret data from graphs showing hormone concentrations during the menstrual cycle.

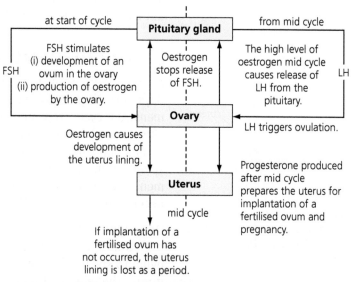

▲ Figure 20.8 Hormones control the menstrual cycle.

KEY TERMS

Follicle-stimulating hormone (FSH) A hormone produced by the pituitary gland that causes an ovum to mature in an ovary and the production of oestrogen.

Follicle A structure in an ovary in which an ovum matures.

Luteinising hormone (LH) A hormone produced by the pituitary gland that stimulates ovulation.

Corpus luteum After ovulation the empty follicle turns into the corpus luteum and releases progesterone.

Follicle-stimulating hormone (FSH) is released by the pituitary gland. This causes an ovum to mature in an ovary. This happens inside a follicle. FSH also causes the ovaries to produce and release oestrogen. This causes the lining of the uterus to thicken in the beginning of the menstrual cycle. The pituitary gland detects this high level of oestrogen and stops producing FSH. Instead the pituitary gland starts to secrete luteinising hormone (LH). This stimulates **ovulation**, releasing an ovum from an ovary. After ovulation the follicle from which the ovum was released develops into the corpus luteum. This releases progesterone, which inhibits the release of both FSH and LH and in doing so maintains the lining of the uterus in the second half of the cycle. If a fertilised ovum settles and implants, the levels of progesterone are maintained, resulting in no menstruation. If an ovum does not implant, the levels of progesterone reduce. This triggers menstruation.

Show you can...

Explain how hormones regulate the menstrual cycle.

Test yourself

8 Where is oestrogen produced?

9 Describe secondary sexual characteristics in women.

10 Describe why the menstrual cycle is an example of negative feedback.

Contraception

Contraception is the name for the methods or devices that stop women becoming pregnant. It is often called birth control or family planning. Some forms of contraception are permanent, while others are only temporary. Contraception is an **ethical issue**. This means that some people disagree with it for religious or moral reasons. Some religions, including perhaps most famously the Roman Catholic Church, of which there are more than a billion members worldwide, only officially accept 'natural family planning'. This is when the male doesn't ejaculate inside the female's vagina during sex, particularly when an ovum is likely to be fertilised (around days 13–17). This is different from abstinence, which is stopping having sex altogether. Other religions such as Judaism, Hinduism and Buddhism have wide-ranging views that allow both 'natural' and artificial contraception.

One method of contraception for men is to have a vasectomy. Here a short and relatively simple operation stops sperm travelling along the **sperm ducts** that link the testes to the penis. The sperm ducts can be tied or cut. This is a **permanent** form of contraception for men. A similar procedure can occur in women. Here the **oviducts** (**fallopian tubes**) are tied or cut, which stops ova reaching the uterus.

Other forms of contraception include barrier methods that stop ova and sperm meeting. **Condoms** are a very widespread form of contraception. They surround an erect penis and stop sperm entering the vagina. They have been used for over 400 years. There are larger female condoms, often called femidoms, which sit inside the vagina and stop sperm in the same way. These are less common. Another contraceptive that forms a barrier is the **diaphragm**. This is a small plastic dome that is inserted into the vagina to cover the cervix. The cervix is the narrow join between the vagina and the uterus. Contraceptive sponges work in the same position. They are also covered in a chemical called a **spermicide**, which kills sperm cells.

Condoms protect against the spread of many **sexually transmitted** diseases (STDs) because they stop the exchange of all bodily fluids. Other forms of contraception are less effective at stopping STDs. The diaphragm and the sponge don't stop their transmission, for example.

All the above methods are **non-hormonal** methods of contraception.

KEY TERMS

Vasectomy A contraceptive medical procedure during which a man's sperm ducts are blocked or cut.

Barrier A contraceptive method that prevents sperm reaching an egg.

TIPS

- Vasectomy and female sterilisation are **surgical** methods of contraception.
- Remember that condoms and diaphragms are **barrier** methods of contraception. They are not permanent, but are easy to use. However, they are less reliable than some other methods.
- Types of contraceptive that use **hormones** work by stopping eggs from maturing and being released.

▲ **Figure 20.9** Condoms are often covered with spermicide.

○ The use of contraceptive hormones

Other forms of contraception involve the use of **hormones** to stop fertilisation of ova. The **oral contraceptive pill** (often known simply as the pill) contains both oestrogen and progesterone. It is a very common form of contraception with over 100 million women worldwide using it. The pill **prevents ovulation** by inhibiting the production of **FSH**.

The woman takes a pill each day at the same time. This can either be for the entire cycle or just the first 21 days followed by a week with no pills. The pill is a very effective contraceptive that prevents pregnancy if used properly. Possible side effects include headaches, and the pill does not stop the transmission of STDs.

Progesterone can also be delivered into a woman's blood in a small device **implanted** under her skin or a **patch** temporarily stuck to it. Implants and patches are as effective at reducing pregnancy as the oral pill. They can stop eggs maturing and being released for a number of months or years.

Intrauterine devices work either by releasing contraceptive hormones or by preventing fertilised eggs from implanting.

TIP

It is important that you can evaluate the use of hormonal and non-hormonal methods of contraception.

KEY TERM

Intrauterine In the uterus. An intrauterine contraceptive device is implanted in the uterus.

▲ **Figure 20.10** The contraceptive patch contains hormones to stop pregnancy.

Test yourself

11 Define the term 'vasectomy'.
12 Describe why contraception is an ethical issue.
13 Describe how the contraceptive pill is taken.

Show you can...

Explain how the contraceptive pill works.

The use of hormones to treat fertility

You learnt in the previous section that the sex hormones oestrogen and progesterone are used to prevent pregnancy. Other related hormones can actually be used to have the opposite effect. That is, they can be used to treat infertility or help a woman become pregnant.

Some women have naturally low levels of the two hormones **FSH** and **LH**. (Look back at Table 20.4 to remind yourself about these hormones.) FSH and LH both work with oestrogen and progesterone to control the menstrual cycle. Some infertile women cannot produce sufficient FSH to begin maturing an ovum in an ovary. 'Fertility drugs' containing FSH and LH can increase the levels in the woman's blood. This may allow her to become pregnant naturally.

KEY TERM

In vitro fertilisation (IVF) Fertilisation of an ovum outside a woman's body.

▲ **Figure 20.11** Nadya Suleman had eight children at the same time following IVF treatment.

If this does not work some woman undergo *in vitro fertilisation (IVF)*. *In vitro* is Latin for 'in glass', which means outside of the body. 'Inside the body' is *in vivo*. Babies born following IVF treatment are often called 'test-tube babies', even though test tubes are not actually used! The first step involves injections of FSH and LH. These hormones stimulate the maturation of several ova in the woman. A small operation removes these ova from the woman's ovaries and they are introduced to a man's sperm in a medical laboratory. The sperm cells sometimes fuse with the ova naturally to fertilise them. Sometimes the nucleus of a sperm cell is injected into an ovum. The fertilised ova then develop into **embryos** (small balls of cells). A second small operation places one or more embryos back into the lining of the uterus. Nine months later the woman has her 'test tube' baby or babies.

What happens to the unused fertilised embryos is an ethical decision. This means that some people disagree with IVF for religious or moral reasons.

These fertility treatments allow women, who otherwise might not be able to, to have a baby of their own. There is often a very natural strong urge to have a baby. The treatments themselves can be **emotionally** and **physically stressful**. Unfortunately the **success rates are not high**. This has led doctors to implant more fertilised ova during IVF and has led to mothers having **larger numbers** of children than they might have wanted.

Test yourself

14 Name the two hormones used in IVF.
15 Give three types of contraception that use hormones.
16 Describe the process of IVF.
17 Describe the uses of hormones in IVF.

Show you can...

Explain how hormones are used to treat infertility.

Negative feedback

The homeostatic regulation of blood glucose levels by the hormones insulin and glucagon is an example of negative feedback control, as discussed earlier in this chapter. The secretion of hormones in the menstrual cycle is another.

▲ **Figure 20.12** This basic sequence applies to all hormonal control.

The formation of **thyroxine** in the thyroid gland is another example of negative feedback control. The thyroid gland is one of your largest glands. It is found in your neck. It produces hormones that control how quickly your body uses energy, and controls your **basal metabolic rate**. It therefore plays an important role in growth and development. Your pituitary gland produces thyroid-stimulating hormone (TSH). This stimulates the release of thyroxine from your thyroid gland. By increasing or decreasing levels of TSH, we can control the amount of thyroxine in the body. In turn, when our thyroxine is at an optimum level, this reduces the amount of TSH produced – another example of negative feedback control.

Adrenaline is a hormone that is secreted by your adrenal glands, which sit just above your kidneys. Unlike the previous examples, the formation of adrenaline is not an example of negative feedback control. Adrenaline is most commonly associated with the '**fight or flight**' response. This response occurs if your brain perceives a threat to you or those around you. Under these circumstances your adrenal glands produce adrenaline. This happens very quickly to avoid any damage from the threat. Adrenaline causes your **heart rate to increase**, providing your muscles and brain with more **glucose** and **oxygen** needed for respiration. This releases the energy you may need to fight or run away. It also increases your blood glucose level, blood pressure and supresses your immune system. All of these are designed to give you a short-term energy boost.

▲ **Figure 20.13** When you are in a dangerous situation, adrenaline causes your heart rate to increase to supply your muscles with more energy so that you can run or stop and fight.

Show you can...

Explain why the effects of adrenaline are important.

Test yourself

18 What does thyroxine control?

19 Where is thyroxine produced?

Chapter review questions

1 What type of organs produce hormones?

2 Describe how hormones move around your body.

3 Name the glands that produce testosterone in men.

4 Define the term 'contraception'.

5 Suggest an advantage of using a condom over a diaphragm or sponge.

6 Give the general name for an organ that a hormone acts upon.

7 Describe what happens when insulin is released into your blood.

8 Describe how people with type 2 diabetes treat the condition.

9 Give the function of oestrogen in the menstrual cycle.

10 Describe what happens in a vasectomy.

11 Describe how the diaphragm and sponge stop pregnancy.

12 Name the two hormones in the contraceptive pill.

13 Explain how the contraceptive patch or implant works.

14 Explain how hormones can be used to treat infertility.

15 Describe the function of thyroid-stimulating hormone (TSH).

16 Explain why you need to be able to store glucose.

17 Explain why the regulation of blood sugar is an example of negative feedback control.

18 Describe the roles of FSH and LH.

19 Describe the process of in vitro fertilisation (IVF).

20 Explain why some people have multiple births following IVF treatment.

Practice questions

1 The human endocrine system consists of structures that can secrete hormones.

a) Describe what a hormone is. [1 mark]

b) Which of the following structures secrete hormones? [1 mark]

A Ventricles   C Glands
B Arterioles   D Ducts

c) Describe how the hormones that regulate the menstrual cycle travel around the body. [1 mark]

2 Figure 20.14 shows the concentrations of blood glucose in the body after a student ate a chocolate bar.

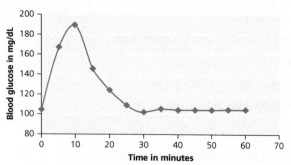

▲ Figure 20.14

a) What is the highest concentration of blood glucose recorded? [1 mark]

b) i) Sketch a graph to show what you would expect the blood glucose concentration to look like if the student had diabetes. [3 marks]

ii) Which hormone can a person with type 1 diabetes not produce? [1 mark]

A Adrenalin   C Glucagon
B Insulin   D Thyroxin

iii) Other than injecting themselves with hormones, what else can people with type 1 diabetes do to manage their condition? [1 mark]

c) i) Why is glucose needed by the body? [1 mark]

ii) Name the organ that monitors blood glucose concentrations. [1 mark]

d) Sometimes blood glucose concentrations drop too low. Describe as fully as you can how blood glucose concentrations can be brought back to normal. [4 marks]

3 Hormones play an important role in regulating the menstrual cycle. Figure 20.15 shows the change in thickness of the lining of the uterus during the menstrual cycle.

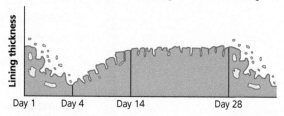

▲ Figure 20.15

a) i) Use Figure 20.15 to explain what happens to the thickness of the uterine lining over the course of the menstrual cycle. [3 marks]

ii) Suggest on which day the ovum is released. [1 mark]

b) Explain how the hormones FSH, oestrogen and LH control the menstrual cycle. [4 marks]

4 The oral pill, the patch and the implant prevent pregnancy by releasing hormones that stop ovulation. Which organ do these hormones act on? [1 mark]

5 The first stage in in vitro fertilisation (IVF) is the fertilisation of an ovum by a sperm.

a) Describe what occurs after the ovum has been fertilised. [2 marks]

b) Often in IVF multiple ova are fertilised.

i) Suggest an advantage of this. [1 mark]

ii) Suggest a disadvantage of using multiple embryos. [1 mark]

Working scientifically:
Scientific thinking

Ethical choices in science

Any new development in science leads to decisions about its use. Since the birth of Louise Brown in 1978, the first 'test-tube baby' born through IVF, there have been ongoing debates about the use of IVF and funding of this procedure on the NHS.

The reasons for the continued interest in this topic are both the ongoing scientific advances in the field and the fact that use of IVF has moral and ethical implications. Morals involve an individual's principles and their judgement of what is right or wrong. Ethics are principles that a group of people or a society agrees are right or wrong.

The Government and hospitals use ethics committees to make ethical decisions. These involve a group of people from different backgrounds who discuss the issues. Usually ethical decisions are based on what leads to the best outcome for the greatest number of people.

In 2013 the guidelines for IVF in the UK were updated to give three full cycles of IVF to women under 40 years old who have not conceived after 2 years of trying. The success rate of IVF depends on a number of factors, including the age of the woman undergoing treatment. Younger women are more likely to have healthier ova, which increases the chances of success. Not everyone can be given IVF and there are waiting lists for the procedure. Fertility clinics need to make decisions about whether a treatment is suitable and ethical.

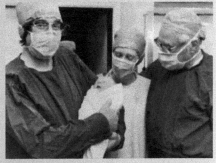

▲ Figure 20.16 The pioneers of IVF, Dr Robert Edwards (left) and Dr Patrick Steptoe (right), celebrate the birth of Louise Brown in 1978, along with the midwife who delivered her.

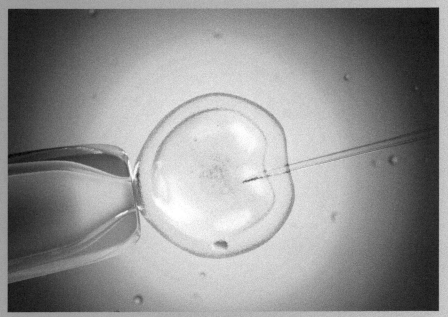

▲ Figure 20.17 On average, one cycle of IVF costs about £5000. However, this varies and there may be additional costs for ongoing medical care, including medicines and consultations.

Questions

1 Imagine you are a member of the ethics committee of a fertility clinic. Read the four case studies below and decide who should be given the three cycles of IVF treatment. You need to justify your reasons.

Couple W are in their late twenties and have been married for 8 years. They have been trying for a child since they were married with no success. They would like IVF treatment.

Couple X are in their mid-thirties and have three daughters already, but they really want a son. They would like to use IVF treatment and pre-select a male embryo to implant.

Couple Y are in their early thirties. One of the couple has an inherited condition called Charcot–Marie–Tooth disease, which causes weakness and wasting of the muscles below the knees and loss of sensation in the fingers. The couple would like IVF treatment in order to pre-select a healthy embryo.

Couple Z have been married for 3 years and they are in their mid-forties. Both have children from their previous relationships. They have been trying for a child of their own for 2 years and would like IVF to assist them in getting pregnant.

2 Give two reasons why some groups of people might be against all IVF treatment.

3 Decisions are not just made about ethics; personal, social, economic and environmental implications need to also be considered. Give an economic reason why some people are against IVF treatment.

4 In February 2015 the UK Parliament voted to approve the use of three-person IVF. Research what this is and the ethical arguments people have for and against its use.

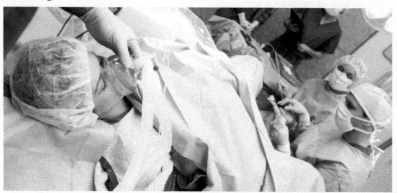

▲ Figure 20.18

21 Reproduction

Imagine cutting off your arm and allowing a genetically identical copy (clone) of yourself to grow from it. This sounds crazy in humans, but other species including plants, worms, fungi, coral and animals like starfish can reproduce in this way. They don't cut themselves up but do reproduce if cut. This is an example of asexual reproduction. Most eukaryotic species like us don't reproduce like this but do so sexually. You were formed from a sperm and an ovum when your parents sexually reproduced.

Previously you could have learnt:

> Heredity is the process by which genetic information is transmitted from one generation to the next.
> DNA is organised into chromosomes, which are subdivided into genes.
> There are similarities and also differences in how animals and plants reproduce.

Test yourself on prior knowledge

1 Name the two gametes in plants.
2 Describe how identical twins are formed.
3 Explain the difference between cross- and self-pollination.
4 Put the following into size order starting with the largest: genes, chromosomes, DNA.

Sexual and asexual reproduction

▲ **Figure 21.1** When a part of a lichen is broken from the parent it will grow into a clone.

Reproduction is the biological process by which a parent or parents have offspring. All life on Earth reproduces, and this is one of the seven life processes. Viruses do not reproduce because they are not alive. They are constructed inside their host cell in a process called replication. There are two types of reproduction: **asexual** and **sexual**.

○ Asexual reproduction

Asexual reproduction involves one parent organism having **genetically identical offspring**. Because of this all offspring are **clones** of the parent. In asexual reproduction, there is no joining (fusion) of sex cells (gametes) such as sperm and ova in humans, for example. So there is no mixing of genetic information (DNA). The only type of cell division involved here is **mitosis**. It is important to remember that organisms that are produced asexually will not all look alike as they can be affected by differences in the environment.

All prokaryotic bacteria reproduce asexually in a process called binary fission. Here the parent bacterium is replaced by two identical daughter bacteria when it divides into two.

Vegetative reproduction occurs in plants and is asexual reproduction, such as by producing runners, bulbs or tubers. Spider and strawberry plants form miniature plantlets on runners, for example, which eventually root near the parent plant and begin to grow. Other plants, such as tulips, form bulbs and some, such as dahlias, form tubers.

KEY TERMS

Asexual reproduction Reproduction involving one parent, giving genetically identical offspring.

Binary fission The asexual reproduction of bacteria.

TIP

It is important that you know that mitosis leads to genetically identical cells being formed. You learnt about this in Chapter 2.

○ Sexual reproduction

Sexual reproduction usually involves two parent organisms, which produce genetically different offspring. The offspring are not clones. Here there is **fusion** (joining) of a **gamete** from each of the male and female organisms. So the offspring inherit DNA from each parent. In animals, the gametes are sperm and ova (egg cells). In plants, the gametes are pollen and egg cells.

Sexual reproduction occurs in eukaryotic organisms. It involves the meiotic production of gametes with half the DNA of the parent. Two gametes then meet and fuse in fertilisation to form a new organism.

A key feature of sexual reproduction is the mixing of genetic information from both parents, which leads to **variety** in the offspring.

Flowering plants all reproduce sexually, although many can reproduce asexually too. In sexual reproduction, they produce pollen as their male gametes and egg cells as their female gametes. A pollen cell must fuse with an egg cell. This forms a zygote, an embryo and then a seed. For example, in insect-pollinated plants, a bee might transfer pollen from one flower to another on the same plant. This is called self-pollination. Alternatively, if the bee visits a flower on another plant and deposits the pollen, then this is called cross-pollination.

Twins

Non-identical twins are formed when a women releases two ova during ovulation. These are then fertilised by two different sperm. Both fertilised ova then embed into the uterus and grow into genetically different organisms. It is possible to have non-identical twins that are different sexes. **Identical twins** are formed when one fertilised ovum splits into two cells, which then grow into two genetically identical twins. Again it is important to remember that identical twins may not always look alike. Environmental variation caused by factors such as diet, scarring, tattoos and haircuts may make them look different from each other.

KEY TERMS ★

Self-pollination When pollen from one plant fertilises egg cells from the same plant.

Cross-pollination When pollen from one plant fertilises egg cells from a different plant.

▲ **Figure 21.2** This honey bee transfers pollen from one flower to another. If this is on a different plant, this is called cross-pollination.

▲ **Figure 21.3** (a) Identical twins are clones of each other but are the result of sexual reproduction, so are not clones of their parents. (b) Non-identical twins are not clones and so can be different sexes.

Meiosis

Mitosis is a type of cell division that copies diploid body cells for growth and repair. There is a second type of cell division, called meiosis, which makes our sex cells or gametes (sperm and ova). These have half of our DNA so that they can join with a second haploid sex cell during fertilisation in sexual reproduction to form a cell with the full amount again. This means that offspring have genetic information (DNA) from two parents.

The four haploid daughter cells produced by meiosis are either sperm or ova in mammals. In plants the haploid cells form pollen and ova. These gametes all have half the DNA of the parent cell. Crucially the gametes of an organism are all slightly different. Unless a person is an identical twin they are not an exact genetic copy of their brother or sister. So their mother's ova must all be slightly genetically different, as must their father's sperm. Meiosis, unlike mitosis, produces genetically different daughter cells.

○ The process of meiosis

Figure 21.4 shows the steps in meiosis. In the first diagram we can see one cell with four chromosomes. The two red chromosomes are from one parent and the blue ones are from the other. There are two pairs of chromosomes, not 23 pairs as in a human cell, to make the diagram a little less confusing.

As in mitosis, the first stage in meiosis is for the nuclear membrane to disappear and for all the chromosomes to shorten and fatten. The chromosomes have already copied themselves completely. If this were a human cell it would have 46 chromosomes and another 46 copies. At this stage each chromosome has changed from a long thin structure into an X-shape. Each X-shape is a long thin chromosome with its new copy.

As in mitosis, the chromosomes and their copies then migrate to the middle of each cell. This is seen in the third stage in the diagram.

At this point half the chromosomes **and** their copies (i.e. one chromosome from each pair of chromosomes) are pulled to one side of the cell and the other half to the other. The cell membrane then starts to pinch inwards and forms two daughter cells. This is the end of the first division. The chromosomes and their copies then line up again in the middle of both the new daughter cells. This is shown in the diagram as the first stage of the second division. The chromosomes split apart from their copies and are pulled to opposite ends of the two cells. The cell membranes then pinch in and form four haploid daughter cells. Crucially these are different from the products of mitosis, because here each cell has half the number of chromosomes of the starting cell and the cells are genetically different from each other, because which of a pair of chromosomes ends up in a particular gamete is random.

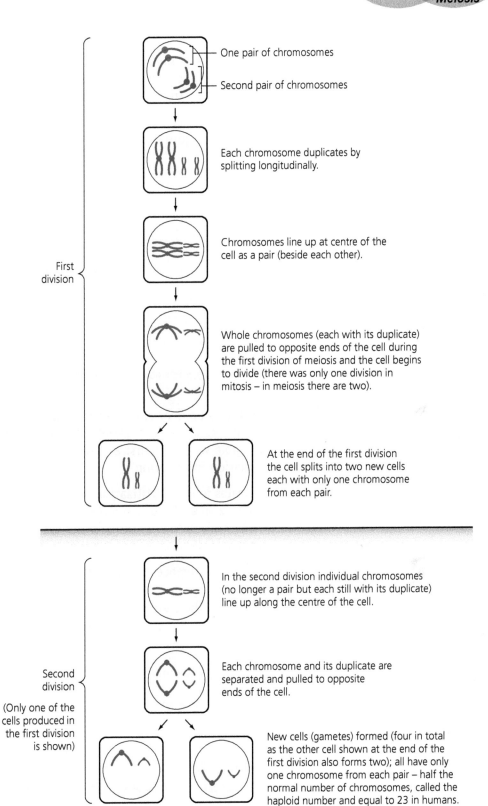

First
division

One pair of chromosomes

Second pair of chromosomes

Each chromosome duplicates by splitting longitudinally.

Chromosomes line up at centre of the cell as a pair (beside each other).

Whole chromosomes (each with its duplicate) are pulled to opposite ends of the cell during the first division of meiosis and the cell begins to divide (there was only one division in mitosis – in meiosis there are two).

At the end of the first division the cell splits into two new cells each with only one chromosome from each pair.

Second
division

(Only one of the cells produced in the first division is shown)

In the second division individual chromosomes (no longer a pair but each still with its duplicate) line up along the centre of the cell.

Each chromosome and its duplicate are separated and pulled to opposite ends of the cell.

New cells (gametes) formed (four in total as the other cell shown at the end of the first division also forms two); all have only one chromosome from each pair – half the normal number of chromosomes, called the haploid number and equal to 23 in humans.

▲ **Figure 21.4** Meiosis.

How does meiosis produce variation?

Humans have 46 chromosomes in 23 pairs. In any one gamete, **either** member of a pair can be present (the key thing is that only one chromosome from each pair is present in a gamete).

Consider the two pairs of chromosomes, with each pair each labelled A and B, in Figure 21.5.

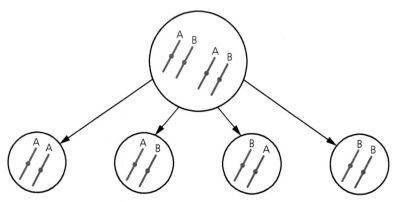

▲ **Figure 21.5** Possible gametes from two pairs of chromosomes at meiosis.

With two pairs of chromosomes, there are four possible arrangements of chromosomes in a gamete. Think how many combinations would be possible with 23 pairs: 2^{23}! No wonder sexual reproduction produces variation in offspring.

Gametes join at **fertilisation** to restore the normal number of chromosomes. Following fertilisation, the zygote (first cell of a new individual) divides by **mitosis** and the number of cells increases. As the embryo develops, cells develop into different types.

The key differences between mitosis and meiosis are shown in Table 21.1.

Table 21.1 The key differences between cell division in mitosis and meiosis.

	Mitosis	Meiosis
Number of cells at beginning	One	One
Type of cell at beginning	Diploid body cell (23 pairs of chromosomes in humans)	Diploid body cell (23 pairs of chromosomes in humans)
Number of cells at end (daughter cells)	Two	Four
Type of cell at end	Diploid body cell (23 pairs of chromosomes in humans)	Haploid gamete (23 chromosomes in humans)
Number of divisions	One	Two
Identical or non-identical cells	Identical	Non-identical
Used for	Growth and repair	Producing gametes
Where it occurs	Everywhere	Sex organs (ovaries and testes in mammals)

TIP ✓

Copy out the table headings in the first row and column and test yourself by filling in the rest of the table from memory. This will help you remember the processes.

Test yourself

1 Give the number of chromosomes in each cell produced by meiosis in humans.
2 How many divisions happen in meiosis and in mitosis?
3 Describe the variation (if any) in daughter cells produced in mitosis and in meiosis.
4 Describe the type of cell produced by meiosis.

DNA and the genome

KEY TERMS ⭐

Double helix The characteristic spiral structure of DNA.

Genome One copy of all the DNA found in an organism's diploid body cells – that is, its entire genetic material.

You may hear scientists or people in the media talk about our genetic code. But what does that mean? Your genetic code is made from DNA, a polymer consisting of two strands that form a double helix. The code tells your body how to make proteins. These are a large group of biological molecules that include enzymes and hormones. Your DNA code tells your body how to make these proteins and others, which define your blood group and eye colour, for example.

DNA is present in chromosomes in the **nuclei** of almost all of your cells. It is not present in your red blood cells. They are full of haemoglobin to bind to oxygen, which they carry to your respiring cells. Only half of your DNA is present in your sex cells – your sperm or ova. But almost all the other billions of cells in your body have one complete copy of all your DNA. This is called your genome.

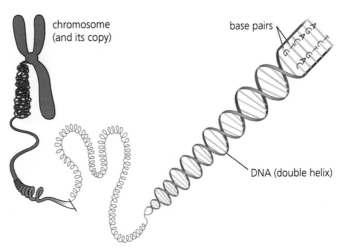

▲ **Figure 21.6** Examine the coiling of DNA to form the chromosome.

○ Chromosomes and genes

The DNA that makes up one genome would stretch to about 2 m long. To fit this much DNA into the nucleus of one of your microscopic cells it needs to be extremely thin and very carefully coiled up. It is coiled into 23 pairs of structures called chromosomes. These come in pairs because you inherit two copies of each chromosome, one from your mother and one from your father. These two copies of each chromosome are different from each other, which makes your genome unique. Unless you are an identical twin, it is highly unlikely that a person has been born with the same genome as you in the past or will be at any time in the future.

Your genome is made from about 20 000 genes. A gene is a short section of DNA (so is part of a chromosome) that provides the code to make a protein. It does this by coding for a sequence of amino acids that forms a protein. It is these proteins that really make us who we are.

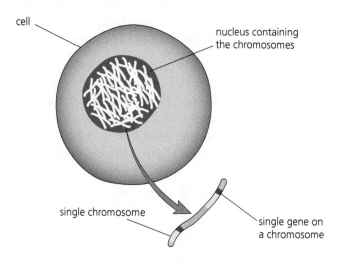

cell

nucleus containing the chromosomes

single chromosome

single gene on a chromosome

▲ **Figure 21.7** Make sure that you can identify the nucleus, chromosomes and the gene locations on the chromosomes.

Test yourself

5   What is a genome?
6   What does DNA code for?

Show you can...

Explain how DNA makes up a genome.

DNA structure

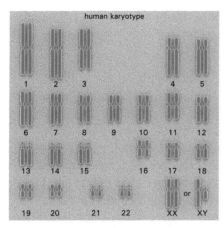

human karyotype

▲ **Figure 21.8** Information gained from work on the Human Genome Project is likely to have great importance for medicine in the future.

○ Understanding the human genome

The Human Genome Project (HGP) was the world's largest collaborative scientific project ever undertaken. It started in 1984 and finished in 2003. Twenty universities in the USA, the UK, Japan, France, Germany and China worked together to identify the sequence of every one of the three billion base pairs that make up our genes.

Since completion of the HGP scientists have continued to work on identifying where genes begin and end in this sequence of bases. Early benefits of the HGP have been genetic tests that show the likelihood of some illnesses developing, including breast cancer and cystic fibrosis, because susceptible people have particular genes. In future many medicines and medical treatments are likely to benefit from follow-up work to the HGP.

The HGP is an ethical concern for some people. This means that some people disagree with it for religious or moral reasons. Some believe that if individuals' genomes were public knowledge some employers might be prejudiced against the genetics of those that they employ. Others worry that health insurance companies that know our genetics may charge some people more than others. Is this fair? Would knowing everyone's genomes make our society better or worse? This is a difficult question to answer.

Knowledge of the human genome also allows us to trace **human migration patterns** in the past. We can do this by comparing similarities and differences among different groups of people.

TIP ✓

It is important that you can understand the importance of the Human Genome Project to the search for genes linked to different types of cancer, the treatment of genetic disorders and the way in which humans originally migrated from Africa.

Show you can...

Explain why some people are concerned by the Human Genome Project.

Test yourself

7 Describe why the Human Genome Project is an ethical issue.

8 Describe the advantages of the Human Genome Project.

Genetic inheritance

KEY TERM ★

Alleles Two versions of the same gene, one from your mother and the other from your father.

Sexual reproduction involves the fusion of two gametes, one from each parent. Therefore the offspring of sexual reproduction have a different genome from either of their parents. Gametes are formed during meiosis and only have half the chromosomes of their parents' body cells. This is so that they can combine and form a new organism with the same number of chromosomes as their parents. So you were made from the 23 chromosomes you inherited from your father in his sperm and the 23 chromosomes you inherited from your mother in her ovum. You therefore have two copies of all 23 chromosomes and two copies of all the genes that are within them. We say that the two copies of any individual gene you have inherited are called alleles. So you have two alleles for eye colour, one from your mother and one from your father.

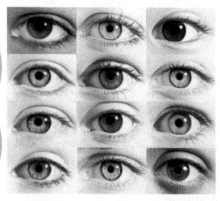

▲ **Figure 21.9** Eye colour is determined by the combination of alleles you inherit from your parents.

○ Eye colour

Many characteristics, such as fur colour in mice and our eye colour, are controlled by one set of genes (a pair of alleles). So you inherit one gene for eye colour from each of your parents. For simplicity we will only talk about blue and brown eyes here. We give letters to represent the colours. Genotypes are given by letters and phenotypes are described in words.

You have inherited a gene from each of your parents for eye colour. (You have therefore inherited a pair of alleles for eye colour.) Brown eyes are dominant over blue eyes. So if you have inherited one brown gene from one parent and a blue one from the other, you will have brown eyes. We give the dominant allele a capital letter. So the three possible genotype combinations for eye colour are BB (phenotype: brown eyes), Bb (phenotype: brown eyes because brown is dominant) and bb (phenotype: blue eyes). Recessive means the opposite of dominant. We say that blue eyes are recessive to brown eyes. These three possible genotypes also have different terms to describe them. Any genotype that is made from two dominant alleles (BB for example) is called is called homozygous dominant. Any genotype made from two recessive alleles (bb for example) is called homozygous recessive. Finally, the genotype made from one dominant and one recessive allele is called heterozygous. This is shown in Table 21.2.

Table 21.2 The three possible allele combinations (genotypes) for eye colour.

Genotype	Phenotype	Terminology
BB	Brown eyes	Homozygous dominant
bb	Blue eyes	Homozygous recessive
Bb	Brown eyes	Heterozygous

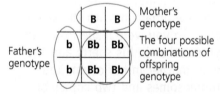

Mother's genotype

Father's genotype

The four possible combinations of offspring genotype

▲ **Figure 21.10** The four possible genotypic offspring of the parents whose genotypes were BB and bb.

We can see how alleles are inherited if we look at a Punnett square diagram (Figure 21.10). This shape shows the four possible genetic combinations of offspring from two gametes. This will only show inheritance of characteristics that are controlled by one set of genes.

A Punnett square can be used to show the possible genotypic offspring from any two parents. The genotype of one parent is placed above and outside the four boxes. The genotype of the other parent is placed to

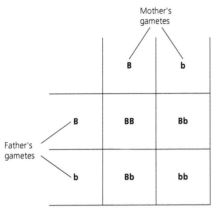

TIP

The genotypes of the mother and father in a Punnett square show the types of gamete the parents can produce.

Mother's gametes

	B	b
B	BB	Bb
b	Bb	bb

Father's gametes

▲ **Figure 21.11** The four possible genotypic offspring of the parents whose genotypes were Bb and Bb.

TIPS

- It is important that you can complete a Punnett Square diagram and interpret the results using direct proportions and simple ratios.
- You should remember that most phenotype features are controlled by multiple genes interacting, rather one gene alone (as described in this chapter for simplicity).

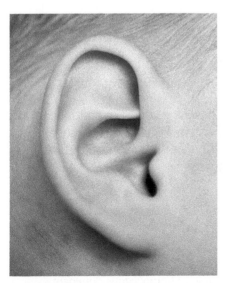

▲ **Figure 21.12** Attached lobes are a homozygous recessive characteristic.

the left and outside the four boxes. In the example in Figure 21.10, the genotype of the first parent is BB and the second bb. The four possible combinations of genotypes in the offspring are then filled in. You can see that the genotypes for the four possible combinations of offspring are all Bb. This means all offspring from these parents will be heterozygous and will have brown eyes.

Other examples are perhaps more interesting. The Punnett square for two parents with the heterozygous genotype Bb is shown in Figure 21.11. This shows that the couple have a 25% chance of having a homozygous dominant baby with the genotype BB and so brown eyes. They have a 50% chance of having a heterozygous baby with the genotype Bb and so, again, brown eyes. They have a 25% chance of having a homozygous recessive baby with blue eyes. These percentages are often given as proportions. It is easiest to give them as proportions of four. So BB would be one in four, Bb two in four and bb one in four. You will be expected to know about ratios. So a chance of one in four, or one to three, would be written as 1:3.

This does not mean that the couple's first baby will be BB, the second two will be Bb and the fourth will be bb. This means that for each baby, the chances of each genotype occurring are as given above. Think about when you toss a coin. The result is not always alternating heads and tails, but you do have a 50% chance of either each time.

Other inherited characteristics can be shown in the same way using different genotypic letters. Whether the lower part of your ear is attached to your head or you have lobes is controlled by a single gene, given the genotypic letters E/e, and whether you can roll your tongue is controlled by a gene given the letters T/t. These are shown in Table 21.3.

Table 21.3 The three possible genotypes and phenotypes for ear lobes and tongue rolling.

Terminology	Ear genotype	Ear phenotype	Tongue genotype	Tongue phenotype
Homozygous dominant	EE	Free lobes	TT	Can roll
Homozygous recessive	ee	Attached lobes	tt	Can't roll
Heterozygous	Ee	Free lobes	Tt	Can roll

▲ **Figure 21.13** The child has inherited his ability to roll his tongue from his father.

It is important that you can **construct your own crosses** as well as interpret the results of those you are given. You need to be familiar with the key terminology used in this chapter to do this. You might be asked to construct a cross for a homozygous dominant parent and a homozygous recessive parent. The results of this are shown in Figure 21.10. You might also be asked to construct a cross for heterozygous parents. This is shown in Figure 21.11. You might be asked to make predictions for outcomes using probability.

○ Family trees

Family trees can show the inheritance of characteristics over multiple generations. Each generation has its own horizontal line and the oldest generations are at the top. Men are shown by squares and women are shown by circles. Those with the characteristic are shown coloured in. Two people connected by a horizontal line are parents, and their children are shown below them.

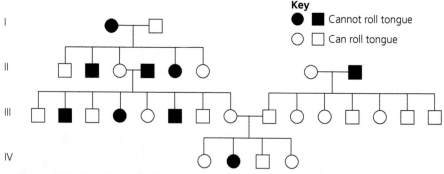

▲ **Figure 21.14** Here is the family tree of a family showing the trait 'inability to roll tongue'.

In the previous section, we have shown how characteristics controlled by one gene (and its two alleles) are inherited. Remember that most characteristics in humans and most other organisms are a result of many genes interacting, rather than just a single gene.

Test yourself

9 Name the term for pairs of genes.
10 Give an example of a phenotype.
11 Describe two phenotypes for eye colour.
12 Calculate the percentage chance of a pair of homozygous dominant parents for brown eye colour having a baby with brown eyes.

Inherited disorders

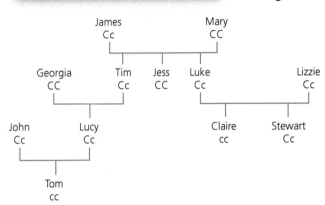

James
Cc

Mary
CC

Georgia
CC

Tim
Cc

Jess
CC

Luke
Cc

Lizzie
Cc

John
Cc

Lucy
Cc

Claire
cc

Stewart
Cc

Tom
cc

▲ **Figure 21.15** The gene for cystic fibrosis is shown by the letter 'c' in this family tree. CC is normal. Cc is a carrier (who doesn't have the disorder). cc is a person with CF.

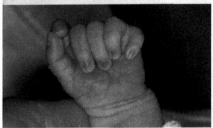

▲ **Figure 21.16** Polydactyly.

Show you can...

Explain how two parents without cystic fibrosis can have a baby that has the disorder.

Many diseases are caused by pathogens and are transmitted from one individual to another. These are called communicable diseases. Those that we inherit from our parents are called inherited disorders.

○ Cystic fibrosis

Cystic fibrosis (CF) affects about 10 000 people in the UK. All of these people inherited two copies of a recessive allele from their parents. If an individual inherits the dominant allele from either of their parents they will not have CF. Thus all those affected are homozygous recessive for this disorder. Heterozygous individuals have one copy of this dominant allele and so do not have the disorder. They can, however, pass it to their children. They are called 'carriers' for the disorder.

People with CF have excessive mucus produced in their lungs. This often gets infected by bacteria and requires treatment with antibiotics. Physiotherapy helps remove much of the mucus on a weekly basis. However, the mucus stops efficient gas exchange in the affected person's lungs. Their ability to exercise is often affected. People with CF sadly have a reduced life expectancy.

○ Polydactyly

Polydactyly is a genetic disorder that results in babies being born with six fingers or toes. This is a very rare condition. Unlike CF, a person with polydactyly inherits a dominant allele coding for polydactyly from one or both parents.

○ Screening for inherited disorders

It is possible to screen embryos (and fetuses) in the uterus for inherited disorders such as cystic fibrosis. Sometimes, the screening process itself can harm the embryo or fetus, or even cause it to be aborted.

If an inherited disorder is identified in the embryo or fetus, the parents may have the opportunity to decide whether to continue with the pregnancy.

Genetic screening raises many ethical issues.

Test yourself

13 What is an inherited disorder?
14 What is the genotype for a carrier of cystic fibrosis?
15 Describe the symptoms of cystic fibrosis.
16 Describe the symptoms of polydactyly.

Sex determination

You learnt previously that your genome is made from 23 pairs of chromosomes and that you inherited one pair of chromosomes from each of your parents. This means you have two copies of every gene, one from each parent. You also learnt that your 23rd and final pair of chromosomes are called your sex chromosomes, and they determine whether you are male or female.

Gametes are sex cells. Sperm are the male gametes and ova are the female gametes. Each of these cell types was formed during meiosis. This process results in four non-identical daughter cells, which have half the number of chromosomes. This means they only have 23 and not 23 pairs (or 46 in total). They are called haploid (not diploid) because of this. All of a man's sperm or a woman's ova (egg cells) are slightly different. If all of a father's sperm and a mother's ova were the same the offspring would be clones and each individual's genome would not be unique.

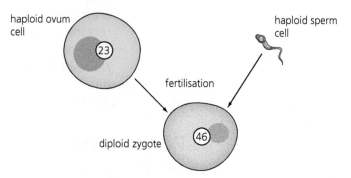

haploid ovum cell

haploid sperm cell

fertilisation

diploid zygote

▲ Figure 21.17 During fertilisation in humans the sperm and the ovum bring together 23 chromosomes from each parent to form the first cell that goes on to produce a new human being. After fertilisation, this cell has the full number of chromosomes (46).

> **TIP** ✓
> It is important that you can carry out a genetic cross to show sex inheritance.

During fertilisation a haploid sperm fuses with a haploid ovum. This means the fertilised ovum now has its full 23 pairs of chromosomes again. It is a diploid cell.

Scientists use letters to represent the different sex chromosomes you can inherit from your parents. All ova have an X chromosome. Half of men's sperm have an X chromosome and the other half have a Y chromosome. If an X ovum is fertilised by an X sperm then the 23rd pair of chromosomes will be **XX** and the baby will be a **girl**. If an X ovum is fertilised by a Y sperm then the pair of chromosomes will be **XY** and the baby will be a **boy**. Because men produce X and Y sperm in approximately equal numbers, the percentages of male and female humans born on Earth are roughly equal.

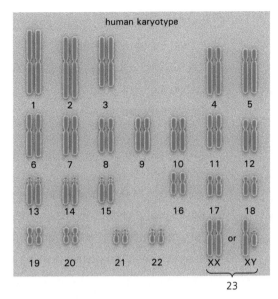

human karyotype

▲ **Figure 21.18** The 23rd pair of chromosomes determine your sex.

Father's genotype
XY

Mother's genotype
XX

		Mother	
		X	X
Father	X	XX	XX
	Y	XY	XY

Offspring   XX   or   XY
           (50%)   (50%)
           Girls   Boys

▲ **Figure 21.19** Possible sex genotypes.

Show you can...

Explain why, in terms of fertilisation, it is not surprising that the population of the world is approximately 50% male and 50% female.

Test yourself

17 Give the two letters used to represent the sex chromosome in a human male.
18 Give the two letters used to represent the sex chromosome in a human female.
19 Describe how you know that every one of a man's sperm and a woman's ova are different.

Chapter review questions

1 Define the term 'genome'.

2 Explain why you have two copies of each chromosome.

3 Define the term 'gene'.

4 Define the term 'ethical decision'.

5 Explain why we call our DNA our genetic code. What specifically is it a code for?

6 Which type of people have identical genomes?

7 Which pair of chromosomes are the sex chromosomes?

8 Describe a major achievement of the Human Genome Project.

9 Explain what collaborative working means.

10 Explain why some people are worried about the information gained from the Human Genome Project.

11 What are proteins made from?

12 Name the organelle in which proteins are produced.

13 Suggest what some future developments of the information from the Human Genome Project might be.

14 Which parental genotypes can give a 3:1 ratio of offspring?

15 Which parental genotypes can give a 1:1 ratio of offspring?

Practice questions

1 In birds, as in humans, chromosomes determine sex. In birds, however, it is due to the inheritance of Z and W chromosomes rather than X and Y. A male bird carries two copies of the Z chromosome, and the female bird carries one copy of Z and one copy of W.

a) Copy and complete the sentences below using appropriate key terms.

Birds, like humans, carry out _____ reproduction. The means there is the fusion of separate male and female _____ . A benefit of this type of reproduction is that it leads to _____ in the offspring. Male birds have two copies of the Z chromosome, therefore they are described as being _____ whereas female birds have one copy of the Z chromosome and one of the W chromosome and so are described as _____. This is different to humans where males have the genotype _____ and females are XX.   [6 marks]

b) Draw a genetic diagram to show the possible sexes produced by mating a male and female bird.   [3 marks]

2 Coat colour in black mice is controlled by a single gene. There are two alleles for this: the dominant black allele (B) and the recessive white allele (b).

a) What is an allele?   [1 mark]

b) Describe the difference between dominant and recessive alleles.   [2 marks]

c) What is the probability of a baby mouse having a white coat if two Bb mice are bred? Draw a genetic diagram to explain your answer.   [4 marks]

3 a) Where in a human cell is DNA found?   [1 mark]

b) DNA exists as a molecule made up of two strands. What is the name given to the structure of this molelecule?   [1 mark]

c) What name is given to a short section of DNA that codes for a protein?

4 Which of the following genotypes would a person with cystic fibrosis have?   [1 mark]

A Cc

B cc

C CC

D cC

Working scientifically:
Scientific thinking

Writing in science

It is vital that scientists can communicate scientific findings clearly and correctly, both to other members of the scientific community and to the general public. Science communication is often used to inform as wide an audience as possible about research findings or developments, to inform decision making, and to gain support for further study.

The main way that scientists communicate to other scientists is by writing scientific papers, which are published in scientific journals. They also communicate their work and findings through poster presentations.

Science is usually communicated to the general public through the media, whether on news programmes, in newspapers, or on social networking sites. There is a growing demand for scientists to be able to communicate with the public themselves, and most universities encourage outreach programmes where researchers and scientists go into schools and speak to the students and teachers.

▲ Figure 21.20 Science is communicated to the public mainly through news programmes and in the newspapers.

Questions

Decide on one of the following titles and produce a piece of writing to allow the general public to understand the science behind it:

- Why have sex? The benefits of sexual reproduction
- Amino acids on asteroids: did life hitchhike to Earth?
- It is a man's world – are sex and gender determined by men?

You will need to research your topic thoroughly, ensuring that you take notes and understand the science yourself to be able to clearly explain it to members of the public.

Make sure you focus on the purpose of and audience for your writing. Remember you are writing to inform the general public of a scientific concept, and they may have a limited background in science. You can present your work as a newspaper article, a blog or a presentation. Don't forget to check your spelling, punctuation and grammar.

A useful tip for helping you structure your writing is to use the Point, Evidence, Explanation writing frame. You might have used this in English or humanities. Using this writing frame for each of your paragraphs will help you make a point, provide evidence to support it and provide an explanation of how the evidence links to the point, wrapping it all together.

22 Variation

There have been over 100 billion people that have ever lived (with seven billion alive today), and you are the only one to look exactly as you do. So why do no two people look exactly identical? If you are an identical twin, you probably look very similar to your twin, but usually people who know you well can tell the difference. These differences are called variations. We can see them in other species too. To many of us, all gorillas look very similar, but they have differences like us, including separate fingerprints.

Specification coverage

This chapter covers specification points 2.1 to 4.6.2.4 and is called Variation.

It covers variation, selective breeding and genetic engineering.

Variation

KEY TERM

Variation The differences that exist within a species or between different species.

▲ **Figure 22.1** Banded snails, showing a wide variation in colour.

TIP

Write a list of all the ways in which you and your sibling or friend vary. Then add next to each one whether it is caused by genetics, the environment or a combination of both.

TIP

Environmental variation is obvious in plants. Plants of the same type often grow very differently at different light levels.

Variation is the sum of all the small or large differences that make one organism different from another of the same species. So the variation between two cats might include the colour of their fur and their size. Variation is also the sum of the differences between two species. Chimpanzees and bonobos are our closest relatives and so the variation that exists between them and us is less than that between us and plants, for example.

○ Causes of variation

Variation can be caused by **genetic** factors. These include the inheritance of genetic disorders like cystic fibrosis. Other common examples of genetic variation in humans are your eye colour and blood group. Whether you have cystic fibrosis, what your eye colour is and what your blood group is are only determined by the genetics you inherited (your genome).

Other variation is caused by **environmental** factors. Your appearance is changed by any scars or tattoos that you have. Your hair colour can be lightened by the Sun. Your skin can be darkened by the Sun, too. All of these examples are not caused by your genetics. They are caused by other factors, which we call environmental.

A third cause of variation is a combination of **genetic and environmental** factors (that is, how your genome interacts with the environment). This tends to be more complicated than either genetic or environmental factors alone. Your weight and height are examples of this. People who are born from tall parents are usually tall themselves. There is a genetic tendency for tall people to have tall children. However, we need calcium in our diet for strong bones and teeth. Children who do not receive enough calcium may have shorter bones. Thus your height can be determined by genetic and environmental factors.

○ Types of variation

All variation can be grouped into two types: continuous and discontinuous. All continuous variation comes in a range of values and the values can have additional values halfway between them. Your height is a good example of this. You can be 150 cm or 151 cm tall. In fact you can be 151.5 cm tall. The results of a survey involving continuous variation are shown in a line graph or histogram.

Other variation can only fall within certain categories or groups, and there is no grouping in between. This is called discontinuous variation. Good examples of this include blood group and eye colour. You can have type A blood or type O blood but you can't be halfway between A and O. This blood group doesn't exist. Similarly, eye colour falls into distinct categories with no groups in between. The results of a survey involving discontinuous variation are shown in a bar chart.

○ Normally distributed variation

The measurement of variation usually involves a survey. Here data are collected from a number of individuals and the relative numbers are recorded and displayed in a line graph or bar chart. A larger survey with more data means that the results are more likely to be reproducible.

Many results of large biological studies looking at variation produce results that are normally distributed. This means that there are a few values at each end of the scale but most come towards the middle. This forms a bell-shaped graph, and we say that this has 'normal' distribution. The more values that are recorded, the closer the graph will be to a bell shape. If you did a survey of height in your class at school then you might not find a perfect bell shape. If you surveyed your whole year group you would be more likely to find a better bell shape. The 'point' of the graph that has the greatest number is the most common value you have measured, while those values at the edges of the bell typically have fewest values. We might call these 'outliers'.

▲ Figure 22.2 Suntans and tattoos are examples of environmental variation.

▲ Figure 22.3 What examples of genetic and environmental variation can you see in this family?

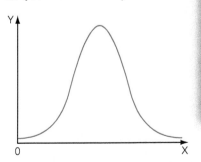

▲ Figure 22.4 Bell-shaped graphs show normally distributed data. The most common value is in the very middle and the least common values at each end.

Activity

Measuring variation

Two medical students were asked to collect data on the average mass of 11 year olds. Their data are shown in Tables 22.1 and 22.2.

Questions

1 Plot the results for student A as a histogram.

2 Do the data show normal distribution?

3 Is the variation continuous or discontinuous?

Table 22.1 Student A's data.

Mass in kg	Frequency
35–39	1
40–44	2
45–49	4
50–54	5
55–59	9
60–64	6
65–70	3

4 What else could the medical students have recorded that would show the other type of variation?

5 If you wanted to know the most common mass for 11 year olds, who has the most accurate data? Can you explain why?

Table 22.2 Student B's data.

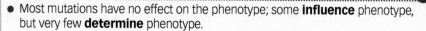

Mass in kg	Frequency
35–39	11
40–44	24
45–49	39
50–54	85
55–59	45
60–64	21
65–70	15

TIPS ✓
- Most mutations have no effect on the phenotype; some **influence** phenotype, but very few **determine** phenotype.
- There is usually a lot of genetic variation within a population of any one species.

○ Mutations

Any change to your DNA is called a mutation. These occur continuously in living organisms. On rare occasions mutations give rise to phenotypic variation. If this is suited to an environment, it can lead to relatively quick changes to species in evolutionary terms. Although sometimes mutations can influence phenotype, usually they have no effect on it.

KEY TERM ⭐

Mutation A change to DNA. Any genetic differences between individuals of one species, or even genetic differences between species, were originally caused by mutations.

Test yourself ⚙

1 Name the three causes of variation.
2 Name the two types of variation.
3 Describe how you would present continuous results.
4 Describe the curve produced by data that are normally distributed.

Show you can...

Explain how mutations can cause variation in a population.

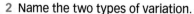
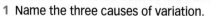

Selective breeding

KEY TERM ⭐

Selective breeding When breeders choose two parents with particular characteristics that they want in the offspring.

For thousands of years before the mechanism of inheritance was understood people had been selectively breeding plants and animals. At its most simple, the principle of selective breeding is that if a big bull is bred with a big cow, then the calves are likely to be big. If this is repeated many, many times then different breeds develop. Selective breeding involves selecting organisms from a population that have a desired variation. These are then bred to produce offspring that share this characteristic.

TIP ✓

It is important that you can explain the benefits and risks of selective breeding.

All dogs belong to the same species (*Canis familiaris*), which is descended from wolves (*Canis lupus*). Selective breeding of big bitches (females) and big dogs (males) has given us large breeds such as the Great Dane. Selective breeding of people-friendly and intelligent bitches and dogs has given us breeds such as the golden retriever. Selective breeding of protective bitches and dogs has given us breeds such as the German shepherd.

▲ **Figure 22.5** These breeds of dogs all belong to the same species, *Canis familiaris.*

Jersey cows have been selectively bred to produce creamy milk. They don't produce much of this, however. Friesian cows have been selectively bred to produce **more milk**, but it is less creamy. Dwarf wheat has been selectively bred to be less easily damaged by bad weather and therefore to increase the farmer's yield.

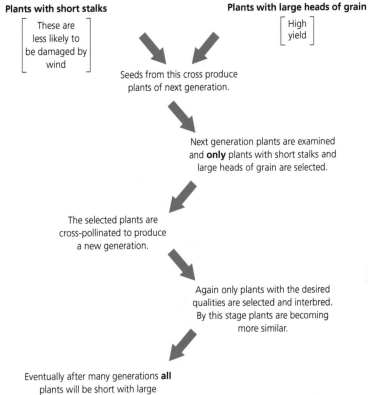

Plants with short stalks
[These are less likely to be damaged by wind]

Plants with large heads of grain
[High yield]

Seeds from this cross produce plants of next generation.

Next generation plants are examined and **only** plants with short stalks and large heads of grain are selected.

The selected plants are cross-pollinated to produce a new generation.

Again only plants with the desired qualities are selected and interbred. By this stage plants are becoming more similar.

Eventually after many generations **all** plants will be short with large heads of grain.

▲ **Figure 22.6** Artificial selection (selective breeding) in wheat.

> **TIP** ✓
>
> What traits have been selectively bred into your dog or that of your neighbours or friends? Researching this will help you remember facts about selective breeding.

▲ **Figure 22.7** Friesian cows have been selectively bred to produce lots of milk.

> **KEY TERM** ★
>
> **Inbreeding** Artificial selection using parents from a small, closely related group, which reduces variation.

Other plants, including orchids and roses, are selectively bred because of their scent or the size, colour and number of their **flowers**. Recent selective breeding is producing crops that are **resistant to disease**.

Selective breeding is also called **artificial selection**. This separates it from natural selection, which is the driver in evolution.

Selective breeding can rapidly reduce the variation in a population. This is often called inbreeding and can result in some genetic weaknesses. Repeated inbreeding can magnify some negative characteristics by mistake alongside the desirable ones. Some breeds of dog suffer from these weaknesses, which include misaligned hips (hip dysplasia). A **gene pool** is a measure of the total set of genes in a population. There are approximately 10 000 pug dogs in the UK. Because of inbreeding they have a gene pool equivalent to that of only 50 animals. Their genetic variation is very low. This means that they might find it difficult to evolve to a changing environment or combat a new communicable disease.

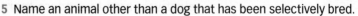

Genetic engineering

KEY TERMS

Genetic engineering A scientific technique in which a gene is moved from one organism to another to give a desired characteristic.

Genetic modification As genetic engineering.

Transgenic Describes a genetically engineered organism.

TIP

It is important that you can explain the potential benefits and risks of genetic engineering and that some people have ethical objections.

▲ **Figure 22.8** Fluorescent mice like these ones only glow in the dark because they have been genetically modified.

TIP

What characteristic or trait would you have liked to be genetically modified into you? (Remember this is illegal!)

▲ **Figure 22.9** Golden rice, which is being grown here on this terraced land, has been genetically engineered to contain carotene.

Genetic engineering is a modern scientific technique that allows us to move one or more genes from one organism into another. It is also called **genetic modification**, and the organisms that are produced are called **transgenic** or genetically modified organisms (GMOs). Genetic engineering is an ethical issue. This means that some people disagree with it for religious or moral reasons.

Genetic engineering is a highly regulated scientific process. That means that licences must be granted before it can be undertaken and organisations must follow specific methods and only work on certain organisms. It is currently illegal to genetically engineer humans.

○ Glow-in-the-dark rabbits

Rabbits do not normally glow in the dark under UV light. Jellyfish do, however, because they possess a gene to make a protein which makes them glow. This gene was cut out from the DNA of a jellyfish using an enzyme. The same enzyme was then used to cut open the DNA of a rabbit embryo. The jellyfish gene was inserted into the genome of the rabbit embryo and sealed into place using a different enzyme. The embryo was therefore genetically engineered. It had a gene from a different species in it. It was a transgenic organism. The embryo was then inserted into the uterus of a rabbit, which from this point onwards had a normal pregnancy. The baby rabbit that was born glowed green under UV light just like the jellyfish.

This was a headline-grabbing example of genetic engineering but not a very useful one. The glow-in-the-dark gene is still used by scientists doing genetic engineering but usually only as a marker to check that other more important genes have been transferred.

○ Genetically engineered crops

Many plant crops have undergone genetic engineering (modification), usually to improve their **yield**. An important example of a genetically engineered crop is **maize**. In many parts of Africa a maize variety that has been genetically modified to be drought tolerant is used. In Europe a genetically engineered form that is resistant to the corn-borer insect is grown. **Golden rice** has been genetically engineered to contain carotene. This reduces the chance of vitamin A deficiency, which causes blindness. **Cotton** has been genetically engineered to be resistant to the boll weevil insect pest. Perhaps the most common example of a genetically engineered crop is **soya**. This has been genetically engineered to be **herbicide resistant**. This means a herbicide can be sprayed all over it, which will kill weeds in the area but not the soya.

Controversy surrounds the production of genetically modified (GM) crops. Many people think that the technology should be used to help those people who currently do not have enough food. Genetically engineered drought-resistant crops can be planted in parts of the world such as Africa and could save many lives. Other people do not like their use. For some, their religion teaches them that humans should not interfere with God's creation. They don't believe genes should be transferred from one species to another. Others think that the genes might spread into the wild gene pool. What would happen if the gene for herbicide resistance spread to weeds? How would we kill dandelions that are resistant to all herbicides? Some people think that the effects on human health of eating GM crops are not yet fully understood.

KEY TERM
Genetically modified (GM) crops
Crops that contain genes from other species.

TIP
GM crops usually show increased yields. They are modified to be of benefit to humans.

○ Genetically modified animals

Transgenic animals can also be genetically engineered to produce molecules that we need. Sheep have been genetically engineered to produce proteins used in medicine. The blood of people with haemophilia does not clot as quickly as that in other people, meaning that they lose more blood every time they are internally or externally injured. Blood-clotting proteins (called factors) have been produced in the milk of genetically engineered sheep and also more recently in genetically modified bacteria.

TIP
Genetically modifying other species to produce products useful to humans as shown here is one stage in genetic modification. Medical research is exploring ways to modify the genome of individuals to eliminate some inherited disorders. This is another form of genetic modification.

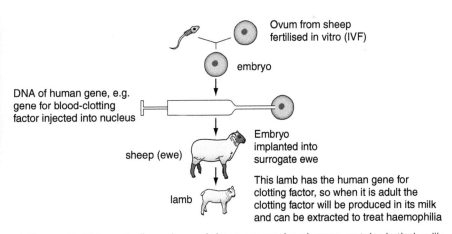

▲ Figure 22.10 Genetically engineered sheep can produce human proteins in their milk.

Show you can...

Explain how glow-in-the-dark rabbits were genetically modified.

Test yourself

9 What has golden rice been genetically modified to include?
10 Describe how maize has been genetically modified.
11 Describe why genetic modification is an ethical issue.

○ Genetically engineering insulin

Human insulin is now produced by genetically engineered bacteria.

Producing human insulin involves a small circle of DNA found in bacteria called a plasmid. This is cut open with the same enzymes that removed the insulin-producing gene from the human cell. The human gene is then inserted into the plasmid and sealed by a second enzyme. The plasmid is inserted into the bacterial cell and produces human insulin. Because the plasmid delivers the human gene to the bacterium we call it a **vector**. **Viruses** can also be used as vectors.

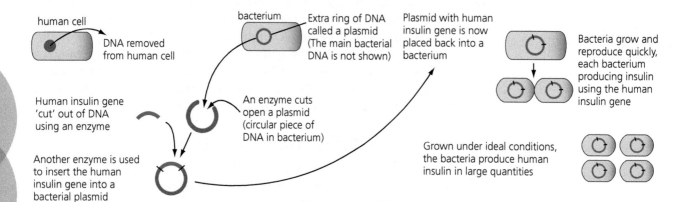

human cell

DNA removed from human cell

Human insulin gene 'cut' out of DNA using an enzyme

Another enzyme is used to insert the human insulin gene into a bacterial plasmid

bacterium

Extra ring of DNA called a plasmid (The main bacterial DNA is not shown)

An enzyme cuts open a plasmid (circular piece of DNA in bacterium)

Plasmid with human insulin gene is now placed back into a bacterium

Bacteria grow and reproduce quickly, each bacterium producing insulin using the human insulin gene

Grown under ideal conditions, the bacteria produce human insulin in large quantities

▲ **Figure 22.11** Bacteria making human insulin as a result of genetic engineering.

Before there was genetically engineered insulin, insulin was obtained from dead livestock at abattoirs. Genetically engineered insulin has the following advantages:

- It is human insulin (not pig or cattle insulin, which is slightly different).
- The production process is more efficient and less time consuming.
- Vegetarians and vegans can use insulin that is not produced from dead animals (which is in line with their principles).

Test yourself

12 Define the term 'plasmid'.
13 Describe how plasmids are used in genetic engineering.

Show you can...

Explain how a bacterium can be genetically modified to produce human insulin.

Chapter review questions

1 Define the term 'variation'.

2 Give the three causes of variation.

3 Give an example of genetic variation.

4 Give an example of environmental variation.

5 Define the term 'genetic engineering'.

6 Give an example of variation that is both genetic and environmental.

7 Your hair has become lightened by the Sun. What type of variation is this?

8 Name the two types of variation.

9 Define the term 'continuous data'.

10 Define the term 'discontinuous data'.

11 Describe the shape of a graph you might expect if your data were normally distributed.

12 Describe the process of selective breeding using large dogs as an example.

13 Give an example of a farmyard animal that has been selectively bred, and specify the trait that it has been bred for.

14 What is another scientific term that means the same as 'selective breeding'?

15 Describe the dangers of inbreeding.

16 Define the term 'transgenic organism'.

17 Explain how scientists genetically engineered a glow-in-the-dark rabbit.

18 Explain how and why golden rice has been genetically engineered.

19 Explain how and why soya has been genetically engineered.

20 Define the term 'plasmid'.

21 Suggest why some people do not like the idea of genetic modification.

22 What substances have been produced in the milk of genetically engineered sheep?

23 Explain how plasmids are used as vectors in genetic engineering.

Practice questions

1 The bar chart in Figure 22.12 shows information on the blood type of 112 blood donors.

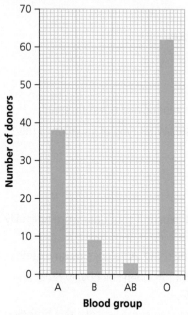

▲ Figure 22.12

a) i) How many donors were blood type B? [1 mark]

ii) To the nearest per cent, what percentage of donors had type O blood? [1 mark]

A 56   C 55
B 50   D 48

b) i) What type of variation do the data on blood type show? [1 mark]

ii) Name the factor that causes variation in blood type. [1 mark]

iii) Give another feature that is only caused by this factor. [1 mark]

2 The Belgian blue is a breed of cattle that has been kept and bred for over 200 years. The cattle have a natural mutation meaning that Belgian blues have a heavily muscled appearance.

a) i) What is the name given to the process of breeding particular characteristics into an animal? [1 mark]

ii) Why do some people disagree with this type of breeding? [1 mark]

b) Describe how the Belgian blue breed was created by farmers. [4 marks]

3 Vitamin A deficiency kills over half a million children under five every year and causes vision problems, including night blindness, for over a million pregnant women. Scientists have genetically engineered a variety of rice called golden rice. It contains several genes from different plants, one of which is the carotene-producing gene from corn, which make the corn yellow. The means that the rice can be eaten as a source of vitamin A.

a) What name is given to an organism that contains the genes of a different species? [1 mark]

b) Why do scientists want to produce golden rice? [1 mark]

c) i) Use Figure 22.13 to explain how golden rice is produced. [5 marks]

ii) Some people are concerned about the use of crops produced through genetic engineering. Explain why. [2 marks]

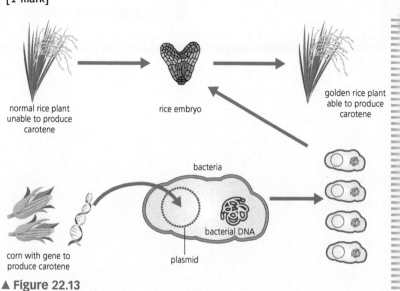

normal rice plant unable to produce carotene

rice embryo

golden rice plant able to produce carotene

corn with gene to produce carotene

bacteria

bacterial DNA

plasmid

▲ Figure 22.13

Working scientifically:
Dealing with data

Data types and graphs

When carrying out an experiment or investigation, observations are made that lead to the collection of data. Data can come in lots of different forms. One form is numerical, involving numbers that represent an amount and have a unit or are a count of something. This type of data is called quantitative data. Another form is descriptive, involving codes, words or sentences all representing a particular category. This type of data is known as qualitative or categoric data.

Questions

1 Look at the data collected in Figure 22.14 and decide which are qualitative and which are quantitative.

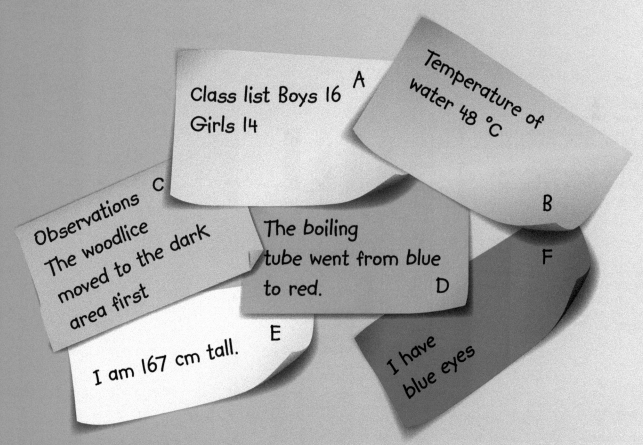

Class list Boys 16 Girls 14 **A**

Temperature of water 48 °C **B**

Observations **C** The woodlice moved to the dark area first

The boiling tube went from blue to red. **D**

I am 167 cm tall. **E**

I have blue eyes **F**

▲ **Figure 22.14** Types of data.

Quantitative data involving quantities or numbers can also be described as being continuous or discontinuous. Continuous data can take any value within a range, such as temperature. Discontinuous data can only take certain values, such as the number of students: you can have whole students but not half a student!

Questions

2 In your class, collect information on the variation seen in height in centimetres. Present your data in a suitable table.

When presenting data, we often use graphs as it makes it easier to read the data and find trends. The type of graph drawn depends on the data you have and what you are trying to find out.

Line graph – this is the most common graph drawn in science. It is used to show how one variable changes with another. It is used when both the independent variable and dependent variable have continuous quantitative data. Data points are first plotted and then a suitable line of best fit is drawn. This is usually a straight line or a curve.	**Temperature in °C** vs **Time in min**
Scatter graph – this is a graph of plotted points. It is used to explore whether a relationship (correlation) exists between two quantitative variables.	**Height in cm** vs **Arm length in cm**
Bar graph – this is used to show comparisons between categories. The x-axis shows discontinuous data (which can be qualitative or quantitative) and the y-axis shows the continuous data. The bars must not touch as this shows the data are discrete and in their own categories.	**Number of students** vs **Chosen liquids/drinks** (Cola, Orange Juice, Water, Lemonade, Milk)
Histogram – this is similar to a bar graph. However, the x-axis in a histogram shows continuous data, not discontinuous data. Because the data are over a range, the bars must touch.	**Frequency** vs ranges (60–64, 65–69, 70–74, 75–79, 80–84, 85–89)

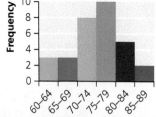

Questions

3 Draw a separate graph to display your results for the variation of height seen in the classroom. Below your graph justify your choice of data presentation.

4 Collect data from 10 students on their hand span and foot length, both in centimetres. Plot a scatter graph of the data and describe the trend seen. Is there any correlation between foot length and hand span?

5 Why was it a good idea to collect the data for foot length in centimetres, not in shoe size?

23 The development of understanding of genetics and evolution

For much of the history of the human race we have not understood how we and all other organisms appeared on Earth. Many of our ancestors believed that all life was created by an all-powerful supernatural being or beings. Many people alive today still believe this. The theory of evolution explains how the nine million species of life on Earth developed from a common ancestor alive around 3.5 billion years ago. If humans had known this all along, would there be more or fewer religious people today?

Specification coverage

This chapter covers specification points 4.6.3.1 to 4.6.3.3 and is called The development of understanding of genetics and evolution.

It covers the theory of natural selection and evolution, and evidence for evolution, including antibiotic-resistant bacteria, fossils and extinction.

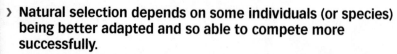

Prior knowledge

Previously you could have learnt:

> Natural selection depends on some individuals (or species) being better adapted and so able to compete more successfully.

> Changes in the environment may leave some individuals within a species, or some species, less well adapted to compete successfully.

> Species that cannot compete successfully may become extinct.

Test yourself on prior knowledge

1 Name the scientist who first described evolution.
2 Describe how polar bears are adapted to their environment.
3 Explain why changes in the environment can lead to extinction.

The theory of evolution

KEY TERM

Common ancestor An organism from which others have evolved.

▲ **Figure 23.1** The biodiversity of life on Earth results from evolution.

▲ **Figure 23.2** Darwin is often called the father of evolution.

Evolution is the theory that explains how millions of different species have developed over billions of years from one common ancestor. It explains how changes in inherited characteristics in a population over time may result in a new species through the process of natural selection.

TIP

The theory of evolution by natural selection proposes that all species of living organism have evolved from a simple life form that first developed more than **three billion** years ago.

We now have approximately nine million different species of life on Earth. Evolution explains how organisms have evolved to live in the hottest deserts, the coldest polar regions and everywhere in between. It explains how the organisms that inhabit these places are highly and specifically adapted to live there.

Charles Darwin (1809–1882) was an English scientist who is most famous for his work on the theory of evolution by a process he named 'natural selection'. A theory is a rational explanation that can be used to form a hypothesis to test.

Over millions of years, natural selection can lead to large changes in species. Species become (or remain) adapted to the environment, because the most highly adapted individuals are those that remain in a population.

In different environments, different adaptations are favoured. For example, characteristics favoured in cold regions will be different to those characteristics favoured in warm regions. This leads to biodiversity and has contributed to the range of species we have on Earth.

TIPS

A characteristic is a phenotypic feature such as length of fur, thickness of skin and height.

A species is a group of organisms that can interbreed to have fertile young. Natural selection can cause new species to form.

Change in species over a long period of time is evolution. Natural selection can account for why species change over time. We will look at an example of natural selection in bacteria in a later section.

Natural selection is the driver of evolution. In effect, natural selection can explain how all species of living thing have evolved from very simple life forms that first developed more than three billion years ago.

Evidence for evolution

Darwin's theory of evolution by natural selection is now widely accepted. In the last 200 or so years most criticism of the theory has come from religious groups. Perhaps the largest group are those that believe in creationism. They believe that the universe and all life within it were created by specific acts of divine creation in seven days. This is one of the main causes of a perceived rift between science and religion. Since Darwin published his theory more evidence for evolution has been discovered, including the relatively new science of **genetic mapping of genomes**. This evidence is still not sufficient for some believers of creationism. **Fossils** and the example of **antibiotic resistance** in bacteria provide further evidence that evolution has taken place and is taking place.

▲ **Figure 23.3** Tollund Man, a remarkably well-preserved body found in a peat bog in Denmark. No oxygen and weakly acidic conditions in bogs prevent decay.

○ Fossils

Fossils are the remains of organisms from millions of years ago, found preserved in rocks. They are formed in three ways.

1 Some fossils are formed when parts of dead organisms **do not decay** because the conditions have kept the specimen preserved. Figure 23.3 shows a human body preserved in a peat bog. In environments like this there is no oxygen and the water has a low pH. This stops decaying microorganisms from breaking down the body.

2 Other fossils are formed when parts of the organism are **replaced by minerals** as they decay. This happens when the organism has been covered with layers of sand, volcanic ash or the silt from the bottom of rivers and seas. The layers above them push down, compressing the organisms. The surrounding water is squeezed out, leaving behind mineral salts, which turn to stone. An exceptionally important fossil formed in this way is shown in Figure 23.4. This species is called *Archaeopteryx* and it shows us that some types of reptile evolved into birds. It has teeth like a reptile but feathers like a bird.

3 The third way in which fossils are formed is when an organism leaves **preserved traces or tracks**. These include dinosaur footprints and eggs. Burrows of some organisms have also been found.

▶ **Figure 23.4** *Archaeopteryx* fossil. How many bird and reptile features can you find?

▲ **Figure 23.5** The Grand Canyon, Arizona, USA. A 500 million year time sequence can be seen in the rock layers.

> **KEY TERM** ⭐
>
> **Fossil record** All of the fossils that have been discovered so far.

The fossil record

The fossil record is the information provided by all fossils that have currently been found all over the world. The date of many fossils has to be estimated by looking at the layers of rock in which the fossils are found. Figure 23.5 shows the Grand Canyon in the USA. Because the rocks have been formed by layers of drying sediments, the oldest rocks are found at the bottom of the canyon.

The fossil record shows the gradual evolution of species over time. Some species do not seem to have changed much at all. Figure 23.6 shows the coelacanth fish, which has lobed fins. It was thought to have been extinct for 65 million years until recently, when it was rediscovered by deep-sea fishermen. Figure 23.7 shows fossils of the jaw and skull of an early human ancestor called *Orrorin tugenisis*.

Fossils tell us how much or how little some species have changed over the years. A good example of this is the evolution of the horse. Fossils show that the horse's hoof evolved over time as a result of the drying of marshes. Originally horses were smaller and had bigger feet to stop them getting stuck in the marshes. As the marshes started to dry up it was the individuals with smaller hooves that had the evolutionary advantage. They could run away from predators faster because they had smaller feet. The fossil record shows this gradual change to smaller hooves.

> **TIP** ✔
>
> The change in horse hoof size is an excellent example of both natural selection and evolution.

▲ **Figure 23.6** This is a photograph of a coelacanth with lobed fins occasionally found by some deep-sea fishermen. Fossils of species similar to this with fin limbs have been found but were were believed to have become extinct 65 million years ago.

▲ **Figure 23.7** Fossils of the jaw and skull of *Orrorin tugenisis* from six million years ago.

TIP ✔

The earliest species when life began (billions of years ago) were soft bodied, and so didn't fossilise. Scientists, therefore, cannot be certain how long ago life began on Earth.

KEY TERMS ⭐

Antibiotic-resistant bacterium
A bacterium that is resistant to antibiotics.

Methicillin-resistant
Staphylococcus aureus (MRSA) A bacterium that has evolved resistance to several antibiotics.

There are **gaps** in the fossil record. Not all fossils have been found yet. Many fossils would have been destroyed in hot volcanic lava. In addition, many organisms died and were not preserved. This may be because their bodies were made of **soft tissue** and not bone. These gaps are slowly being filled in as new fossils are found, but they make it difficult for scientists to be specific about when life first appeared on Earth and how it evolved early on.

◯ Antibiotic-resistant bacteria

You have already learnt that evolution is a gradual process that takes multiple generations. In recent years scientists have begun the study of evolution in species that have a much shorter lifespan and so can be seen evolving. This began in the fruit fly *Drosophila melanogaster* in 1980. The fruit fly lives for only about 10 days and so several generations can be studied in a month. Other long-term experiments look at bacteria.

The development of antibiotic-resistant bacteria also provides evidence for evolution. This is where a mutation arises that makes bacteria resistant to antibiotics. The bacteria carrying the mutation survive antibiotic treatment and reproduce, passing their genetic advantage on to the next generation. Eventually all of the bacteria are resistant to the antibiotics and treatment no longer works. The most common strain of bacterium that has developed resistance to antibiotics is methicillin-resistant *Staphylococcus aureus* (MRSA). Methicillin is a specific antibiotic that is related to Fleming's penicillin. MRSA bacteria cannot be killed by a range of current antibiotics.

MRSA is a communicable pathogen. It is particularly serious in hospital patients. These people often have open wounds that make their normally effective first line of defence less good at stopping the entry of bacteria. They may also have a weakened immune system. The occurrence of MRSA in many hospitals is being reduced through:

- strict hygiene measures
- isolation of patients
- less liberal use of antibiotics.

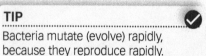
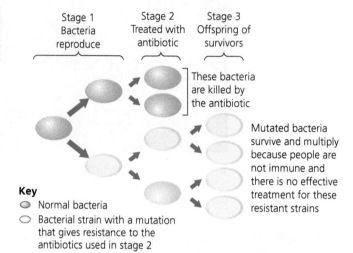

▲ **Figure 23.8** Antibiotic resistance in action.

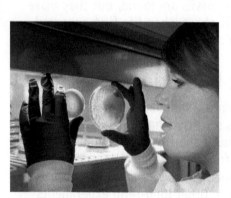

▲ **Figure 23.9** Research into MRSA continues, to try to find alternatives to antibiotics.

The use of antibiotics is currently being tightened. Some people think that in recent years antibiotics have been too readily prescribed by doctors and used in other ways, such as being giving to farm animals in their food. This **overuse or misuse of antibiotics** has helped speed up antibiotic resistance. Many doctors now think that we should not be using so many antibiotics so widely. Perhaps they should not be given to treat **non-serious infections** such as mild throat infections. It is also extremely important that, if antibiotics are prescribed, **patients take all the tablets** and don't stop as soon as they feel better. Completing the course of antibiotics makes it less likely that resistant bacteria will survive. We have managed to find one new antibiotic in the last 30 years. What happens when a pathogen first becomes resistant to all our antibiotics? The development of new drugs is a slow and expensive procedure and is unlikely to keep up with the emergence of new antibiotic-resistant strains.

Some countries (but not the UK) add antibiotics to the food of livestock even though they are not ill. This stops the livestock becoming ill and acts as a **growth promoter**. This practice has the effect of encouraging antibiotic resistance to develop in bacteria. If strains of bacteria that affect humans develop antibiotic resistance, this is a problem. Many countries, therefore, ban the adding of antibiotics to livestock feed. Antibiotics should only be used for animals that are ill.

Antibiotic resistance is a particular concern for two main reasons:

- To develop new antibiotics is very **expensive** and takes **many years**.
- Antibiotic-resistant strains are evolving so quickly that we cannot produce new, effective antibiotics fast enough.

MRSA infection rates

The data in Table 23.1 show the rate of MRSA infection from 1993 to 2011.

Questions

1 Plot the data as a line graph.
2 What is the trend shown by the data? What has happened to infection rates?
3 Why do you think the infection rates changed after 2005?
4 Explain how MRSA-resistant bacteria can arise.

Table 23.1

Year	Rate of infection per million of the population
1993	1
1996	5
1999	8
2002	15
2005	26
2008	17
2011	5

Show you can...

Explain how MRSA provides evidence for evolution.

Test yourself

1 Name the fossil that proved birds developed from dinosaurs.
2 What are MRSA resistant to?
3 Describe how antibiotic resistance has been sped up.
4 Describe how hospitals are now trying to stop MRSA and other infections linked to antibiotic resistance.

Extinction

KEY TERM

Extinct When no members of a species remain alive.

▲ **Figure 23.10** We are not really sure what the extinct dodo actually looked like.

The theory of evolution explains how organisms that are well adapted to their environment are more likely to reach reproductive maturity, have offspring and pass on their characteristics. The reverse of this is that those organisms or species that are not well adapted are less likely to survive and reproduce. If this continues a species will become extinct (die out). Scientists estimate that over 99% of all species that have ever existed on Earth have already become extinct.

The rate at which extinctions occur is increasing together with the increasing numbers of humans on Earth. Our impact on our environment is like that of no other species in our planet's history. We are rapidly cutting down rainforests to plant crops, creating bigger cities and urban areas and will soon be under pressure to drill for oil or mine other precious materials in some of the world's most remote places. This will inevitably result in more extinctions.

Extinctions are also increasing as a result of natural changes to the environment over time. For example, the end of the last ice age would have resulted in many extinctions.

The introduction of **new predators** can also cause extinctions. When Dutch sailors first landed on Mauritius in the Indian Ocean in 1598 they found the flightless dodo bird. This bird had no predators until they arrived. The last dodo was seen in 1662.

KEY TERM ⭐

Mass extinction A large number of extinctions occurring at the same time (humans are the latest cause of a mass extinction).

New diseases can cause extinctions. Extinctions can also be caused when a more successful **competitor** is introduced. This is not always a predator. Introduced species sometimes outcompete existing species for resources such as food or territory.

Approximately 66 million years ago an asteroid hit the east coast of Mexico, making a crater over 110 miles wide. This asteroid was estimated to be over 6 miles wide itself. This forced a super-heated cloud of dust, ash and steam into the sky. Its impact triggered global volcanic eruptions and earthquakes. Sunlight was blocked from the Earth, which cooled its surface. Plants found photosynthesis difficult. Animals further up the food chains that depended on those plants had less food. Many species, including the dinosaurs, became extinct. This is called a mass extinction.

Volcanic eruptions can have similar devastating effects to asteroid collisions.

In summary, extinction can happen for a number of reasons. These include:

- **natural disasters**, such as volcanic eruptions
- **climate change**, such as ice ages
- **new diseases**
- **the activity of humans** in introducing new species to countries that previously did not have them, removing natural habitats (e.g. deforestation) and causing pollution.

Test yourself

5 What percentage of species on Earth are now extinct?
6 Name the species that became extinct on Mauritius in the 17th century.
7 Describe what a mass extinction is and give an example.

Show you can...

Explain how the dinosaurs became extinct.

Chapter review questions

1 Name the key scientist who developed the theory of evolution.
2 Define the term 'species'.
3 Define the term 'fossil record'.

4 Define the term 'common ancestor'.
5 Define the term 'natural selection'.
6 Describe how fossils provide evidence for evolution. Give a specific example.
7 Explain why animal remains are often preserved in peat bogs.
8 Explain why the fossil record is not complete.
9 Define the term 'mass extinction'.

10 Explain the link between natural selection and evolution.
11 Explain how the evolution of horses' hooves is linked to a changing environment.
12 Explain how MRSA bacteria provide evidence for evolution.
13 Explain the difference between natural selection and selective breeding (artificial selection).

Practice questions

1 Figure 23.11 shows the fossilised remains of a small shoal of extinct fish.

▲ Figure 23.11

a) Define the term 'extinct'. [1 mark]

b) Give two possible causes for the extinction of the fish. [2 marks]

c) Explain how the fossils of the fish could have been formed. [3 marks]

2 Charles Darwin formulated his theory of evolution by natural selection in his book *On the Origin of Species*.

Copy and complete the sentence using one of the words below.

complex   adapted   simple

In this theory Darwin suggested that all species of living things evolved from _____ life forms. [1 mark]

24 Classification of living organisms

We are automatically very good at putting things into groups. Before you were able to form memories (usually at the age of three or four) you probably lined up your toys in order of their size, or put them in groups according to their colour or shape. These tendencies continue into later life. Many people collect stickers during childhood, and many adults have collections of objects too. Several hundred years ago we started to find more of the nine million species of life on Earth and scientists started to put them into groups. This is classification.

Specification coverage

This chapter covers specification point 1.5.4 and is called Classification of living organisms.

It covers traditional binomial classification and modern methods of classification.

Classification

▲ **Figure 24.1** Linnaeus is known as the father of classification, or taxonomy.

KEY TERMS ★

Kingdom The largest group of classifying organisms, e.g. the animal kingdom.

Genus The second smallest group of classifying organisms.

Species The smallest group of classifying organisms, all of which are able to interbreed to produce fertile offspring.

○ Carl Linnaeus and binomial classification

Carl Linnaeus (1707–1778) was a Swedish botanist. He is known as the father of modern classification. This is the putting of similar organisms into groups and giving them names. Linnaeus put organisms into these groups based upon their structure and characteristics. In 1735 he published an important book called *Systema Naturae* in which he introduced the binomial method of naming organisms. This system gave two names to each organism. In the following years he revised his work until the 12th edition, which was the last he wrote. Throughout these editions his system for classification became more and more detailed.

Only 10 000 species were ever listed in *Systema Naturae*, which is well below the nine million we think exist today.

Linnaeus established three large groups of organisms and called them **kingdoms**. These included the animal kingdom and the plant kingdom. He classified organisms within these large groups into ever smaller groups inside them that are now called **phylum**, **class**, **order**, **family**, **genus** and **species**. The classification system that Linnaeus established is the basis of the classification system we use today.

Linnaeus put each related group of organisms or species of organism into a larger group with those other species that are similar. He called this a genus (plural genera). So binomial classification gives each organism a two-part name. The first is its genus and the second its species. Our genus is *Homo* and our species *sapiens*. There are currently no other species of human alive, but previously we lived with *Homo erectus* and *Homo neanderthalensis*. Perhaps easier to remember are the five species of the big cat genus *Panthera*. Can you name them? They are *Panthera leo* (lion), *P. tigris* (tiger), *P. onca* (jaguar), *P. pardus* (leopard) and *P. uncia* (snow leopard). Many organisms have their own common name, which may or may not be related to the organism's binomial name. *Leo* is close to lion but *onca* is not like jaguar. There is one organism with the same common and binomial name: *Boa constrictor*.

▲ **Figure 24.2** The five members of the *Panthera* genus.

Scientists write binomial names in italics with a capital letter only for the genus. We also shorten the genus section of the name to its first letter when we write it more than once, so *Salmonella enterica* becomes *S. enterica*.

Since Linnaeus began his work, scientific equipment such as **microscopes** and techniques such as **genome mapping** have developed. These technological developments have made classification much easier. We now classify all organisms into five kingdoms. These are **animals**, **plants**, **fungi**, **protists** and **prokaryotes**. Remember that viruses are not alive and don't fit into any of these five. Table 24.1 shows one example of an organism in each kingdom and how it is classified into these groups. The final row shows the common name.

Table 24.1 An example of one organism from each kingdom and how the examples are classified.

Kingdom	Animals	Plants	Fungi	Protists	Prokaryotes
Phylum	Chordates	Angiosperms	Basidomycota	Apicomplexa	Proteobacteria
Class	Mammals	Monocots	Agaricmycetes	Aconoidasida	Gammaproteobacteria
Order	Primates	Asparagales	Agaricales	Haemosporida	Enterobacteriales
Family	Hominidae	Orchidaceae	Physalacriaceae	Plasmodiidae	Enterobacteriaceae
Genus	*Gorilla*	*Paphiopedilum*	*Armillaria*	*Plasmodium*	*Escherichia*
Species	*beringei*	*rothschildianum*	*solidipes*	*falciparum*	*coli*
Common name	Mountain gorilla	Slipper orchid	Honey fungus	None	None

TIPS ✓
- You do not need to remember the detail in this table. You do need to remember the order of classification groups from kingdom to species. Some people remember it by making a sentence from the first letters, such as **Keeping Precious Creatures Organised For Grumpy Scientists.**
- As you go up the groups from 'species', the number of species in each group increases.

TIP ✓
You should be able to extract and interpret information from charts, graphs, tables and diagrams such as evolutionary trees.

TIPS ✓
- The purpose of evolutionary trees is to show relationships between groups. The further apart the groups, the more distantly they are related.
- Evolutionary trees for extinct organisms are worked out from fossils.

The relationship between organisms can be shown graphically in a representation a little like a family tree, but over many more generations. Figure 24.3 shows this **evolutionary tree** for the vertebrates (animals with a backbone). There are five classes: amphibians, mammals, reptiles, birds and fish. Figure 24.4 shows the evolution of humans from other primates. You can see in Table 24.1 that the mountain gorilla also belongs to the primate order.

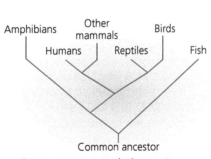

▲ **Figure 24.3** An evolutionary tree showing relatedness of some animals with backbones. Trace those that have scales.

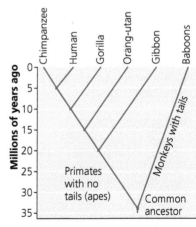

▲ **Figure 24.4** This tree is based on DNA similarities.

◯ Problems with classification

The definition of a species is a group of organisms that can interbreed to produce fertile offspring. A lion cannot mate with a zebra, so they are different species. A lion can mate with a tiger to produce a liger, but this is infertile. So lions and tigers are different species. Salamanders (Figure 24.5) are an example where this rule fails.

▲ **Figure 24.5** Salamanders can form fertile hybrids (offspring from interbreeding between two different species).

Show you can...

Explain how vertebrates are classified.

Test yourself

1 Give the hierarchy of classification beginning with kingdom.
2 Describe why ligers do not disrupt the definition of a species.
3 Describe how the three domain system is different from previous classifications.

Carl Woese (1928–2012) was an American microbiologist. He pioneered the use of chemical and molecular analysis in classification. The results of this are used to further refine classification, and often undo groups based simply on the way organisms look or behave. In particular, Woese looked at some molecules linked to the genome. By looking at the molecules in bacteria, Woese was able to refine their classification. He came up with the '**three domain system**'. His three domains are:

1 **Eukaryota** (with all animals, plants, fungi and protists)
2 **Bacteria**
3 **Archaea** (primitive bacteria usually found in extreme environments such as hot springs).

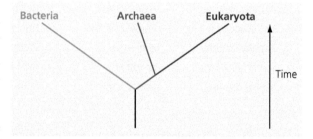

▲ **Figure 24.6** The three-domain system proposed by Woese.

Activity

Boloceroides daphneae

In 2006 a new species of marine creature with 2-metre tentacles was discovered in the deep sea (Figure 24.7). Scientists thought that it was a giant sea anemone and it was named *Boloceroides daphneae*.

Questions

1 Compare the picture of *Boloceroides daphneae* with another member of the *Boloceroides* genus shown in Figure 24.8. What features do you think they have in common that led scientists to group them together?

2 In 2014 the American Museum of Natural History examined the DNA of *Boloceroides daphneae* along with more than 112 species of anemones from the world's oceans. Their data revealed that *Boloceroides* was not a sea anemone at all, but instead classified it as being the first of its own order, and it was renamed *Relicanthus daphneae*.

▲ **Figure 24.7** Is this *Boloceroides daphneae*?

▲ **Figure 24.8** Another species of *Boloceroides*.

a) Suggest why scientists prefer to use DNA evidence to classify animals rather than rely on observations of anatomical features.
b) Why is it important that scientists are willing to change their views when new evidence is obtained?

Chapter review questions

1 Describe the process of classification.

2 Define the term 'species'.

3 Name the five kingdoms we use in classification today.

4 Give the binomial name for humans.

5 Name the diagram in which we can see the relationship between organisms over time.

6 Name the scientist that first developed the binomial system.

7 What are the two groupings of organisms in the binomial system?

8 Define the term 'kingdom' in classification and give an example of a kingdom.

9 Describe how scientists write binomial names.

10 Describe the groupings in how scientists classify organisms.

11 Name a level of hierarchy that includes mammals and reptiles. Give the other three groupings in your answer.

12 Which recent developments have allowed us to refine classification?

13 Explain why ligers (half lion, half tiger) do not conflict with our definition of a species.

14 Name two other species belonging to the *Homo* genus that are now extinct.

15 Explain why we need binomial names for species.

16 Describe the new system of classification developed by Carl Woese.

17 Define the term 'archaea'.

Practice questions

1 In 2014 around 18 000 new species were discovered and named. One of these was the cartwheeling spider, named for how it moves over the sand dunes in its home environment.

As with all species, the cartwheeling spider has a binomial name.

a) i) Define the term 'species'. [2 marks]

   ii) What does the first word of the binominal name indicate? [1 mark]

b) Which scientist is credited with developing the binominal classification system? [1 mark]

2 Polar bears (*Ursus maritimus*) and brown bears (*Ursus arctos*) are related.

Copy and complete Table 24.2 to show how a polar bear and brown bear would be classified. [8 marks]

Table 24.2

Level of classification	Polar bear	Brown bear
Kingdom		
	Chordata	Chordata
Class	Mammalia	
Order	Carnivora	Carnivora
Family		Ursidae
	Ursus	*Ursus*
Species		

3 Attenborough's pitcher plant (*Nepenthes attenboroughii*) is a large plant found on the island of Palawan in the Philippines and is named after the TV presenter and naturalist Sir David Attenborough.

▲ **Figure 24.9** Attenborough's pitcher plant.

a) *Nepenthes attenboroughii* has been classified into the kingdom Plantae, or plants.

   i) What features would be present in its cells that would classify it as a plant? [2 marks]

   ii) Plants are one of the five kingdoms. Name **three** others. [3 marks]

b) In 1977 Carl Woese proposed a new three domain system to replace the five kingdom model.

   i) What are the three domains he proposed? [3 marks]

   ii) Discuss why classification systems change over time. [2 marks]

4 a) Using Figure 24.10 what is the binomial name for the European otter? [1 mark]

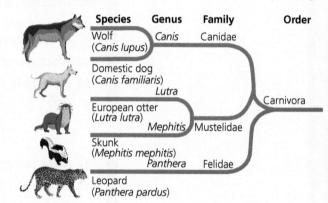

▲ **Figure 24.10**

b) From the diagram, which species is most closely related to the European otter? [1 mark]

   A Leopard (*Panthera pardus*)

   B Skunk (*Mephitis mephitis*)

   C Dog (*Canis familiaris*)

   D Wolf (*Canis lupus*)

c) From the diagram, which is the most distantly related of all the species shown? [1 mark]

   A Leopard (*Panthera pardus*)

   B Skunk (*Mephitis mephitis*)

   C Dog (*Canis familiaris*)

   D Wolf (*Canis lupus*)

Working scientifically:
Scientific thinking

Naming conventions

As you have seen, Carl Linnaeus devised a system of naming organisms. It is called the binomial system because each organism is given a two-part name. The first part of the name refers to the genus that the organism belongs to; it always starts with a capital letter. The second part refers to the species name; it does not have a capital. To make it stand out the binomial name is written in italics, or if handwritten it is underlined.

Today of the nine million species predicted to be alive, 1.9 million have been given a scientific name.

So why bother? One of the issues is due to differences in language, region and knowledge. The same species may go by many different common names, or the same name may be used to refer to several different species. For example, in the UK we think of a daddy longlegs as being a crane fly (Figure 24.11a) but in the USA that common name refers to a harvestman spider (Figure 24.11b).

The cat species in Figure 24.12 is a cougar (*Puma concolor*), but it is also known as a mountain lion, puma or catamount. Having a scientific name avoids confusion, as everyone is clear exactly which organism is being referred to.

▲ Figure 24.11 (a) *Tipula oleracea* is a daddy longlegs in the UK. (b) *Opiliones* spp. are called daddy long legs in the USA.

Questions

1 What type of animal do you think the following species names are referring to?
   a) *Octopus vulgaris*
   b) *Sphyraena barracuda*
   c) *Crocodylus niloticus*
   d) *Elephas indicus*
   e) *Panthera tigris*

2 What about these? Have a guess, and then use the internet to find out the answers.
   a) *Hippocampus hippocampus*
   b) *Gallus gallus*
   c) *Macropus giganteus*
   d) *Vulpes vulpes*
   e) *Tursiops truncates*

3 Once a new species has been discovered and verified, the scientist that discovered it must select a name and write a description. The name given can be based on a description or the species' geographical location, or be commemorative (named after a person), or a combination of these.

   Use Table 24.3 to choose an animal genus, and then choose a species name which describes your organism. Now draw a picture of it based on the descriptions and write its binomial name underneath.

Table 24.3

Animals (genus)	Description (species)
Bufo – toad	*helix* – spiral
Glis – dormouse	*lepisma* – scale
Talpa – mole	*lestes* – robber
Apis – bee	*salticus* – dancer
Lumbricus – worm	*satrapes* – a ruler
Panthera – large cat	*nyctalus* – sleepy

▲ Figure 24.12 *Puma concolor*: is this a cougar, mountain lion, puma or catamount?

25 Adaptations, interdependence and competition

What makes a good rhino? Or what makes a good example of any organism? The answer is the one that is best suited to live in its habitat in a delicate balance alongside the other species there. That would make it the best adapted and most likely to be successful competing with others from its own and other species. So a good rhino will be adapted to live on the grassy plains of Africa together with the lions, zebras and all the other species. If, as is nearly the case with rhinos, they are all hunted then this delicate balance can be quickly destroyed with devastating knock-on effects for other species as well. How long will it be before many species of African mammals are hunted to extinction and what will happen to the other species that live in balance with them?

Specification coverage

This chapter covers specification points 4.7.1.1 to 4.7.1.4 and is called Adaptations, interdependence and competition.

It covers competition within communities, interdependence and adaptations.

Previously you could have learnt:

> Species differ from each other in a range of ways.
> Organisms (and species) in an ecosystem are dependent on other organisms (and species) in many ways, including as a food source.

Test yourself on prior knowledge

1 Give the term used to describe the differences between species.
2 Explain why photosynthetic organisms are at the bottom of most food chains.

Communities

KEY TERMS ⭐

Population The total number of all the organisms of the same species that live in a particular geographical area.

Community A group of two or more populations of different species that live at the same time in the same geographical area.

Competition The contest between organisms for resources such as food and shelter.

A population is the **total number** of all the organisms of the same species that live in a particular geographical area. The overall human population is the total number of humans alive. The geographical area for the human population is the entire planet.

A community is a group of two or more populations of different species that live at the same time in the same geographical area. Populations and their interactions with each other and their environment are studied by ecologists.

Competition is the contest between organisms within a community. There are two types. **Interspecific** competition is between different species in the community. **Intraspecific** competition is between organisms of the same species.

Some form of competition exists between all species of life on Earth, including microorganisms. Charles Darwin described this as the 'struggle for existence'. Competition is the driver for evolution. Without competition evolution slows or stops completely. What biological competition now exists for humans? Are we still evolving?

○ Plant competition

Imagine the death of a mature tree in the rainforest. This might be 30 m tall and form part of the tree canopy. The canopy is the primary layer of the forest, which forms a relatively horizontal 'roof'.

As soon as this tree falls a plant race begins. Seedlings that have already germinated and started to grow suddenly speed up their growth. They race for the light above. The gap caused by the old tree falling has provided more light, which means more photosynthesis, which means more growth. Eventually one seedling will outgrow the others. In time this will form part of the canopy. As well as **light** and **space**, plants also compete for **water** and **nutrients** from the soil. Many plants have evolved ways of dispersing their seeds far away from the parent plant or plants. This reduces competition between them.

This example is in the rainforest, but competition between plants

TIP ✓

Plants compete for many resources, including light, space, water and minerals in the soil.

▲ **Figure 25.1** When this large tree dies it will be replaced by the fastest growing seedling.

▲ **Figure 25.2** Hair grass is adapted to live in the Antarctic.

exists wherever they grow. A similar situation exists in the conifer forests of the Arctic (the region around the north pole), the grass plains of Africa, the hair grass that grows in the Antarctic (the south pole), and everything in between. Competition between two hair grass plants in the Antarctic is an example of intraspecific competition. If it is between a hair grass plant and a pearlwort plant (the only other plant species found there) then it is interspecific competition.

Looking at a woodland

Figure 25.3 shows a woodland in May (a) and July (b).

Questions

1 Look at the photos. What resources are the bluebells competing for?

2 Why do you think bluebells have adapted to flower in early summer before the trees come fully into leaf?

3 Some scientists are concerned that global warming is making trees in woodlands like this come into leaf earlier every year and that bluebells are flowering early to compensate. They are concerned that if bluebells come into flower too early it might lead to them not successfully reproducing. Suggest why you think this is the case.

▲ **Figure 25.3** Bluebells can be seen carpeting the floors of many British woodlands in early summer, but they quickly disappear as summer progresses.

◯ **Animal competition**

Animal competition may be easier to observe than competition in plants. Perhaps it is because it can happen over a shorter timescale. Imagine lions and hyenas on the grassy plains of Africa. Surprisingly, the hyena, which is known as a scavenger of prey caught by lions, actually kills more prey itself. The lions scavenge more hyena kills than *vice versa*. These two species are competing for food.

Now visualise a troop of mountain gorillas. The family is dominated by the alpha male, called a silverback. He is usually the oldest and strongest gorilla and has matured with a silver stripe across his back. There may be other male gorillas in the troop. These are almost certainly children of the alpha male. The alpha male will want to mate with all the female gorillas and he will want to stop the other males from doing so in his place.

As well as competing for **food** and **mates**, animals often compete for **territory**. Many of us live in towns or cities, or at least have close neighbours. If a number of people in any one residential area have cats as pets there will be territorial fights. Competition is particularly fierce between rival male cats. We have found that the territories of domestic cats often change as a result of competition and in many cases overlap.

TIP

Animals compete for many things, including food, territory and mates.

▲ **Figure 25.4** Lions will fight to keep their kill.

▲ **Figure 25.5** The competition for space is intense in this gannet colony.

Interdependence

Interdependence means that all the species that live in a community depend upon each other. For example, in a marine ecosystem sharks depend upon seals and smaller fish for food.

A **stable community** is one in which there is much interdependence – that is, all the species are in balance with each other. There are an appropriate number of predators and prey. As the numbers of prey increase, so do the numbers of predators after a short lag phase. As the numbers of predators increase, the number of prey decrease as they are being eaten. Eventually the reduced number of prey means that the numbers of predators will fall because there is not enough food. This is called **predator–prey** cycling and is shown in Figure 25.6. While the numbers of predators and prey cycle naturally, interdependence means that they will almost always remain within a stable number range.

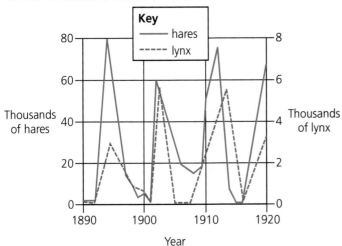

▲ **Figure 25.6** Predator–prey cycling in the Canadian lynx and snowshoe hare.

Abiotic factors

TIPS

- It is important that you can explain how a change in an abiotic factor would affect a given community.
- Temperature is another important abiotic factor in polar regions.

▲ **Figure 25.7** These trees are close to the tree line at the north pole.

▲ **Figure 25.8** *Hosta* plants growing in shade.

▲ **Figure 25.9** *Hydrangea* plants have pink flowers in alkaline soil and blue flowers in acidic soil.

Abiotic factors are the non-living parts of the environment. These can be chemical or physical, but not biological.

Light intensity is an important abiotic factor. Because the polar regions are at the poles of our planet they face away from the Sun and therefore they receive a much lower light intensity than the area around the equator. Without this light plants find it particularly difficult to grow in the polar regions. If travelling towards the north pole, you will come to an area where there are no more trees. This is called the tree line. Beyond this point there is not sufficient light intensity to support trees, so none grow.

Other plants require less light. Plants that are used to growing in shady areas like those within the *Hosta* genus usually have big, dark green leaves. The leaves are dark green because they contain lots of chlorophyll and are big to absorb as much light as possible. If you moved a *Hosta* from the shade into the middle of your lawn it would wilt and die very quickly.

Water (moisture level) is another abiotic factor. Cacti are plants famous for surviving in dry deserts with very high light intensity. Moving a cactus from the desert to a country like the UK would kill it. The excess water would kill the cactus. It would 'drown' in too much water. It is often said that more houseplants are killed by overwatering than by underwatering.

The **pH of the soil** and its **mineral content** are other abiotic factors. Some plants are adapted to live in nutrient-poor soils. Carnivorous plants have evolved to catch insects to supplement the low levels of nutrients in the boggy soils or rocky outcrops they grow upon.

Some plants are very sensitive to soil pH. Species of *Azalea* are known specifically for only growing in acidic soil. Gardeners add peat to their soil to make it more acidic and suitable for these plants.

Farmers spread manure on to their fields and gardeners add compost to their plots to increase the mineral content of their soils. This provides nutrients but also makes the soil more acidic. Farmers often then add lime to reduce the acidity.

Interestingly, the *Hydrangea* plant can grow in both acidic and alkaline soils. Its flowers actually change colour depending upon the type of soil it grows in. If it is found in an acid soil it has blue flowers. If in an alkaline soil its flowers are pink. In this way it acts just like universal indicator paper!

Other abiotic factors that affect a community are **wind intensity** and **direction**. Close to the headlands in many coastal regions in the UK

▲ **Figure 25.10** This tree has been battered by the winds and so cannot grow straight upwards.

are trees growing at very odd angles and in strange shapes. These have spent their entire lives growing against strong coastal winds.

Temperature is also an important abiotic factor. It affects both plant and animal life.

A common abiotic factor that affects plants is the level of available **carbon dioxide**. This is a limiting factor in photosynthesis so plants generally prefer higher levels.

In a similar way, the dissolved **oxygen** levels in water act as an abiotic factor affecting the distribution of aquatic animal life. More oxygen means that more life will be able to use it to respire.

Activity

The distribution of lichens

Lichens are made from two organisms (an alga and a fungus) living in a mutually beneficial partnership. They are often found growing on rocks or tree trunks. Design an experiment to compare the distribution of lichens on the north- and south-facing sides of a tree.

Show you can...

Explain how abiotic factors might affect the growth of a plant.

Test yourself

8 What are abiotic factors?
9 What do farmers add to their fields to reduce the pH?
10 Describe how the *Hydrangea* plant is unusual.
11 Describe why trees cannot grow beyond the tree line in the Arctic.

Biotic factors

KEY TERM

Biotic factors The living parts of the environment.

TIP

It is important that you can explain how a change in a biotic factor would affect a given community.

▲ **Figure 25.11** The cane toad is a highly invasive species in Australia.

Biotic factors are those that are living or related to living organisms. Perhaps the most obvious biotic factor is the amount of **food** available. The number of populations of organisms in a community depends reasonably heavily on the amount of food available.

○ Introducing predators

The introduction of a **new predator** is a biotic factor that can drastically affect a community. The relationship between predators and prey in a stable community is finely balanced. The numbers of each organism rise and fall in a predator–prey cycle but remain fairly constant within this. If a new predator was introduced into this community the number of prey might reduce extremely quickly. This would also affect other organisms that depend upon either the predator or the prey.

An example of this is the cane toad. This is native to Central and South America but has spread to Australia. In Central and South America it is eaten by several predators. These predators are not found in Australia,

▲ **Figure 25.12** Red squirrels are outcompeted by the larger grey squirrels.

so the cane toad populations have increased massively there. The toad is now an invasive species in these areas.

○ Introducing diseases

The introduction of **new diseases** (pathogens) to an area can have devastating effects on the populations of animals and plants. Plants, as well as animals, can be infected by pathogens. Dutch elm disease is caused by a fungus that is spread by a species of beetle. This began in Asia and has now spread to Europe. It has killed over 25 million elm trees in the UK alone.

○ Competing species

One species, when introduced, can simply **outcompete** another. The red squirrel (*Sciurus vulgaris*) is native to the UK and Europe. In the UK the numbers of red squirrels have reduced dramatically in recent years. Fewer than 140 000 are left, most of which are in Scotland. In the 1870s the larger grey squirrel (*Sciurus carolinensis*) was introduced. There are now more than two million grey squirrels. This was done on purpose, as grey squirrels were thought to be a fashionable addition to many large country estates. Together with the cutting down of trees, the grey squirrel has caused the massive reduction in numbers of red squirrels in the UK. Because grey squirrels are larger they can store up to four times more fat than red squirrels. This means they are more likely to survive harsh winters. They can also produce more young than red squirrels. Finally, grey squirrels are immune to a pox virus that has killed many entire red squirrel populations.

As a result of the issues described above, the transport of many plants and animals into countries is often illegal or restricted by licences and permits.

Test yourself

12 Define the term 'biotic factor'.
13 Give an example of a biotic factor.
14 Describe why the numbers of native red squirrels are reducing in the UK.
15 Describe why the introduction of the cane toad to Australia is causing problems.

Show you can...

Explain how biotic factors might affect a herd of zebras.

Ecosystems

Previous sections discuss the effect of abiotic and biotic factors on communities – essentially, what an ecosystem is.

Adaptations

TIP

It is important that you can explain how organisms are adapted to live in their natural environment.

All organisms are **adapted** to the environment they live in. Without a number of these adaptations any organism finds itself severely disadvantaged against others within its species (intraspecific competition) and against other species (interspecific competition). Imagine a rabbit with small ears. It is less likely to hear a predator coming and more likely to be eaten. This means it will be outcompeted by other rabbits with more effective ears. It also means that other herbivores such as deer will outcompete it as well.

Adaptations allow organisms to outcompete others and provide them with an evolutionary advantage. Without adaptations and competition, there would be no evolution.

KEY TERMS

Structural adaptation An advantage to an organism as a result of the way it is formed, like the streamlining seen in fish.

Behavioural adaptation An advantage to an organism as a result of its behaviour.

▲ **Figure 25.13** Look at the adaptations of the thorn bush in this photo. What adaptations do you need to eat thorn bush leaves? A tongue up to 45 cm long to strip the leaves while avoiding the thorns, and a leathery mouth, allow the giraffe to deal with the plant's defences.

Adaptations can be structural, behavioural or functional. Structural adaptations are the physical features that allow an organism to compete. Examples include the sharpness of a tiger's teeth to allow it to kill its prey and the eyesight of a bird of prey to allow it to hunt better. Some animals mimic the structural adaptations of others. Hoverflies have evolved black-and-yellow banding on their body similar to that of wasps. This is shown in Figure 25.14. This type of adaptation is called mimicry.

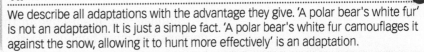
TIP

We describe all adaptations with the advantage they give. 'A polar bear's white fur' is not an adaptation. It is just a simple fact. 'A polar bear's white fur camouflages it against the snow, allowing it to hunt more effectively' is an adaptation.

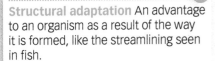

▲ **Figure 25.14** (a) What visual stimulus makes you avoid wasps? (b) This is a harmless hoverfly. Why might insect predators avoid the hoverfly?

Behavioural adaptations explain how specific behaviours benefit an organism. Crows have evolved to use sticks to poke into holes as probes to find food. When under threat, a hedgehog rolls itself into a ball. Only its spines are exposed, making it very difficult for predators to kill and eat it. This is a very effective behavioural adaptation, protecting the hedgehog.

KEY TERM ⭐

Functional adaptation An advantage to an organism as a result of a process, such as the production of poisonous venom.

▲ **Figure 25.15** A chimpanzee using a small branch as a tool.

Functional adaptations are processes that organisms complete to help them survive. A good example is the production of venom. Many organisms produce venom to deter predators from hunting and killing them. Other animals, including many snakes and some spiders, produce poisons to help them hunt. These can be injected by a bite or a sting. Plants have evolved functional adaptations by producing poisons as well. *Atropa belladonna* is commonly known as deadly nightshade. If eaten this can kill humans in severe cases. Lilies are poisonous to domestic cats.

Plants produce hormones in their roots and shoots that allow them to turn towards water and light, respectively. This is an example of a functional adaptation.

Activity

Odd adaptations

All animals and plants have adaptations to aid survival, but some adaptations are bizarre. Choose one of the following and research how it is adapted, to discover its odd adaptation:

- basilisk lizard (*Basiliscus basiliscus*)
- treehopper (*Cyphonia clavata*)
- vampire squid (*Vampyroteuthis infernalis*)
- pistol shrimp (e.g. *Alpheus bellulus*)
- toad fish (*Tetractenos hamiltoni*)
- mimic octopus (*Thaumoctopus mimicus*)

KEY TERM ⭐

Extreme environment A location in which it is challenging for most organisms to live.

○ Extreme environments

An **extreme environment** is one that is challenging for most organisms to live within. Imagine somewhere you would find it hard or impossible to live in, and that is probably an extreme environment. These are places that may have highly **acidic** or **alkaline** environments. They may be extremely **hot** or **cold**. They may not possess much oxygen or water. The majority of the moons and planets in our solar system are extreme environments.

Polar regions

The region around the north pole is called the Arctic. The region around the south pole is the Antarctic. Polar bears are only found in the Arctic and penguins mainly in the Antarctic. Both regions are obviously very cold. The temperature in Antarctica has reached −89 °C.

Few animals and plants live in the polar regions. In Antarctica perhaps the most famous are the emperor penguins. The male birds look after the eggs on their feet, huddled into a tight circle to keep warm during the cold winter months. This is an example of a behavioural adaptation. Polar bears in the Arctic have a small head and ears to reduce heat loss. They have a thick layer of fat (up to 11 cm) under their thick fur to keep warm. Each of these organisms is highly adapted to survive in this extreme environment.

▲ **Figure 25.16** Polar bears are adapted to survive in cold conditions.

Deep-sea hydrothermal vents

A deep-sea hydrothermal vent is a small gap or fissure on the bottom of the sea or ocean. Here, volcanic activity below the sea or ocean bed provides an area where the water is warmer. On land these are called hot springs or geysers. The volcanic activity underneath them releases huge plumes of black or white 'smoke' into the water. This is full of dissolved minerals. These minerals precipitate out forming solid chimneys that stretch like stalagmites into the water.

The vents are under extreme **pressure** from all the water above them and around them. Their plumes are **superheated** by the volcanic activity but several metres away the temperature drops to almost freezing. No light can reach the bottom of these vents. Yet life exists here. Any life which exists in an extreme environment, such as the polar regions or around deep-sea vents, is called an extremophile.

We originally thought that all life required light from the Sun. We thought that an organism that photosynthesised was at the beginning of every food web on Earth. Then life was discovered at the bottom of the ocean surrounding hydrothermal vents. Several metres away from the vent there is very little life at all, if any. Bacteria have evolved to feed on the chemicals released by the vents. In particular they feed on sulfur compounds (e.g. hydrogen sulfide), which are toxic to almost all other life.

▲ **Figure 25.17** Could anything survive in a hydrothermal vent with temperatures close to 100°C and pressures high enough to crush a submarine?

▲ **Figure 25.18** These species of worms and fish may only exist around this one hydrothermal vent.

Test yourself

16 What is an extreme environment?
17 Describe a structural adaptation of a hoverfly.
18 Describe a behavioural adaptation of an emperor penguin.

Show you can...

Explain what is unique about hydrothermal vents.

Chapter review questions

1 Define the term 'population'.

2 Define the term 'community'.

3 Describe what is unusual about the *Hydrangea* plant.

4 What is competition?

5 Explain what an adaptation is.

6 Define the term 'extremophile'.

7 Give two examples of extreme environments.

8 Describe how cacti are adapted to live in the desert.

9 What are the two factors that make an ecosystem?

10 Define the term 'abiotic factor'.

11 Suggest two abiotic factors that gardeners might be aware of.

12 Define the term 'biotic factor'.

13 Give two examples of biotic factors.

14 What biotic factor often correlates with the amount of food available in a community?

15 Suggest some factors that plants compete for.

16 Suggest some factors that animals compete for.

17 Define the term 'interdependence'.

18 Describe what happens to individual organisms that are not well adapted.

19 Name the three types of adaptation.

20 Suggest an example of structural adaptation in a shark.

21 Suggest an example of functional adaptation in a rattlesnake.

22 Describe the behavioural adaptation of male emperor penguins.

23 Describe how polar bears are adapted to live in the Arctic.

24 Describe the conditions around a deep-sea hydrothermal vent.

25 Explain why light is an abiotic factor.

26 Describe an experiment in which you use pond weed and a lamp to investigate the effects of light on the rate of photosynthesis.

27 Name a plant species that grows in acidic soil.

28 Name the only two plant species that compete at the south pole.

29 Describe a graph of predator–prey cycling.

Practice questions

1 In a rainforest which two of the following would plants most likely compete for? [2 marks]

  A Light   C Mineral nutrients
  B Water   D Carbon dioxide

2 The native white-clawed crayfish, found in waterways throughout the UK, is currently under threat from a larger and more aggressive invasive species called the American signal crayfish. Crayfish eat plants and animals, and will eat any organic matter they find.

  a) i) Name an abiotic factor that could affect the distribution of the white-clawed crayfish in its habitat. [1 mark]

      ii) Name two biotic factors that could affect the distribution of the white-clawed crayfish. [2 marks]

  b) i) Define the term 'invasive'. [1 mark]

      ii) Explain in as much detail as possible how the introduction of the American signal crayfish could cause the loss of the white-clawed crayfish from British waterways. [2 marks]

3 The red squirrel is a native species and is one of the most threatened of UK mammals. It was once found throughout the UK but has suffered a marked population loss over the last 50 years. In contrast the grey squirrel is an introduced species whose population is increasing throughout the UK.

  a) i) Define the term 'population'. [1 mark]

      ii) Suggest **two** reasons for the decline in the red squirrel population. [2 marks]

  b) Table 25.1 gives information on the two species of squirrel. Scientists are concerned that grey squirrels are outcompeting red squirrels.

Table 25.1

	Red squirrel	Grey squirrel
Average body length nose to tail in cm	40	51
Average mass in g	255–340	396–567
Breeding habits	Once a year, 1–8 per litter	Twice a year, 1–8 per litter
Food preferences	Scots pine nuts Hazelnuts Spruce cones	Beech mast Oak acorns Sycamore seeds Sweet chestnuts Scots pine nuts Hazelnuts Spruce cones
Number in UK	140000	2500000

      i) Define the term 'outcompeting'. [1 mark]

      ii) What evidence from the table suggests that grey squirrels would outcompete red squirrels? [2 marks]

      iii) What is the name for the type of competition between red and grey squirrels? [1 mark]

  c) Calculate the percentage of squirrels in the UK that are red squirrels. Show your working. [2 marks]

  d) Give an adaptation that both squirrels have to living in a forest habitat. [1 mark]

4 Wildebeest are prey animals that are hunted by lions in savannah grasslands. They have several adaptations to avoid being caught.

▲ Figure 25.19

  a) i) Describe a structural adaptation wildebeest have to avoid predation. [1 mark]

      ii) Looking at Figure 25.19, describe a behavioural adaptation that the wildebeest have to avoid predation. [2 marks]

  b) When a baby wildebeest is born the calf can stand and run within minutes of birth. Explain how this adaptation also helps avoid predation. [2 marks]

Working scientifically:
Experimental skills

Evaluating experiments

Emperor penguins have several structural adaptations to allow them to survive in the extreme cold and wind-chill of Antarctica. They have a thick layer of blubber, or fat, under the skin, and outer feathers that interlock to prevent water penetrating their inner feathers, which trap air and insulate them (Figure 25.20).

▲ **Figure 25.20** Emperor penguins live in freezing conditions in Antarctica.

You are going to carry out an experiment to see whether the huddling behaviour that penguins exhibit is a behavioral adaptation to help emperor penguins keep warm. As you carry out the experiment note down any issues you have with the method or errors that were made.

Method

1 Collect 11 test tubes and divide them so you have one on its own, three in a group, and seven in another group, as in Figure 25.21. Use an elastic band to secure each group.

2 Fill the single test tube so it is three-quarters full of hot water from a recently boiled kettle.

3 Place the single test tube in a beaker of cold tap water, immediately record the starting temperature and start a stopwatch. Record the temperature of the water in the test tube every minute for the next 10 minutes.

4 Repeat the experiment with the group of three and the group of seven test tubes in a 'huddle', ensuring each tube is filled with the same volume of hot water.

Working scientifically: Experimental skills

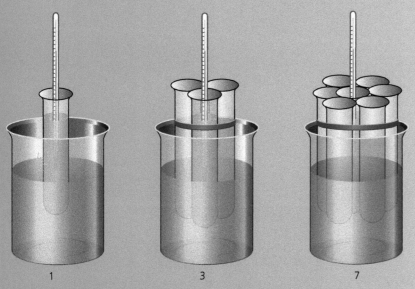

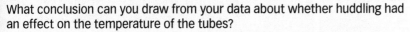

▲ Figure 25.21 The equipment used to investigate huddling behaviour.

Questions

What conclusion can you draw from your data about whether huddling had an effect on the temperature of the tubes?

You are going to write an evaluation for this practical. There are three things that need to be evaluated: the method, the results and the conclusion. Use the following question prompts to help you write an evaluation of your method. Don't just answer the questions; instead write your answers in clear paragraphs.

Evaluating your method

1 Were there issues with the method or equipment used?
2 Was your experiment a fair test? Were all control variables kept constant?
3 Was the range of independent and dependent measurements sufficient to show a clear trend?
4 How could you improve the method, range or equipment?

Evaluating the data

1 How good are your data? Can you identify any anomalous results?
2 If you have anomalous results can you explain what may have caused them?
3 Is there any evidence for random or systematic error in your data? If so can you identify the cause?
4 Is your data repeatable or reproducible? If not, what could you have done to ensure it was?

Evaluating your conclusion

1 Did your results show a clear trend?
2 Is there enough evidence to support your conclusion?
3 What are the areas of weakness in your method or data that make you less sure of your conclusion? What could be done to reduce these?
4 What further evidence is needed to fully support your conclusion?

26 Organisation of an ecosystem

You are made from thousands of billions of cells. Each of these is made from millions of billions of atoms. These numbers are probably too big for most of us to understand. Many of these atoms are carbon, which forms a 'cellular skeleton' for all life on Earth. The carbon atoms in your cells were not always yours. They are recycled when we die and form other living or non-living substances. When you die some of your carbon atoms might make up carbon dioxide, some might be trapped in chalk rock, others might become a part of animals or plants. Scientists estimate that each of us has some of the carbon atoms that once made up William Shakespeare (or any other long dead person you might prefer). How many of your carbon atoms came from Charles Darwin?

Specification coverage

This chapter covers specification points 4.7.2.1 to 4.7.2.2 and is called Organisation of an ecosystem.

It covers levels of organisation, the carbon and water cycles, and the impact of environmental change.

Previously you could have learnt:

› In an ecosystem, organisms are interdependent.
› All life on Earth depends on organisms that photosynthesise, producing food and maintaining levels of carbon dioxide and oxygen in the atmosphere.
› The carbon cycle is an important cycle that involves living organisms.

Test yourself on prior knowledge

1 Define the term 'interdependence'.
2 Describe how respiration alters the carbon cycle.
3 Explain why there is less carbon dioxide in the atmosphere in summer.

Levels of organisation

This section describes some ways in which we can find out more about the organisms living around us.

KEY TERMS

Sampling The process of recording a smaller amount of information to make wider conclusions.

Quadrat A square frame used in biological sampling.

○ Sampling

Sampling is a process by which scientists look at a part of a habitat and draw conclusions about the whole of it. For example, if you wanted to know whether polar bears were becoming less common, you could either count them all or you could count a smaller but representative sample of them and make judgements using these data alone. Counting them all would take too long and might mean that the numbers had changed by the time you thought you had finished counting.

○ Quadrats

Quadrats are used in many forms of sampling. A quadrat is simply a square of wire or wood frame that is placed on the ground and the organisms that are within it are recorded. This is obviously most useful for plants or small, static or slow-moving animal species. Often $0.25\,m^2$ quadrats are used. If you were looking to sample large organisms such as bushes or shrubs you would definitely use quadrats that are much bigger than this.

The recording of organisms happens in one of three ways. Sometimes the **number** of individual organisms is counted. If you wanted to see whether there were more daisies or dandelions, you would have to count them both. Other times the number of different species is counted. This gives you an indication of how biodiverse an area is.

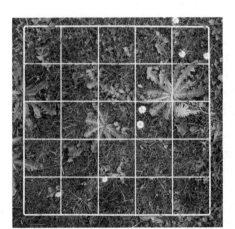

▲ **Figure 26.1** A quadrat on a lawn with weeds.

Sometimes it is difficult to see where one plant stops and another starts. Imagine trying to count grass plants in a garden! In this situation, we work out the **percentage cover**. This is the percentage of the quadrat covered by a particular type of plant. It is impossible to calculate exactly the percentage cover of a particular species, therefore we round the figure to the nearest 10%, as shown in Figure 26.2.

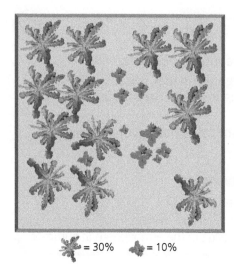

= 30%   = 10%

▲ **Figure 26.2** Using a quadrat to measure percentage plant cover.

○ **Random sampling using quadrats**

Imagine that you are trying to estimate the number of daisy plants in a field. It is common mistake to think that you stand in one place, close your eyes, spin around, throw your quadrat and sample where it lands. How often is it likely to land at your feet? How often is it likely to land beyond the reach of your throw? It never will. Most of the sampling that we do involves the random placing of the quadrat. This is important to remove all forms of bias from your results.

So you choose a fixed location from where you will always start. It is usually the corner of the field or area. From here you use a pair of random numbers from a table in a book or the internet as coordinates. The first random number tells you how far to walk forwards. You then turn to one side and the second random number tells you how far to walk in that direction. It is here that you put your quadrat down and count the daisies in it. You then move back to the start and repeat using the next two random numbers.

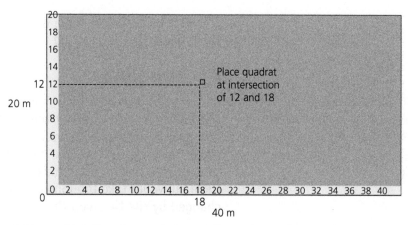

▲ **Figure 26.3** Place the quadrat according to the random numbers.

If you only completed two 0.25 m² quadrats you would not be able to make appropriate conclusions about the number of daisies in the field. Sampling must be **representative**. So you may have to complete 30 quadrats to get a representative sample of the field. The more quadrats you sample the more confident you can be in your results, but the longer it will take.

After you have completed a set number of quadrats, you calculate an average per quadrat. You then measure the area of the field. Next you work out how many quadrats fit into it. If you know that 250 quadrats fit into the whole field and that you have found an average of three daisies per quadrat, you multiply the numbers to estimate the total number in the field.

This sort of sampling is often used to compare the number of organisms in two different areas. If you wanted to find out whether there were more dandelions in sunny parts of the field, you would divide the field into sunny and shady areas. You would repeat the process just described in these two separate areas and compare the results.

◯ Systematic sampling using quadrats

There is another method of using quadrats, called **systematic sampling**. Instead of randomly placing your quadrat you would place it in a systematic (or regular) way. You would do this only when you wanted to check whether the distribution of an organism changed in an area.

Imagine you are on a rocky seashore. You could try random sampling to estimate the total number of crabs on the seashore. You would do this using random numbers as described in the first part of the previous section. However, if you want to look at how seaweed is distributed on the seashore you could use systematic sampling. Here you would place a line called a transect from the top of the shore to the bottom. You would systematically place your quadrat on this line, say every metre, and record the number of seaweed species.

Scientists often record abiotic factors along transects to compare their biological data with. This helps them draw conclusions about **why** the organisms are distributed the way they are along a transect. For example, if you wanted to check how the number of a particular species of plant changes as you progress from grassland to the edge of a pond, you would run a transect from the grassland to the pond. You could measure soil moisture (an abiotic factor) at intervals along the transect to see whether soil moisture correlates with the number of plants of the particular species. This would help you draw conclusions about why the plant distribution changes.

The type of sampling you undertake depends upon the question you are trying to answer. Is it just an estimation of a population? If so, then you will place the quadrats using random numbers. Is it comparing the numbers of organisms in two areas? If so, then you will place the quadrats using random numbers separately within these areas. Or are you trying to see how the distribution of an organism changes with differences in the habitat? If so, then you will place the quadrats systematically along a transect.

▲ **Figure 26.4** Students completing systematic sampling using a quadrat on a transect.

Test yourself ⚙

1 Name the square device that is often used in sampling.
2 Name the line upon which you might sample.
3 Describe why it is necessary to sample habitats.
4 Describe how you would place a quadrat in random sampling.

Show you can...

Explain when you would use random and when you would use systematic sampling.

Measure the population size of a common species in a habitat. Use sampling techniques to investigate the effect of a factor on the distribution of this species

This practical has two parts. First you will work out an estimation for the population size of a species in a given area and then you will investigate how an abiotic factor affects its distribution within this area.

Part 1: method

1   When you are given a quadrat the first thing you need to do is to work out the area of your quadrat in m². It is likely that you have been given either a 1 m quadrat or a 0.5 m quadrat.

   area = length × width

> **Example**
>
> For a 1 m quadrat:
>
> area = 1 × 1, so the area is 1 m²
>
> For a 0.5 m quadrat:
>
> area = 0.5 × 0.5, so the area is 0.25 m²

2   Using your method of random sampling, find your first sampling point and place the quadrat down. Determine either the number or the percentage cover of your chosen plant in your first quadrat. If some plants are half inside the quadrat, remember to only class the plants on two sides of the quadrat as being in and the plants on the other two sides as being out (Figure 26.5).

3   Repeat this process for the remaining nine quadrats.

4   Determine a mean for the plant you studied across all your quadrats. If you had a 1 m² quadrat do this by adding up all your data and dividing by 10. If you had a 0.25 m² quadrat you need to do the same but then multiply your final answer by 4 to give your estimation of plants per m².

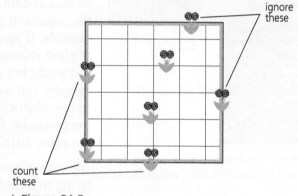

▲ Figure 26.5

5   Your teacher will provide you with the approximate area for the total site studied. Use this to estimate the abundance of the plant across the site (total area m² × average abundance of plant = abundance of plants per m²).

> **Example**
>
> Using the data in Table 26.1:
>
> Area = 5060 m²
>
> So you can estimate that there are:
>
> 8.6 × 5060 = 43 516 daisies

Table 26.1

Quadrat	Number of daisies
1	10
2	5
3	6
4	8
5	7
6	11
7	15
8	7
9	8
10	9
Total daisies in all quadrats (1 m²)	86
Mean number of daisies per m²	8.6

6   Repeat the process by taking 10 more random samples.

  a) What did this do to your mean?

  b) How did this change your estimation for the abundance of your chosen plant on the field?

7   Combine all your class data and use this to determine a mean abundance of the plant per m² for your class. Use this to estimate the abundance of the plant across the site.

  a) Why does taking more samples lead to a more valid estimation?

  b) Why do you think that when ecologists are measuring abundance they usually only measure at most 10% of the area?

Part 2: method

1   Your teacher will tell you which abiotic factor you will be studying, the length of the transect and the distance between your quadrat sampling points.

2   Within the chosen habitat, identify a sampling site that shows variation in the abiotic factor being studied. Position a tape measure to guide your transect.

3   Starting at the 0 m end of the tape, lay your quadrat down so that the left-hand bottom corner is level with the 0 m mark. Identify the abundance of your chosen species within the quadrat and record in a suitable table, as in Table 26.2.

Table 26.2

Distance in metres	Abundance of grass in %	Light intensity in arbitrary units
0	2	1
2	4	2
4	15	4
6	53	6
8	74	8
10	100	9

4   Measure and record the abiotic value for the quadrat.

5   Move the quadrat to the next sampling point and determine and record the abundance of the species and record the value for the abiotic factor.

6   Repeat this until you reach the end of the transect.

Questions

1   Describe how the abundance of your chosen species changed as the abiotic factor changed.

2   Suggest a reason to explain the difference in abundance.

3   Why could you be more confident that you had a valid trend if you had carried out four more transects?

Producers, consumers and decomposers

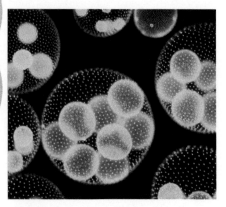

▲ **Figure 26.6** *Volvox aureus*, a photosynthesising alga that lives in fresh water.

▶ **Figure 26.7** The energy use of a cow. Compare the food energy eaten with the amount built into body tissue. Look at the large amount of energy left in the faeces. What organisms can use this?

There are three ways in which organisms obtain the energy store they need to live. They can produce the energy store they need. They can consume other living organisms that have already produced this energy store. Or they can obtain it from the bodies of dead organisms.

⚪ Producers

Anything that photosynthesises is a producer. It is able to 'produce' its own food using light energy from the Sun. It is often helpful to remember that they produce biomass (living tissue) from inorganic compounds.

All **plants** photosynthesise. They harness the energy from the Sun transferred by light and convert it into chemical energy stored in glucose. **Algae** can photosynthesise too. In fact, about 70% of the oxygen made each day by photosynthesis comes from photosynthetic algae and not plants.

Photosynthesising plants and algae are the only organisms on Earth that are able to use the energy transferred from the Sun in this way. Surprisingly they can only use (or capture) about 1% of the energy from light. Even though this does not sound like very efficient energy transfer, almost all life on Earth is dependent upon this.

⚪ Consumers

Any organism that obtains its energy by eating another is a consumer. One group of consumers are **herbivores**. Any animal that eats any plant is a herbivore. Common examples are sheep and rabbits. Herbivores are primary consumers. Primary consumers eat producers.

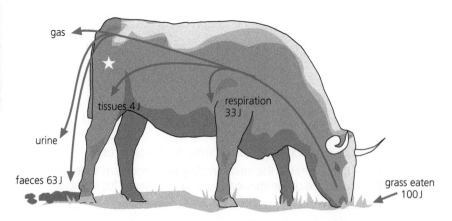

A second group of consumers are **carnivores**. These are any animals that eat meat, either from herbivores or from other carnivores. Examples are lions and sharks. A predator is any animal that hunts and kills prey for food. An animal that feeds on a primary consumer is a secondary consumer, and an animal that feeds on a secondary consumer is a tertiary consumer.

Feeding relationships can be represented by **food chains**. All food chains begin with a **producer** that is able to use light energy to make glucose by photosynthesis.

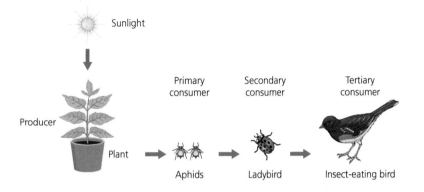

Sunlight

Producer

Plant

Primary consumer

Aphids

Secondary consumer

Ladybird

Tertiary consumer

Insect-eating bird

▲ **Figure 26.8** A food chain.

Show you can...

Describe the flow of energy in a food chain.

Test yourself

5 Give an example of an aquatic producer.

Materials cycling

All materials in the living world are recycled to provide the building blocks for future organisms.

TIP ✔

It is important that you can understand graphs used to model these cycles. It is also important that you can explain the processes in diagrams of the carbon cycle.

○ **The carbon cycle**

Carbon is the key element for all known life. The complex molecules that make up all life (such as proteins) are made when carbon chemically bonds with other elements, especially oxygen, nitrogen and hydrogen.

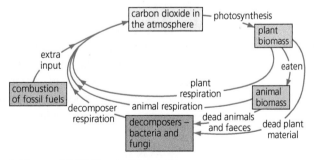

carbon dioxide in the atmosphere — photosynthesis

plant biomass

extra input

eaten

plant respiration

combustion of fossil fuels

animal respiration

animal biomass

decomposer respiration

decomposers – bacteria and fungi

dead animals and faeces

dead plant material

▲ **Figure 26.10** The carbon cycle.

The carbon cycle is shown in Figure 26.10. It can be simplified into four separate processes called photosynthesis, respiration, combustion and decay.

The equation for photosynthesis is:

$$\text{carbon dioxide and water} \xrightarrow{\text{light}} \text{glucose and oxygen}$$

Carbon is converted from the carbon in carbon dioxide to the carbon in glucose.

The equation for respiration is:

$$\text{glucose and oxygen} \xrightarrow{\text{energy}} \text{carbon dioxide and water}$$

So the conversion here is between carbon in glucose and carbon in carbon dioxide.

Combustion is the release of energy from burning a fuel. The reaction for burning natural gas (like when you light a Bunsen burner) is:

$$\text{methane and oxygen} \xrightarrow{\text{energy}} \text{carbon dioxide and water}$$

So the conversion here is between carbon in methane and carbon in carbon dioxide. The percentage of carbon dioxide in the atmosphere is 0.04%. This sounds very low but is the highest it has been in 800 000 years.

Decay occurs when a living organism is broken down by decomposing bacteria and fungi. These organisms help biological material decompose. When plants die and rot they release their carbon back into the atmosphere as carbon dioxide waste from the respiration of the microorganisms causing decomposition. The carbon was incorporated into the plant tissue when the plants photosynthesised. When animals die and rot they release carbon dioxide into the atmosphere, too. Decay also releases mineral ions back to the soil.

> **TIP** ✓
>
> Figure 26.10 shows how many processes put carbon dioxide back into the atmosphere, but only one process (photosynthesis) takes it out.

> **KEY TERM** ★
>
> **Combustion** Burning.

> **TIP** ✓
>
> Microorganisms are important in returning carbon to the atmosphere as carbon dioxide in the carbon cycle, and in returning mineral ions to the soil.

Test yourself

6 Name a key process in the carbon cycle.
7 What is the percentage of carbon dioxide in the atmosphere?
8 Describe the conversion of carbon during combustion.
9 Describe the conversion of carbon during respiration.

Show you can...

Explain the processes by which carbon is cycled.

KEY TERMS ⭐

Precipitation Rain, snow, hail, sleet.

Evaporation When water changes state to form water vapour (a gas). (This is the opposite of condensation.)

○ The water cycle

Water is the major liquid component of all cells and therefore all life.

We find water on, above and below the surface of the Earth. It can exist as a solid in the polar ice caps, a liquid in the oceans and a gas as the water vapour you breathe out. The movement of water from one place to another and from one state of matter to another is called the water cycle. The water cycle is shown in Figure 26.11.

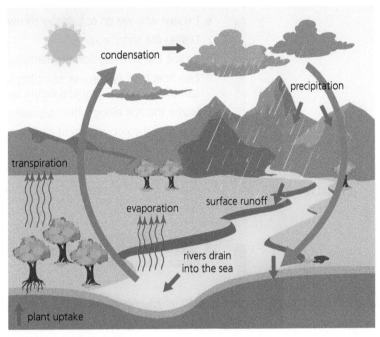

condensation

precipitation

transpiration

evaporation

surface runoff

rivers drain into the sea

plant uptake

▲ **Figure 26.11** The water cycle.

Precipitation is the scientific name for rain. This obviously occurs when condensed water vapour falls to the surface of our planet. Precipitation is therefore preceded by **condensation**. Snow, hail and sleet are all types of precipitation too.

Water in oceans, lakes and rivers returns to water vapour in the air and in clouds, because of **evaporation**. This change of state requires energy, which usually comes from the Sun.

Transpiration is evaporation from plants' leaves of up to 90% of the water those plants take in through their roots. Here water is absorbed into a plant as a liquid before being released into the atmosphere as a gas.

Test yourself

10 Name a process in the water cycle.
11 Describe the different types of precipitation.

Show you can...

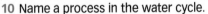

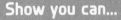

Explain the processes in the water cycle.

Chapter review questions

1 Define the term 'sampling'.
2 Name the square frames using in sampling.
3 Explain why we do not sample animals using quadrats.
4 Name the three main processes in the carbon cycle.
5 Name the main processes in the water cycle.

6 Explain why we do not simply throw quadrats and sample where they land.
7 Define the term 'ecology'.
8 Define the term 'systematic sampling'.
9 Describe how systematic sampling using a quadrat is different from random sampling. Explain why you would use systematic sampling.
10 Name the line along which a quadrat is placed during systematic sampling.
11 Describe the conversion of carbon in photosynthesis.
12 Describe the conversion of carbon in respiration.
13 Describe the conversion of carbon in combustion.
14 Describe what happens to the carbon in dead plants and animals.

15 Suggest why, on a rocky seashore, you might want to sample systematically rather than randomly.

Practice questions

1 Figure 26.12 below shows the carbon cycle.

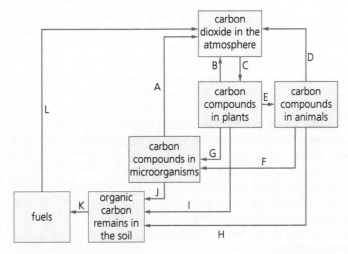

▲ Figure 26.12

a) i) In which form does carbon exist in the atmosphere? [1 mark]

ii) Line A represents respiration; give two other lines that also represent this. [2 marks]

b) Line L represents combustion of fuels. In 1998 5.4 billion tonnes of carbon was released into the atmosphere. In 2014 it was 9.9 billion tonnes of carbon. Calculate the percentage increase in the amount of carbon released. Show your working. [2 marks]

c) An increase in levels of carbon dioxide in the atmosphere is thought to lead to which process? [1 mark]

2 If you wanted to estimate the population of a plant species in a field which of the following methods would you use? [1 mark]

A Systematic sampling using a transect
B Random sampling using quadrats
C Field notes

3 Nadine carried out an experiment to examine decay in leaf litter in two woodland habitats: a pine forest (coniferous) and an oak woodland (broadleaf). She removed 500 g of leaf litter from three forest floors and placed it into black bin bags. She then recorded the starting mass of each bag and sealed the bags and left each of them in a separate bin. The bins were sealed but not airtight; however, water could not get in. Nadine re-weighed the bags every month for 6 months.

The results of the experiment are shown in Table 26.3.

Table 26.3

Month	Mass of leaf litter in g	
	Pine forest	Oak woodland
0	500	500
1	430	400
2	421	386
3	401	372
4	382	363
5	387	359
6	362	346

a) i) The leaf litter from which woodland decayed more? [1 mark]

ii) How much mass was lost as it decayed? [1 mark]

iii) In the table there is an anomalous result. What does this mean? [1 mark]

b) Suggest why litter lost most mass within the first month. [1 mark]

c) Name two types of organisms that can bring about decay. [2 marks]

Working scientifically:
Experimental skills

Sampling and bias

Ecology is often referred to as the 'study of distribution and abundance'. Putting it simply, this means how many of a particular type of organism are in an area and where are they found.

Because of the complexity of environments and habitats it is almost impossible to measure the distribution and abundance of organisms in a particular area, as it would take too long and cost too much to do. Instead, ecologists need to make an estimate by sampling. They will want their estimate to be the best estimate that it can be. This means that their results need to be accurate and precise.

Questions

1 Define the terms 'accurate' and 'precise'.
2 How can ecologists ensure that their results are precise?

Ecologists also need to ensure there is no bias in their results. Bias is when evidence is shifted in one direction. Bias can occur because a person wants something to be proven and their presentation of material reflects this, or it can be due to errors in estimating values or collecting data.

Questions

3 Why might a pharmaceutical company be biased in reporting data on the effectiveness of a new drug?

Ensuring that the method used to collect data truly reflects what is occurring can reduce measurement bias. To do this, measurements need to be taken accurately and the experiment needs to be a fair test with controls kept constant.

Sampling bias is introduced when the sample used is not representative or is inappropriate for the aim of the investigation. For example, if ecologists were studying limpets (a type of marine mollusc) along a rocky shore and they only examined two quadrats – one on some dry rocks and one in a sandy area – then their estimate for abundance would be much lower than the true abundance.

To reduce the impact of sampling bias, ecologists must ensure their sample is big enough to calculate a best estimate for their mean. They also must ensure their samples are random. A random sample is one in which every potential sample plot within a studied area has an equal chance of being chosen. To carry out random sampling usually a random number table (found in a statistical book) or a random number generator is used.

Figure 26.13 shows a section of rocky shore with the distribution of the three species that live there. The section represented is split into 36 equally sized quadrats. Using dice, you are going to select random quadrats to sample. To do this, roll two dice. The numbers that come up show you the coordinates to pick. If you rolled a 5 and a 1 you would count five across and one up. Repeat this process five more times to pick six random samples. Count the number of each species present and record your results in a table like Table 26.4.

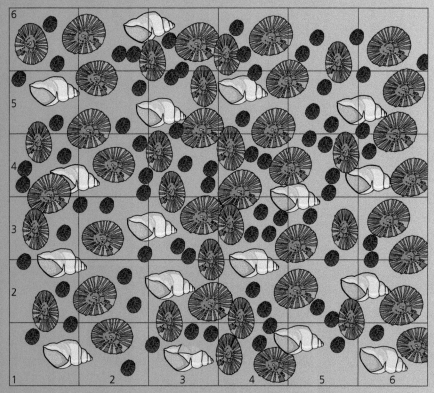

▲ Figure 26.13 A section of rocky shore divided into quadrats.

Table 26.4

Coordinates for the quadrat	Number of limpets	Number of barnacles	Number of dog whelks
1 (,)			
2 (,)			
3 (,)			
4 (,)			
5 (,)			
6 (,)			
Total			

Questions

4 Which species showed the highest abundance?

5 Compare your estimation for abundance for the number of dog whelks with the actual abundance (count all of them). Do you think your sample size was large enough to not be biased?

6 A dog whelk is a predator. Explain why its abundance is lower than that of the barnacles.

27 Biodiversity and the effect of human interaction on ecosystems

For much of our early existence humans lived as hunter-gatherers. We moved from place to place looking for food, shelter and water. We didn't damage our environment because we needed it to survive. We lived in balance with the other species around us. In recent years human activities, especially those that release pollution, have caused tremendous and often irreversible damage to our environment. Will we ever return to living more naturally? Or will we pollute our planet beyond repair before we learn not to?

Specification coverage

This chapter covers specification points 4.7.3.1 to 4.7.3.6 and is called Biodiversity and the effect of human interaction on ecosystems.

It covers biodiversity, waste management, land use, deforestation, global warming, and maintaining diversity.

Previously you could have learnt:

> It is important to maintain biodiversity.
> Organisms are affected by their surroundings in many ways, including by the build-up of toxic materials.
> Humans are responsible for adding high levels of carbon dioxide to the atmosphere. This is thought to be leading to climate change.

Test yourself on prior knowledge

1 Define the term 'biodiversity'.
2 Describe the impact of global warming on the seas and oceans.
3 Explain why toxic materials accumulate in top predators.

Biodiversity

KEY TERM

Biodiversity A measure of the different species present in a community.

Biodiversity is a measure of how many different species of organism live in the same geographical area. All species of life are included in biodiversity, including animals, plants, fungi and microorganisms.

Biodiversity is not the same all over the world. Areas with high biodiversity include tropical rainforests and ancient oak woodlands. Here we find a huge number of different species of plant, animal and microorganism. Areas with low biodiversity include deserts and the polar regions. Here fewer species are found. Sadly some areas of high biodiversity, such as the rainforest, are being destroyed by human activities, including **deforestation** for farming. These can reduce biodiversity significantly. Only relatively recently have humans taken some measures to stop this.

High biodiversity ensures the **stability** of an ecosystem. It **reduces the dependency** of species on each other for food, shelter and the maintenance of the environment.

Biodiversity is often highest in the tropical regions. It is in these regions that tropical rainforests are found. A good example of this is the island of Borneo in South East Asia. It has just under 20 million people, but over 15 000 species of plants and nearly 1500 species of amphibians, birds, fish, mammals, reptiles and insects. Borneo is so biodiverse because it is made up of many different types of habitat. The different types of forests include tropical rainforests, mangrove trees (near the coast), peat-swamp forests and high forests on Mount Kinabalu. In each of these different habitats different species have evolved to live.

Show you can...

Explain why some countries, for example Borneo, are biodiverse.

Test yourself

1 Give an example of an area of high biodiversity.
2 Give an example of an area of low biodiversity.

Waste management

TIP ✔

It is important that you can explain the impact that waste has on biodiversity.

▲ **Figure 27.1** Rats are thought to have spread the Black Death.

▲ **Figure 27.2** Transporting clean water.

▲ **Figure 27.3** Attempting to clean up a horrific oil spill.

The total population of humans on Earth is increasing at a staggering rate and is now over seven billion. Estimations show that the population might rise to 16 billion by the end of this century. Where will these people live, what will they eat and where will their fresh water come from? The last time the population didn't increase continually was at the end of the Great Famine and Black Death in 1350.

Food shortages are not the only consequence of an increasing population. More people means communicable diseases could increase. Other consequences include **pollution**, **deforestation** and **global warming**. These kill plants and animals, which can reduce biodiversity.

○ **Water pollution**

Water is vital for all known forms of life. Over half of your body is made from water. As well as for drinking directly ourselves, we use water for growing our food, washing, transportation (up and down rivers and canals and across lakes and oceans), and for finding food. As a consequence, water pollution is a serious challenge facing the world today and often results in loss of life or poor health.

Water can be polluted by **pathogens** that cause communicable diseases. These include species of *Salmonella* bacteria, the *Norovirus* and parasitic worms. These are often found in water as a result of contamination from sewage. Urban areas of developed countries have sewage treatment centres to which sewage travels in underground sewers. This poses little threat to drinking water. In many other parts of the world, sewage is not transported away by underground sewers. Open sewers travel through parts of many cities in the developing world. These can run straight into streams and rivers and then concentrate in lakes or by the sea.

Chemical pollution

Other contaminants include organic **chemicals** such as **pesticides** and waste from factories and industry.

Some farmers overuse **fertilisers**, which then wash off fields and concentrate in slow-moving water such as ponds and lakes. Here the fertiliser can speed up the growth of algae, which can form a 'bloom'. This can cover an entire lake in a few days. At this point the plants below the bloom do not receive enough light for photosynthesis. They die shortly afterwards and quickly begin to rot. Microorganisms then feed on the dead organisms, and their respiration uses up much of the oxygen in the lake. Without oxygen, animal life, including fish and invertebrates such as insects, begin to die. In a reasonably short period of time a fresh-water lake with many different species existing in a stable ecosystem can be completely destroyed.

Oil spills

Pollution of our oceans is also occurring on a grand scale. In recent years there have been many **oil spills** from tankers or the drills from which oil is pumped.

A recent oil spill at an oil rig called Deepwater Horizon occurred in spring 2010 in the Gulf of Mexico. It resulted from a leak of the well on the ocean floor. The US Government estimates that 4.9 million barrels of oil were lost. Much of this oil washed up on the shores of Mexico and the USA. In this area over 8000 different species live.

▲ **Figure 27.4** This white swan is covered in black oil.

▲ **Figure 27.5** Shanghai has some heavy smogs.

○ **Air pollution**

Just as water is a key component of life on Earth, so is the air that contains the oxygen we use. Pollution of the air is therefore another very serious problem. Air pollution is often caused by waste gases from vehicles or factories, but it can also be caused by particles of solids or liquids. These are called particulates and form the smogs that are often seen cloaking large cities in some parts of the world.

Sulfur dioxide is produced when petrol and other fossil fuels are burnt, and so the increase in motor vehicles in recent years has resulted in more of this pollutant being released. Sulfur dioxide reacts with water vapour and sunlight, to form sulfuric acid. This can lower the pH of water and result in the formation of acid rain. This can destroy whole forests of trees, affecting entire ecosystems. It also often damages stone buildings and statues.

Carbon monoxide is a colourless, poisonous gas without a smell. This makes it difficult to detect. It is again produced in large volumes by vehicles. It can also be produced during incomplete combustion of natural gas in faulty boilers and incomplete combustion of other fuels such as coal and wood.

Carbon dioxide is produced when fuels are burnt completely. This is a key greenhouse gas and is responsible for global warming.

○ **Land pollution**

The volume of waste produced has increased with the rapid rise in the population of humans. Much of this waste goes straight into huge **landfill** sites, where it is buried. These landfill sites attract vermin, including rats, which can spread communicable diseases. As the items themselves break down they can produce **toxic liquids** and release a large volume of methane, which is also a greenhouse gas. It is particularly important that we dispose of some items such as batteries correctly and that they don't go into landfill. Batteries contain heavy metals and other toxic chemicals, which can easily pollute local soil and water.

These days more waste is being burnt in incinerators. This reduces the need for landfill and so commonly occurs in smaller, more densely

> **KEY TERMS**
>
> **Acid rain** Precipitation that is acidic as a result of air pollution.
>
> **Incomplete combustion** The burning of fuel without sufficient oxygen, which produces poisonous carbon monoxide.

KEY TERM ⭐

Recycle Changing a waste product into new raw materials to make another.

▲ **Figure 27.6** It is very important to recycle all batteries to stop heavy metals going into landfill.

▲ **Figure 27.7** Paper at a recycling centre.

Show you can...

Explain how fertilisers can cause water pollution.

populated countries such as Japan. The thermal energy released is often used to generate electricity, which of course is a positive benefit. However, some examples of air pollutants have been found in higher concentrations close to incinerators.

To reduce the need for landfill and incineration it is very important that we all **reduce**, **reuse** and recycle as much as we can. These are often called the three 'R's.

Reducing means simply having less or going without. Do we really need the latest mobile phone or most fashionable item of clothing?

'Reuse' simply means using again. This can involve using the item for the purpose it was made; a jam jar can be reused to store homemade jam. Reuse also means reusing objects for different purposes; the same jam jar can be used to store many other things too. Clothing banks and charity shops are a great way of allowing others to reuse clothes you no longer want.

'Recycle' means changing a waste product into new raw materials to make another. As with reuse this can be the same as the original product; paper is often recycled into more paper. A waste product can also be used to make other objects; plastic bottles can be used to make fleece clothing. Glass, plastic, paper and electronic goods such as mobile phones are often recycled. In the UK many items are collected for recycling alongside weekly or fortnightly rubbish collections. Other items need to be taken to recycling centres.

It is worth remembering that every time someone drops a piece of litter, this is another example of land pollution. If only everyone made a point of putting all litter in a bin.

Test yourself

3 Explain the main problems with landfill.
4 How can we reduce landfill?
5 Describe the effects of acid rain.

Land use

Land use is simply the way that we use land. This can be fields for **farming** crops or livestock. It can be managed woods to produce timber. Land can also be used for human habitation (urban). Imagine life about 10 000 years ago. This was towards the end of the Stone Age. Small groups of humans (*Homo sapiens*) roamed around, hunting and gathering. The number of people per unit of land area was very low. The population density was probably around one person per square mile. They lived very sustainable lives and their impact on the environment was very limited. Now because the population density has increased to about 120 people per square mile, our lives are less sustainable and the negative impact that we have on the environment has increased.

KEY TERM ⭐

Sustainable Describes an activity that can continue without damaging the environment.

▲ **Figure 27.8** Some fields are now so large that the crops can only be treated efficiently by plane.

TIP

Humans reduce the amount of land available for other species by building, quarrying, farming and dumping waste (e.g. in landfill sites). These all have a negative effect on biodiversity.

▲ **Figure 27.9** This rare peat bog contains many species that grow in no other habitats.

Some human activities are very damaging to land. These include **quarrying** and **mining**. One of the largest mines in the world is called Bingham Canyon Mine in Utah in the USA. This is so large that it can be seen from space.

Some of our **cities** have also grown over huge areas. Beijing in China is the largest, with a total area of more than 16 000 square kilometres. London is currently about 40th in this list, with an area of 1600 square kilometres. This means that Beijing is the size of 10 Londons!

○ **Peat bogs**

A bog is an area of wet soil without trees, where many species of moss grow. The water at ground level is acidic and low in nutrients. Despite this, bogs are very important areas of biodiversity, often containing species that are not found anywhere else. They are generally found in areas of high rainfall and cool temperatures. This, combined with the acidity, results in slow plant growth and also slow decomposition. This means that remains found in peat bogs are very well preserved. Because of these conditions peat accumulates. Peat is made up of partially decayed vegetation.

Peat is often the first step in the formation of fossil fuels. So it is a very valuable source of fuel in certain parts of the world. Burning peat as fuel releases this carbon as **carbon dioxide**, which causes global warming. Peat used to be used regularly by farmers and gardeners to improve the quality of their soil. In recent years, this use has reduced as people have learnt about the destruction that this practice causes to wetlands.

TIP

Conflict exists between the need for compost to increase food production and the need to conserve peat bogs and peat lands, both to increase biodiversity and to reduce carbon dioxide emissions.

Test yourself

6 Give the current human population density on our planet.
7 Give an example of human activity that can be seen from space.
8 Describe the environmental conditions under which peat bogs develop.

Show you can...

Explain how we have changed our environment in the UK since the Stone Age.

Deforestation

TIP

It is important that you can explain the impact that deforestation has on biodiversity.

▲ **Figure 27.10** (a) When this village moves on, the area will recolonise from seeds blown in from the surrounding forest. (b) Vast tracts of deforested land do not recolonise as quickly.

Deforestation is the clearance of trees from an area that will then be used for other purposes, commonly **farming** or **urban** use. Many trees are also felled for their timber. A large number of people in developing countries still collect firewood each day to burn to cook and keep warm. Deforestation by small communities of people who manage areas of land where they live is probably not a major cause for concern. However, in recent years large areas of forests have been cut down by large companies to provide land to feed vast **herds of cattle** or plant huge fields of crops such as **rice** in **tropical areas**. This now includes crops used to make a type of fuel called **biodiesel.**

Deforestation of the rainforests is often more publicised than that of other regions, perhaps because of the increasing rate at which the rainforests are disappearing. We have now cut down about half of the rainforest that existed 75 years ago. Some predictions suggest that only 10% will be left in 2030 unless we take drastic action to reduce this.

Trees take in carbon dioxide for photosynthesis, and this reduces the amount of carbon dioxide in the atmosphere. Fewer trees mean less photosynthesis and less carbon dioxide removed from the atmosphere. Trees are often burnt when cut down, which further releases carbon dioxide into the atmosphere. This increases the greenhouse effect and therefore global warming.

Deforestation massively reduces the number of different living organisms in an area and so reduces biodiversity. We think that about 80% of the world's biodiversity could be in our tropical rainforests. We are losing over 100 different species each day as a result of rainforest deforestation. There is no way that these species will ever be seen again.

TIP

In tropical areas, large-scale deforestation has occurred to provide land for cattle and rice fields, and to grow crops for biofuels.

Test yourself

9 What proportion of the rainforests have we cut down in the last 75 years?
10 What is the expected proportion of biodiversity found in rainforests?
11 Explain why people cut down trees.
12 Describe a negative effect of biodiesel.

Show you can...

Explain the effects of deforestation on the environment.

Global warming

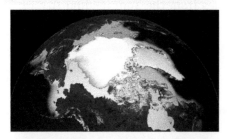

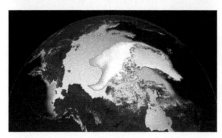

▲ **Figure 27.11** These two photographs show how much the Arctic sea ice retreated between 1980 and 2012 as a result of global warming.

Global warming is the gradual increase in the average temperature of the Earth. This includes its land and oceans. The temperature of the Earth has fluctuated over time. We have had ice ages and periods when animals and plants found in the tropics were present in the UK. These changes happened naturally and not as a result of human activity.

For many years scientists have argued about whether the current change in our climate is a result of human activity or just natural causes. In recent years scientists have become certain that global warming is a result of increased emission of greenhouse gases by us.

The effects of global warming are seen the world over. In Arctic regions we are seeing **polar ice-caps melting**. The total area of sea ice is reducing, which means it is ever harder for polar bears to hunt. Throughout the world more freak instances of weather (**climate change**) are thought to result from global warming. These include heavy rainfall leading to flash floods, as well as heat waves leading to droughts. The melting of sea ice is causing the sea and ocean levels to rise. Many of the world's major cities are near the coast and will be lost to the sea if this continues. These include Venice, Amsterdam, New York and London.

Species are migrating away from regions they once lived in to those that have temperatures they are more suited to and where they will find their food. If global warming continues those species found in northern Africa will migrate to southern Europe. Then those found in southern Europe will migrate to northern Europe. Currently the climate in the UK is too cold for tropical mosquitos to survive and so malaria is not present. If global warming continues, how long will it be before they can survive here? Such species migration may have hugely damaging effects on ecosystems. This is a **biological consequence** of global warming.

Many scientists are now predicting that global warming is a big threat to our food security. Perhaps positive changes to reduce global warming will happen faster when our food is threatened.

Activity

Energy wastage

Mobile phone chargers left plugged in are responsible for wasting more than £60 million of electricity and releasing a quarter of a million tonnes of carbon dioxide every year. Working as a small group, design a survey to find out what energy-wasting habits students in your year group have. Then design a poster to raise awareness of them and the impact they are having on the planet.

What causes global warming? If you stand in a greenhouse (or sit in a car with the windows up) on a sunny day you will quickly get very hot. This is because there is no way that the thermal radiation from the Sun can leave the greenhouse (or the car). This is a model for the effect that increased greenhouse gases (mainly **carbon dioxide** and **methane**) have in trapping more energy in our atmosphere. The Sun's radiation initially penetrates our atmosphere and about 50% is absorbed by the Earth's surface. The rest is reflected back. Before the last few hundred years most of this radiation would have left the atmosphere and only some would have been retained. This low level of the greenhouse effect actually supports life on Earth and is essential for it to continue.

Recent increases in greenhouse gases released by burning fossil fuels and compounded by deforestation have meant that more greenhouse gases are found in our atmosphere. Consequently, the Earth is retaining more of the energy that comes from the Sun, leading to global warming.

▲ **Figure 27.12** Fossil-fuel-powered power plants are responsible for releasing carbon dioxide into the atmosphere.

Show you can...

Explain how an increase in the greenhouse effect leads to global warming

Test yourself

13 Name three fossil fuels.
14 Which country currently releases the most carbon dioxide as a percentage of total emissions?
15 Describe why we are producing more greenhouse gases now than in past centuries.
16 Describe what effect global warming will have on species distribution.

Maintaining biodiversity

TIP ✓

It is important that you can explain the conflicting pressures of maintaining biodiversity.

KEY TERM ★

Breeding programme Activity of zoos to breed captive animals together to increase their numbers and the gene pool.

▲ **Figure 27.13** *Tyrannosaurus rex* became extinct along with the rest of the dinosaurs in a mass extinction event.

Biodiversity is a measure of the number of different species of plants, animals and microorganisms that live in an area. Habitats with particularly high biodiversity include coral reefs and rainforests. These two habitats are threatened by rising ocean temperatures as a result of global warming and by deforestation, respectively.

Fossils have helped us to look at biodiversity since the earliest forms of life on Earth. We have seen at least five major mass extinctions throughout this time. The most recent mass extinction event is happening now as a result of human activity. This is the first time extinctions have happened as a result of the activity of one particular species.

Many zoos carry out breeding programmes in which the reproduction of animals is carefully managed to promote the widest gene pool possible. This stops inbreeding and the genetic disorders that come with it. Perhaps the most famous animals undergoing a breeding programme are the giant pandas. They are notoriously difficult to breed. Horticultural groups have breeding programmes for some species of plant as well.

▲ **Figure 27.14** Giant pandas are notoriously difficult to breed in zoos.

○ Conservation efforts

Conservation is one way in which countries, organisations and charities, and individual people are helping to promote biodiversity. The formation of national parks is one way in which governments are doing this. Currently, there is only one male northern white rhino left in the world. It is continuously protected by armed guards to prevent poaching.

National parks and nature reserves are being used to protect many areas of the world. The Great Barrier Reef is the world's largest coral reef system, off the coast of Australia. It can been seen from outer space and is the largest structure on Earth made by living organisms. It was selected as a World Heritage Site in 1981 and a large part of it is protected by a national park. This is an example of the **protection and regeneration of rare habitats**.

A hedge or **hedgerow** is a line of trees or shrubs that have grown together to form a boundary. These separate areas belonging to different people or keep animals in fields. Hedgerows are home to many species of animals, plants and microorganisms. For many years farmers have been removing hedgerows to make bigger fields, which are easier to tend using machines such as combine harvesters. The removal of these hedgerows has reduced biodiversity. Farmers are now encouraged to reintroduce hedgerows and to leave **field margins** where wild plants can grow.

▲ **Figure 27.15** Hedgerows are far more biodiverse than the fields that surround them.

Conservation also involves **reducing deforestation** and **recycling resources** rather than using landfill. Maintaining biodiversity is arguably hardest for developing countries. Here there are real tensions between the needs of local people who in some cases are in direct conflict with endangered animals. It must be very difficult for an African farmer to not trap and kill leopards that are continually killing livestock.

This tension is also seen at a regional and national level. Many years ago, our ancestors cut down many of the trees that made huge forests that covered the UK. Are we now in an ethical position to ask countries in the developing world with a lower standard of living than ours to not do the same?

There are also a number of things you can do as an individual to indirectly and directly increase biodiversity. These include:

- not using pesticides and fertilisers in your garden
- leaving areas to become wild (perhaps with a stack of wood for creatures to hide in)
- reducing, reusing and recycling
- growing wild flowers
- putting waste food into compost, not landfill.

Activity

Conservation debate

Many organisms that are endangered cannot now survive without the help of humans. In order to try to conserve a species, various people's viewpoints need to be considered. As a class, imagine you are at a meeting to decide whether to make an area of forest a reserve to protect the Sumatran tiger (the world's most endangered tiger). Hold a class debate to decide whether you think this should happen. Some of you should take on the role of people against the reserve, such as local farmers, loggers working in the forest, and people concerned about tigers leaving the area and harming people. Others should take on the role of people in favour, such as scientists, representatives from a conservation charity, and local hotel managers and business owners who want the reserve to increase tourism.

Ensure that you think about your character's role in the debate and back your view up with arguments and evidence.

Test yourself

17 Give an example of an area with low biodiversity.
18 Give an example of an area with high biodiversity.
19 Describe why hedgerow removal damages biodiversity.
20 Define the term 'biodiversity'.

Show you can...

Explain how you can directly increase biodiversity.

Chapter review questions

1 Suggest some consequences of the increasing population of humans.

2 Explain why we should recycle batteries and not put them into landfill.

3 What do the three 'R's stand for?

4 Give three examples of materials that are commonly recycled.

5 Define the term 'deforestation'.

6 Explain why deforestation is often considered a bad thing.

7 Define the term 'biodiversity'.

8 Describe the formation of acid rain.

9 Give an advantage and a disadvantage of using an incinerator over landfill.

10 Define the term 'sustainable'.

11 Explain why reusing is more sustainable than recycling.

12 Explain why the rate of decay is low in peat bogs.

13 Describe the effects of global warming.

14 Define the term 'food security'.

15 Name two common greenhouse gases.

16 Define the term 'breeding programme'.

17 Define the term 'conservation'.

18 Suggest why national parks are important for biodiversity.

19 Explain why hedgerow removal is a bad thing in terms of ecology.

20 Why is the preservation of peat bogs so important?

21 Explain why some people still do not believe that global warming is being caused by increased carbon dioxide levels in the atmosphere.

Practice questions

1 Figure 27.16 shows information on how much different groups of organisms are currently under threat of extinction.

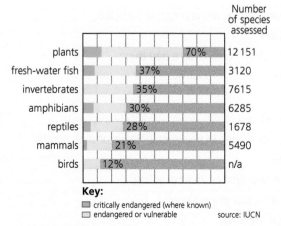

Key:
▢ critically endangered (where known)
▢ endangered or vulnerable   source: IUCN

▲ **Figure 27.16**

a) i) What does the term 'endangered' mean?   [1 mark]

ii) Looking at Figure 27.16, which group of animals is most endangered?   [1 mark]

iii) Thirty per cent of amphibians are endangered; calculate from the number assessed how many this is. Show your working.   [2 marks]

b) There are fewer than 40 Amur leopards left in the wild in Siberia, and this species is critically endangered.

i) Suggest two reasons why Amur leopards might be critically endangered.   [2 marks]

ii) There are a number of conservation strategies used in order to help protect the species from further loss, including captive breeding programmes and educating the local population to prevent poaching and habitat destruction. For each of these two strategies give an advantage and disadvantage of its use.   [4 marks]

2 Figure 27.17 shows two fields separated by a river.

▲ **Figure 27.17**

a) i) The farmer sprays fertilisers on his crops. Why does he do this?   [1 mark]

ii) Name another chemical the farmer may spray on his crops.   [1 mark]

b) The local water company is concerned that there has been an increase in nitrates from fertiliser in the water supply. The farmer was careful to not spray the area of the field near the river and only sprayed on days with no wind. Suggest how the nitrates could have got in the water.   [2 marks]

c) Over time the number of fish in the river has declined and in the summer many are found dead on the banks of the river. Explain how an increase in spraying fertiliser could have led to the death of these fish.   [4 marks]

3 a) The global temperature is increasing due to global warming. The increase in greenhouse gases is accelerating this process. Which two gases are known to contribute most to the greenhouse effect?   [2 marks]

A Chlorofluorocarbons (CFCs)

B Carbon monoxide

C Carbon dioxide

D Methane

b) Which of the following is not a consequence of global warming?   [1 mark]

A Glaciers retreating

B Reduction in sea ice

C Heavy rainfall leading to flash floods

D Reduced sea levels

Working scientifically:
Dealing with data

Correlation and causation

A correlation is a relationship between two variables. If one variable changes, so does the other. There are different types of correlation: positive correlations, where the variables increase together, and negative correlations, where as one variable increases the other decreases.

It is often difficult to spot a correlation and whether it is positive or negative simply by looking at data in a table, so scientists draw scatter graphs, or scatter plots. These help them see the trends more clearly, identify any anomalous results and determine the strength of the correlation.

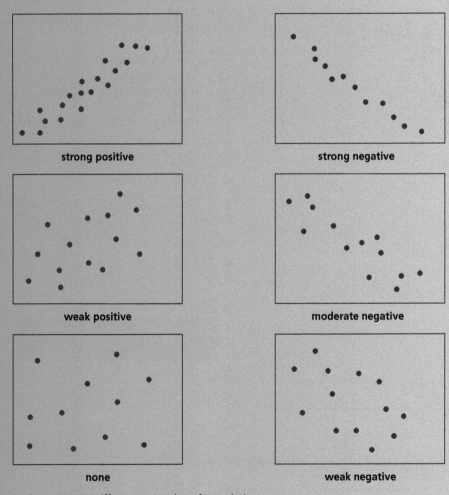

strong positive

strong negative

weak positive

moderate negative

weak negative

▲ Figure 27.18 Different examples of correlation.

Sometimes you can have a spurious correlation. This is an apparent correlation between two variables when the relationship is not directly between them but from a different variable that affects both of them. For example, the number of churches in a town might show a positive correlation with the number of pubs. This does not mean that because there are more churches more pubs were built, or that because there are more pubs more churches are needed! It is that both are affected by another variable in a similar way – in this case, the population. The bigger the population, the more churches and pubs it supports.

It is important to remember that correlation is not causation. That is, just because two variables show a correlation it does not mean that one thing causes the other. Once scientists find a correlation they must then identify the cause to prove it is not just spurious.

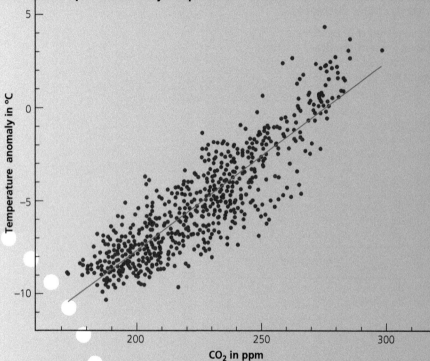

▲ **Figure 27.19** The relationship between carbon dioxide concentration and temperature anomaly. (Temperature anomaly is how much the temperature (at time of measurement) differs from the long term average.)

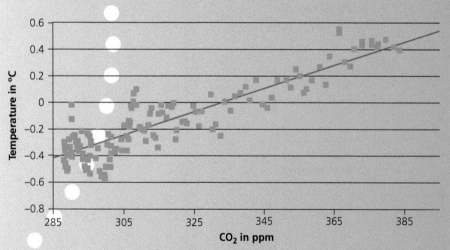

▲ **Figure 27.20** The relationship between carbon dioxide concentration and temperature.

Questions

1 Look at Figure 27.19. What trend can you see?

2 What type of correlation is this?

3 Why do scientists believe that increasing carbon dioxide can cause an increase in temperature?

4 Why do some people disagree that increasing carbon dioxide levels are causing global warming?

5 There is concern that human population growth is accelerating the rate of species extinction. Plot the data from Table 27.1 as a scatter graph and add a trend line.

6 What is the trend shown in your data?

7 What type of correlation does it show?

8 What are the possible causes for increasing an number of humans leading to increasing rates of extinction?

Table 27.1

Population in millions	Number of estimated extinctions
1800	2000
1900	2000
1950	2000
2000	2000
2100	2000
2500	2500
2900	7000
4000	9000
6600	30000

28 The rate and extent of chemical change

The explosion of dynamite is a very fast reaction. The rusting of steel is a very slow reaction. Some reactions completely react, but in others the products can turn back into the reactants. Being able to control the speed of a reaction and reducing the amount of products turning back into reactants are very important and are studied in this chapter.

Specification coverage

This chapter covers specification points 5.6.1.1 to 5.6.2.7 and is called The rate and extent of chemical change.

It covers the rate of reactions, reversible reactions and dynamic equilibrium.

Rate of reaction

○ Measuring the rate of reaction

Some chemical reactions are fast while others are slow. Reactions that are fast have a high rate of reaction. Reactions that are slow have a low rate of reaction.

Measuring the mean rate of reaction

The mean rate of a reaction can be found by measuring the quantity of a **reactant used** or a **product formed** over the time taken:

$$\text{mean rate of reaction} = \frac{\text{quantity of reactant used}}{\text{time}}$$

or

$$\text{mean rate of reaction} = \frac{\text{quantity of product formed}}{\text{time}}$$

The time is typically measured in seconds.

The quantity of a chemical could be measured as:

- mass in grams (g) or
- volume of gas in cubic centimetres (cm^3) or
- amount in moles (mol).

The rate of a reaction changes during a reaction. Most reactions are fastest at the beginning, slow down and then eventually stop. As the rate is constantly changing, the rate at one moment is likely to be different to the rate at another. This means that when we calculate the rate of reaction over a period of time, we are actually working out the mean rate of reaction over that time.

Example

In a reaction where some calcium carbonate reacts with hydrochloric acid, 40 cm³ of carbon dioxide was produced in 10 seconds. Calculate the mean rate of reaction during this time in cm³/s of carbon dioxide produced.

Answer

$$\text{rate of reaction} = \frac{\text{quantity of product formed}}{\text{time}} = \frac{40}{10} = 4.0\,\text{cm}^3/\text{s}$$

Example

Magnesium reacts with sulfuric acid. In a reaction, 0.10 g of magnesium was used up in 20 seconds. Calculate the mean rate of reaction during this time in g/s of magnesium used.

Answer

$$\text{rate of reaction} = \frac{\text{quantity of reactant used}}{\text{time}} = \frac{0.10}{20} = 0.0050\,\text{g/s}$$

Example

Copper carbonate decomposes when it is heated. In a reaction, 0.024 moles of copper carbonate had decomposed in 30 seconds. Calculate the mean rate of reaction during this time in mol/s of copper carbonate decomposing.

Answer

$$\text{rate of reaction} = \frac{\text{quantity of reactant used}}{\text{time}} = \frac{0.024}{30} = 0.00080\,\text{mol/s}$$

Test yourself

1 Carbon dioxide gas is formed when sodium carbonate reacts with ethanoic acid. In a reaction, 75 cm³ of carbon dioxide was produced in 30 seconds. Calculate the mean rate of reaction during this time in cm³/s of carbon dioxide produced.

2 When nickel carbonate is heated it decomposes. During an experiment it was found that 0.25 g of nickel carbonate decomposed in 40 seconds. Calculate the mean rate of reaction during this time in g/s of nickel carbonate decomposing.

3 When sodium thiosulfate solution reacts with hydrochloric acid, one of the products is solid sulfur. In a reaction, 0.010 moles of sulfur was formed in 50 seconds. Calculate the mean rate of reaction during this time in mol/s of sulfur formed.

Show you can...

Some copper carbonate was placed in a crucible and weighed. It was heated to decompose it for 40 seconds and reweighed, and the results recorded in the table below. Carbon dioxide is formed and escapes from the crucible as the copper carbonate decomposes.

Table 28.1

| Mass of copper carbonate and crucible before heating in g | 18.1 |
| Mass of copper carbonate and crucible after 40 seconds heating in g | 16.9 |

Calculate the mean rate of reaction during this time in g/s of carbon dioxide produced.

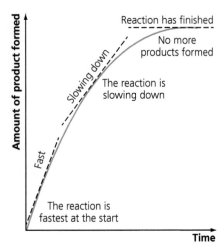

▲ **Figure 28.1** Graph showing the progress of a reaction.

TIP ✓

When drawing tangents, to help you judge the angle, use a see-through ruler and move the ruler along the line, changing the angle of the ruler as you move along.

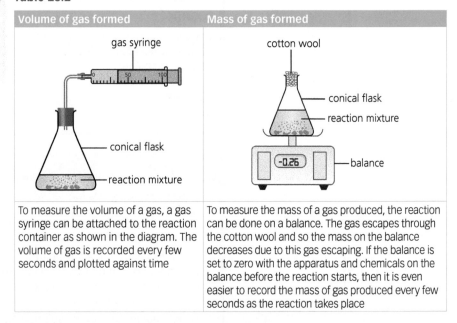

At the time shown (T), the slope for reaction A is steeper than the slope for reaction B. This means that reaction A is faster than reaction B

▲ **Figure 28.2** Comparing reaction rates with tangents.

Reaction rate graphs

Graphs can be drawn to show how the quantity of reactant used or product formed changes with time. The shapes of these graphs are usually very similar for most reactions. The slope of the line represents the rate of the reaction. The steeper the slope, the faster the reaction.

The relative slope can be judged by simply looking at the line but it can be useful to draw tangents to the curve (Figure 28.1). At any point on a curve, a straight line is drawn with a ruler that just touches the curve and that we judge to be at the same slope as the curve at that point. The steeper the tangent, the faster the rate of reaction.

In an individual reaction, the line is steepest at the start when the reaction is at its fastest. The line becomes less steep as the reaction slows down and eventually becomes horizontal when the reaction stops.

In reactions that produce a gas, it is easy to measure the volume or mass of gas formed over time (Table 28.2) and draw a graph of this type.

Table 28.2

Volume of gas formed	Mass of gas formed
To measure the volume of a gas, a gas syringe can be attached to the reaction container as shown in the diagram. The volume of gas is recorded every few seconds and plotted against time	To measure the mass of a gas produced, the reaction can be done on a balance. The gas escapes through the cotton wool and so the mass on the balance decreases due to this gas escaping. If the balance is set to zero with the apparatus and chemicals on the balance before the reaction starts, then it is even easier to record the mass of gas produced every few seconds as the reaction takes place

The rate of one reaction can be compared to the rate of another using these graphs. The steeper the slope of the line, the greater the rate of reaction. For example, in the graph in Figure 28.2, reaction A is faster than reaction B. We can see that the line for reaction A is steeper, but this is confirmed if we draw a tangent to each curve near the start.

Finding the rate of reaction from the gradient of tangents on reaction rate curves

The rate of reaction at any given moment can be found by measuring the gradient of the tangent to the curve at that point.

To calculate the rate of reaction by drawing a tangent to a graph:

1 Select the time at which you want to measure the rate.

2 With a ruler, draw a tangent to the curve at that point (the tangent should have the same slope as the graph line at that point).

3 Choose two points a good distance apart on the tangent line and draw lines until they meet the axes.

4 Find the slope using this equation: slope $= \dfrac{\text{change in } y\text{-axis}}{\text{change in } x\text{-axis}}$

In Figure 28.3, the volume of carbon dioxide produced when calcium carbonate reacts with acid was plotted against time. The rate of reaction was found at 18 and 40 seconds by drawing tangents to the curve and calculating the slope. The rate of reaction was $1.11\,\text{cm}^3/\text{s}$ at 18 seconds. The rate was lower at 40 seconds where it was $0.53\,\text{cm}^3/\text{s}$.

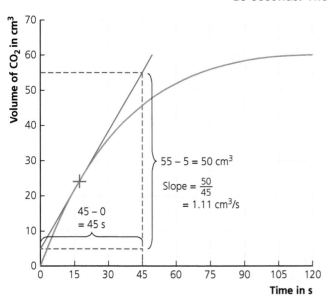
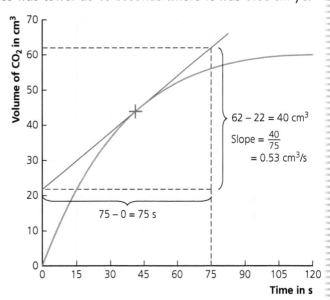

▲ **Figure 28.3** Using tangents to work out the rate of reaction.

Test yourself

4 Figure 28.4 shows how the mass of carbon dioxide formed when sodium carbonate reacts with hydrochloric acid varies over time.
   a) At what point on the graph is the rate of reaction greatest? Explain your answer.
   b) At what point on the graph does the reaction stop? Explain your answer.
   c) Is the rate of reaction faster at point B or E? Explain your answer.
5 Look at Figure 28.5 for reactions P, Q and R, which all produce a gas.
   a) Which reaction is the fastest?
   b) Which reaction is the slowest?
6 Hydrogen gas is formed when magnesium reacts with sulfuric acid. Table 28.3 shows how the volume of hydrogen changed with time when some magnesium was reacted with sulfuric acid.

Table 28.3

Time in s	0	10	20	30	40	50	60	70	80	90	100
Volume of hydrogen in cm^3	0	30	55	75	88	98	102	104	104	104	104

   a) Given that $30\,\text{cm}^3$ of hydrogen has formed after 10 seconds, calculate the mean rate of reaction during the first 10 seconds in cm^3/s of hydrogen formed.
   b) Plot a graph of the volume of hydrogen against time.
   c) Calculate the rate of reaction at 20 seconds by drawing a tangent to the curve.
   d) Calculate the rate of reaction at 40 seconds by drawing a tangent to the curve.
   e) Write a word equation for this reaction.
   f) Write a balanced equation for this reaction.

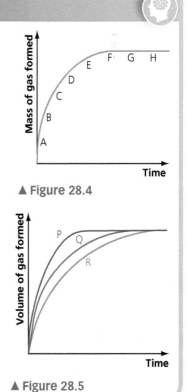

▲ Figure 28.4

▲ Figure 28.5

KEY TERM

Activation energy The minimum energy particles must have to react.

Collision theory

For a chemical reaction to take place particles of the reactants must collide with enough energy to react. The minimum amount of energy particles need to react is called the activation energy.

Successful collisions (i.e. ones which result in a reaction) take place when reactant particles collide with enough energy to react (Figure 28.6). Unsuccessful collisions, ones which do not result in a reaction, take place when reactant particles collide but do not have enough energy to react.

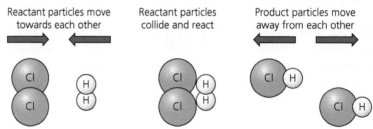

▶ **Figure 28.6** A successful collision.

The rate of a reaction depends on the frequency of successful collisions.

Factors affecting the rate of reactions

There are several factors that affect the rate of chemical reactions.

Temperature

The **higher the temperature**, the faster a reaction. This is because the particles have **more energy** (they are more energetic) and so more of the particles have enough energy to react when they collide. The particles are also **moving faster** and so collide more frequently. This means that there are more frequent successful collisions.

An everyday example of the effect of temperature on chemical reactions is the use of fridges (Figure 28.7). When food goes off, chemical reactions are taking place. We can slow down the rate at which these reactions take place and so slow down the rate at which food goes off by cooling food down and putting it in a fridge.

▲ **Figure 28.7** Food is kept in a fridge to keep it cool and slow down chemical reactions that make food go off.

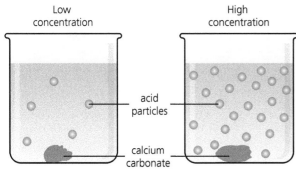

Low concentration   High concentration

acid particles

calcium carbonate

▲ **Figure 28.8** Acid reacting with calcium carbonate.

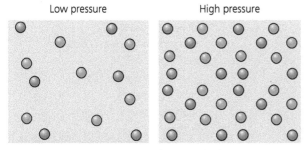

Low pressure   High pressure

▲ **Figure 28.9** Reactions are faster at higher pressure.

> **TIP** ✓
> Remember that changing pressure only affects the rate of a reaction if one or more of the *reactants* are gases. If the only gases present in a reaction mixture are products then changing the pressure has no effect.

▶ **Figure 28.10** The effect of changing surface area.

> **TIP** ✓
> Remember that if a solid is broken up into smaller pieces it has a bigger surface area.

Concentration of reactants in solution

Many chemical reactions involve reactants that are dissolved in solutions. The concentration of a solution is a measure of how much solute is dissolved. The higher the concentration, the more particles of solute that are dissolved (Figure 28.8).

The **higher the concentration** of reactants in solution, the greater the rate of reaction. This is because there are more reactant particles in the solution and so there are more frequent successful collisions.

Pressure of reacting gases

Some chemical reactions involve reactants that are gases. The greater the pressure of a gas, the closer the reactant particles are together.

The **higher the pressure** of reactants that are gases, the greater the rate of reaction. This is because the reactant particles are closer together and so there are more frequent successful collisions (Figure 28.9).

Surface area of solid reactants

Some chemical reactions involve reactants that are solids. The surface area of a solid is increased if it is broken up into more pieces (Figure 28.10). When a solid is made into a powder it has a massive surface area. The **greater the surface area** of a solid reactant, the greater the rate of reaction. This is because there are more particles on the surface that can react and so there are more frequent successful collisions.

Large piece of solid – low surface area

Only reactant particles at the surface can collide with particles of the other reactant

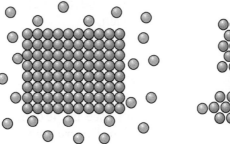

Smaller pieces of solid – bigger surface area

If the solid is broken up into more pieces, there are more reactant particles that can be collided with

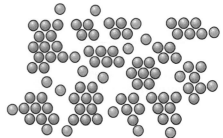

A solid cube with sides of length 2 cm has a total surface area of 24 cm². If it is broken up into eight smaller cubes with sides of length 1 cm then its surface area increases to 48 cm². This means that the surface area has increased while the total volume stays the same (Table 28.4).

The surface area to volume ratio can also be considered to explain this. The single cube with 2 cm sides has a surface area to volume ratio of 3 : 1. When it is broken up into eight smaller cubes with 1 cm sides, the surface area to volume ratio is greater at 6 : 1 (Table 28.4).

Table 28.4 Surface area to volume ratios in a cube.

Cube	One larger cube	Broken up into smaller cubes
	2cm × 2cm × 2cm	1cm × 1cm × 1cm
Surface area	Surface area of each side = $2 \times 2 = 4\,cm^2$ There are six sides, so Total surface area = $6 \times 4 = 24\,cm^2$	Each cube: Surface area of each side = $1 \times 1 = 1\,cm^2$ There are six sides, so Surface area of each cube = $6 \times 1 = 6\,cm^2$ For all eight cubes: Total surface area = $8 \times 6 = 48\,cm^2$
Volume	Volume = $2 \times 2 \times 2 = 8\,cm^3$	Each cube: Volume = $1 \times 1 \times 1 = 1\,cm^3$ For all eight cubes: Total volume = $8 \times 1 = 8\,cm^3$
Surface area : volume ratio	$24:8 = 3:1$	$48:8 = 6:1$

KEY TERM

Catalyst A substance that speeds up a chemical reaction.

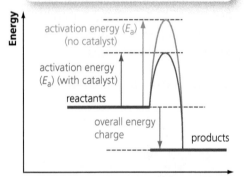

▲ **Figure 28.11** A catalyst lowers the activation energy of a reaction.

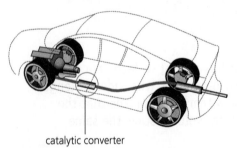

catalytic converter

▲ **Figure 28.12** Catalytic converters reduce air pollution.

The presence of catalysts

A **catalyst** is a substance that speeds up a chemical reaction but does not get used up. A catalyst works by providing a different route (pathway) for the reaction that has a lower activation energy (Figure 28.11).

Different reactions have different catalysts. Some examples of catalysts include:

- iron in the production of ammonia from reaction of hydrogen with nitrogen (the Haber process)
- nickel in the production of margarine from reaction of vegetable oils with hydrogen
- platinum in catalytic converters to remove some pollutants from the exhaust gases of cars (Figure 28.12).

The catalyst is not used up in a reaction. The amount of catalyst left at the end of the reaction is the same as there was at the start. As the catalyst is not used up, it does not appear in the chemical equation for the reaction. For example, in the reaction to remove nitrogen monoxide (NO) from car exhaust gases, the catalyst platinum (Pt) does not appear in the equation. Sometimes the catalyst is written on top of the arrow in the equation:

$$2NO \rightarrow N_2 + O_2 \quad \text{or} \quad 2NO \xrightarrow{Pt} N_2 + O_2$$

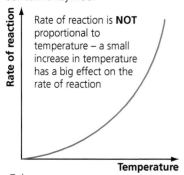

▲ **Figure 28.13** Biological detergents contain enzymes.

Rate of reaction is **NOT** proportional to temperature – a small increase in temperature has a big effect on the rate of reaction

Temperature

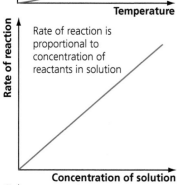

Rate of reaction is proportional to concentration of reactants in solution

Concentration of solution

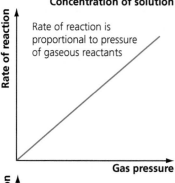

Rate of reaction is proportional to pressure of gaseous reactants

Gas pressure

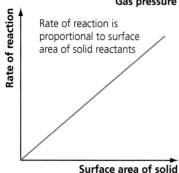

Rate of reaction is proportional to surface area of solid reactants

Surface area of solid

▲ **Figure 28.15** The relative effects of different factors on reaction rates.

Enzymes are molecules that act as catalysts in biological systems. For example:

- Amylase is an enzyme that catalyses the breakdown of starch into sugars.
- Protease is an enzyme that catalyses the breakdown of protein into amino acids.

Enzymes are produced in living organisms and are vital for most life processes such as respiration and digestion. We can also make artificial use of enzymes. For example, biological washing powders contain protease enzymes to break down proteins in dirt on clothes (Figure 28.13).

The relative effect of factors affecting rates

Most factors affecting rates (including the concentration of reactants in solution, the pressure of reactant gases and the surface area of solid reactants) usually have a proportional effect on the rate of reaction. For example, if any of these factors are doubled, the reaction rate doubles. This is because in each case if the factor is doubled, the number of reactant particles available for collisions doubles and so the frequency of successful collisions will double (Figure 28.14).

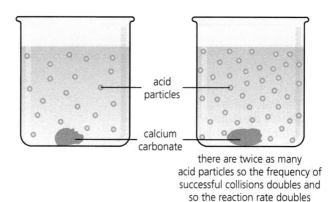

acid particles

calcium carbonate

there are twice as many acid particles so the frequency of successful collisions doubles and so the reaction rate doubles

▲ **Figure 28.14** The effect of doubling the concentration of a reactant in solution.

However, changes in **temperature** have a **much greater effect**. A small increase in temperature causes a big increase in rate of reaction (Figure 28.15). This is because a small increase in temperature leads to many more particles having the activation energy.

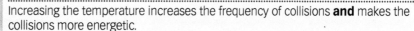

TIP ✓

In a proportional relationship, any change made to one factor has the same effect on the other factor. For example, if one factor is doubled, the other factor doubles. If one factor is made 10 times bigger, the other factor gets 10 times bigger. If one factor is made 3 times smaller, the other factor becomes 3 times smaller.

TIP ✓

Increasing the temperature increases the frequency of collisions **and** makes the collisions more energetic.

7 Hydrogen peroxide decomposes very slowly into water and oxygen. The reaction is much faster if a catalyst such as manganese(IV) oxide is used.

a) What is a catalyst?

b) How do catalysts work?

c) Figure 28.16 shows the reaction profile for the decomposition of hydrogen peroxide with and without a catalyst. Which arrow represents the activation energy for the reaction that takes place with a catalyst?

d) The rate of reaction can also be increased by adding an enzyme. What are enzymes?

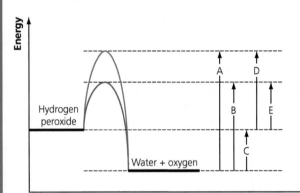

▲ Figure 28.16

8 Nitrogen gas reacts with hydrogen gas to make ammonia. The reaction is carried out at 200 atmospheres pressure. Explain why the reaction is faster at higher pressures.

9 Sodium thiosulfate solution reacts with hydrochloric acid to form solid sulfur. As the sulfur is formed, the solution becomes cloudy. The reaction can be carried out in a conical flask on top of a piece of paper with a cross drawn on it. The time it takes for the solution to become so cloudy that the cross can no longer be seen can be measured and used to compare reaction rates under different conditions. Table 28.5 below shows the results of some experiments carried out in this way.

Table 28.5

Experiment	Concentration of sodium thiosulfate solution in g/dm³	Temperature in °C	Time taken to become too cloudy to see cross in s
A	16	20	75
B	32	20	37
C	48	20	25
D	32	30	18

a) Which experiment was fastest?

b) What is the effect of changing the concentration on the rate of reaction? Which experiments did you use to work this out?

c) Explain why changing concentration affects the rate of reaction.

d) What is the effect of changing the temperature on the rate of reaction? Which experiments did you use to work this out?

e) Explain why changing temperature affects the rate of reaction.

10 Calcium carbonate reacts with hydrochloric acid and produces carbon dioxide gas. An experiment was done to compare the rate of reaction using large pieces, small pieces and powdered calcium carbonate. In each case the same mass of calcium carbonate was used.

Table 28.6

Type of calcium carbonate	Time taken to produce 100 cm³ of carbon dioxide in s
Large pieces	75
Small pieces	48
Powder	5

a) Which type of calcium carbonate had the greatest surface area?

b) Calculate the mean rate of reaction for each experiment in cm³/s of carbon dioxide produced.

c) What is the effect of changing the surface area of the calcium carbonate on the rate of reaction?

d) Explain why changing the surface area affects the rate of reaction.

11 This question compares the surface area to volume ratio of a single cube with 10 cm sides to 1000 smaller cubes with 1 cm sides that can be made from the bigger cube.

a) Calculate the surface area of a cube with 10 cm sides.

b) Calculate the surface area to volume ratio of a cube with 10 cm sides.

c) Calculate the total surface area of 1000 cubes with 1 cm sides.

d) Calculate the surface area to volume ratio of 1000 cubes with 1 cm sides.

Investigating how changes in concentration affect the rates of reactions by measuring the volume of a gas produced

An experiment was carried out to determine the effect of changing the concentration of hydrochloric acid on the rate of the reaction between magnesium and hydrochloric acid. 1.0g of magnesium turnings was reacted with excess hydrochloric acid of different concentrations and the volume of gas produced recorded every minute. The apparatus used is shown in Figure 28.17.

Questions

1 Name the piece of apparatus labelled A.

2 Explain why the bung must be inserted immediately once the magnesium is added to the acid.

3 Name one other piece of apparatus which must be used in this experiment.

4 Write a balanced symbol equation for the reaction between magnesium and hydrochloric acid.

5 What is observed in the flask during the reaction?

6 How would you ensure that this experiment is a fair test?

7 State one source of error in this experiment.

The results from this experiment are plotted on a graph (Figure 28.18).

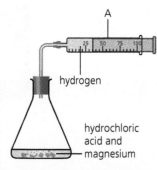

▲ **Figure 28.17** Measuring the volume of hydrogen gas produced.

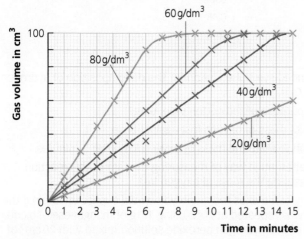

▲ **Figure 28.18** The hydrogen produced when different concentrations of hydrochloric acid react with magnesium.

8 Use the graph to determine the time at which each reaction ended for acid of concentration 80g/dm³ and 60g/dm³.

9 Are there any results which are anomalous? Explain your answer.

10 Which concentration of acid produced the slowest rate of reaction?

11 What is the relationship between the concentration of acid and the rate of reaction?

12 Explain this relationship in terms of collision theory.

Investigating the rate of decomposition of hydrogen peroxide solution by measuring the loss in mass

Hydrogen peroxide decomposes in the presence of solid manganese(IV) oxide to produce water and oxygen.

hydrogen peroxide → water + oxygen

The apparatus shown in Figure 28.19 was used to investigate the rate of decomposition of hydrogen peroxide solution. 20 cm³ of hydrogen peroxide solution were added to 1.0 g of solid manganese(IV) oxide at 20°C.

20 cm³ of hydrogen peroxide + 1.0 g manganese(IV) oxide

cotton wool

conical flask

reaction mixture

-0.26

balance

▲ Figure 28.19

The following results were obtained.

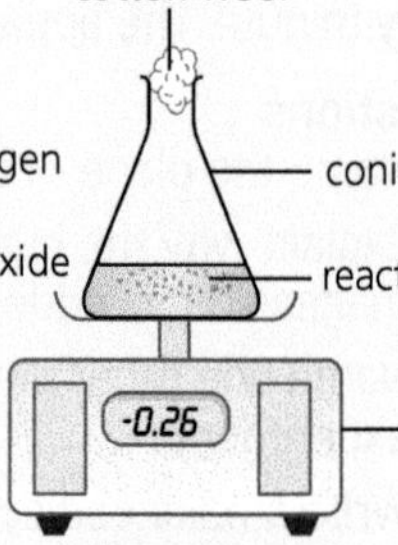

Table 28.7

Time in minutes	Mass of oxygen lost in g
1	0.23
2	0.34
3	
4	0.45
5	0.47
6	0.48
7	0.48

Questions

1 a) Write a balanced symbol equation for the decomposition of hydrogen peroxide (H_2O_2).

 b) What is the purpose of the cotton wool plug?

 c) Plot a graph of mass of oxygen lost against time.

 d) Use the graph to state the mass of oxygen lost after 3 minutes.

 e) Suggest an alternative way of measuring the rate of this reaction without measuring the mass of oxygen lost.

 f) Sketch on the same axes the graph you would expect to obtain if the experiment was repeated at 20°C using 1.0 g of manganese(IV) oxide with 20 cm³ of this hydrogen peroxide solution mixed with 20 cm³ of water.

2 At the end of this experiment the manganese(IV) oxide can be recovered.

 a) Draw a labelled diagram of the assembled apparatus which could be used to recover the manganese(IV) oxide at the end of the experiment.

 b) How would you experimentally prove that the manganese(IV) oxide was not used up in this experiment?

 c) Complete this sentence:
 In this experiment the manganese oxide is acting as a _____.

Required practical 11b

Investigating how changes in concentration affect the rates of reactions by a method involving a change in colour or turbidity

Sodium thiosulfate solution ($Na_2S_2O_3$) reacts with dilute hydrochloric acid according to the equation:

$$Na_2S_2O_3(aq) + 2HCl(aq) \rightarrow 2NaCl(aq) + S(s) + SO_2(g) + H_2O(l)$$

In an experiment 25 cm³ of sodium thiosulfate was placed in a conical flask on top of a piece of paper with a cross drawn on it (Figure 28.20). Hydrochloric acid was added and the stopwatch started. A precipitate was produced that caused the solution to become cloudy. The stopwatch was stopped when the experimenter could no longer see the cross on the paper through the solution, due to the precipitate formed. The precipitate causes turbidity, which is cloudiness in a fluid caused by large numbers of individual solid particles.

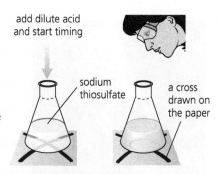

▲ **Figure 28.20** Measuring reaction rates using changes in turbidity.

Questions

1 Look at the equation and identify the product which causes the solution to become cloudy.

The experiment was repeated using different concentrations of sodium thiosulfate. The results are recorded below.

Table 28.8

Experiment	Concentration of sodium thiosulfate in g/dm³	Time taken for cross to disappear in s	Rate of reaction in s⁻¹ (1/time)
1	60	105	0.0095
2	128	79	0.0127
3	192	54	0.0185
4	256	32	

> **KEY TERM**
> **Turbidity** The cloudiness of a solution.

2 Calculate the value for the rate of reaction for experiment 4 using the equation: rate = $\frac{1}{time}$. Give your answer to 3 significant figures.

3 Identify three variables that must be kept constant to make this a fair test.

4 From the results of the experiment state the effect of increasing the concentration of sodium thiosulfate solution on the rate of the reaction.

5 Explain in terms of collision theory why increasing the concentration of sodium thiosulfate solution has this effect.

6 State and explain one change which could be made to this experiment to give more accurate results.

Reversible reactions and dynamic equilibrium

○ Reversible reactions

Some chemical reactions are reversible. This means that once the products have been made from the reactants, the products can react to reform the reactants. The $\rightleftharpoons$ arrows are used to show the reaction is reversible.

$$\text{reactants} \underset{\text{reverse reaction}}{\overset{\text{forward reaction}}{\rightleftharpoons}} \text{products}$$

(a)

(b)

▲ **Figure 28.21** (a) Reaction of anhydrous copper sulfate with water; (b) heating hydrated copper sulfate.

▲ **Figure 28.22** Heating ammonium chloride.

For example, anhydrous copper sulfate (which is white) reacts with water to form hydrated copper sulfate (which is blue). However, this reaction is reversible as when the blue hydrated copper sulfate is heated it breaks back down into white anhydrous copper sulfate and water (Figure 28.21).

anhydrous copper sulfate	+ water	$\rightleftharpoons$	hydrated copper sulfate
$CuSO_4$	$+ 5H_2O$ $\rightleftharpoons$		$CuSO_4.5H_2O$
white solid	colourless liquid		blue solid

When ammonium chloride (NH_4Cl) is heated it breaks down into ammonia (NH_3) and hydrogen chloride (HCl). However, when cooled the ammonia can react with the hydrogen chloride to reform ammonium chloride (Figure 28.22).

ammonium chloride	$\rightleftharpoons$ ammonia	+ hydrogen chloride
NH_4Cl	$\rightleftharpoons$ NH_3	+ HCl
white solid	colourless gas	colourless gas

Energy changes in reversible reactions

If a reversible reaction is exothermic in one direction, then it will be endothermic in the other (Table 28.9). The amount of energy transferred will be the same in each direction. For example, if the forward reaction is exothermic with an energy change of −92 kJ, then the reverse reaction will be endothermic with an energy change of +92 kJ. This is due to the law of conservation of energy.

Table 28.9 Energy changes in reversible reactions.

A reversible reaction where the forward reaction is exothermic	A reversible reaction where the forward reaction is endothermic
Exothermic (−92 kJ) $\longrightarrow$	Endothermic (+58 kJ) $\longrightarrow$
nitrogen + hydrogen $\rightleftharpoons$ ammonia	dinitrogen tetroxide $\rightleftharpoons$ nitrogen oxide
N_2 + $3H_2$ $\rightleftharpoons$ $2NH_3$	N_2O_4 $\rightleftharpoons$ $2NO_2$
$\longleftarrow$ Endothermic (+92 kJ)	$\longleftarrow$ Exothermic (−58 kJ)

Dynamic equilibrium

If a reversible reaction takes place in a closed system, that is apparatus where no substances can get in or out, a dynamic equilibrium is reached. This happens when both the forward and reverse reactions are taking place simultaneously and at exactly the same rate of reaction.

A good analogy of a system in dynamic equilibrium is an athlete walking up an escalator that is moving down (Figure 28.23). The athlete on the escalator is in a state of dynamic equilibrium if she walks up the escalator at the same rate as the escalator moves down.

KEY TERMS

Closed system A system where no substances can get in or out.

(Dynamic) equilibrium System where both the forward and reverse reactions are taking place simultaneously and at the same rate.

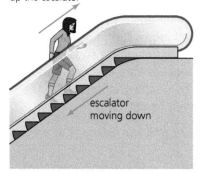

person walking up the escalator

escalator moving down

▲ **Figure 28.23** An athlete in a state of dynamic equilibrium with the escalator.

Test yourself

12 The following reaction is reversible and reaches a state of dynamic equilibrium in a closed system.

hydrogen iodide $\rightleftharpoons$ hydrogen + iodine

a) The forward reaction is endothermic. Will the reverse reaction be endothermic or exothermic?

b) What is a closed system?

c) What is happening to the reactions at dynamic equilibrium?

d) Given the formulae below, write a balanced equation for this reaction. (hydrogen iodide = HI, hydrogen = H_2, iodine = I_2)

13 Ethanol can be made in the following reaction which is reversible and reaches a state of dynamic equilibrium in a closed system. The energy change for the forward reaction is −42 kJ.

ethene + steam $\rightleftharpoons$ ethanol

a) Is the forward reaction endothermic or exothermic?

b) Will the reverse reaction be endothermic or exothermic?

c) What is the energy change for the reverse reaction?

Show you can...

a) Draw a reaction profile for the reaction
$N_2 + 3H_2 \rightleftharpoons 2NH_3$
The energy change for the forward reaction is −92 kJ mol^{-1}
Label:
the activation energy for the forward reaction;
the activation energy for the reverse reaction;
the energy change for the forward reaction.

b) Under what condition does this reaction reach dynamic equilibrium?

○ Le Châtelier's principle

The position of an equilibrium

The athlete on the escalator can be in state of dynamic equilibrium wherever she is on the escalator, whether she is near the top, middle or bottom (Table 28.10). In a similar way, in a chemical reaction at dynamic equilibrium there could be mainly reactants, mainly products or a similar amount of each. The relative amount of reactants and products can be described by the position of the equilibrium.

Table 28.10 Equilibrium positions.

Equilibrium position	Lies to the left	Lies somewhere in the middle	Lies to the right
reactants $\rightleftharpoons$ products	Means there are *more reactants* than products in the mixture of chemicals at equilibrium	Means there is a *similar amount of reactants and products* in the mixture of chemicals at equilibrium	Means there are *more products* than reactants in the mixture of chemicals at equilibrium
Escalator analogy	position lies to the left	position lies somewhere in the middle	position lies to the right

Changing the position of an equilibrium

The relative amounts of reactants and products present at equilibrium (i.e. the position of an equilibrium) depend on the conditions. If the conditions are changed, the position of the equilibrium changes and so the relative amounts of reactants and products change.

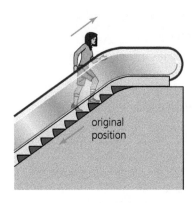

▲ **Figure 28.24** A new equilibrium is established with its position further to the left after the speed of the escalator was increased.

Imagine the athlete on the escalator. If the speed of the escalator moving down was increased then the system would no longer be in equilibrium and the athlete would start moving down the escalator. If the athlete responded to this change by walking or running faster, a new equilibrium would be reached when she matches the speed of the escalator (Figure 28.24). By the time she reaches this new equilibrium, her position will have moved and she will be nearer the bottom of the escalator and so her position will have moved to the left.

Chemical reactions at dynamic equilibrium respond to changes as well. **Le Châtelier's principle** states that:

> If a change is made to the conditions of a system at equilibrium, then the position of the equilibrium moves to oppose that change in conditions.

The effect of changing concentration

If the concentration of a chemical in an equilibrium is changed, the position of the equilibrium will move to oppose that change. If more of a chemical is added, the equilibrium position moves to remove it. If some of a chemical is removed, the equilibrium position moves to make more of it.

Table 28.11 illustrates how changes in concentration affect the position of an equilibrium of the form:

$$\text{reactants} \rightleftharpoons \text{products}$$

Table 28.11 The effect of changes in concentration on the position of an equilibrium.

	Increase concentration of a reactant	Decrease concentration of a reactant	Increase concentration of a product	Decrease concentration of a product
How the system responds to oppose the change	Equilibrium position moves right to reduce concentration of reactant	Equilibrium position moves left to increase concentration of reactant	Equilibrium position moves left to reduce concentration of product	Equilibrium position moves right to increase concentration of product

Example

Ammonia is made by reacting nitrogen with hydrogen. What would happen to the amount of ammonia formed if the concentration of nitrogen was increased?

Answer

nitrogen + hydrogen $\rightleftharpoons$ ammonia

$N_2(g) + 3H_2(g) \rightleftharpoons 2NH_3(g)$

If the concentration of nitrogen is increased:

- The equilibrium position moves right to reduce the concentration of added nitrogen.
- More ammonia is produced because the equilibrium position moves right.

Example

Hydrogen can be made by reacting methane with steam. What would happen to the amount of hydrogen formed if the carbon monoxide was removed as it was formed?

Answer

methane + steam $\rightleftharpoons$ hydrogen + carbon monoxide

$CH_4(g) + H_2O(g) \rightleftharpoons 3H_2(g) + CO(g)$

If the carbon monoxide is removed:

- The equilibrium position moves right to increase the concentration of carbon monoxide.
- More hydrogen is produced because the equilibrium position moves right.

The effect of changing temperature

If the temperature of an equilibrium is changed, the position of the equilibrium will move to oppose that change. In order to increase the temperature, the equilibrium position moves in the direction of the exothermic reaction. In order to decrease the temperature, the equilibrium position moves in the direction of the endothermic reaction (Table 28.12).

Table 28.12 The effect of changes in temperature on the position of an equilibrium.

Equilibrium	Forward reaction is exothermic		Forward reaction is endothermic	
	reactants $\xrightleftharpoons[\text{endothermic}]{\text{exothermic}}$ products		reactants $\xrightleftharpoons[\text{exothermic}]{\text{endothermic}}$ products	
Change	Increase temperature	Decrease temperature	Increase temperature	Decrease temperature
How the system responds to oppose the change	Equilibrium position moves left in endothermic direction to lower the temperature	Equilibrium position moves right in exothermic direction to increase the temperature	Equilibrium position moves right in endothermic direction to lower the temperature	Equilibrium position moves left in exothermic direction to increase the temperature

However, reactions where more product is formed at lower temperatures are not usually carried out at low temperature. This is because they would be far too slow. A compromise temperature is usually used that gives a reasonable yield of product but at a fast rate.

Example

Ammonia is made by reacting nitrogen with hydrogen. The forward reaction is exothermic. What would happen to the amount of ammonia formed if the temperature was increased?

Answer

nitrogen + hydrogen $\rightleftharpoons$ ammonia

$N_2(g) + 3H_2(g) \rightleftharpoons 2NH_3(g)$

If the temperature is increased:

- The equilibrium position moves left in the endothermic direction to reduce the temperature.
- Less ammonia is produced because the equilibrium position moves left.

Example

Hydrogen can be made by reacting methane with steam in an endothermic reaction. What would happen to the amount of hydrogen formed if the temperature was increased?

Answer

methane + steam $\rightleftharpoons$ hydrogen + carbon monoxide

$CH_4(g) + H_2O(g) \rightleftharpoons 3H_2(g) + CO(g)$

If the temperature is increased:

- The equilibrium position moves right in the endothermic direction to reduce the temperature.
- More hydrogen is produced because the equilibrium position moves right.

The effect of changing pressure

The more molecules that are present in a gas, the greater the pressure of the gas (Figure 28.25).

If the pressure of an equilibrium containing gases is changed, the position of the equilibrium will move to oppose that change (if it can). In order to increase the pressure, the equilibrium position moves to the side with the most gas molecules. In order to decrease the pressure, the equilibrium position moves to the side with fewer gas molecules (Table 28.13).

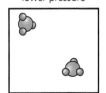 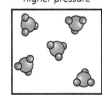

lower pressure   higher pressure

▲ **Figure 28.25** The more gas particles present, the higher the pressure.

Table 28.13 The effect of changes in pressure on the position of an equilibrium.

Equilibrium	More gas molecules of reactants than products e.g. 2A(g) ⇌ B(g)		Same number of gas molecules of reactants and products e.g. A(g) ⇌ B(g)		More gas molecules of products than reactants e.g. A(g) ⇌ 2B(g)	
Change	Increase pressure	Decrease pressure	Increase pressure	Decrease pressure	Increase pressure	Decrease pressure
How the system responds to oppose the change	Equilibrium position moves right to side with less gas molecules to reduce pressure	Equilibrium position moves left to side with more gas molecules to increase pressure	Equilibrium does not move because there are same number of gas molecules of reactants and products		Equilibrium position moves left to side with less gas molecules to reduce pressure	Equilibrium position moves right to side with more gas molecules to increase pressure

The use of high pressure is very expensive due to the high energy cost of compressing gases and the high cost of pipes to withstand that pressure. In reactions where higher pressure gives more product, the value of the extra product formed is sometimes less than the cost of those higher pressures. This means that sometimes the actual pressure used in a process is not as high as might be predicted because it is not cost effective. There is often a compromise between the amount of product formed and the cost of using higher pressure.

Example

Ammonia is made by reacting nitrogen with hydrogen. What would happen to the amount of ammonia formed if the pressure was increased?

Answer

If the pressure is increased:

- The equilibrium position moves right to the side with fewer gas molecules to decrease the pressure.
- More ammonia is produced because the equilibrium position moves right.

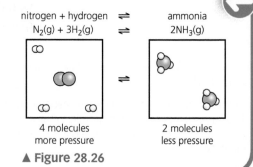

nitrogen + hydrogen ⇌ ammonia
$N_2(g) + 3H_2(g)$ ⇌ $2NH_3(g)$

4 molecules
more pressure

2 molecules
less pressure

▲ Figure 28.26

Example

Hydrogen can be made by reaction of methane with steam. What would happen to the amount of hydrogen formed if the pressure was increased?

Answer

If the pressure is increased:

- The equilibrium position moves left to the side with fewer gas molecules to decrease the pressure.
- Less hydrogen is produced because the equilibrium position moves left.

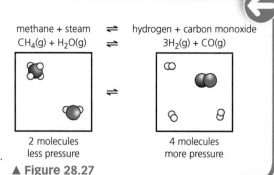

methane + steam ⇌ hydrogen + carbon monoxide
$CH_4(g) + H_2O(g)$ ⇌ $3H_2(g) + CO(g)$

2 molecules
less pressure

4 molecules
more pressure

▲ Figure 28.27

Example

Hydrogen gas reacts with iodine gas to form hydrogen iodide gas. What would happen to the amount of hydrogen iodide formed if the pressure was increased?

Answer

If the pressure is increased:

- The equilibrium position does not move because there are the same number of gas molecules on both sides of the equation.
- The amount of hydrogen iodide formed remains the same.

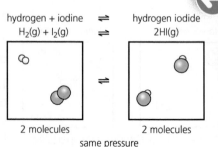

hydrogen + iodine ⇌ hydrogen iodide
$H_2(g) + I_2(g)$ ⇌ $2HI(g)$

2 molecules   2 molecules
same pressure

▲ Figure 28.28

The effect of catalysts

Catalysts have no effect on the position of an equilibrium. However, the catalyst does increase the rate of forward and backward reactions, both by the same amount. This means that the system reaches equilibrium faster and the product is formed faster.

Test yourself

14 Look at the following four reactions that all make substance **X**. The energy change for the forward reaction is shown in each case.

Reaction 1: $A(g) + B(g) \rightleftharpoons X(g)$ exothermic
Reaction 2: $C(g) + 2D(g) \rightleftharpoons X(g)$ endothermic
Reaction 3: $E_2(g) \rightleftharpoons 2X(g)$ exothermic
Reaction 4: $F(g) + G(g) \rightleftharpoons 2X(g)$ exothermic

a) In which of the reactions, if any, would the amount of **X** formed increase as the temperature increases?

b) In which of the reactions, if any, would the amount of **X** formed increase as the pressure increases?

c) In which of the reactions, if any, would the amount of **X** formed remain the same as pressure increases?

15 The colourless gas dinitrogen tetroxide (N_2O_4) forms an equilibrium with the brown gas nitrogen dioxide (NO_2). An equilibrium mixture is a pale brown colour. The energy change for the forward reaction is +58 kJ. What will happen to the colour of the equilibrium mixture if it is heated up? Explain your answer.

$N_2O_4(g) \rightleftharpoons 2NO_2(g)$

16 In aqueous solution, pink cobalt ions (Co^{2+}) react with chloride ions (Cl^-) to form blue cobalt tetrachloride ions ($CoCl_4^{2-}$). This reaction is reversible and forms a dynamic equilibrium.

$Co^{2+} + 4Cl^- \rightleftharpoons CoCl_4^{2-}$
pink   blue

a) State Le Châtelier's principle.

b) An equilibrium mixture was a purple colour as it contained a similar amount of the pink Co^{2+} ions and the blue $CoCl_4^{2-}$ ions. What would happen to the colour if more chloride ions were added? Explain your answer.

c) When this purple mixture is cooled down it goes pink. Is the forward reaction exothermic or endothermic? Explain your answer.

17 Gas R is made when gases P and Q react:
$P(g) + Q(g) \rightleftharpoons R(g)$

a) Table 28.14 below shows how the percentage of R in the equilibrium mixture varies with temperature. Plot a graph to the show the percentage of R against temperature.

Table 28.14

Temperature in °C	100	200	300	400	500
Percentage of R in equilibrium mixture	58	42	30	21	16

b) Is the reaction to form R endothermic or exothermic? Explain your answer.

Show you can...

Chlorodifluoromethane reacts to produce tetrafluoroethene in an endothermic reaction.

chlorodifluoromethane $\rightleftharpoons$ tetrafluoroethene + hydrogen chloride

$2CHClF_2(g)$ $\rightleftharpoons$ $C_2F_4(g)$ + $2HCl(g)$

Explain in terms of reaction rate and equilibrium whether:

a) a high or low pressure should be used
b) a high or low temperature should be used.

Chapter review questions

1 Carbon monoxide reacts with steam to form carbon dioxide and hydrogen in a reversible reaction. This reaches a state of dynamic equilibrium in a closed system. The energy change for the forward reaction is −42 kJ.

carbon monoxide + steam $\rightleftharpoons$ carbon dioxide + hydrogen

a) What is a dynamic equilibrium?

b) Is the forward reaction endothermic or exothermic?

c) What is the energy change for the reverse reaction?

d) This reaction is often done using a catalyst. What is a catalyst?

2 Hydrogen (H_2) can react explosively with oxygen (O_2) to make water. However, a mixture of hydrogen and oxygen does not explode unless a flame or spark is brought near the mixture.

hydrogen + oxygen $\rightarrow$ water

a) Define the term 'activation energy'.

b) Explain why hydrogen and oxygen molecules may not react with each other even when they collide unless a flame or spark is present.

c) Write a balanced equation for this reaction.

3 a) A reaction produced 50 cm³ of hydrogen gas in 40 seconds. Calculate the mean rate of reaction in cm³/s of hydrogen produced in these 40 seconds.

b) A reaction produced 0.16 g of oxygen gas in 20 seconds. Calculate the mean rate of reaction in g/s of oxygen produced in these 20 seconds.

4 Calcium carbonate reacts with hydrochloric acid and forms carbon dioxide gas. A student carried out a series of experiments to investigate how concentration of acid, temperature and surface area of calcium carbonate affected the rate. Table 28.15 gives details about each reaction (the acid was in excess in each case).

Figure 28.29 shows how the volume of carbon dioxide gas formed changed with time. Match experiments 2, 3 and 4 in Table 28.15 with lines P, Q and R. Explain your reasoning.

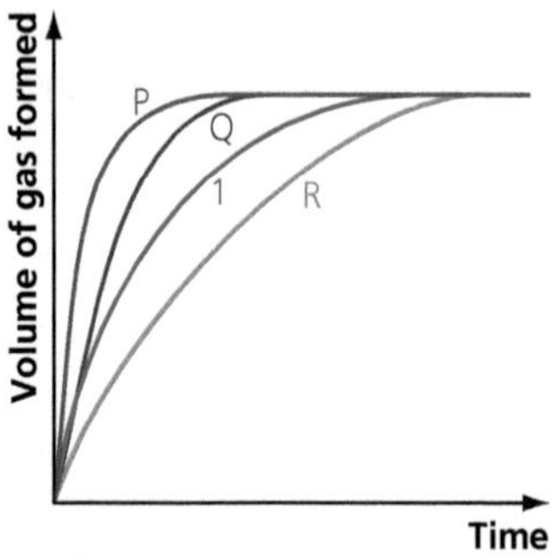

Table 28.15

Experiment	Concentration of acid in g/dm³	Temperature in °C	Type of calcium carbonate
1	80	20	Small pieces
2	40	20	Small pieces
3	80	30	Small pieces
4	80	20	Powder

▲ Figure 28.29

5 The metal cobalt reacts slowly with acid. However, it reacts faster if the temperature and the concentration of acid are increased.

a) Explain why increasing the temperature increases the rate of reaction.

b) Explain why increasing the concentration of acid increases the rate of reaction.

c) Which has the biggest relative effect – increasing the temperature or increasing the concentration of acid?

6 Methanol is a useful fuel. Two ways in which it can be made are shown below. Both reactions are exothermic. Method 1 is typically done with a catalyst at 250°C and at 80 atmospheres pressure.

Method 1: $CO(g) + 2H_2(g) \rightleftharpoons CH_3OH(g)$

Method 2: $CO_2(g) + 3H_2(g) \rightleftharpoons CH_3OH(g) + H_2O(g)$

a) Explain why a high pressure is used for method 1.

b) Explain why pressure higher than 80 atmospheres is not used for method 1.

c) Explain why the rate of reaction in method 1 increases as the pressure increases.

d) According to Le Châtelier's principle, more methanol would be produced in method 1 at temperatures lower than 250°C. Explain why temperatures lower than 250°C are not used.

7 In aqueous solution, the yellow chromate(VI) ion, CrO_4^{2-}, forms an equilibrium with the orange dichromate(VI) ion, $Cr_2O_7^{2-}$

$$2CrO_4^{2-} + 2H^+ \rightleftharpoons Cr_2O_7^{2-} + H_2O$$
yellow　　　　　　　orange

a) What would happen to the colour of the mixture if sulfuric acid was added? Explain your answer.

b) What would happen to the colour of the mixture if sodium hydroxide was added? Explain your answer.

8 Rhubarb stalks contain some ethanedioic acid. If rhubarb stalks are placed in a purple solution of dilute acidified potassium manganate(VII), the solution goes colourless as the ethanedioic acid reacts with it. The results from two experiments are shown below. In the first experiment one piece of rhubarb stalk was used in the shape of a cuboid measuring 5 cm × 1 cm × 1 cm. In the second experiment, a similar piece of rhubarb stalk was chopped up into five cubes each measuring 1 cm × 1 cm × 1 cm.

Table 28.16

	Size of rhubarb stalk used	Time to go colourless in s
Experiment 1	One piece of rhubarb stalk in the shape of a cuboid measuring 5 cm × 1 cm × 1 cm	75
Experiment 2	Five pieces of rhubarb stalk in the shape of cubes each measuring 1 cm × 1 cm × 1 cm	52

a) Show that the same amount of rhubarb was used in each experiment.

b) Calculate the surface area of rhubarb in each experiment.

c) Calculate the surface area to volume ratio of rhubarb in each experiment.

d) What does this experiment show about the effect of surface area on rate of reaction?

e) Explain why changing surface area has the effect that it does.

f) What variables should be controlled between experiments 1 and 2 to ensure that this is a fair test?

9 Carbon dioxide gas is formed when calcium carbonate reacts with hydrochloric acid. Table 28.17 shows how the volume of carbon dioxide changed over time when this reaction was done.

Table 28.17

Time in s	0	10	20	30	40	50	60	70	80	90	100
Volume of carbon dioxide in cm^3	0	22	35	43	48	52	55	57	58	58	58

a) Plot a graph of the volume of carbon dioxide over time.

b) Calculate the rate of reaction at 20 seconds by drawing a tangent to the curve.

c) Calculate the rate of reaction at 60 seconds by drawing a tangent to the curve.

d) Write a word equation for this reaction.

e) Write a balanced equation for this reaction.

Practice questions

1 Which one of the following is used to increase the rate at which ammonia is produced from hydrogen and nitrogen? [1 mark]

A catalyst   B oxidising agent

C reducing agent   D reduced temperature

2 In which one of the following experiments will the rate of reaction be quickest at the start of the reaction? [1 mark]

A zinc powder reacting with an excess of 40 g/dm³ HCl at 20°C

B zinc powder reacting with an excess of 40 g/dm³ HCl at 30°C

C zinc powder reacting with an excess of 80 g/dm³ HCl at 30°C

D zinc granules reacting with an excess of 40 g/dm³ HCl at 30°C

3 In the laboratory, preparation of oxygen from hydrogen peroxide using manganese(IV) oxide as catalyst, the mass of manganese(IV) oxide was measured at various times. Which one of the graphs in Figure 28.30 best shows the experimental results? [1 mark]

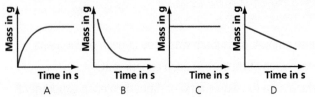

▲ Figure 28.30

4 In a laboratory experiment 0.5 g of magnesium ribbon was reacted with excess dilute hydrochloric acid at room temperature. The volume of gas produced was noted every 10 seconds. The results were plotted in Figure 28.31 below as graph C.

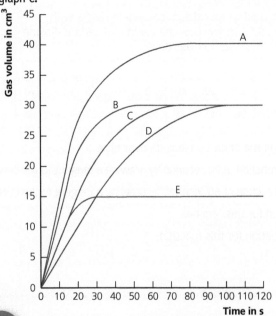

Time in s   ◄ Figure 28.31

a) At what time did reaction C finish? [1 mark]

b) Calculate the mean rate of reaction of experiment C over the time of the whole reaction. [1 mark]

The experiment was repeated using different conditions and the results obtained given in graphs A, B, D and E.

c) State which of the graphs A, B, D or E would have been obtained if the 0.5 g of magnesium ribbon was replaced by 0.5 g of magnesium powder. Give a reason for your answer. [2 marks]

d) State which of the graphs A, B, D or E would have been obtained if the 0.5 g of magnesium ribbon was added to excess dilute hydrochloric acid at a temperature below room temperature. Give a reason for your answer. [2 marks]

e) State which of the graphs A, B, D or E would have been obtained if 0.25 g of magnesium ribbon was reacted with excess dilute hydrochloric acid at room temperature. Give a reason for your answer. [2 marks]

f) State and explain in terms of collision theory the effect of increasing the concentration of hydrochloric acid on the rate of the reaction between hydrochloric acid and magnesium. [3 marks]

5 Nitrogen(IV) oxide (NO₂) is a brown gas which can form from the colourless gas dinitrogen tetroxide N₂O₄ in a dynamic equilibrium. The energy change for the forward reaction is exothermic.

$$2NO_2(g) \leftrightharpoons N_2O_4(g)$$

a) What is meant by the term 'dynamic equilibrium'? [2 marks]

b) Explain what is observed when the pressure on the equilibrium mixture is decreased. [3 marks]

c) Explain what is observed when the temperature of the equilibrium mixture is reduced. [3 marks]

Working scientifically:
Presenting information and data in a scientific way

Recording results

During experimental activities you must often record results in a table. When drawing tables and recording data ensure that:

▶ the table is a ruled box with ruled columns and rows

▶ there are headings for each column or row

▶ there are units for each column and row – usually placed after the heading – for example 'Temperature in °C'

▶ there is room for repeat measurements and averages – remember the more repeats you do the more reproducible the data.

Table 28.18 Example of a table to record experimental results.

Times in minutes	Mass of oxygen lost in g	Mass of oxygen lost in g	Average mass of oxygen lost in g
1			
2			
3			
4			
5			

Questions

1 50 cm³ of hydrogen peroxide and 1.0 g of manganese dioxide were allowed to react at 25°C. The volume of oxygen collected from the reaction at 10 second intervals was:

after 10 seconds, 30 cm³; after 20 seconds, 49 cm³; after 30 seconds, 59 cm³; after 40 seconds, 63 cm³; and after 50 seconds, 63 cm³

On repeating the experiment the volume of gas obtained at each time interval was 32, 51, 59, 63, 65 cm³ respectively.

a) Present these results in a suitable table with headings and units.

b) Calculate the mean volume of gas produced and include this in your table.

2 Solubility is measured in g/100 g water. The solubility of potassium sulfate in 100 g water was measured at different temperatures with the following results:

at 90°C, 22.9 g/100 g dissolves; at 50°C, 16.5 g/100 g dissolves; at 70°C, 19.8 g/100 g dissolves; at 10°C, 9.3 g/100 g dissolves; and at 30°C, 13.0 g/100 g dissolves.

a) Present these results in a suitable table with headings.

b) What is the trend shown by the results?

Plotting graphs

When carrying out experiments you must be able to translate data from one form to another. Most often this involves using data from a table to draw a graph. A graph is an illustration of how two variables relate to one another.

When drawing a graph remember that:

▶ The independent variable is placed on the *x*-axis, while the dependent variable is placed on the *y*-axis (Figure 28.32).

▶ Appropriate scales should be devised for the axes, making the most effective use of the space on the graph paper. The data should be critically examined to establish whether it is necessary to start the scale(s) at zero.

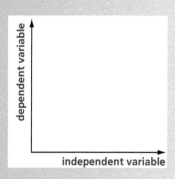

▲ Figure 28.32

▶ Axes should be labelled with the name of the variable followed by 'in' and the unit of measurement. For example the label may be 'Temperature in °C'.

▶ A line of best fit should be drawn. When judging the position of the line there should be approximately the same number of data points on each side of the line. Resist the temptation to simply connect the first and last points. The best fit line could be straight or a curve. Ignore any anomalous results when drawing your line.

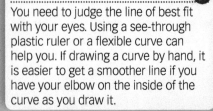

TIP

You need to judge the line of best fit with your eyes. Using a see-through plastic ruler or a flexible curve can help you. If drawing a curve by hand, it is easier to get a smoother line if you have your elbow on the inside of the curve as you draw it.

Questions

3 In an experiment some calcium carbonate and acid were placed in a conical flask on a balance and the balance reading recorded every minute. The results were recorded and the graph shown below was drawn.

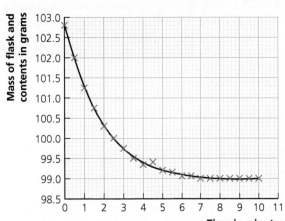

▲ **Figure 28.33** Changes in mass over time when calcium carbonate and acid react.

a) Are there any results which you would ignore when drawing a best fit curve?

b) What is the independent variable in this experiment?

c) At what time does the reaction stop?

d) What is the mass of the flask and contents at time 2 minutes?

29 Organic chemistry

Organic chemistry is the study of compounds containing carbon. The number of organic compounds that exist is greater than all other compounds added together. This is because carbon is able to form bonds to other carbon atoms very easily and form chains and rings.

Many medicines, detergents, clothing fibres and solvents are organic molecules and many of these are made from chemicals that we find in crude oil. Our own bodies and food also contain many organic molecules.

Specification coverage

This chapter covers specification point 5.7.1.1 to 5.7.1.4 and is called Organic chemistry.

It covers carbon compounds as fuels and feedstock, fractional distillation, hydrocarbons, cracking and alkenes.

Previously you could have learnt:

> Crude oil, coal and natural gas are fossil fuels.
> Fuels are substances that burn in oxygen releasing a lot of thermal energy.
> Non-metal atoms bond to each other by sharing electrons; one covalent bond is made up of two shared electrons.
> Carbon atoms make four covalent bonds in molecules.
> Mixtures of miscible liquids with different boiling points can be separated by fractional distillation.
> Polymers are long chain molecules.

Test yourself on prior knowledge

1 What are fuels?
2 Give three examples of fossil fuels
3 What type of bonds are made when carbon atoms bond with hydrogen atoms?
4 How many covalent bonds do carbon atoms make?
5 What is a polymer?
6 How are mixtures of miscible liquids separated?

Crude oil, hydrocarbons and alkanes

○ What is crude oil?

Crude oil is a fossil fuel that is found underground in rocks. It was formed over millions of years from the remains of sea creatures. These creatures were mainly plankton that were buried in mud at the bottom of the oceans (Figure 29.1).

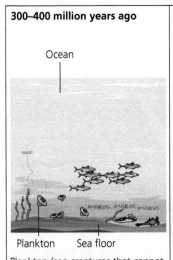

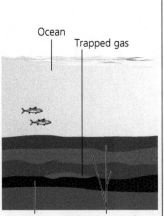
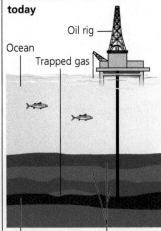

300–400 million years ago

Ocean

Plankton   Sea floor

Plankton (sea creatures that cannot swim against a current, including algae, bacteria, protists and some animals) died and fell onto the sea floor.

Ocean

Mud   Dead plankton

The dead plankton were covered in mud.

Ocean   Trapped gas

Oil in porous rocks   Layers of rock

Over millions of years, more and more sediment built up. The high temperature and pressure turned the dead plankton into oil and gas.

today

Oil rig

Ocean   Trapped gas

Oil in porous rocks   Layers of rock

Today we drill down through rock to reach the oil and bring it up to the surface.

▲ **Figure 29.1** The formation of crude oil.

Crude oil is a form of ancient biomass as it was made from the remains of creatures that lived many years ago. Crude oil is a **finite resource** because we cannot replace it as we use it up.

Crude oil is a mixture of many different compounds. Most of these compounds are hydrocarbons. Hydrocarbons are compounds that contain hydrogen and carbon **only**. Most of the hydrocarbons in crude oil are **alkanes**.

○ Alkanes

Alkanes are a family of saturated hydrocarbons. Saturated molecules are ones that only contain single covalent bonds. The structures of the first four alkanes are shown in Table 29.1. The table includes the displayed formula which shows all the atoms and all the bonds in each molecule.

Table **29.1** The first four alkanes.

Alkane	Ball and stick structure	Displayed structure	Molecular formula
Methane			CH_4
Ethane			C_2H_6
Propane			C_3H_8
Butane			C_4H_{10}

All the alkanes have a molecular formula of the form C_nH_{2n+2}. For example, if there are 3 carbon atoms ($n = 3$), then there are 8 hydrogen atoms ($2n + 2 = (2 \times 3) + 2 = 8$). This and other examples are shown in Table 29.2.

Table **29.2** The first four alkanes.

	Methane	Ethane	Propane	Butane
Number of C atoms (n)	$n = 1$	$n = 2$	$n = 3$	$n = 4$
Number of H atoms ($2n + 2$)	$2n + 2 = (2 \times 1) + 2 = 4$	$2n + 2 = (2 \times 2) + 2 = 6$	$2n + 2 = (2 \times 3) + 2 = 8$	$2n + 2 = (2 \times 4) + 2 = 10$
Molecular formula	CH_4	C_2H_6	C_3H_8	C_4H_{10}

The names of organic compounds are made up of two parts. The first part of the name indicates the number of carbon atoms and the second part indicates which homologous series the molecule belongs to (Figure 29.2).

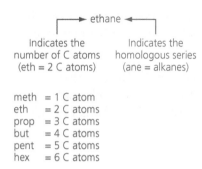

ethane

Indicates the
number of C atoms
(eth = 2 C atoms)

Indicates the
homologous series
(ane = alkanes)

meth = 1 C atom
eth = 2 C atoms
prop = 3 C atoms
but = 4 C atoms
pent = 5 C atoms
hex = 6 C atoms

▲ **Figure 29.2** The alkanes as a homologous series.

KEY TERMS

Fractional distillation A method used to separate miscible liquids with different boiling points.

Fraction A mixture of molecules with similar boiling points.

The alkanes are an example of a homologous series. A homologous series is a family of compounds with:

- the same general formula
- similar chemical properties.

⃝ **Fractional distillation of crude oil**

For crude oil to be useful, the hydrocarbons it contains have to be separated. The hydrocarbons have different boiling points and this difference is used to separate them by fractional distillation at an oil refinery. This process separates the hydrocarbons into fractions. A fraction is a mixture of molecules with similar boiling points. In each fraction, the hydrocarbons contain a similar number of carbon atoms.

The crude oil is heated and vaporised (**evaporated**). The vaporised crude oil enters the fractionating tower which is hotter at the bottom and cooler at the top. The hydrocarbons cool as they rise up the tower and **condense** at different heights because they have different boiling points. Hydrocarbons with large molecules are collected as liquids near the bottom of the tower while those with small molecules collect at the top (Figure 29.3).

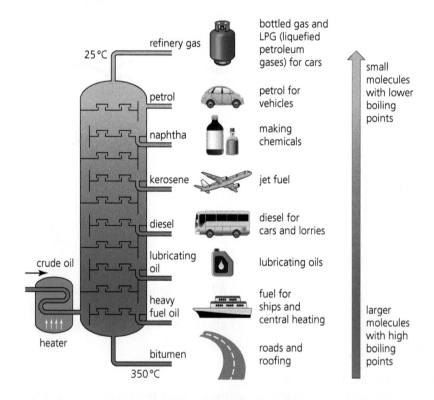

▶ **Figure 29.3** Fractional distillation of crude oil.

The fractions from crude oil have different uses because they have different properties. Many are used as fuels, such as liquefied petroleum gases (cars and gas heaters), petrol (cars), kerosene (aeroplanes), diesel (cars and lorries) and heavy fuel oil (ships and heating). Other fractions are used as feedstock for processes to make useful substances such as medicines, detergents, lubricants and polymers.

The petrochemical industry is a huge industry that deals with the fractional distillation of crude oil and provides fuels and other substances made from crude oil. Modern life would be very different without these fuels and other substances produced from crude oil.

○ The use of hydrocarbons as fuels

The main use of hydrocarbons from crude oil is as fuels. Hydrocarbons are good fuels because they release a lot of energy when they burn.

When hydrocarbons burn, they react with oxygen. Complete combustion takes place if there is a good supply of oxygen from the air. The carbon atoms are oxidised, combining with the oxygen to form carbon dioxide. The hydrogen atoms are also oxidised, combining with the oxygen to form water.

> Complete combustion: hydrocarbon + oxygen → carbon dioxide + water

When writing balanced equations for the complete combustion of alkanes:

- Balance the C atoms: for each C atom in the hydrocarbon (alkane) there will be one CO_2 molecule formed.
- Balance the H atoms: for every two H atoms in the hydrocarbon (alkane) there will be one H_2O molecule formed.
- Count the number of O atoms in the CO_2 and H_2O: the number of O_2 molecules will be half this number.
- If the number of O_2 molecules has a half in it, double all the balancing numbers to get rid of the half.

Example

Write a balanced equation for the complete combustion of methane, CH_4.

Answer

Word equation: methane + oxygen → carbon dioxide + water

Unbalanced equation: $CH_4 + O_2 → CO_2 + H_2O$

CO_2:   there is 1 C atom in CH_4 so there will be 1 CO_2 formed

H_2O: there are 4 H atoms in CH_4 so there will be 2 H_2O formed

O_2: $CO_2 + 2H_2O$ contains 4 O atoms and so 2 O_2 needed

Balanced equation: $CH_4 + 2O_2 → CO_2 + 2H_2O$

Example

Write a balanced equation for the complete combustion of butane, C_4H_{10}.

Answer

Word equation: butane + oxygen → carbon dioxide + water

Unbalanced equation: $C_4H_{10} + O_2 \rightarrow CO_2 + H_2O$

CO_2:   there are 4 C atoms in C_4H_{10} so there will be 4 CO_2 formed

H_2O: there are 10 H atoms in C_4H_{10} so there will be 5 H_2O formed

O_2: $4CO_2 + 5H_2O$ contains 13 O atoms and so 6½ O_2 needed

As there is a half in the equation we will double all the values

Balanced equation: $2C_4H_{10} + 13O_2 \rightarrow 8CO_2 + 10H_2O$

KEY TERMS

Flammability How easily a substance catches fire; the more flammable, the more easily it catches fire.

Viscosity How easily a liquid flows; the higher the viscosity the less easily it flows.

The properties of hydrocarbons depend on the size of the molecules and this affects their use as fuels (Figure 29.4). The flammability of a fuel is how easily it catches fire. A flammable fuel catches fire easily. The viscosity of a liquid is how easily it flows. A runny liquid has a low viscosity while a thick, slow moving liquid (e.g. syrup) has a high viscosity.

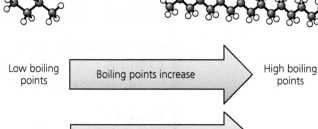

▲ **Figure 29.4** The properties of hydrocarbons (alkanes).

TIP

To help remember the properties of hydrocarbons, look at Figure 29.3 and compare petrol and bitumen.

This means that shorter hydrocarbons are more in demand as fuels because they flow more easily, are more flammable and burn with a cleaner flame. Although larger hydrocarbons are less easy to use, some are used as fuels. For example, fuel oil is used in ships and in some central heating systems.

Test yourself

1 a) Describe how crude oil was formed.
   b) Explain why crude oil can be described as an ancient source of biomass.
   c) Crude oil is a finite resource. Explain what this means.

2 Crude oil is a mixture of hydrocarbons. They are separated by fractional distillation at an oil refinery.
   a) What is a hydrocarbon?
   b) What property of the hydrocarbons allows them to be separated in this way?
   c) Describe how fractional distillation separates the hydrocarbons.

3 Hexane is an alkane containing six carbon atoms.
   a) Alkanes are saturated hydrocarbons. What does saturated mean in this context?
   b) Give the molecular formula of hexane.
   c) Draw the displayed formula of hexane.

4 Pentane and decane are both alkanes. Pentane has the formula C_5H_{12}. Decane has the formula $C_{10}H_{22}$.
   a) Which one has the highest boiling point?
   b) Which one is the most flammable?
   c) Which one is most viscous?

5 Write a balanced equation for the complete combustion of the following alkanes:
   a) pentane, C_5H_{12}
   b) ethane, C_2H_6

Show you can...

A hydrocarbon P (C_xH_y) burns in excess air as shown in the equation below:

$$C_xH_y + 5O_2 \rightarrow 3CO_2 + 4H_2O$$

a) Explain why C_xH_y is a hydrocarbon.
b) Determine the values of x and y using the equation given above.
c) Name hydrocarbon P and draw its displayed formula.

Cracking and alkenes

 Cracking

Shorter hydrocarbons (alkanes) are in very high demand as fuels but longer ones are in less demand. This means that there is a surplus of the longer hydrocarbons from the fractional distillation of crude oil. These longer alkanes can be broken down into the shorter molecules that are in higher demand by a process called cracking. This process also produces unsaturated hydrocarbons called **alkenes** which can be used as a starting material to make many other substances such as polymers and medicines (Figure 29.5).

> **KEY TERM**
>
> Cracking The thermal decomposition of long hydrocarbons (alkanes) into shorter alkanes and alkenes.

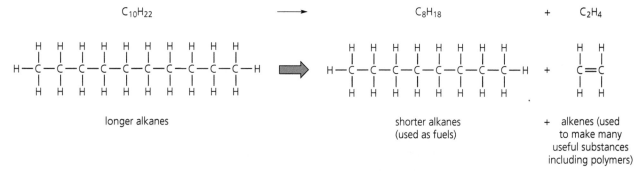

▲ **Figure 29.5** An example of a cracking reaction.

Cracking can be done in a number of ways. Two methods of cracking are:

- **Catalytic cracking**: heat the alkanes to vaporise them and then pass them over a hot catalyst.
- **Steam cracking**: heat the alkanes to vaporise them, mix them with steam and then heat them to very high temperature.

Cracking is a thermal decomposition reaction because the alkanes are broken down into smaller molecules by heating them.

Alkenes

KEY TERMS

Alkenes A homologous series of unsaturated hydrocarbons with the general formula C_nH_{2n}.

Unsaturated (in the context of organic chemistry) A molecule that contains one or more double covalent bonds.

The **alkenes** are a homologous series of **unsaturated** alkenes. They are unsaturated because they contain a carbon-carbon double bond (C=C).

There is no alkene with one carbon atom because at least two carbon atoms are needed in alkenes as they all contain a C=C double bond.

Alkenes are much **more reactive** than alkanes, because they contain a C=C bond.

TIP

The reaction of alkenes with bromine water is the test to identify alkenes.

Reaction of alkenes with bromine (Br_2)

Bromine (Br_2) reacts readily with alkenes. The C=C double bond opens up and one bromine atom adds onto each of the carbon atoms in the double bond, as shown in Figure 29.6.

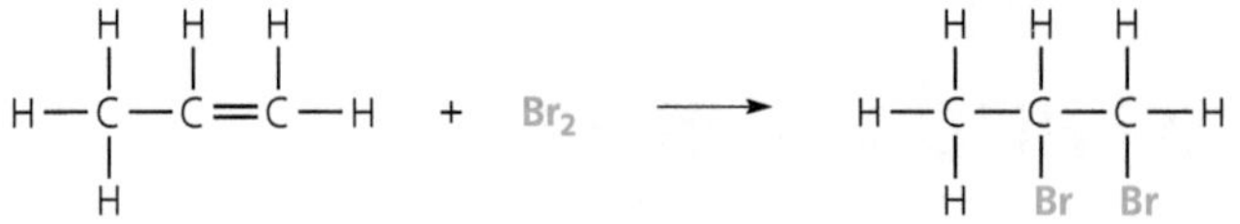

▶ **Figure 29.6** Reactions of alkenes with bromine.

The reaction with **bromine** is used to test for the presence of C=C double bonds in compounds. Bromine water is a solution of bromine in water and has an orange colour due to the dissolved Br_2 molecules. When the bromine water reacts with the C=C double bond, the Br_2 molecules are used up and so the reaction mixture goes colourless (Table 29.3).

TIP

When bromine water reacts with alkenes, it goes colourless. It is wrong to say that it goes clear. Clear does not mean the same thing as colourless. Colourless means there is no colour but clear means see-through. Bromine water itself is an example of a clear, coloured solution. It is a clear, orange solution.

Table 29.3 Using bromine to test for alkenes.

Type of compound	Saturated compound	Unsaturated compound
C=C double bonds	No C=C double bonds	Contains C=C double bond(s)
Reaction with bromine water	Stays orange 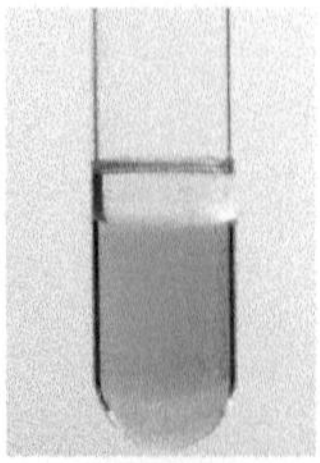	Orange → colourless 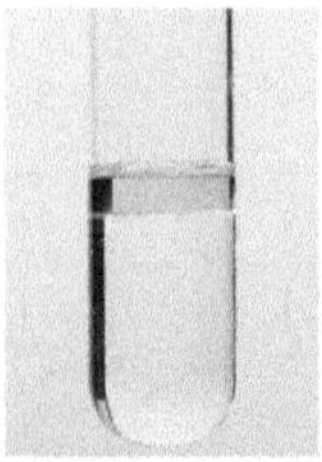

KEY TERM

Polymer A long chain molecule in which lots of small molecules (monomers) are joined together.

Alkenes are used to produce **polymers**. For example, many subunits (monomers) of ethene can be linked together to form polyethene (polythene).

Chapter review questions

1   Hexane is a *saturated hydrocarbon*. Hexene is an *unsaturated* hydrocarbon.

   a) Define the three words in italics.

   b) These two compounds can be distinguished using bromine water. What would happen to orange bromine water if it was added to:

   i)   hexane?

   ii) hexene?

2   a) Crude oil is split into fractions by fractional distillation. Explain how this process separates crude oil into fractions.

   b) Fuel oil is one of the fractions produced from crude oil. It is made of long alkane molecules.

   i)   Give two reasons why long alkanes are in less demand as fuels than short alkanes.

   ii) Name the process used to produce short alkanes from long alkanes.

   iii) Describe one way in which this process is done.

   iv) What other type of compound is produced in this process besides short alkanes?

   v)   Balance the following equation for this reaction that takes place in this process.

   $C_{18}H_{38} \rightarrow C_{10}H_{22} + C_3H_6 + C_2H_4$

Practice questions

1 Which one of the following molecular formulae represents a compound which is a member of the same homologous series as C_2H_6? [1 mark]

A C_2H_4

B C_3H_6

C C_4H_8

D C_4H_{10}

2 Crude oil is the source of hydrocarbons such as alkanes.

a) i) What is crude oil? [1 mark]

ii) How are alkanes obtained from crude oil? [1 mark]

iii) Name two fuels which are obtained from crude oil. [2 marks]

b) Octane, C_8H_{18}, is an alkane which is a constituent of petrol.

i) Octane is a saturated hydrocarbon. What is meant by the terms 'saturated' and 'hydrocarbon'? [2 marks]

ii) What is the general formula for the alkanes? [1 mark]

c) Alkenes, such as propene, can be obtained from larger alkanes such as octane.

i) What name is given to the process of forming alkenes from large alkanes? [1 mark]

ii) Write an equation for the formation of propene and an alkane from octane. [2 marks]

iii) Describe a chemical test for an unsaturated hydrocarbon such as propene. [2 marks]

iv) Write an equation for the reaction involved in the test described in part (iii). [1 mark]

3 Fractional distillation of crude oil is used to produce hydrocarbon fuels. The diesel fraction contains heptadecane molecules which have the formula $C_{17}H_{36}$.

a) State which has the higher boiling point, heptadecane or decane ($C_{10}H_{22}$). [1 mark]

b) Cracking of long chain hydrocarbons produces shorter chain hydrocarbons. State two different conditions which could be used for cracking. [2 marks]

4 a) Explain the respective roles of evaporation and condensation in fractional distillation. [2 marks]

b) State how the size of a hydrocarbon affects:

● boiling point

● viscosity

● flammability [3 marks]

5 a) Ethane has a molecular formula C_2H_6. Draw the displayed structure of ethane. [1 mark]

b) The displayed structure of propane is shown.

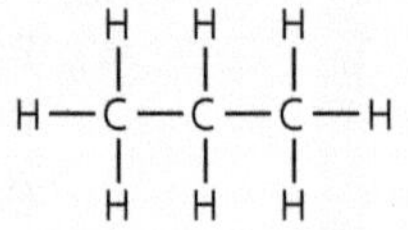

What is the molecular formula of propane?

Working scientifically: Assessing risk in science experiments

Whenever experiments and investigations are carried out in the laboratory, you need to decide if the experiment is safe by carrying out a risk assessment. A risk assessment is a judgement of how likely it is that someone might come to harm if a planned action is carried out and how these risks could be reduced.

A good risk assessment includes:

1 A list of all the hazards in the experiment.

2 A list of the risks that the hazards could cause.

3 Suitable control measures you could take which will reduce or prevent the risk.

For chemicals there should be a COSHH hazard warning sign on the container.

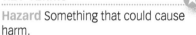

KEY TERMS

Hazard Something that could cause harm.

Risk An action involving a hazard that might result in danger.

Dangerous to the environment	Toxic	Gas under pressure
Corrosive	Explosive	Flammable
Caution – used for less serious health hazards like skin irritation	Oxidising	Longer term health hazards such as carcinogenicity

▲ Figure 29.7

TIP

An oxidising substance does not burn. It should be kept away from flammable substances though as it will provide the oxygen allowing the flammable substance to burn fiercely.

The hazards for each chemical can be found by looking up the CLEAPSS student safety sheets or hazcards. These should be recorded in your risk assessment; for example, 'when using pure ethanol it should be labelled **HIGHLY FLAMMABLE (Hazcard 60)**.' Hazcards will tell you about control measures and list when a fume cupboard needs to be used, when to wear gloves and the suitable concentrations and volumes for a safe experiment.

Table 29.4 Some examples of hazards, risks and control measures.

Hazard	Risk	Control measure
Concentrated sulfuric acid	Corrosive	Handle with gloves; use small volumes; wear eye protection
Ethanol	Flammable	Keep away from flames; no naked flames to be used; use water bath to heat
Bromine	Toxic	Handle with gloves; use dilute solutions; wear eye protection; use in fume cupboard
Cracked glassware	Could cause cuts	Check for cracks before use
Hot apparatus	Could cause burns	Allow apparatus to cool before touching
Bags and stools	Could be tripping hazard	Tuck stools under benches; leave bags in bag store
Chemicals being heated in test tubes	Chemicals could spit out of test tubes	Wear eye protection; point tubes away from people
Beaker being heated on a tripod and gauze	Could fall over spilling hot liquid	Keep apparatus away from edge of bench; work standing up; wear eye protection
Long hair	Could catch fire	Tie back long hair

Questions

1 Figure 29.8 shows the apparatus used to heat some hydrated copper sulfate in a boiling tube. Copy and complete the table to give a risk assessment for this experiment.

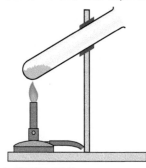

▲ **Figure 29.8**

Hazard	Risk	Control measure

2 To prepare a sample of an ester, 5 cm³ of ethanol and 5 cm³ of ethanoic acid were mixed in a test tube with 5 drops of concentrated sulfuric acid and warmed. Write a risk assessment for this experiment.

3 The instructions to react sodium with ethanol were:
'Place a piece of sodium in a test tube of ethanol and observe the reaction.' Rewrite these experimental instructions in view of health and safety precautions.

4 What hazards and risks are associated with carrying out the electrolysis of a solution in the laboratory?

30 Chemical analysis

Blood is a mixture of many substances. It is very important that medics can identify which substances are in blood and how much of each substance is present. Analysis is a key area of chemistry and there are many tests and techniques that can be used to identify, measure and test the purity of substances.

Specification coverage

This chapter covers specification points 5.8.1.1 to 5.8.2.4 and is called Chemical analysis.

It covers purity, formulations, chromatography and the identification of common gases.

Previously you could have learnt:

› **A pure substance is a single element or compound.**

› **An element is a substance containing one type of atom only.**

› **A compound is a substance containing atoms of different elements joined together.**

› **While a pure substance is melting, freezing, boiling or condensing, the temperature remains the same.**

› **Coloured dyes can be analysed and separated by paper chromatography.**

› **There are simple tests for the gases oxygen, hydrogen and carbon dioxide.**

› **Ionic substances are compounds with metals combined with non-metals.**

Test yourself on prior knowledge:

1 Which of the following are pure substances?
air, sea water, tap water, mineral water, diamond, a strawberry, carbon dioxide

2 Here is a list of some substances: Ag, S, H_2O, CH_3COOH, CO, Co, Br_2, SiO_2
   a) Which of these substances are elements?
   b) Which of these substances are compounds?

3 Which of the following are ionic substances?
hydrogen sulfide (H_2S), carbon dioxide (CO_2), aluminium oxide (Al_2O_3), calcium bromide ($CaBr_2$), sodium nitrate ($NaNO_3$), ethanol (C_2H_6O)

Purity, formulations and chromatography

○ Pure substances

What is a pure substance?

In everyday language, a pure substance is regarded as a natural substance that has had nothing added to it. For example, 'pure orange juice' is considered to be the juice taken from oranges with nothing else, such as colourings or sweeteners, added; 'pure soap' is considered to be soap with nothing else, such as perfumes, added; milk may be regarded as a pure substance because it is taken straight from the cow and nothing else is added. However, a scientist uses the word 'pure' in a different way and would not consider the orange juice, soap or milk to be pure substances.

A **pure substance** is a single element or compound. For example:

● Diamond (C) is a pure substance because it contains only carbon atoms.

● Oxygen (O_2) is a pure substance because it contains only oxygen molecules.

● Glucose ($C_6H_{12}O_6$) is a pure substance because it contains only glucose molecules.

KEY TERM

Pure substance A single element or compound that is not mixed with any other substance.

▲ **Figure 30.1** A scientist would not consider this orange juice to be 'pure'.

A mixture contains more than one substance. For example:

- Orange juice is a mixture of water molecules, citric acid molecules, vitamin C molecules, glucose molecules, etc. (Figure 30.1).
- Soap is a mixture of several salts made from different fatty acids.
- Milk is a mixture of several substances including water, animal fats, emulsifiers, minerals, etc.
- Mineral water is a mixture because it contains water molecules, calcium ions, magnesium ions, nitrate ions and many other ions.
- Air is a mixture because it contains nitrogen molecules, oxygen molecules, argon atoms, etc.

Melting and boiling points of pure substances and mixtures

Pure substances melt and boil at specific temperatures. For example, water has a boiling point of 100°C and a melting point of 0°C. While a pure substance changes state, the temperature remains constant at these values. For example, while pure water is boiling, the temperature stays at 100°C. When pure water is freezing, the temperature stays at 0°C.

However, mixtures change state over a range of temperatures. Some everyday examples of this are shown in Figure 30.2.

> **TIP** ✓
>
> The melting point of a substance is the same temperature as the freezing point. Melting is a solid turning into a liquid. Freezing is the opposite, a liquid turning into a solid. For example, the melting point of water is the same as the freezing point, both being 0°C.

The car radiator contains a mixture of antifreeze and water so that the mixture freezes below 0°C.

Salt is put on roads to create a mixture with water that freezes below 0°C.

Petrol is a mixture of hydrocarbons that boil over a range of temperatures from about 60 to 100°C.

► **Figure 30.2** Some very useful mixtures.

- **Salt water** freezes at a range of temperatures below 0°C, for example between –5°C and –10°C depending how much salt is dissolved. Salt is used to stop ice forming on roads in winter. Grit which contains salt is put on roads to prevent water freezing if the temperature drops below 0°C. Salt water boils at temperatures above 100°C, for example between 101°C and 103°C depending how much salt is dissolved.

- **Antifreeze** is mixed with water in car radiators to stop the water freezing in cold weather. The melting point of antifreeze is −13°C and that of water is 0°C. The mixture typically has a melting point range between −30°C to −40°C depending how much antifreeze is used.
- **Petrol** is a mixture of hydrocarbons and boils over a range of temperatures from about 60°C to 100°C. Each individual substance in petrol has its own specific boiling point, but the mixture boils over a range of temperatures.

◯ Formulations

A formulation is a **mixture** that has been designed as a useful product. It is made by mixing together several different substances in **carefully measured quantities** to ensure the product has the required properties. Some examples are listed in Table 30.1.

Table 30.1 Examples of formulations.

Product	Comments
Alloys	• Alloys are specific mixtures of metals with other elements (e.g. steel is a mixture of iron and carbon; brass is a mixture of copper and zinc) • Alloys are harder than pure metals • There are many different alloys • Each alloy is designed to have the specific properties required for its use
Fertilisers	• Most fertilisers contain specific mixtures of different substances (e.g. many fertilisers contain ammonium nitrate, phosphorus oxide and potassium oxide) • There are many different types of fertiliser with different amounts of different substances in them that are suitable for different plants and/or different soil types
Fuels	• Many fuels are specific mixtures (e.g. petrol and diesel are complex and carefully controlled mixtures of hydrocarbons designed to burn well and power a car engine)
Medicines	• Many medicines are specific mixtures of substances (e.g. aspirin tablets contain several other substances besides aspirin, including corn starch which is there to bind the tablet together; Calpol contains paracetamol in malitol liquid so that the medicine can be taken off a spoon)
Cleaning agents	• Cleaning agents are specific mixtures of many different substances (e.g. some dishwasher tablets contain detergents, alkalis, bleaches, rinse aid, etc.; some toilet cleaners contain bleaches, alkalis, detergents, etc.; some oven cleaner sprays contain sodium hydroxide to react with dirt, butane as a propellant, etc.)
Foods	• Many foods are very specific mixtures (e.g. margarine is a mixture of vegetable oils, water, emulsifiers, salt, etc.; tomato ketchup is a mixture of tomatoes, vinegar, sweeteners, spices, salt, etc.; vegetable soup is a mixture of water, vegetables, spices, etc.)
Paints	• Paints are mixtures whose contents include a solvent (water for emulsion, water or hydrocarbons for gloss), pigments (for the colour) and binder (to hold the pigments in place when the paint dries)

TIP

A formulation is a special type of mixture: the substances in the formulation are mixed in carefully measured (not random) quantities.

Test yourself

1 Some people may say that a bottle of mineral water is pure water.
   a) Explain why some people may regard this as being pure.
   b) Explain why a scientist would not say that it is pure.
2 A sample of water was found to freeze between −2°C and −4°C. Was this water pure? Explain your answer.
3 Some students made samples of aspirin in the laboratory. The melting point of aspirin is 136°C. Which student(s) made pure aspirin? Explain your answer.

Student	A	B	C	D	E	F
Melting point in °C	137–138	130–132	136	139–144	136–137	131–136

4 a) What is a formulation?
   b) Give four examples of formulations.

▲ **Figure 30.3** Many everyday products are formulations.

Show you can...

Table 30.2 shows the melting points and boiling points of some metallic elements and alloys named by letters A to C.

Table 30.2

	Melting point in °C	Boiling point in °C
A	−34	356
B	420	913
C	1425–1540	2530–2545

a) What state does substance A exist in at room temperature (20°C) and pressure?
b) What state does substance B exist in at room temperature (20°C) and pressure?
c) What state does substance C exist in at room temperature (20°C) and pressure?
d) What is an alloy?
e) Classify the substances in the table as elements or alloys and explain your answer.

Chromatography

What happens in paper chromatography?

Chromatography is a very useful technique that can be used to separate and analyse mixtures. There are several types of chromatography including paper, thin layer, column and gas chromatography.

Paper chromatography is often used to analyse coloured substances. In paper chromatography:

1 A pencil line is drawn on the chromatography paper near the bottom. Pencil is used as it will not dissolve in the solvent.

2 Small amounts of the substances being analysed are placed in spots on the pencil line.

3 The paper is hung in a beaker of the solvent. The pencil line and spots must be above the level of the solvent so that the spots do not dissolve into the solvent in the beaker.

4 Over the next few minutes, the solvent soaks into and moves up the paper.

5 When the solvent is near the top, the paper is taken out of the solvent and the level that the solvent reached is marked. This is known as the solvent front.

6 The paper is left to dry.

A pure substance can only ever produce one spot in chromatography, whatever solvent is used. Mixtures will usually produce more than one spot, one for each substance in the mixture (Figure 30.4). It is possible two substances in a mixture will move the same distance and appear as a single spot in some solvents.

In the example in Figure 30.4, Y is a mixture of two substances as it produces two spots. We can see by comparing the spots to those for A, B and C, that Y is a mixture of substances A and C.

TIP
Chromatography can be used to separate mixtures and also to identify substances.

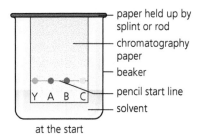
at the start

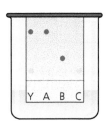
after the solvent has soaked up the paper

▲ **Figure 30.4** Paper chromatography.

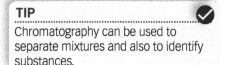

How chromatography works

In each type of chromatography there is a mobile phase and a stationary phase. In paper chromatography, the **stationary phase** is the piece of chromatography paper and the **mobile phase** is a solvent.

How far each substance moves depends on its relative attraction to the paper and the solvent. Substances that have a stronger attraction to the solvent move quickly and travel a long way up the paper. Substances that have a stronger attraction to the paper move slowly and only travel a short distance up the paper. In Figure 30.5 it can be seen that substance Q moves the greatest distance and so has a stronger attraction to the solvent than the paper, while substance P moves the shortest distance and so has a stronger attraction to the paper than Q has.

R_f values

The ratio of the distance a substance moves to the distance moved by the solvent is called the R_f value (Figure 30.5). The distance is measured to the centre of the spot.

$$R_f = \frac{\text{distance moved by substance}}{\text{distance moved by solvent}}$$

The R_f value for a substance is always the same in the same solvent. However, substances will have different R_f values in different solvents. R_f values can be used to identify substances.

> **TIP**
> You must mark the position of the solvent front when you stop the investigation, because the position may be difficult or impossible to identify when the solvent evaporates.

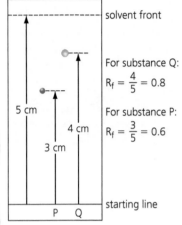

For substance Q:
$R_f = \frac{4}{5} = 0.8$

For substance P:
$R_f = \frac{3}{5} = 0.6$

▲ **Figure 30.5** Finding R_f values.

> **TIP**
> An R_f value cannot be more than 1.

Test yourself

5 Food colouring S was analysed by paper chromatography and compared with substances 1–6 (Figure 30.6). The samples were placed on a pencil line on a piece of chromatography paper which was hung in a solvent.

a) Explain why the starting line was drawn in pencil.

b) Explain why the level of the solvent must be below the level of the pencil line.

c) Which of the substances 1 to 6 appear to be pure substances? How can you tell?

d) How many substances are in colouring S?

e) Which substances are in colouring S?

f) Calculate the R_f values for each spot in colouring S. Give your answer to 2 significant figures.

g) Which colour spot moved slowest during the experiment?

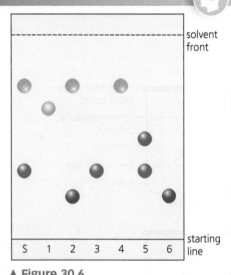

▲ **Figure 30.6**

Test yourself

6 A substance was analysed by chromatography in two different solvents (see Figure 30.7).

   a) Calculate the R_f value in each solvent. Give your answer to 2 significant figures.

   b) Explain why the substance moved further in solvent 2 than solvent 1.

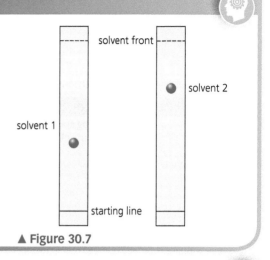

▲ Figure 30.7

Investigating how paper chromatography can be used to separate and tell the difference between coloured substances

When freezing water for ice rinks, it is important that the water does not contain too many dissolved metal ions as this makes poor quality ice for skating. To determine if a sample of water contained some dissolved metal ions a chromatography experiment was carried out using the water sample (A) and known metal ion solutions B (containing copper(ɪɪ) ions), C (containing iron(ɪɪ) ions) and D (containing iron(ɪɪɪ) ions). The method used is shown below.

Method

1 Draw a base line on the chromatography paper 1.5 cm from the bottom using a pencil.

2 Place a concentrated drop of each solution to be tested on the base line.

3 Place the chromatography paper into a chromatography tank containing water at a depth of 1 cm.

4 After the water soaks up the paper, dry the paper and spray with sodium hydroxide solution.

Figure 30.8 shows spots that appeared on the paper.

Questions

1 Why is it necessary to draw a pencil line as a base line 1.5 cm from the bottom?

2 Why is the water solvent at a depth of only 1 cm?

3 How is a concentrated drop of each solution added to the chromatography paper?

4 The chromatogram obtained is shown here. Name two metal ions which are present in the water sample A.

5 Write the formula of the compound formed when spot C reacts with sodium hydroxide.

6 Write an equation which is used to calculate R_f value and calculate the R_f value for spot B.

7 If the experiment was repeated using a different solvent would the R_f value be the same?

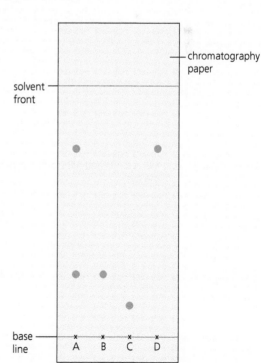

▲ Figure 30.8 Paper chromatography.

Identification of common gases

The gases oxygen, hydrogen, carbon dioxide and chlorine are common gases. There is a simple chemical test to identify each one (Table 30.3).

Table 30.3 Identifying some common gases.

Gas	Test		Result
Oxygen, O_2	Insert a glowing splint (one that has just been blown out) into a tube of the gas		The splint relights
Hydrogen, H_2	Insert a burning splint into a tube of the gas		The hydrogen burns rapidly with a squeaky pop sound
Carbon dioxide, CO_2	The gas is shaken with or bubbled through limewater (a solution of calcium hydroxide in water)		Limewater goes milky (cloudy)
Chlorine, Cl_2	Insert damp (red or blue) litmus paper into the gas		Litmus paper is bleached (it loses its colour and turns white)

Test yourself

7   Magnesium reacts with hydrochloric acid to produce a gas. This gas was tested and found to give a squeaky pop with a burning splint.
    a)   Identify the gas produced.
    b)   Write a word equation for the reaction between magnesium and hydrochloric acid.
    c)   Write a balanced equation for the reaction between magnesium and hydrochloric acid.

8   Copper carbonate reacts with nitric acid to produce a gas. This gas was tested and found to turn a solution of limewater cloudy.
    a)   Identify the gas produced.
    b)   What is limewater?
    c)   Write a word equation for the reaction between copper carbonate and nitric acid.
    d)   Write a balanced equation for the reaction between copper carbonate and nitric acid.

9   The electrolysis of a solution of sodium chloride was carried out. A gas was produced at the positive electrode that turned damp blue litmus paper white. Another gas was produced at the negative electrode which gave a squeaky pop with a burning splint.
    a)   Identify the gas produced at the positive electrode.
    b)   Identify the gas produced at the negative electrode.
    c)   Write a half equation for the process at the positive electrode and state whether this is reduction or oxidation.
    d)   Write a half equation for the process at the negative electrode and state whether this is reduction or oxidation.

10 The electrolysis of a solution of magnesium sulfate was carried out. A gas was produced at the positive electrode that was found to relight a glowing splint. Another gas was produced at the negative electrode which gave a squeaky pop with a burning splint.
    a)   Identify the gas produced at the positive electrode.
    b)   Identify the gas produced at the negative electrode.
    c)   Write a half equation for the process at the positive electrode and state whether this is reduction or oxidation.
    d)   Write a half equation for the process at the negative electrode and state whether this is reduction or oxidation.

Show you can...

A student carried out four tests on a gas X and recorded the results in the Table 30.4.

Table 30.4

Test	Observations
Test 1: damp red litmus paper	Stays red
Test 2: bubble into limewater	Stays colourless
Test 3: lighted splint	Squeaky pop
Test 4: bubble into bromine water	Stays orange

a) What does the result of test 1 tell you about gas X?
b) What does the result of test 2 tell you about gas X?
c) Use the results of tests 3 and 4 to decide if gas X is hydrogen or ethene.

Chapter review questions

1 Figure 30.9 shows some particles in air.

a) Is air a pure substance? Explain your answer.

b) Give the formula of each of the elements shown in the diagram.

c) Give the formula of each of the compounds shown in the diagram.

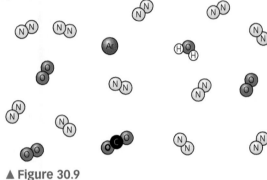

▲ Figure 30.9

2 Identify gases A, B, C and D using the results in the Table 30.5.

Table 30.5

Test	Gas A	Gas B	Gas C	Gas D
Effect on damp red litmus paper	No effect	Paper goes white	No effect	No effect
Effect on burning splint	Flame continues to burn	Flame goes out	Squeaky pop	Flame goes out
Effect on limewater	No effect	No effect	No effect	Goes cloudy
Effect on glowing splint	Flame relights	No effect	No effect	No effect

3 A food colouring K was analysed by paper chromatography. A spot of K was placed on a pencil line on a piece of chromatography paper along with spots of dyes W, X, Y and Z. The paper was placed in a beaker of water and left for a few minutes. The paper was removed from the beaker, the level of the solvent marked and the paper left to dry. The results are shown in Figure 30.10.

a) How many substances are in food colouring K?

b) Which of the substances W, X, Y and Z are in food colouring K?

c) Which of the substances W, X, Y and Z are likely to be:

i) pure substances

ii) mixtures?

d) Why was the line drawn in pencil?

e) Calculate the R_f value for the spots in food colouring K. Give your answer to 2 significant figures.

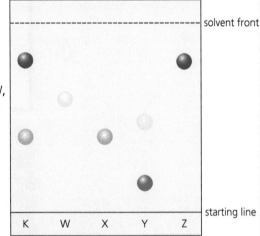

▲ Figure 30.10

f) What would happen to the R_f value of each spot if a different solvent was used?

4 Sun cream is used by many people to protect themselves from harmful UV rays from the Sun. Sun cream is a formulation. What is meant by the term 'formulation'?

Practice questions

1 A solid is thought to be pure benzoic acid. Which of the following is the best way to test its purity? [1 mark]

A determine the density

B determine the pH

C determine the melting point

2 Some ice was taken out of a freezer and allowed to warm up. Table 30.6 shows the temperature over the next 15 minutes as the ice melted.

Table 30.6

Time in min	0	1	2	3	4	5	6	7	8	9	10	11	12	13	14	15
Temperature in °C	−20	−15	−10	−5	−4	−3	−2	3	8	13	18	20	20	20	20	20

a) Plot a graph to show the temperature change over time. [3 marks]

b) Was the ice pure? Give the reason for your answer. [1 mark]

c) Why do you think the temperature stopped rising when it reached 20 °C? [1 mark]

3 a) In chromatography, what is meant by the term 'solvent front'? [1 mark]

b) Give the reason why an R_f value cannot exceed 1. [1 mark]

c) A particular mixture is thought to contain three different dyes, X, Y and Z. Explain how you could use chromatography to show that the mixture does contain dyes X, Y and Z. [3 marks]

Working scientifically:
Recording observations

Making and recording observations is an important skill in chemistry. Qualitative observations are what we see and smell during reactions. Important types of observations in chemistry and notes on how to record these are shown in Table 30.7.

Table 30.7 Making and recording observations.

Type of observation	Notes on recording observations	Examples	
Colour change	Always state the colour of the solution before the reaction and after	Bubbling an alkene into bromine water – the colour change is orange solution to colourless solution	
Bubbles produced	If a gas is produced, then bubbles are often observed in the liquid	When sodium carbonate reacts with an acid, the observation is bubbles	
Temperature change	Often in a reaction the temperature changes – in an exothermic reaction it increases and in an endothermic reaction it decreases	When acids react with alkalis, the temperature increases. When acids react with sodium hydrogencarbonate, the temperature decreases. When water reacts with anhydrous copper sulfate, the temperature increases	
Precipitate produced	When two solutions mix, an insoluble precipitate may form. Ensure you use the word precipitate in your observation. Also state the colour of the precipitate and the colour of the solution before adding the reagent	When barium chloride solution is added to a solution containing sulfate ions a white precipitate is formed from the colourless solution	
Solubility of solids	When a spatula of a solid is added to water, the solid may dissolve. If it dissolves, the observation is often that the solid dissolves to form a solution. Make sure you state the colour of the solution formed. If it does not dissolve, state that it is insoluble	Copper(ɪɪ) sulfate crystals dissolve in water to produce a blue solution	
Solubility of liquids	When a liquid is added to water always record if it is miscible or immiscible with water	Ethanol and water are miscible (left-hand tube). The liquids in the right-hand test tube are immiscible	

TIP
Remember that clear means see-through and does not mean colourless.

TIP
Effervescence is another term to describe the production of bubbles in a reaction.

TIP
Note that writing that carbon dioxide is formed is not an observation.

Questions

1 In a laboratory experiment a student added some magnesium metal to some copper sulfate solution and recorded the observation that 'the magnesium became covered in copper and magnesium sulfate was formed.'

   a) Write a word equation for the reaction occurring.

   b) State and explain if the observations given by the student are correct.

2 Some hydrochloric acid was placed in a conical flask and a spatula of calcium carbonate powder was added.

   a) Write a balanced symbol equation for this reaction.

   b) State the observations which occurred.

3 Copy and complete the table.

Reaction	Observations
ethanoic acid + sodium carbonate	
potassium iodide solution + silver nitrate solution	
bromine water + alkene	
hydrochloric acid + magnesium	
acidified barium chloride + sulfuric acid	

31 Chemistry of the atmosphere

There is air all around us even though we cannot see it. This air contains oxygen which is vital for life, but there was no oxygen in the atmosphere when the Earth was young. This chapter looks at what air is and how the atmosphere has changed over time. It also looks at how the oxygen in the air is used to burn fuels, and how this can produce gases that pollute our atmosphere.

Specification coverage

This chapter covers specification points 5.9.1.1 to 5.9.3.2 and is called Chemistry of the atmosphere.

It covers the composition and evolution of the Earth's atmosphere, carbon dioxide and methane as greenhouse gases, and common atmospheric pollutants and their sources.

The composition and evolution of the Earth's atmosphere

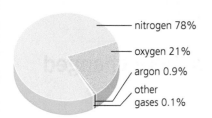

▲ **Figure 31.1** The gases in dry air.

nitrogen 78%
oxygen 21%
argon 0.9%
other gases 0.1%

○ The atmosphere today

The Earth is about **4.6 billion (4600 million) years old**. For the last **200 million years** or so, the composition of the air has been much the same. Most of the air is nitrogen (about four-fifths) and oxygen (about one-fifth), with small amounts of noble gases (mainly argon) and carbon dioxide (Figure 31.1). There is also a small amount of water vapour but this does vary with weather conditions.

○ The early atmosphere of the Earth

Scientists believe that the Earth's atmosphere was very different when it was young. For example, there is evidence that there was little or no oxygen in the atmosphere. However, there is much uncertainty about the Earth when it was young and there are many theories about what the atmosphere was like and how and when it changed.

Venus and Mars have very similar atmospheres to each other. The atmospheres on both planets are mainly carbon dioxide with some nitrogen but with little or no oxygen (Figure 31.2). Given that the Earth is between Venus and Mars, it may have been that the early atmosphere of the Earth was like that on Venus and Mars. It is thought that the evolution of life on Earth changed our atmosphere, but this did not happen on Venus or Mars.

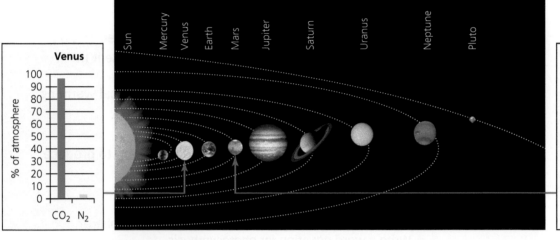

▲ Figure 31.2 The atmospheres of Venus and Mars.

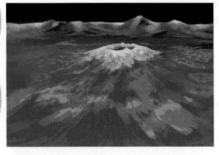

▲ Figure 31.3 There is intense volcanic activity on Venus today.

▲ Figure 31.4 The oceans may have been formed as water vapour released by volcanoes cooled and condensed.

> **TIP** ✓
> Make sure you can explain how volcanic activity may have changed the early Earth's atmosphere.

One theory about the Earth is that it was very hot when it was formed and there was intense **volcanic activity** on the planet. Volcanoes release gases from the inside of the planet and the gases in the early atmosphere could have been released from volcanoes in this way (Figure 31.3).

The volcanoes on Earth may have released:

- carbon dioxide (CO_2) – which is likely to have been the main gas in the early atmosphere
- nitrogen (N_2) – this may have gradually built up in the atmosphere over time
- methane (CH_4) – probably in small amounts only
- ammonia (NH_3) – probably in small amounts only
- water vapour (H_2O) – this could have condensed and formed the oceans as the Earth cooled down (Figure 31.4).

○ How the Earth's atmosphere changed

Where did the oxygen come from?

Life evolved on Earth and it is thought that the oxygen in the atmosphere was formed by **photosynthesis** in living creatures. During photosynthesis, carbon dioxide reacts with water to form glucose and oxygen.

$$\text{carbon dioxide} + \text{water} \rightarrow \text{glucose} + \text{oxygen}$$
$$6CO_2 + 6H_2O \rightarrow C_6H_{12}O_6 + 6O_2$$

▲ **Figure 31.5** Algae.

Algae were present on Earth from about 2.7 billion years ago. Soon after this, oxygen appeared in the atmosphere.

Over the next billion years, more complex life forms including plants evolved. As the number of organisms that photosynthesised increased on Earth, the amount of oxygen in the air gradually increased and reached a point where animals could evolve.

Where did the carbon dioxide go?

There were two main ways in which carbon dioxide was removed from the atmosphere (Figure 31.6). Much of the carbon dioxide was:

● **dissolved in the oceans** or
● **used in photosynthesis**.

Some of the carbon dioxide that dissolved in the oceans reacted to form insoluble carbonate compounds, such as calcium carbonate, that became sediment. Some produced compounds in the oceans that became part of the shells and skeletons of sea creatures. When these creatures died, their shells and skeletons fell into sediment on the sea floor. Over millions of years, all this sediment became sedimentary rocks such as limestone, which is mainly calcium carbonate.

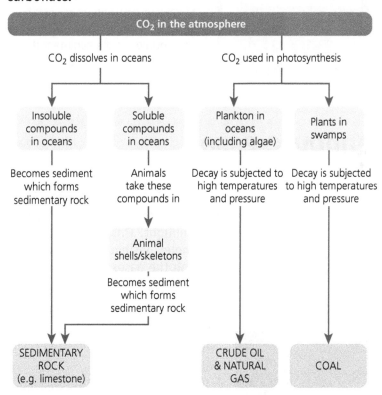

▶ **Figure 31.6** Carbon dioxide has been removed from the atmosphere in many ways.

A lot of carbon dioxide was absorbed by algae and plants for photosynthesis. In the oceans, as algae and other plankton died, their remains were buried in the mud on the sea floor and compressed. Over millions of years this formed crude oil and natural gas that was trapped under rocks. In swamps, the remains of plants were buried and compressed and formed coal, which is a sedimentary rock, over millions of years.

Show you can...

Table 31.1 shows suggested percentages of different gases found in the atmosphere from early times to present the day.

Table 31.1

	% in the early atmosphere	% in today's atmosphere
Carbon dioxide	85	0.04
Nitrogen	Very small	78.08
Oxygen	0	20.95

Compare the percentage of each of the following gases in the early atmosphere with the percentage in today's atmosphere. Explain the changes in the percentages of each gas.

a) carbon dioxide
b) nitrogen
c) oxygen

Test yourself

1 a) Draw a bar chart to show the main two gases that make up about 99% of the present atmosphere on Earth.
  b) What gases make up the other 1% of the present atmosphere?
2 a) Where may the gases that formed the Earth's early atmosphere have come from?
  b) How may the oceans have been formed?
3 a) What process produced the oxygen in the atmosphere?
  b) Write a word equation for this process.
  c) Write a balanced equation for this process.
4 A lot of the carbon from the carbon dioxide in the Earth's early atmosphere has ended up in the fossil fuels coal, natural gas and crude oil.
  a) Outline how coal was formed.
  b) Outline how natural gas and crude oil were formed.
5 Limestone is a sedimentary rock that is mainly calcium carbonate. Give two sources of the sediment that forms limestone.

Greenhouse gases

○ What are greenhouse gases?

The Sun gives off radiation. Some of this reaches the Earth and we call this sunlight. This sunlight contains electromagnetic radiation in the ultraviolet, visible and infrared regions. Much of this radiation passes through the gases in the atmosphere and reaches the surface of the Earth. This provides energy and warmth to the planet.

The Earth also gives off radiation. As the Earth is cooler than the Sun, the radiation given off by the Earth has a longer wavelength than the Sun's radiation. The radiation given off by the Earth is in the infrared region.

Some of the gases in the atmosphere absorb infrared radiation given off by the Earth but do not absorb the radiation coming in from the Sun. These are known as greenhouse gases and are important for keeping the Earth warm (Figure 31.7). Many occur naturally and include **water vapour** (H_2O), **carbon dioxide** (CO_2) and **methane** (CH_4) (Figure 31.8).

KEY TERM

Greenhouse gas A gas that absorbs long wavelength infrared radiation given off by the Earth but does not absorb the Sun's radiation.

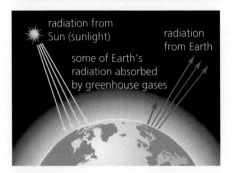

▲ **Figure 31.7** Radiation entering and leaving the Earth's atmosphere.

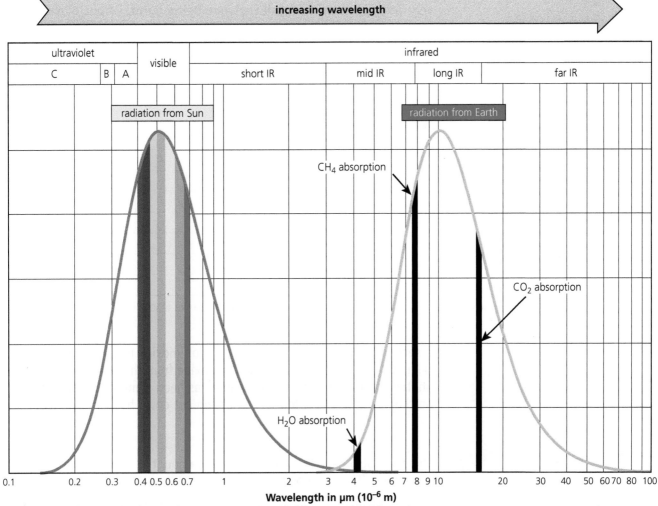

▲ **Figure 31.8** Wavelengths of radiation entering the Earth's atmosphere from the Sun and being given off by the Earth.

⃝ The increase in the amount of greenhouse gases in the atmosphere

The amount of water vapour in the atmosphere varies with weather conditions, and human activities do not have much impact. However, human activities are leading to an increased amount of other greenhouse gases, such as carbon dioxide and methane, in the atmosphere.

The amount of carbon dioxide in the atmosphere

There is a lot of accurate data about the amount of carbon dioxide in the air. It shows clearly that the amount is steadily increasing (Figure 31.9).

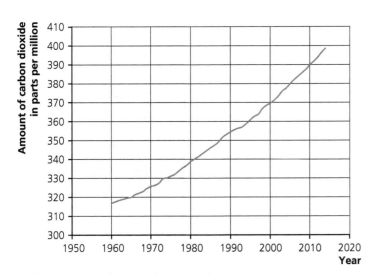

▲ **Figure 31.9** CO_2 levels in the atmosphere.

One reason why the amount of carbon dioxide is increasing is that very large quantities of **fossil fuels are being burnt**. Carbon dioxide is produced when fossil fuels are burnt. Figure 31.10 shows how the amount of fossil fuels being burnt has increased in recent years. This increase matches the increase in the amount of carbon dioxide in the atmosphere.

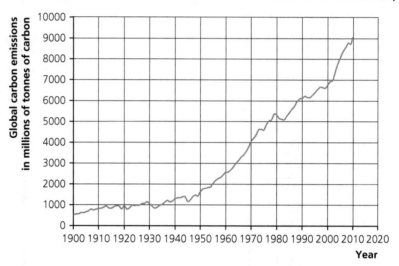

► **Figure 31.10** Global carbon emissions from burning fossil fuels.

A second reason for the increase in carbon dioxide is deforestation (Figure 31.11). Trees remove carbon dioxide from the air for photosynthesis. In recent years, there has been significant **deforestation** on the planet as many forests have been cut down but not replaced.

The amount of methane in the atmosphere

There are also data that show that the amount of methane in the air is increasing (Figure 31.12).

Animal farming produces a lot of methane. Methane is a product of normal digestion in animals. Most of this methane comes from cattle due to the nature of their digestive system (Figure 31.13). It is also produced when manure decomposes. The amount of animal farming has increased in recent years and this is responsible for some of the increase in methane in the atmosphere.

▲ **Figure 31.11** There has been large scale deforestation in the last century.

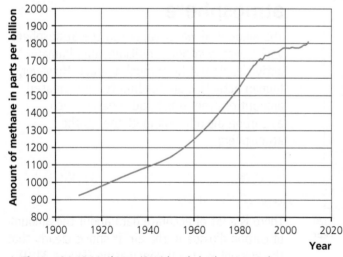

▲ **Figure 31.12** Methane (CH_4) levels in the atmosphere.

▲ **Figure 31.13** Cattle produce a lot of methane.

▲ **Figure 31.14** Methane gas is produced by rubbish decomposing in landfill sites.

A large amount of waste is buried in the ground in **landfill sites** (Figure 31.14). Underground the waste decomposes and this also produces a lot of methane gas.

Test yourself

6 What is different, in terms of wavelength, about the radiation the Earth receives from the Sun and the radiation the Earth gives off?

7 a) What does a greenhouse gas do?

 b) Name three important greenhouse gases in the Earth's atmosphere.

8 a) What is happening to the amount of carbon dioxide released into the atmosphere?

 b) Give two reasons for this change.

9 a) What is happening to the amount of methane released into the atmosphere?

 b) Give two reasons for this change.

Show you can...

Look again at Figure 31.9. Which of the following statements can be made using the information from this graph alone?

A) The concentration of carbon dioxide has risen steadily since 1960.

B) There has been no decrease in carbon dioxide concentration over the past 50 years.

C) The concentration of carbon dioxide has increased very rapidly over the last 20 years.

D) The carbon dioxide concentration increases as the amount of fossil fuels burnt increases.

○ Global warming and climate change

Scientific opinions on global warming

Scientists publish their research in scientific journals for other scientists and the general public to see (Figure 31.15). Before it can be published, this work is peer reviewed. This means that it is examined by other scientists who are experts in the same area of science to check that it is scientifically valid.

Based on published peer-reviewed evidence, many scientists believe that the Earth will become warmer due to the human activities that are increasing the amount of greenhouse gases in the atmosphere. They believe that an increase in the temperature at the Earth's surface, known as global warming, will result in global climate change. However, models that predict the climate are simplifications as there are so many factors that affect the climate and some of these factors are not fully understood. This means that it is very difficult to model the climate fully and there is some uncertainty in this area. For example, there are some scientists who do not believe that the increasing levels of greenhouse gases will make the Earth warmer.

Due to the uncertainty of climate models and significant public interest, there is much speculation and a wide range of opinions

KEY TERMS

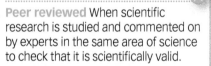

Peer reviewed When scientific research is studied and commented on by experts in the same area of science to check that it is scientifically valid.

Global warming An increase in the temperature at the Earth's surface

▲ **Figure 31.15** Scientists publish their work in scientific journals.

are presented in the media on climate change. Some of this may be biased, for example being put forward by industries that use fossil fuels or that promote renewable energy sources. Also, some may be based on incomplete evidence, perhaps using evidence from one scientific study on its own that does not agree with the majority of published studies.

Global warming

There is a great deal of evidence from many sources that the temperature at the Earth's surface has increased in recent years, possibly by about 0.5°C in the last 30 years (Figure 31.16). Most scientists believe that this is largely due to the increased amount of greenhouse gases in the atmosphere.

Climate change and its effects

It is very difficult to predict the effects of the increasing surface temperature of the Earth. Some possible effects are given below.

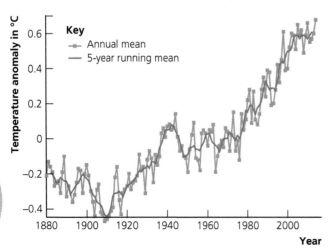

▲ **Figure 31.16** Global mean surface temperature change.

Sea level rise

As the Earth becomes warmer, sea levels are expected to rise. This may be due to polar ice caps and glaciers melting (Figure 31.17), leading to an increased volume of water in the oceans. Another significant factor is that as the water in the oceans becomes warmer, it expands and so has a greater volume. It is difficult to predict by how much sea levels may rise, but there are estimates it could rise by as much as a metre over this century.

(a)

(b)

▲ **Figure 31.17** Glaciers (ice rivers) are in retreat as they are melting (Mer de Glace, Chamonix, France). Photographs taken in (a) 1875 and (b) 2013.

Height above sea level (m)

▲ **Figure 31.18** Some areas at sea level are at risk of flooding.

As sea levels rise, some parts of the world are at risk of flooding (Figure 31.18). For example, there are some Pacific islands that could be completely submerged. Large areas of countries such as Vietnam, Bangladesh and the Netherlands could be flooded. Some major cities such as London and New York could be at risk in tidal surges if sea levels are higher.

Increased sea levels will also increase coastal erosion. This is where the sea wears away the rock along the coast. Figure 31.19 shows the effect of coastal erosion in Sussex.

Storms

As the Earth warms and the climate changes, there may be more frequent and severe storms (Figure 31.20). However, this is uncertain and there is mixed evidence that this is happening. For example, while there is evidence that there is more rain falling in the most severe storms, there is mixed evidence as to whether the number of hurricanes is increasing or not.

▲ **Figure 31.19** Coastal erosion (Birling Gap, Sussex). Photographs taken in a) 1906 and b) 2012.

▲ **Figure 31.20** Hurricane Katrina led to massive flooding in New Orleans, USA, in 2005.

Rainfall

It is expected that global warming will affect the amount, timing and distribution of rainfall. This could increase rainfall in some parts of the world but decrease it in others. For example there has been more rainfall in northern Europe in recent years but there has been less rainfall in regions near the equator. The number of heavy rainfall events has increased in recent years, but so has the number of droughts, especially near the equator.

Temperature and water stress

As the temperature and climate changes, it is expected that this will have an impact on the availability of fresh water. This causes water stress, which is a shortage of fresh water and affects all living creatures (Figure 31.21). There have been more droughts in areas near the equator in recent years and this leads to water shortages for humans but also for plants and animals affecting food chains.

> **KEY TERM**
>
> **Water stress** A shortage of fresh water.

▲ **Figure 31.21** Water stress has impact on humans, other animals and crops.

Wildlife

Changes in the climate affect wildlife in many ways. For example, plants may flower earlier, birds may lay eggs earlier and animals may come out of hibernation earlier. Some species may migrate further north as temperatures rise. For example, there are species of dragonfly now being seen in the south of the UK that had only been seen in warmer countries previously. The number of polar bears is expected to fall significantly as the temperature increases as there are fewer suitable places for them to live (Figure 31.22).

Food production

The weather has a very significant impact on crop production. Changes in the climate may affect the capacity of some regions to produce food due to changes in rainfall patterns, drought, flooding, higher temperatures and the type and number of pests in the region.

▲ **Figure 31.22** Climate change could reduce the population of polar bears.

Test yourself

10 Before scientists can publish their research it is peer reviewed. What is peer review in this context?
11 a) What is global warming?
   b) What do most scientists believe is causing global warming?
12 One impact of global warming could be rising sea levels.
   a) Give two reasons why global warming would cause sea levels to rise.
   b) Why are rising sea levels a problem?
   c) Rising sea levels could increase coastal erosion. What is coastal erosion?
13 a) What is water stress?
   b) Why might global warming cause water stress?
14 a) In what ways could global warming affect global climate?
   b) Why could global warming affect food production?
15 In what ways could global warming affect wildlife?

○ Carbon footprint

What is a carbon footprint?

The amount of carbon dioxide and other greenhouse gases in the air is increasing and likely to be causing global warming and climate change. Therefore it is important that we monitor the amount of greenhouse gases released by different activities. A carbon footprint is defined as the amount of carbon dioxide and other greenhouse gases given out over the full life cycle of a product, service or event.

For example, the carbon footprint of a plastic bag made from poly(ethene) could include carbon dioxide released as fuels are burnt to provide energy/thermal energy to:

- drill for crude oil
- heat and vaporise crude oil in fractional distillation
- heat alkanes to crack them to make ethene
- provide high temperatures and pressure to polymerise ethene to make poly(ethene)
- transport plastic bags to shops
- transport used plastic bags to waste disposal sites.

In addition, if the bag is burnt in an incinerator after disposal then more carbon dioxide is produced.

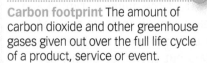

KEY TERM

Carbon footprint The amount of carbon dioxide and other greenhouse gases given out over the full life cycle of a product, service or event.

Ways to reduce a carbon footprint

In order to prevent the amount of greenhouse gases in the atmosphere increasing further, it is important to reduce the carbon footprint of products, services and events. Some ways in which this can be done are shown in Table 31.2.

Table 31.2 Reducing a carbon footprint.

Method of reducing carbon footprint	Comments
Increase the use of alternative energy sources	• We can reduce the use of fossil fuels and increase the use of **alternative energy sources** such as wind turbines, solar cells and nuclear power. These alternative sources have much lower carbon footprints than burning fossil fuels
Energy conservation	• There are many things we can do in a more **energy efficient** way that conserves energy. Some examples include: – using more energy efficient engines in cars and other vehicles – increasing insulation in homes – using more energy efficient boilers in heating systems – using low energy (LED) light bulbs instead of filament or halogen light bulbs – using better detergents so clothes can be washed at lower temperatures – switching off electrical devices instead of leaving them on standby
Carbon capture and storage (Figure 31.23)	• A significant amount of the carbon dioxide produced by burning fossil fuels in a power station can be **captured** to prevent it being released into the atmosphere. This carbon dioxide can then be moved by pipeline and stored deep underground in rocks. This is a new technology that is being developed but some plants are in operation and it is thought it could reduce carbon dioxide emissions from power stations by up to 90%
Carbon off-setting	• **Carbon off-setting** is where something is put in place to reduce the emission of greenhouse gases to compensate for the release of greenhouse gases elsewhere. Some examples of carbon off-sets include: – setting up renewable energy projects, e.g. wind farms – setting up plants to prevent the emission into the atmosphere of methane from landfill sites – planting more trees
Carbon taxes and licences	• Governments can put a **carbon tax** on the use of fossil fuels. For example, in many countries there is a large tax on petrol and diesel for cars. As this makes the fuel very expensive for drivers, car-makers are producing more fuel efficient engines so that less fuel is needed to meet the demands of car owners. Carbon licences can also be used whereby the amount of greenhouse gases a company can release is limited
Carbon neutral fuels (Figure 31.24)	• There are fuels and processes that are **carbon neutral**. This means that their use results in zero net release of greenhouse gases to the atmosphere. For example, ethanol made from fermentation of crops can be used as a fuel – it is carbon neutral because it releases the same amount of carbon dioxide when it burns as the crops it was made from took in for photosynthesis as they grew

Problems of reducing carbon footprints

There is a need to reduce carbon footprints, but there are some problems that make the reduction in greenhouse gas emissions difficult (Table 31.3).

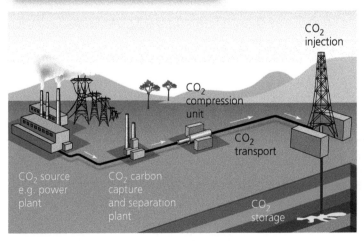

▲ **Figure 31.23** Carbon capture and storage.

▲ **Figure 31.24** 85% of the fuel in E85 fuel is ethanol, which is a carbon neutral fuel.

Table 31.3 Problems with reducing a carbon footprint.

Problem	Explanation
Scientific disagreement	• **Scientists do not all agree** about the causes and consequences of global climate change. For example, there are some scientists who do not believe that the extra greenhouse gases in the air are responsible for global warming. This makes efforts to reduce greenhouse gas emissions more difficult as some scientists say there is no need to do so
Economic considerations	• Many of the methods of reducing carbon footprints are **expensive**. For example, carbon capture and storage significantly increases the cost of generating electricity by burning fossil fuels. There is some reluctance to put these costly methods into place while there is some disagreement about the cause of climate change. The extra cost would be particularly difficult for poor countries in the developing world
Incomplete international co-operation	• The **Kyoto Protocol** is a major international treaty agreed by many nations to reduce emissions of greenhouse gases. This has been successful in some countries with, for example, greenhouse gas emissions in the UK falling by about 20% since 1990 (Figure 31.25). • However, not all countries are keen to do this. For example: – Some countries did not sign up. – Developing nations such as China and India (countries where over a third of the world's population live) are exempt. – The USA, which emits more greenhouse gases than any other nation, has not agreed to this which means it is not required to cut its emissions. – Canada has withdrawn from the agreement. – Greenhouse gas emissions have actually increased in some countries that agreed to the treaty. • The original Kyoto Protocol expired in 2012, but the **Paris Agreement** was negotiated by many countries in 2015 to cut greenhouse gas emissions. It is hoped that enough countries will sign up to this so that it has an impact.
Lack of public information and education	• Many people are confused and **do not understand** the issues of greenhouse gases and climate change. This problem is even greater in developing countries. The better educated and informed people are about the issues, the more likely that effective action will be taken to reduce the problem
Lifestyle changes	• The world's **energy consumption is rapidly increasing**. This is partly because there are more people on the planet but also due to our greater demand for energy as our standard of living increases and lifestyle changes. For example: – We are using more electrical devices (e.g. smart phones, tablets, dishwashers, tumble driers). – We expect to be warmer in our homes than in the past. – We are travelling more (e.g. longer commutes to work, travelling further on holiday, making more overseas business visits).

> **TIP** ✅
> To reduce carbon footprints mainly involves reducing emissions of carbon dioxide and methane.

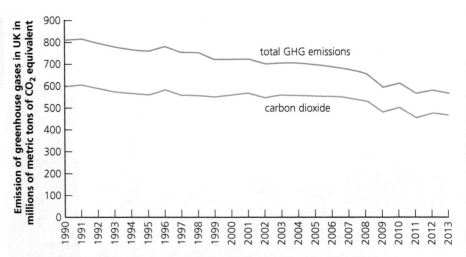

▲ **Figure 31.25** Greenhouse gas (GHG) emissions in the UK 1990–2013.

Test yourself

16 What is meant by the term 'carbon footprint'?

17 There are many ways to reduce our carbon footprint.

    a) One way is to use more alternative energy sources. Give some examples of these.

    b) Another way is to be more energy efficient. Give some examples of ways in which this can be done.

    c) Carbon capture and storage could reduce our carbon footprint a great deal. What happens in this process?

    d) Carbon off-setting can be used. Give some ways in which this can be done.

    e) Explain how carbon taxes on fuel for cars has led to cars with more fuel efficient engines being produced.

    f) Biodiesel is a carbon neutral fuel made from crops such as rapeseed. Explain what it means to be carbon neutral.

18 Explain why disagreement among scientists about the causes of global warming may limit steps to lower greenhouse gas emissions.

19 Worldwide energy consumption is increasing rapidly. Why might our improving lifestyles contribute to this increase?

Common atmospheric pollutants and their sources

◯ What is formed when fuels burn?

Most fuels, including coal, oil and gas, contain carbon and/or hydrogen and may also contain some sulfur.

When fuels burn, they are oxidised which means that they react with oxygen and the atoms in their molecules combine with the oxygen. For example, when fuels containing hydrogen burn, the hydrogen combines with oxygen to form water vapour. When fuels containing sulfur burn, the sulfur combines with oxygen to form sulfur dioxide.

For fuels containing carbon, complete combustion takes place in a good supply of oxygen to form carbon dioxide. In a poor supply of oxygen, incomplete combustion takes place which gives carbon monoxide and/or soot (a form of carbon).

Table 31.4 gives some equations which are examples of reactions taking place when fuels burn.

Table 31.4 The combustion of fuels.

Substance in fuel	Products from combustion	Combustion reactions
Hydrogen, H_2 (an important fuel)		$2H_2 + O_2 \longrightarrow 2H_2O$ hydrogen + oxygen $\longrightarrow$ water
Methane, CH_4 (the main compound in natural gas)		$CH_4 + 2O_2 \longrightarrow CO_2 + 2H_2O$ (complete combustion) methane + oxygen $\longrightarrow$ carbon dioxide + water $2CH_4 + 3O_2 \longrightarrow 2CO + 4H_2O$ (incomplete combustion) methane + oxygen $\longrightarrow$ carbon monoxide + water $CH_4 + O_2 \longrightarrow C + 2H_2O$ (incomplete combustion) methane + oxygen $\longrightarrow$ carbon + water
Ethanol, C_2H_6O (a common alcohol used as a fuel)		$C_2H_6O + 3O_2 \longrightarrow 2CO_2 + 3H_2O$ (complete combustion) ethanol + oxygen $\longrightarrow$ carbon dioxide + water $C_2H_6O + 2O_2 \longrightarrow 2CO + 3H_2O$ (incomplete combustion) ethanol + oxygen $\longrightarrow$ carbon monoxide + water $C_2H_6O + O_2 \longrightarrow 2C + 3H_2O$ (incomplete combustion) ethanol + oxygen $\longrightarrow$ carbon + water
Carbon disulfide, CS_2 (small amounts are found in oil)		$CS_2 + 3O_2 \longrightarrow CO_2 + 2SO_2$ (complete combustion) carbon disulfide + oxygen $\longrightarrow$ carbon dioxide + sulfur dioxide $2CS_2 + 5O_2 \longrightarrow 2CO + 4SO_2$ (incomplete combustion) carbon disulfide + oxygen $\longrightarrow$ carbon monoxide + sulfur dioxide $CS_2 + 2O_2 \longrightarrow C + 2SO_2$ (incomplete combustion) carbon disulfide + oxygen $\longrightarrow$ carbon + sulfur dioxide

○ Pollution from burning fuels

The burning of fuels is a major source of pollutants in the atmosphere. Table 31.5 shows some of these, how they are formed and how the problems they cause could be reduced.

Table 31.5 Combustion of fuels and pollution.

Product of combustion	How it is formed	Problems it causes	How the problem could be reduced
Carbon dioxide (CO_2)	Most fuels contain carbon and this produces carbon dioxide on complete combustion	It is a greenhouse gas and is thought to be causing global warming and climate change	Reduce the use of fossil fuels by using alternative energy sources and improving energy efficiency
Carbon monoxide (CO)	Most fuels contain carbon and this can produce carbon monoxide on incomplete combustion	It is toxic because it combines with haemoglobin in blood **reducing** the ability of the blood to carry **oxygen** It is also **colourless** and has no smell so can be hard to detect	Ensure there is a good supply of oxygen when fuel is burnt
Soot (C) (Figure 31.26)	Most fuels contain carbon and this can produce soot (carbon particulates) on incomplete combustion	These carbon particulates pollute the air and blacken buildings They also cause **global dimming** which reduces the amount of sunlight reaching the Earth's surface They can damage the lungs and cause **health problems** to humans	Ensure there is a good supply of oxygen when fuel is burnt
Water vapour (H_2O)	Many fuels contain hydrogen and this produces water vapour on combustion	It is not a problem as although it is a greenhouse gas, human activity does not appear to affect the amount of water vapour in the air as weather patterns prevent the amount of water vapour changing over time	It is not a problem
Sulfur dioxide (SO_2)	Many fuels contain some sulfur and this produces sulfur dioxide on combustion	This is a cause of **acid rain** which damages plants and stonework It also causes **respiratory problems** for humans	Remove the sulfur dioxide from the waste gases before they reach the atmosphere
Nitrogen oxides (NO and NO_2)	Nitrogen in the air reacts with oxygen in air at very high temperatures when fuels are burnt	This is a cause of **acid rain** It also causes **respiratory problems** in humans	Remove the nitrogen oxides from the waste gases before they reach the atmosphere (this is done using catalytic converters in cars) Adjust the way the fuel is burnt to reduce the amount produced
Unburnt hydrocarbons	When hydrocarbon fuels are burnt (e.g. petrol, diesel), some molecules may not burn and so unburnt hydrocarbons are released into the atmosphere	This wastes fuel and they are greenhouse gases	Ensure efficient burning so that all the fuel is burnt

▲ **Figure 31.26** Tower Bridge has been cleaned to remove layers of soot caused by air pollution.

Show you can...

Burning petrol in car engines can lead to the formation of acid rain.

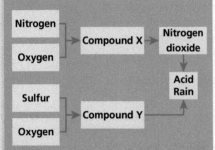

▲ Figure 31.27

a) What is the formula of compound X?
b) What is the formula of nitrogen dioxide?
c) What is the source of the nitrogen and oxygen that react together to form compound X?
d) What does nitrogen dioxide react with to form acid rain?
e) What is the formula of compound Y?
f) State two ways in which acid rain can be prevented.

Test yourself

20 Petrol is an important fuel used in cars.
  a) One of the chemicals in petrol is isooctane (C_8H_{18}). Under what conditions can the following gases be formed when isooctane burns?
    i) carbon dioxide
    ii) carbon monoxide
    iii) soot
    iv) water vapour
  b) Explain why some sulfur dioxide may be formed when petrol burns.
  c) Explain why some nitrogen monoxide and nitrogen dioxide may be formed when petrol burns.
  d) What two problems do the pollutants sulfur dioxide and nitrogen oxides cause in the air?
  e) Some unburnt isooctane can be given off in exhaust fumes from cars. Why might this be a problem?
21 a) Carbon monoxide can be formed on the incomplete combustion of methane (CH_4).
    i) What is incomplete combustion?
    ii) Why is carbon monoxide toxic?
    iii) Why is it difficult to detect carbon monoxide?
  b) Soot can be formed on the incomplete combustion of methane (CH_4).
    i) What is soot?
    ii) What problems can soot cause?

Investigating the products of combustion of a fuel

Practical

The experiment shown in Figure 31.28 was carried out to investigate the products of the combustion of a hydrocarbon candle wax. In the experiment the products of the combustion were drawn through the apparatus by the vacuum pump.

Questions

1 Explain why a colourless liquid forms at the bottom of the U tube.

2 The colourless liquid in the U tube changes white anhydrous copper sulfate to blue hydrated copper sulfate. Suggest the name of the colourless liquid.

3 Liquid A was clear and colourless at the start of the experiment and it slowly became cloudy. Identify liquid A and explain why it became cloudy.

4 From the results of this experiment identify two of the combustion products formed when the hydrocarbon burns completely.

5 Some solid black particles were found on the apparatus at the end of the experiment. Suggest the identity of these particles and suggest how they are formed.

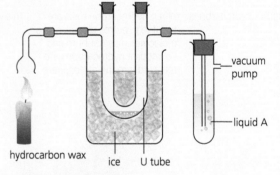

▲ Figure 31.28

6 The experiment was repeated using a different fuel, which contained sulfur as an impurity.
  a) Name the gas formed when sulfur is burned in air.
  b) This gas is acidic. Suggest what happens to it as it passes through the apparatus.
  c) Suggest why burning sulfur-containing fuels is a problem for the environment.

Chapter review questions

1 Copy the paragraph below and complete the gaps.

The Earth is thought to be 4.6 _____ years old. There is _____ evidence about the atmosphere of the Earth when it was young. One theory is that there was a lot of _____ activity when the Earth was young and this gave out the gases that formed the early atmosphere.

2 a) The air today contains 78% nitrogen, 21% oxygen and 1% other gases. Draw a bar chart to show the composition of air today.

   b) What happened on Earth that led to the formation of oxygen in the atmosphere?

3 a) What is a greenhouse gas?

   b) Name three greenhouse gases and state where each one comes from.

   c) The amount of greenhouse gases in the air is increasing. What impact is this having on the Earth's temperature?

   d) State three ways in which this change in temperature could affect the global climate.

4 Some scientists believe that the gases that formed the atmosphere of the Earth when it was young came from volcanoes. Table 31.6 shows the percentages of gases given off by a present-day volcano.

Table 31.6

Gas	Percentage
Water vapour	38
Carbon dioxide	50
Sulfur dioxide	10
Other gases	2

   a) Draw a pie chart to show the gases given off by the present-day volcano.

   b) Some scientists believe that the oceans were formed from water vapour given off by ancient volcanoes. How could the oceans have been formed from this water vapour?

5 Scientists believe that the main gas in the atmosphere when it was young was carbon dioxide. The evolution of life is thought to have changed the Earth's atmosphere when algae and plants evolved that photosynthesise.

   a) Which two planets in the solar system have atmospheres that are mainly carbon dioxide?

   b) Write a word equation and balanced equation for photosynthesis.

6 Copy the paragraph below and complete the gaps.

Limestone and coal are both _____ rocks that contain carbon. Limestone is mainly the compound _____. Limestone was formed from the skeletons and _____ of marine organisms that died and fell into sediment.

7 When petrol burns in an engine, several pollutants are formed. For each one, explain how it is formed and one problem it can cause.

   a) sulfur dioxide (SO_2) b) nitrogen oxides (NO and NO_2)

   c) carbon dioxide (CO_2) d) carbon monoxide (CO)

   e) soot (C)

8 a) Methane (CH_4) is a greenhouse gas. Describe two ways in which human activities are increasing the amount of methane in the atmosphere.

b) Carbon dioxide (CO_2) is a greenhouse gas. Describe two ways in which human activities are increasing the amount of carbon dioxide in the atmosphere.

9 There is a lot of work taking place to reduce the carbon footprint of many processes.

a) What does the term 'carbon footprint' mean?

b) Describe three ways in which carbon footprints can be reduced.

c) Describe three problems that can prevent attempts to reduce carbon footprints.

10 Write a balanced equation for the complete combustion of each of the following fuels:

a) hydrogen, H_2

b) heptane, C_7H_{16}

c) methanol, CH_3OH

11 There is a small amount of diethyl sulfide ($C_4H_{10}S$) found in the petrol fraction of crude oil used as a fuel in cars. Write a balanced equation for the complete combustion of diethyl sulfide.

12 The Sun's radiation is in the ultraviolet, visible and near infrared region of the electromagnetic spectrum. The radiation given out by the Earth is in the infrared region.

a) How does the wavelength of the Earth's radiation compare to the Sun's?

b) Why are carbon dioxide, methane and water vapour greenhouse gases?

Practice questions

1 Which one of the following makes up about 20% of the Earth's atmosphere? [1 mark]

A carbon dioxide B helium

C nitrogen D oxygen

2 A company has decided to reduce the amount of sulfur dioxide from the waste gases which it emits from its factories. Which one of the following substances in solution would most effectively remove the sulfur dioxide from the waste gases? [1 mark]

A calcium chloride B calcium hydroxide

C sodium chloride D sulfuric acid

3 Mars is often called the red planet due to the presence of haematite, which contains iron(III) oxide, on its surface. A recent study of the Huygens Crater on Mars has also shown the presence of iron(III) hydroxide and calcium carbonate.

a) i) Calcium carbonate and iron(III) hydroxide undergo thermal decomposition. What is meant by the term 'thermal decomposition'? [2 marks]

ii) Write a balanced equation for the thermal decomposition of iron(III) hydroxide into iron(III) oxide and water. [2 marks]

b) 'Atmosphere' is the term used to describe the collection of gases that surround a planet. The suggested composition of the atmosphere of Mars is shown in Table 31.7. Compare the composition of the Earth's atmosphere today, with that of the planet Mars. [4 marks]

Table 31.7

Gas	Composition in %
Carbon dioxide	95.0
Nitrogen	3.0
Noble gases	1.6
Oxygen	Trace
Methane	Trace

c) Changes in the atmosphere of the Earth occurred slowly over millions of years due to photosynthesis and other processes.

i) The equation for the production of glucose ($C_6H_{12}O_6$) in photosynthesis is shown below. Balance this equation. [1 mark]

$$\_\_CO_2 + \_\_H_2O \rightarrow C_6H_{12}O_6 + \_\_O_2$$

 ii) Photosynthesis is an endothermic reaction. Explain why this reaction is endothermic in terms of breaking and making bonds. [3 marks]

iii) State one other process which caused the composition of the Earth's atmosphere to change. [1 mark]

4 Table 31.8 shows some information about different fossil fuels.

Table 31.8

Fossil fuel	Appearance	% Carbon	% Moisture	Thermal energy released when burnt
Natural gas	Colourless gas	75	0	High
Crude oil	Black liquid	80	0	High
Peat	Soft brown fibrous solid	20	70	Low
Lignite	Soft dark solid	55	35	Medium
Coal	Hard shiny black solid	90	0	High

a) Using Table 31.8 state the relationship between thermal energy released when burnt and:

i) % carbon [1 mark]

ii) % moisture [1 mark]

b) State three differences between lignite and coal. [3 marks]

c) Name two products formed when coal burns incompletely. [2 marks]

d) If the coal contained sulfur, name one other product that may be formed during combustion, and state two environmental problems which may occur. [3 marks]

e) Some lignite is burnt in a test tube and a gas given off. The gas can be condensed to give a liquid by passing the gas into a cooled test tube. The test tube is cooled in a beaker of iced water.

i) Draw a labelled diagram to illustrate the assembled apparatus described above. [3 marks]

ii) Name two products of the complete combustion of lignite. [1 mark]

5 a) There is much international concern that an increase in atmospheric concentrations of carbon dioxide and methane may lead to global warming and climate change.

i) Give the formulae of carbon dioxide and methane. [2 marks]

ii) What type of radiation is absorbed by carbon dioxide and methane molecules? [1 mark]

iii) What is meant by global warming? [1 mark]

iv) Suggest why scientists are more concerned about carbon dioxide as a greenhouse gas than methane. [1 mark]

b) Scientists in the Antarctic have measured the concentration of carbon dioxide in air bubbles in the ice there, and this has allowed them to estimate the atmospheric concentration of carbon dioxide over many thousands of years, as shown in Figure 31.29. Figure 31.30 shows the change in average temperature of the Earth's surface over the same time period as Figure 31.29.

Do the graphs show that an increase in atmospheric carbon dioxide concentration increases global warming? Explain your answer. [2 marks]

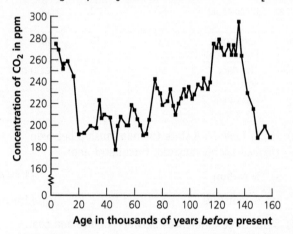

▲ Figure 31.29 Changes in atmospheric carbon dioxide over time.

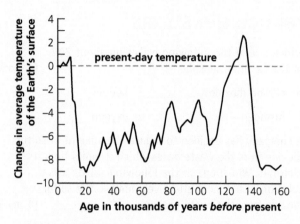

▲ Figure 31.30 Changes in temperature at the Earth's surface over time.

c) Chemists are developing methods to minimise climate change. State two methods which have been used to minimise climate change by removing carbon dioxide. [2 marks]

d) Governments have set targets to reduce emissions of carbon dioxide. One way of doing this is by issuing carbon licences to businesses. Describe how carbon licences could be used to reduce carbon dioxide emissions. [1 mark]

Working scientifically:
Communicating scientific conclusions

When drawing conclusions from data about burning fossil fuels and air pollution, it is important to realise that a number of different influences can affect the data and these need to be taken into consideration. There are many things which a scientist must do to produce valid results and conclusions.

▶ A scientist may take a set of measurements or make some observations and draw conclusions from them. However to make valid conclusions the experiment must be repeated and similar data obtained.

▶ Getting other people to successfully repeat the experiment or achieving the same results with a slightly different method means that the results are reproducible. This increases the validity of the conclusion.

▶ Scientists publish their results in journals. Before their work can be published, it is checked and evaluated by other experts in the same field. This process is known as peer review. Scientists also present their ideas at conferences where other scientists can review and evaluate their ideas.

Cold fusion

Nuclear fusion in a test tube developed by Utah professors – *Financial Times*

Scientists pursue endless power source – *The Times*

Scientists claim techniques to control nuclear fusion – *Boston Globe*

If you had read the newspaper headlines shown above in March 1989, you might have believed that the world's energy generation problems could be over for ever. Two scientists, Stanley Pons and Martin Fleischmann, called a press conference to announce that they had successfully caused nuclear fusion to happen using simple laboratory electrolysis apparatus at room temperature. This became known as 'cold fusion'.

Nuclear fusion involves two atomic nuclei overcoming the repulsion between them and joining together to make a larger nucleus. It is only known to happen at very high temperatures. This fusion releases huge amounts of energy and is the energy source in stars (Figure 31.31) where it takes place at about 15 million °C. Reports that fusion was occurring at room temperature received worldwide media attention and raised hopes of a cheap and abundant source of energy.

Pons and Fleischmann did not follow the normal route of publishing scientific work. They announced their findings at a press conference before

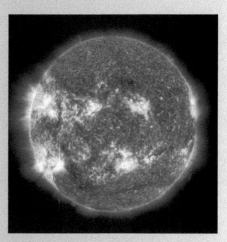

▲ **Figure 31.31** Fusion reactions take place at high temperature inside the Sun.

other scientists had reviewed them or tried to repeat their work. It was only after their announcements that details of the experimental method were published. Other scientists tried to replicate the work of Fleischmann and Pons but they were unable to produce the same results. Soon, the work of Fleischmann and Pons was discredited and the hope for cold fusion gone.

Questions

1 What was unusual about the way in which Pons and Fleischmann announced their results?

2 What procedures should scientists follow before presenting their findings?

3 The concentration of particulates in the air of a town centre was measured over several days. The number of patients seeking medical treatment for asthma was recorded over the same days. The results were plotted in the graph shown (Figure 31.32).

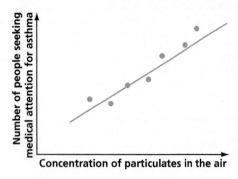

▲ Figure 31.32

a) What correlation does the graph show?

b) Evaluate the validity of the data.

c) The results of this experiment were presented to a newspaper. The headline claims:

Asthma is caused by particulates in the air.

i) How much confidence can be placed in the newspaper claim?

ii) The particulates that pollute the air are made of carbon. Explain how carbon particulates get into the air.

32 Using the Earth's resources

GLASS METAL PAPER

The planet we live on provides all the raw materials for making the products we use in everyday life. Many of these products are made from resources such as rocks, crude oil, the oceans and the air. Some of these resources will run out if we keep using them as we do. This chapter looks at some ways in which we make use of the Earth's resources and how we can use them in a more sustainable way.

Specification coverage

This chapter covers specification points 5.10.1.1 to 5.10.2.2 and is called Using the Earth's resources.

It covers using the Earth's resources and obtaining potable water, life cycle assessment and recycling.

Using the Earth's resources

○ The Earth's resources

Everything that we need for life is provided by the Earth's resources. These resources include:

- rocks in the ground
- fuels such as coal, oil and natural gas found underground
- plants and animals (produced by agriculture)
- fresh water and sea water
- air
- sunlight
- wind.

Table 32.1 shows some examples of how we use these resources to provide energy and essential substances for everyday life.

Table 32.1 Using the Earth's resources.

What we need	Where it comes from
Oxygen	Air
Water	Rain
Food	Plants and animals
Clothes	Fibres made from chemicals in oil (e.g. nylon, polyesters) or from natural fibres from plants (e.g. cotton) or animals (e.g. wool)
Shelter / buildings	Building materials include: stone, sand, bricks made from clay, cement made from limestone and clay, timber from trees
Warmth	Burning fuels (e.g. coal, oil, gas, biofuels from plants)
Electricity	Generated using fuels (e.g. coal, oil, gas, biofuels from plants), wind, sunlight, waves, decay of radioactive substances, etc.
Fuel for transport	Burning fuels (e.g. oil, natural gas, biofuels from plants)
Medicines	Mainly made from chemicals found in oil and/or plants
Fertilisers	From plant and animal waste or made from nitrogen in the air, water, natural gas and minerals from the ground
Metals	Made from ores in the ground
Polymers (plastics)	Mainly made from chemicals found in oil

Many substances, such as metals, medicines, plastics and some clothing fibres, are made from natural resources by chemical reactions. In fact, chemistry is all about the production of useful substances from the Earth's natural resources.

As time goes on, chemistry is playing an important role in improving agricultural and industrial processes. For example the following are being made:

- new products with new uses (e.g. new medicines)
- new products to supplement natural ones (e.g. polyesters, nylon, Gore-Tex® and microfibre for clothing fibres to add to natural fibres such as wool, cotton and silk) and
- new products or improving processes (e.g. better fertilisers or more efficient processes to make them).

○ Sustainable development

Many of the Earth's resources are finite. This means that we cannot replace them once we have used them. For example, supplies of crude oil and many metal ores are limited and if we continue to use them as we do now we will run out. In contrast to this, some resources are renewable, which means we can replace them once we have used them. Biofuels, which are fuels made from plants, are good examples of renewable resources. Examples of biofuels include ethanol and biodiesel, which can be made from crops and so can be replaced once used.

Sustainable development is where we use resources to meet the needs of people today without preventing people in the future from meeting theirs. Table 32.2 shows some ways in which we can meet our needs today in more sustainable ways.

Table 32.2 Unsustainable and sustainable use of resources.

	Unsustainable way	Sustainable alternative
Metals	Throw away metals and then use metal ores to provide more metal	Recycle metals
Fuels for transport	Use fossil fuels (e.g. petrol and diesel from crude oil)	Use biofuels (e.g. biodiesel and ethanol made from crops)
Electricity generation	Use fossil fuels (e.g. coal and natural gas)	Use renewable energy sources (e.g. solar, wind, tidal) (Figure 32.1)

▲ **Figure 32.1** Using the wind to generate electricity is sustainable.

547

Test yourself

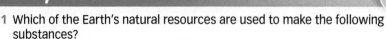

1 Which of the Earth's natural resources are used to make the following substances?
   a) metals
   b) plastics
   c) clothes
2 a) What is a finite resource?
   b) Name two finite resources.
   c) What is a renewable resource?
   d) Name two renewable resources.
3 a) What is sustainable development?
   b) The use of natural gas as a fuel for electricity generation is not sustainable. Suggest how we could generate electricity in a sustainable way.

○ Ways to reduce the use of resources

We use large amounts of metals, glass, building materials, clay ceramics and many plastics, but they are all produced from limited raw materials. Much of the energy used in their production comes from resources that are also limited, such as fossil fuels. Many of the raw materials used to make these products, such as rocks and ores, are obtained by mining or quarrying which also have environmental impact.

If we, as the end-users of a product, reuse or recycle products then we make processes more sustainable as it will reduce:

- the use of the Earth's limited resources
- energy consumption and the use of fuels
- the impact of manufacturing new materials on the environment
- the impact of waste disposal on the environment.

Reuse

Some products, such as glass bottles, can be reused. For example, most glass milk bottles are collected, washed and reused (Figure 32.2). Some glass beer bottles are also reused in a similar way.

Recycling

Many products cannot be reused, but the materials they are made of can be recycled. This usually involves melting the products and then remoulding or recasting the materials into new products. Some examples are shown in Table 32.3.

▲ Figure 32.2 Milk bottles are reused.

Table 32.3 Recycling and remoulding.

Materials	How they are recycled
Metals	1 Separated (e.g. iron/steel can be removed with a large magnet) 2 Melted 3 Recast/reformed into new metal products
Glass	1 Separated into different colours 2 Crushed 3 Melted 4 Remoulded into new glass products
Plastics	1 Separated into different types 2 Melted 3 Remoulded into new plastic products

Materials for some uses have to be of very high quality. For example, glass used for making pans for cooking and metals used for making tools must be of very high quality to have the required properties. For other uses, such as plastic for making rulers, the quality of the material is less important. In general, when using recycled products to produce materials, the better the separation of the materials when they are recycled, the higher the quality of the final material produced.

When metals, glass or plastics are recycled, the different types are separated. For example:

- iron/steel, aluminium and copper are common metals that are separated from other metals
- colourless, brown and green glass are separated
- plastics such as high-density poly(ethene), low density poly(ethene), PET and PVC are separated (Figure 32.3).

| PET | HDPE | PVC | LDPE | PP | PS |
| PET | high density poly(ethene) | PVC | low density poly(ethene) | poly(propene) | polystyrene |

▶ **Figure 32.3** Some symbols on plastic products to help separation when recycling.

A good example of recycling is the use of scrap steel to make new steel (Figure 32.4). Steel is made using mainly iron which is extracted from iron ore in a blast furnace. By adding scrap steel to some iron from a blast furnace, the amount of iron that needs to be extracted to make steel is reduced.

▲ **Figure 32.4** Some scrap steel can be used to make new steel.

○ **Life cycle assessment**

A life cycle assessment is carried out to assess the impact of a product on the environment throughout its life. This includes the **extraction of raw materials**, its **manufacture**, its **use** and its **disposal** at the end of its useful life. Factors that are considered include the

- use of and sustainability of raw materials (including those used for the packaging of the product)
- use of energy at all stages
- use of water at all stages
- production and disposal of waste products (including pollutants) at all stages
- transportation and distribution at all stages.

Simple life cycle assessments of the use of plastic and paper to make shopping bags are shown in Table 32.4 for comparison.

Table 32.4 Life cycle assessments for plastic bags and paper bags.

		Plastic shopping bags	Paper shopping bags
Raw materials	What they are	Crude oil	Trees
	Sustainability	Not sustainable – crude oil cannot be replaced	Sustainable – more trees can be planted (but take a long time to grow)
	Obtaining raw materials	Extracting crude oil uses lots of energy	Habitats are destroyed as trees are cut down
	Transporting raw materials	Transport of crude oil uses up fuels and causes some pollution; potential damage to environment from spillages	Transport of logs uses up fuels and causes some pollution
Manufacture of bag		Much energy used to separate crude oil into fractions, for cracking and for polymerisation	Uses a lot of water; uses some harmful chemicals (leaks would damage the environment)
Use of bags	Transport to where used	Transport of bags uses up fuels and causes some pollution	Transport of bags uses up fuels and causes some pollution
Disposal options	Landfill	Does not rot and so remains in ground for many years	Rots releasing greenhouse gases including methane
	Incinerator	Gives off greenhouse gas carbon dioxide when burnt	Gives off greenhouse gas carbon dioxide when burnt
	Recycled	Transport of bags uses up fuels and causes some pollution; melting of plastic uses energy	Transport of bags uses up fuels and causes some pollution; relatively easy to recycle

KEY TERM ⭐

Value judgement An assessment of a situation that may be subjective, based on a person's opinions and/or values.

In a life cycle assessment, the use of raw materials, energy and water plus the production of some waste can be quite easy to quantify. However, it can be difficult to quantify the effects of some pollutants. For example, it is hard to judge how much damage to the environment is caused by the release of the greenhouse gas methane from the rotting of paper bags in landfill sites. This means that a life cycle assessment involves some personal judgements to make a value judgement to decide whether use of a product is good or not. This means that the process is not completely objective and so judgements about the benefits or harm of the use of a product may be a matter of opinion.

Life cycle assessments can be produced that do not show all aspects of a product's manufacture, use and disposal. These could be misused, for example by the manufacturer of a product to support the use of their product.

Show you can...

Table 32.5 shows data about plastic and glass milk bottles.

Table 32.5

	Energy needed: non-reusable plastic milk bottle in MJ	Energy needed: reusable glass milk bottle in MJ
Manufacture	4.7	7.2
Washing, filling and delivering	2.2	2.5

a) How many times must the glass bottle be reused before there is an energy saving compared with the plastic milk bottle?
b) Create a life cycle assessment for the glass bottle and for the plastic bottle.

Test yourself

4 a) Glass bottles can be reused or recycled. What is the difference?
  b) What are the stages in the recycling of glass?
5 Describe how some scrap steel can be used in the manufacture of steel using iron from a blast furnace.
6 How does the way in which waste plastics are separated before melting down to make recycled plastic affect the usefulness of the recycled plastic formed?
7 What is a life cycle assessment?
8 Create a life cycle assessment for copper water pipes that will be recycled at the end of their useful life.
9 The manufacturer of a product included a life cycle assessment on their website. Why should we be cautious about this life cycle assessment?

The use of water

○ Producing potable water

Types of water

Water is essential for life and humans need water that is safe to drink (Figure 32.5). Water that is safe to drink is called potable water. Potable water is not pure water as it contains some dissolved substances, but these are at low levels that are safe. There should also be safe levels of microbes in potable water and ideally none.

Potable water and other types of water are described in Table 32.6.

Table 32.6 Different types of water.

Type	Description	Contents	
		Dissolved substances	Microbes
Pure water	Water that contains only water molecules and nothing else	✗	✗
Potable water	Water that is safe to drink	✓ low levels	✗ (or very low levels)
Fresh water	Water found in places such as lakes, rivers, the ice caps, glaciers and underground rocks and streams	✓ low levels	✓ (very low from some sources)
Ground water	Fresh water found in underground streams and in porous rocks (aquifers)	✓ low levels	✓
Sea water (salty)	Water in the seas and oceans	✓ high levels	✓
Waste water	Used water from homes, industry and agriculture	✓ high levels	✓

▲ **Figure 32.5** Humans need water that is safe to drink.

Water treatment

In the UK we produce our potable water from fresh water that comes from rain. This rain collects in lakes, rivers and reservoirs. It also collects in the ground in underground streams and in the pores of some rocks, which are called aquifers.

This fresh water can be treated to produce potable water. In some parts of the UK, fresh water comes mainly from rivers and reservoirs, but in other areas much water comes from ground water sources. There are many stages in making this water safe to drink, but two of the main stages are shown in Table 32.7.

Table 32.7 Making water safe to drink.

Stage	What it does	Details
Filtration	Removes solids	The water is passed through **filter beds**. These filter beds are usually made of sand. As the water flows through the sand, any solids in the water are removed.
Sterilisation	Kills microbes	The most common way is to use small amounts of **chlorine**. However, the microbes can also be killed by treating the water with **ozone** or by passing **ultraviolet** light through the water.

In some parts of the world, such as Saudi Arabia, the United Arab Emirates (UAE) and parts of Spain, there is little fresh water but there is lots of (salty) sea water. Potable water can be made from sea water by desalination. Two methods of doing this are **distillation** and **reverse osmosis** (Table 32.8).

Table 32.8 Methods of desalination to produce potable water.

Method	How it works	Diagram
Distillation	Sea water is heated so that it boils. The water molecules are turned to steam leaving behind the dissolved substances. The water vapour is then cooled and condensed	
Reverse osmosis	Sea water is passed through a semipermeable membrane using pressure. The water molecules pass through the membrane but many of the dissolved substances cannot. This is the opposite of normal osmosis where water would move in the opposite direction	

Both methods require a lot of **energy**. In distillation the energy is needed to boil the water. In reverse osmosis the energy is needed to create the pressure. Reverse osmosis is becoming more popular in places with sea water but little fresh water as it is cheaper than distillation. However, due to the high energy costs both methods are more expensive than producing potable water from fresh water.

Analysis and purification of water samples

Different experimental techniques can be used to analyse water samples. The pH can be measured or, if dissolved solids are present, then evaporating the water will leave them behind. The purer the water, the less solid residue will be left. Pure water samples will boil at 100°C. If impurities are present, the boiling point increases; the greater the amount of dissolved substances present, the more the boiling point changes from the exact boiling point of water.

Sea water has been purified by distillation to form drinking water since at least 200 AD, when the process was clearly described by the Greek philosopher Aristotle. The distillation apparatus used to purify water from different sources, such as salt solution (sea water), is shown in Figure 32.6.

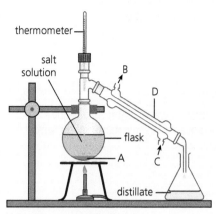

▲ **Figure 32.6** Distillation.

Questions

1 Name the solute in salt solution.

2 Name the solvent in salt solution.

3 What is solid A? Suggest its purpose.

4 What labels should be inserted at positions B and C?

5 Name the piece of apparatus labelled D and state its purpose.

6 What is the temperature on the thermometer when distillation is occurring? What does this suggest?

7 State and explain what happens to the salt dissolved in the solution during this process.

8 Distillation involves two processes: evaporation and condensation.

   a) What is meant by evaporation? Where in the apparatus does evaporation take place?

   b) What is meant by condensation? Where in the apparatus does condensation take place?

9 What advantages does the apparatus in Figure 32.6 have over the apparatus in Figure 32.7?

10 What is the purpose of the beaker of water in Figure 32.7?

11 Name the pieces of apparatus X, Y and Z.

12 Describe how you would test the purity of the distillate.

13 Explain why obtaining water from sea water is not good for the environment.

14 State how you would determine the pH of the water before and after distillation.

15 How would you prove that a sample of mineral water contained dissolved salts?

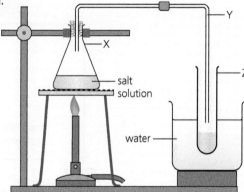

▲ **Figure 32.7** A simpler method of distillation.

○ **Waste water treatment**

Large amounts of waste water are continuously produced. This waste water comes from domestic, industrial and agricultural uses of water and has to be treated before the water can safely be returned to the environment. The waste water is collected in underground pipes called sewers and taken for treatment. In some places, all waste water is collected together in combined sewers, but in other places there are separate systems for different types of waste water. These separate systems are more efficient as different types of waste water require different treatment.

Sewage is an example of waste water and includes water used to flush toilets that contains human waste and toilet paper. It also includes water from sinks, baths and showers. Sewage needs to be treated to remove any objects in the water, the organic matter (the human waste) and harmful microbes. Waste water from agricultural and industrial sources also needs treating as it may also contain organic matter, harmful microbes and some harmful chemicals.

Key stages in the treatment of waste water include (Figure 32.8):

- **Screening** and **grit removal** – this removes large solids from the waste water.
- **Sedimentation** – this separates the human waste from the rest of the water which is called effluent.
- **Aerobic treatment** of effluent (from sedimentation) – air is passed through the effluent in aeration tanks which leads to good bacteria killing harmful bacteria.
- **Anaerobic treatment** of sludge (from sedimentation) – in the absence of air, bacteria produce methane from sludge.

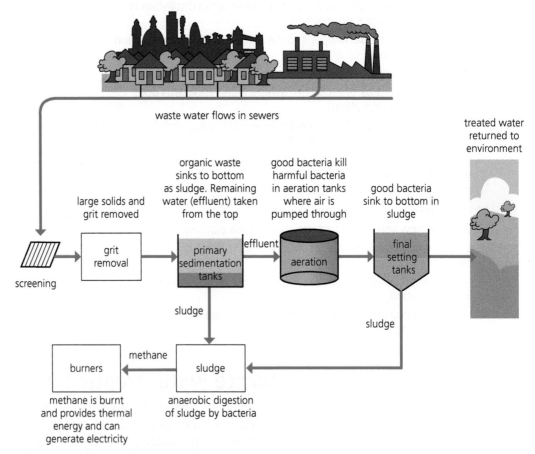

▲ **Figure 32.8** Waste water treatment.

Test yourself

10 What is potable water?

11 a) In the UK, potable water is produced from fresh water. What is fresh water?

   b) A key step in the production of potable water is filtration. Why and how is this done?

   c) Another key step in the production of potable water is sterilisation. Why is this done? Give three ways in which this is done.

12 a) What is desalination?

   b) More potable water is produced by desalination in Saudi Arabia than in any other country. Why is water produced this way in Saudi Arabia?

   c) Describe how potable water is produced from sea water by distillation.

   d) Describe how potable water is produced from sea water by reverse osmosis.

13 a) Give three sources of waste water.

   b) For each of the following stages of sewage treatment, describe what happens:
      i)   screening
      ii) sedimentation
      iii) aerobic treatment of effluent
      iv) anaerobic treatment of sludge

Show you can...

Table 32.9

	Waste water	Ground water	Salt water
Method of producing potable water			Desalination by distillation or reverse osmosis

a) Copy and complete Table 32.9 to give details of the methods of converting the different types of water into potable water. One column has been completed for you.

In 2010 the Thames Water desalination plant, the first and only desalination plant in the UK, opened. It is able to produce 1.5% of the UK's water requirement. Israel has many desalination plants and 40% of its water requirement is supplied in this way. Singapore, once heavily dependent on Malaysia for its water supply, now recycles sewage and waste water and uses it as a primary supply of potable water.

b) State and explain which method of obtaining potable water is primarily used in the UK.

c) Suggest reasons why there is a difference in the percentage use of desalination in the UK and in Israel.

d) Suggest why using waste water has been important for Singapore.

Alternative methods of extracting metals

Metals are very important materials that have many uses. For example:

- Cars, trains and planes are made from metal.
- Large buildings have metal frameworks.
- All electrical devices have metal circuits and wires.
- Tools are made from metal.

Metals are extracted from compounds found in rocks (see Chapter 12). The metal compounds are found in rocks are called ores. An ore is a rock from which a metal can be extracted for profit.

KEY TERM

Ore A rock from which a metal can be extracted for profit.

Extraction of copper

The Earth's resources of metal ores are limited. We have already reached a point with copper metal where there are only low-grade ores remaining. Low-grade ores are ones which only contain a small percentage of metal compounds. Due to this, scientists are having to develop new methods to extract copper.

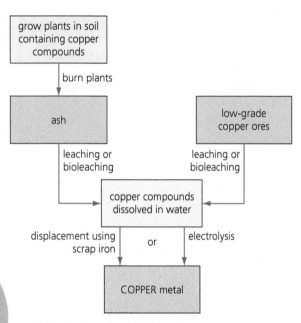

▲ **Figure 32.9** Phytomining.

One way of doing this is to use phytomining to extract copper from soil that contains copper compounds (Figure 32.9). Some plants are very good at absorbing copper compounds from the ground through their roots. These plants can be grown in land containing copper compounds. These plants can then be burnt leaving ash that is rich in copper compounds.

Copper can be extracted from this plant ash or from remaining low-grade ores using bioleaching or leaching. Bioleaching uses bacteria whereas leaching uses an acid such as sulfuric acid. In each case, insoluble copper compounds react to produce a solution containing soluble copper compounds. This solution is known as a leachate. Some bacteria are very efficient at this and so bioleaching can be very effective.

The copper can be extracted from the solution containing copper compounds. This can be done by electrolysis or using scrap iron in a displacement reaction (Table 32.10).

Traditional methods of metal extraction involve mining with digging, moving and disposing of large amounts of rock. These new methods for copper extraction are very different and do not involve this and so are better for the environment.

Table 32.10 Using electrolysis and displacement reactions to extract copper.

	Electrolysis	Displacement reaction
What happens	Two inert electrodes (e.g. graphite) are placed in the solution of copper compounds in an electrolysis circuit – copper forms at the negative electrode	Scrap iron/steel is placed in the solution of copper compounds – iron is more reactive than copper and so the copper is displaced by the iron
Equations	Negative electrode: $Cu^{2+}(aq) + 2e^- \rightarrow Cu(s)$	$Cu^{2+}(aq) + Fe(s) \rightarrow Cu(s) + Fe^{2+}(aq)$

Test yourself

14 a) What is a low-grade ore?

b) Why are new methods being developed to extract copper?

15 Copper can be extracted from low-grade ores by bioleaching. What is bioleaching?

16 Leachate solutions containing copper compounds can be produced by phytomining in areas with soil containing copper compounds. Describe how these leachate solutions are produced in these areas.

17 Two ways to extract copper metal from leachate solutions are by electrolysis and by displacement using scrap iron.

a) In the electrolysis process:

   i) At which electrode is the copper formed?

   ii) Why is copper formed at this electrode?

   iii) Write a half equation for the process at this electrode.

   iv) Explain why the formation of copper in this process involves reduction.

b) In the displacement process:

   i) Explain why iron will displace copper.

   ii) Write an ionic equation for the process.

   iii) Write a half equation for the formation of copper in this process.

   iv) Explain why the formation of copper in this process involves reduction.

Show you can...

Extracting copper metal from leachate solutions by electrolysis and by displacement are processes which both involve reduction of copper ions. Explain this statement using equations.

Chapter review questions

1 Here is a list of some of the Earth's raw materials.

ores, crude oil, sea water, air, plants

From which of these raw materials is each of the following products made?

a) oxygen gas   b) plastics   c) iron   d) biofuels

2 State whether each of the following raw materials is finite or renewable.

a) crude oil   b) ores   c) rain water

d) cotton plants   e) limestone

3 a) What is meant by a process that is sustainable?

b) Suggest a way in which each of the following products could be made in a more sustainable way than the way described below:

i) extracting aluminium from the ore bauxite

ii) using petrol to fuel a car

iii) making a glass wine bottle

iv) generating electricity by burning natural gas (methane)

c) Some products can be reused while others can be recycled. Explain the difference.

4 a) Water that is safe to drink is called potable water. Name two important stages in the production of potable water from fresh water in the UK. For each stage, explain why it is done.

b) In some countries potable water is produced from sea water by desalination. Outline how this can be done by:

i) distillation   ii) reverse osmosis

c) Explain why distillation and reverse osmosis are expensive ways to produce potable water.

5 Copper is a very important metal used for making water pipes and electrical cables. There is a shortage of copper-rich ores and new methods are being developed to extract copper from low-grade ores.

a) Solutions containing copper compounds can be produced from low-grade ores by bioleaching. What is used to bioleach copper compounds in this way?

b) Copper metal can be extracted from solutions containing copper compounds by electrolysis or a displacement reaction with scrap iron.

i) Write a half equation to show the formation of copper at the negative electrode in the electrolysis of copper sulfate.

ii) Write a balanced equation for extraction of iron from copper sulfate solution using iron.

c) Copper can be produced by phytomining. Outline what happens in phytomining.

6 Waste water from domestic, industrial and agricultural uses of water has to be treated before the water can safely be returned to the environment. Describe what happens in each of the following stages:

a) screening   c) aerobic treatment of effluent

b) sedimentation   d) anaerobic treatment of sludge

Practice questions

1 Which of these methods could be used to obtain a sample of pure water from sea water? [1 mark]

A chromatography   B crystallisation

C distillation   D filtration

2 In the UK 45 million mobile phone users discard about 15 million handsets every year. Only about 2% of these handsets are recycled, with the remainder going to landfill dumps. Recycling mobile phones means an overall reduction in carbon dioxide emissions as well as saving precious metals such as gold, silver and copper. Suggest why it is important to recycle phones to recover metals such as gold, silver and copper. [1 mark]

3 There are different types of water. Some of these are listed below:

fresh water, ground water, potable water, waste water

Match each of these types of water to the descriptions below. You should use each answer once.

a) used water from homes, industry and agriculture [1 mark]

b) water found in underground streams and in porous rocks [1 mark]

c) water found in places such as lakes, rivers, the ice caps, glaciers and underground rocks and streams [1 mark]

d) water that is safe to drink [1 mark]

4 a) In the treatment of waste water, explain what is meant by sedimentation. [2 marks]

b) In the treatment of waste water at sewage works, explain the roles of anaerobic treatment of sludge and aerobic treatment of effluent. [4 marks]

5 Copper can be found in the Earth's crust as ores containing copper sulfide. Copper ores are becoming scarce and many quarries are exhausted. Large areas of land around exhausted quarries contain low percentages of copper sulfide.

a) i) State one reason why extracting copper from this land using traditional methods is too expensive. [1 mark]

ii) State why extracting copper from this land would have a major environmental impact. [1 mark]

b) New methods such as phytomining and bioleaching can be used to extract copper from land containing low percentages of copper sulfide.

i) Describe the process of phytomining. [4 marks]

ii) State two advantages of phytomining over traditional methods of extracting copper. [2 marks]

c) Bioleaching uses bacteria to produce a solution of copper sulfate. It is possible to extract copper from copper sulfate solution using scrap iron.

i) Suggest why it is economical to extract copper using scrap iron. [1 mark]

ii) Write a symbol equation for the reaction of iron and copper sulfate to produce copper. Explain why this reaction occurs. [3 marks]

Working scientifically:
Evaluating results and procedures

In everyday life, we are constantly evaluating situations and drawing conclusions. For example when baking muffins, we might evaluate how the finished muffins look and decide how to improve on our method. If the muffins are:

▶ **too pale** – then they probably should have been left in the oven for a longer time.

▶ **burnt around the edges** – then a cooler oven should have been used or perhaps the muffins should have been removed from the oven sooner.

▶ **not moist enough** – then perhaps some extra liquid ingredients should have been added or maybe less flour should have been used.

▶ **very heavy** – then it is likely that the ingredients should have been stirred for longer to introduce more air to the mixture.

Evaluating an experimental procedure allows us to assess its effectiveness, to plan for future modifications and to judge whether an alternative method might be more suitable. Part of the evaluation process includes asking questions such as those in Table 32.11.

▲ Figure 32.10

Table 32.11 Evaluating an experimental procedure.

Questions to consider	Explanation
Are the results accurate?	Accurate results are those which are close to the true value.
Are the results repeatable?	Results are repeatable if similar results are obtained when the same person carries out the same experiment several times.
Are the results reproducible?	Results are reproducible if similar results are obtained using a different technique or by someone else doing the same experiment.
Are the results precise?	Precise results are those in which there is little spread around the mean value, i.e. all the results are close to the mean value.
Are there any significant random errors?	There will be some small variations every time an experiment is done that lead to slightly different results each time. In a well designed, well carried out experiment, these random errors will be small. The impact of random errors is reduced by doing several repeats and finding a mean value.
Are there any systematic errors?	Systematic errors give results that differ from the true value by a similar amount each time. This is because the same problem is affecting the result each time. Finding a mean value does not reduce the impact of systematic errors.
Are the results valid?	Valid results are those from a fair test in which only one variable is changed.
Was the method suitable?	Was the method easy to carry out? Did it give results that were accurate, repeatable, reproducible and precise, with minimal random errors and no systematic errors?

KEY TERMS

Accurate A measurement or calculated value that is close to the true value.

Repeatable Measurements are repeatable when similar results are achieved when an experiment is repeated by the same person.

Reproducible Measurements are reproducible when similar results are achieved when an experiment is repeated by another person or a different method is used.

Precise Measurements which have little spread around the mean value.

Random error Errors that cause a measurement to differ from the true value by different amounts each time.

Systematic error Errors that cause a measurement to differ from the true value by a similar amount each time.

Valid results Results from a fair test in which only one variable is changed.

Use the questions above to help you evaluate the following experiments.

Questions

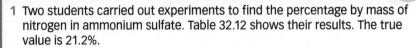

1 Two students carried out experiments to find the percentage by mass of nitrogen in ammonium sulfate. Table 32.12 shows their results. The true value is 21.2%.

Table 32.12

| Student A | 21.4% | 21.2% | 21.1% | 21.3% | 21.4% |
| Student B | 22.5% | 22.4% | 22.6% | 22.5% | 22.5% |

a) Calculate the mean value for each student, including the uncertainty as a ± value.

b) Comment on the accuracy of the result for each student.

c) Comment on the repeatability of the result for each student.

d) Why did the results differ when each student repeated the experiment?

e) Did either of the students have a systematic error in their experiment? Explain your answer.

2 Susie measured out $25\,cm^3$ of sodium hydroxide solution into a conical flask using a $25\,cm^3$ measuring cylinder. She added three drops of indicator and added sulfuric acid rapidly from a burette until the colour of the indicator changed. She found it difficult to see the colour of the indicator in the conical flask placed on the bench. The experiment was repeated several times. The results were 23.3, 27.8, 25.4, 25.9 and $22.6\ cm^3$.

Evaluate the procedure used, calculate the mean volume (including the uncertainty as a ± value) and comment on the precision of the measurements.

3 In an experiment $25\,cm^3$ of $0.1\,mol/dm^3$ hydrochloric acid was measured into a flask, placed on a pencil-drawn cross on a piece of paper and $25\,cm^3$ of sodium thiosulfate solution added. A stopwatch was started and the time taken for the cross to disappear from sight was recorded. The experiment was repeated using five different concentrations of hydrochloric acid. A different pencil cross was used each time.

Evaluate the experimental procedure.

4 In an experiment 2.0g of calcium carbonate chips was added to hydrochloric acid in a conical flask. A stopper with a delivery tube was placed in the flask, a stopwatch started and the volume of carbon dioxide collected under water in an upturned measuring cylinder was recorded every 2 minutes. The experiment was repeated using 2.0g of calcium carbonate powder.

State any source of error in this experiment and state any improvements which could be made.

The photograph shows the Millau Road Bridge in the South of France. At a height of 270 metres, this is the highest road bridge in the world. Engineers who build bridges have a detailed understanding of forces. You will meet different types of force in this chapter.

Prior knowledge

Previously you could have learnt:

> A force is a push or a pull.
> A force can squash or stretch an object.
> A force can twist or turn an object.
> The resultant force acting on an object is the sum of the forces acting on the object.
> Turning moment = force × perpendicular distance from the pivot
> Pressure = $\dfrac{\text{force}}{\text{area}}$

Test yourself on prior knowledge

1 Name three common forces.
2 What is a resultant force?

Forces and their interactions

○ Scalars and vectors

A scalar quantity is one that only has a size or magnitude.

A vector quantity is one that has a size and a direction.

Some examples of scalar quantities are: mass (for example, 3 kg of apples); temperature (for example, 27 °C); energy (for example, 200 joules). These quantities do not have a direction, they only have a size.

Force is an example of a vector quantity. You can push with a force of 250 N to the right or with a force of 250 N to the left. Each force has the same magnitude, but the effect is in a different direction.

Velocity and speed

Throughout this chapter, the words velocity and speed are both used to describe how fast something is moving. The difference between these two quantities is that **speed** is a **scalar** quantity and **velocity** is a **vector** quantity.

So when we say a car moves with a speed of 30 m/s we are just interested in how fast it is travelling. However, when we say that an aeroplane travels with a velocity of 180 m/s due east, we are saying two things: the direction the plane is travelling and how fast it is travelling.

KEY TERMS ⭐

Scalar quantity A quantity with only magnitude (size) and no direction.

Vector quantity A quantity with both magnitude (size) and direction.

KEY TERM ⭐

Velocity A speed in a defined direction.

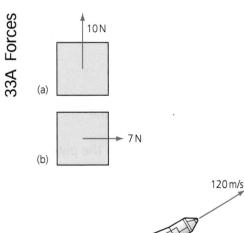

(a)

(b)

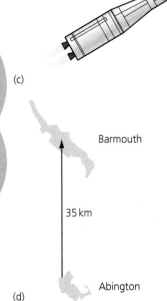

(c)

Barmouth

35 km

Abington

(d)

▲ **Figure 33.1** Examples of vectors.

Distance and displacement

You are used to describing a distance as a scalar quantity: for example, 'the ruler is 1 m long', or 'I ran 13 km this afternoon'. However, we are often interested in a direction too. For example, Bradford is 13 km due west of Leeds. When we add a direction to a distance, its correct name is a **displacement**.

Drawing vectors

You can represent a vector quantity by using an arrow. The length of the arrow represents the magnitude of the vector. The direction of the arrow shows the direction of the vector quantity. Some examples are given in Figure 33.1:

a) A force of 10 N acts upwards.

b) The box has a force of 7 N acting on it to the right. Note that the arrow here is shorter than the arrow in (a).

c) The rocket has a velocity of 120 m/s. The direction of the velocity tells us that the rocket is climbing as well as moving past a point on the ground.

d) When a car travels from Abington to Barmouth, its displacement is 35 km due north.

Test yourself

1 a) Which one of these is a scalar quantity?
   velocity   energy   displacement
 b) Which one of these is a vector quantity?
   mass   speed   force
2 What is wrong with this statement? 'The girl ran with a velocity of 8 m/s.'
3 a) A man walks 4 km due north. Make a drawing to show this displacement.
 b) The man in part (a) then walks 4 km due south. What is his displacement from his starting point? What distance did he travel?
 c) The next day the man in parts (a) and (b) walks 4 km east, then 3 km north.
   i) Draw a diagram to show his displacement from his starting point.
   ii) How far has the man walked?
 d) Explain the difference between distance and displacement.

Show you can...

Write a paragraph to explain the difference between a vector and a scalar quantity. Include some examples.

KEY TERMS ⭐

Displacement A distance travelled in a defined direction.

Force A push or a pull.

○ **What is a force?**

A **force** is a push or a pull. Whenever you push or pull something you are exerting a force on it. The forces that you exert can cause three things:

● You can change the *shape* of an object. For example, you can stretch or squash a spring and you can bend or break a ruler.
● You can change the *speed* of an object. For example, you can increase the speed of a ball when you throw it and you decrease its speed when you catch it.
● A force can also change the *direction* in which something is travelling. For example, we use a steering wheel to turn a car.

KEY TERMS

Contact force A force that can be exerted between two objects when they touch.

Non-contact force A force that can sometimes be exerted between two objects that are physically separated.

▲ **Figure 33.2** The shot putter pushes the shot.

▲ **Figure 33.3** The archer pulls the string of the bow.

TIP

1 kN = 1 000 N (1 × 10³ N)

1 MN = 1 000 000 N (1 × 10⁶ N)

TIP

Force is a **vector** quantity – it has both size and direction.

Contact forces

The forces described so far are called contact forces. Your hand touches something to exert a force.

Here are some further examples of contact forces:

- **Friction** is the contact force that opposes objects moving relative to each other. Friction acts between two sliding surfaces. Friction can also act to stop something beginning to move.
- **Air resistance**, or drag, is a force that acts on an object moving through the air. You can feel air resistance if you put your hand out of a car window when the car is moving. A boat experiences drag when it moves through water.
- **Tension** is the name we give to the force exerted through a rope when we pull something.
- **Normal contact force** is the force that supports an object that is resting on a surface such as a table or the floor.

Non-contact forces

There are also non-contact forces. These are forces that act between objects that are physically separated. **Gravitational**, **electrostatic** and **magnetic** forces are examples of non-contact forces.

Gravity

The Earth pulls you downwards whether you are in contact with it or not. The Sun's gravity acts over great distances to keep the Earth and other planets in orbit.

Electrostatic forces

These are forces that act between charged objects. By rubbing a balloon you can charge it and stick it to the wall.

Magnetic forces

A magnet can attract objects made from iron or steel towards it.

The size of forces

The unit we use to measure force is the **newton** (N). Large forces may be measured in kilonewtons (kN) or meganewtons (MN). A few examples of the size of various forces are given below.

The pull of gravity on a fly = 0.001 N

The pull of gravity on an apple = 1 N

The frictional force slowing a rolling football = 2 N

The force required to squash an egg = 50 N

The tension in a rope towing a car = 1000 N (1 kN)

The frictional force exerted by the brakes of a car = 5000 N (5 kN)

The push from the engines of a space rocket = 1 000 000 N (1 MN)

Example 1

Sakhib, who has a mass of 50 kilograms, weighs:

$$W = 50 \times 9.8 = 490 \text{ newtons}$$

Example 2

What is the weight of a 50 kg mass on the Moon's surface?

Answer

$$W = m \times g$$
$$= 50 \times 1.6$$
$$= 80 \text{ N}$$

TIP

In calculations involving weight and gravity, the value of the gravitational field strength will be given.

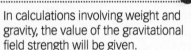

▲ **Figure 33.4** What is the weight of the apple?

KEY TERM

Centre of mass The point through which the weight of an object can be taken to act.

○ Weight

Weight is the name that we give to the pull of gravity on an object. Large objects such as the Earth produce a gravitational field, which attracts masses. (Smaller objects exert gravitational forces on each other, but these are too small to notice.)

● Near the Earth's surface the gravitational field strength, g, is 9.8 N/kg.
● This means that each kilogram has a gravitational pull of 9.8 N.
● The weight of 1 kg on the Earth's surface is 9.8 N.

To calculate weight use this equation:

$$W = mg$$
weight = mass × gravitational field strength

where weight is in newtons, N
  mass is in kilograms, kg
  gravitational field strength is in newtons per kilogram, N/kg.

Since the value of g is constant (9.8 N/kg near the Earth's surface) the weight of an object and the mass of an object are directly proportional.

weight ∝ mass

So if your mass in kilograms goes up or down, your weight in newtons will go up or down in the same proportion.

The mass of an object remains the same anywhere in the universe. The mass of an object is determined by the amount of matter in it. However, a 50 kg mass would have a different weight on the Moon, where the gravitational field strength is 1.6 N/kg.

The weight of an object can be considered to act through a single point. This point is called the centre of mass of the object.

The weight of an object can be measured with a spring balance, which is calibrated in newtons.

○ Resultant forces

A force is a vector, so we can represent the size and direction of a force with an arrow.

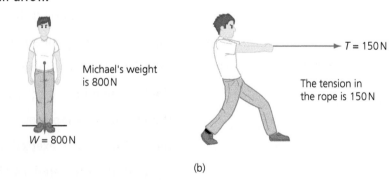

Michael's weight is 800 N

$W = 800$ N

$T = 150$ N

The tension in the rope is 150 N

(a)   (b)

▲ **Figure 33.5** Two examples of forces acting on Michael: (a) his weight (the pull of gravity on him) is 800 N. (b) A rope with a tension of 150 N pulls him forwards.

When two forces act in the same direction, they add up to give a larger resultant force.

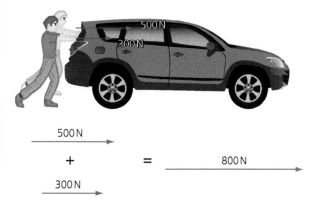

▲ **Figure 33.6** Two people push the car, one with a force of 300N, and the other with a force of 500N. The resultant force acting on the car is now 800N to the right.

When two forces act in opposite directions, they produce a smaller resultant force. If the person pushing with 300N in Figure 33.6 moves to the front of the car and exerts the same force, the resultant force will now be: 500N – 300N = 200N to the right.

Resultant force and state of motion

Figure 33.7 shows two more examples of how forces add up along a line.

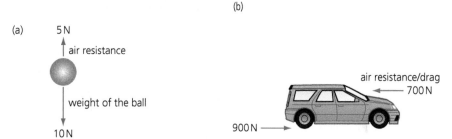

▲ **Figure 33.7** (a) A ball is falling downward. The air resistance exerts an upwards force on the ball, and the weight of the ball exerts a downwards force on it. The resultant force of 5N speeds up the ball. (b) A car is moving forwards and increases its speed. The push from the road is greater than the air resistance, so an unbalanced force of 200N helps to speed up the car.

Test yourself

4 a) What is a contact force?
   b) Which one of the following is an example of a contact force?
      friction gravity magnetic
5 a) What is a non-contact force?
   b) Which one of the following is an example of a non-contact force?
      air resistance   electrostatic   tension
6 Explain how would you demonstrate that the magnetic force between two magnets is stronger than the gravitational pull on the magnets.
7 A girl has a mass of 57kg. Calculate her weight on Earth.
8 Figure 33.8 shows some boxes. In each case, state what the resultant force is on the box. Give both the magnitude and direction of the force.

▶

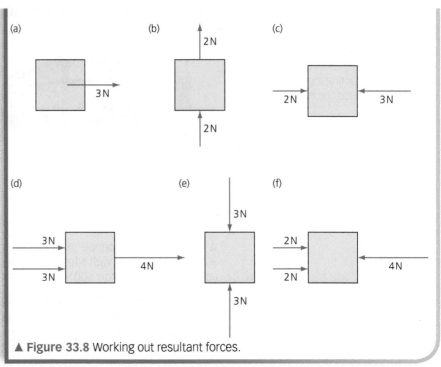

▲ **Figure 33.8** Working out resultant forces.

Show you can...

Show you understand what a resultant force is by completing this task.

Draw vector diagrams to show how two forces, acting along the same line, can produce a) a larger resultant force, b) a smaller resultant force.

Free body diagrams

Free body diagrams are used to show the magnitude and direction of all of the forces that act on an object.

Figure 33.7(a) is an example of a free body diagram. Two forces act on the ball: the air resistance exerts an upward force on the ball, and the pull of gravity (the ball's weight) exerts a downwards force on it. The sum of these two forces is 9.8 N − 5 N = 4.8 N downwards, so the ball continues to speed up.

Figure 33.9 shows a more complicated example of a free body diagram. Here, four forces are acting on Michael. Note that in drawing free body diagrams we treat the object as a point – so all the forces act through Michael's centre of gravity.

- In the vertical direction his weight, W, of 800 N is balanced by the normal contact force, R, from the floor of 800 N. So the resultant force upwards is zero.
- In the horizontal direction the pull from the rope, T, is 150 N and the frictional force, F, from the floor is 50 N. So the resultant horizontal force on Michael is 100 N to the right.

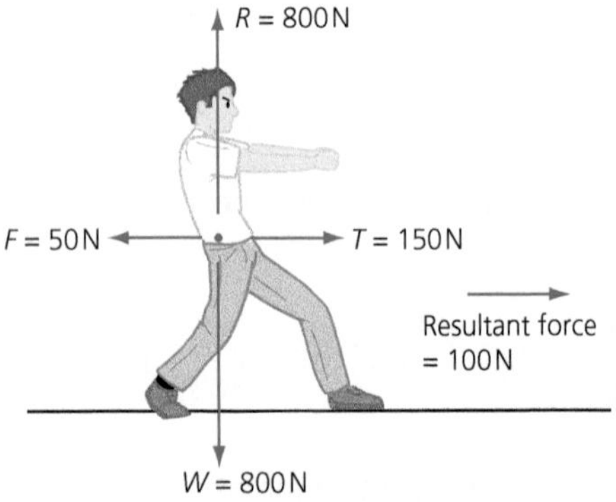

▲ **Figure 33.9** Forces acting on Michael.

Resolving forces

Figure 33.10(a) shows a box which is being pulled along the floor with a force of 10 N. However, the force is applied at an angle of 37° to the horizontal. How much of the force is being used to pull the box along the ground? The answer to the question is illustrated in Figure 33.10(b).

Figure 33.10(b) shows that the force can be resolved into two components, one vertical and one horizontal. The force is resolved in this way.

(a)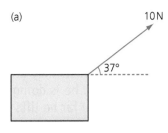

(b)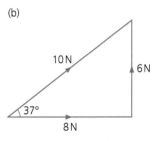

▲ **Figure 33.10** (a) A box is pulled along the floor at an angle. (b) The pulling force on the box can be resolved into two components.

TIP ✓

Resolving a force.
A single force can be split into two components acting at right angles to each other. The two component forces together have the same effect as the single force.

▲ **Figure 33.12**

Show you can...

Show that you understand what weight is by completing this task.

Explain why your weight would change but your mass would remain the same if you moved to a different planet.

- Draw the force to scale as shown by the red arrow in Figure 33.10(a), at an angle of 37°.
- Draw the horizontal component from the lower end of the red vector.
- Draw the vertical component down from the higher end of the red vector to meet the horizontal component.

In this example the horizontal component of the force is 8 N and the vertical component is 6 N. The horizontal component of 8 N is used to overcome frictional forces exerted on the box by the floor, and the vertical component of the force acts to reduce the normal contact force between the box and the floor.

Test yourself

9 An astronaut has a mass of 120 kg in his spacesuit. Calculate his weight on Mars, where the gravitational field strength is 3.7 N/kg.

10 Draw a free body diagram for a tennis ball that is travelling mid-air in a horizontal direction. Include the drag acting on it.

11 Resolve each of the forces shown in Figure 33.11 into vertical and horizontal components.

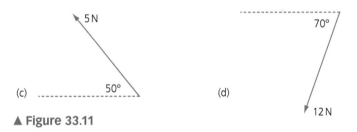

▲ **Figure 33.11**

12 The box shown in Figure 33.10(a) is stationary. It has a weight of 18 N.
   a) State the frictional force on the box.
   b) State the normal contact force acting on the box.
   c) Draw a free body diagram to show the forces acting on the box.

13 Figure 33.12 shows a box on the ground; it is sliding to the left.
   a) Draw a free body diagram to show the three forces acting on the box: friction, the box's weight and the normal contact force.
   b) Use the free body diagram to explain why the box slows down.

▲ **Figure 33.13** This man does work when he lifts the wheelbarrow, and when he pushes it forwards.

○ Work done and energy transfer

A job of work

Tony works in a supermarket. His job is to fill up shelves when they are empty. When Tony lifts up tins to put them on the shelves, he is doing some work. The amount of work Tony does depends on how far he lifts the tins and how heavy they are.

The amount of work done can be calculated using the equation:

$$W = F\,s$$

work done = force × distance moved in the direction of the force

where work done is in joules, J
force is in newtons, N
distance is in metres, m.

Work is measured in **joules** (J). **One joule** of work is done when a force of **1 newton** moves something through a distance of **1 metre** in the direction of the applied force.

$$1\,J = 1\,N \times 1\,m$$

so, 1 joule = 1 newton-metre

Large amounts of work are measured in kJ (1000 J or 10^3 J) and MJ (1 000 000 J or 10^6 J).

Example

How much work does Tony do when he lifts a tin weighing 20 N through a height of 0.5 m?

Answer

$$W = F \times s$$
$$= 20 \times 0.5$$
$$= 10\,J$$

Tony would do the same amount of work if he lifted a tin weighing 10N through 1m.

You will notice that the unit of work, J, is the same as that used for energy. This is because when work is done, energy is transferred.

- In the case of the tins, the work done by Tony is transferred to the tins as gravitational potential energy.
- If you do work dragging a box along the ground, you are working against frictional forces. This causes the **temperature** of the box and the ground to rise.
- If you push-start a car, the work you do is transferred to the kinetic energy of the car.

Does a force always do work?

A force does not always do work. In Figure 33.14, Martin is helping Salim and Teresa to give the car a push start. Teresa and Salim are pushing from behind and Martin is pushing from the side. Teresa and Salim are doing some work because they are pushing in the right direction to get the car moving. Martin is doing nothing useful to get the car moving. Martin does no work because he is pushing at right angles to the direction of movement.

◄ **Figure 33.14** Teresa and Salim are doing work but Martin is not.

In Figure 33.15, Samantha is doing some weight training. She is holding two weights but she is not lifting them. She becomes tired because her muscles transfer energy, but she is not doing any work because the weights are not moving. To do work you have to move something, for instance lift a suitcase or push a car along the road.

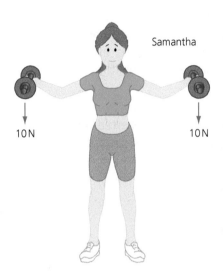

▲ **Figure 33.15** Samantha is not doing any work because the weights are not moving.

Test yourself

14 In which of the following cases is work being done? Explain your answers.
  a) A magnetic force holds a fridge magnet on a steel door.
  b) You pedal a bicycle along a road.
  c) A crane is used to lift a large bag of sand.
  d) You hold a bag of shopping without moving it.
15 Calculate the work done in each case.
  a) You lift a 20 N weight through a height of 2.5 m.
  b) You drag a box 8 m along a floor using a force of 75 N, parallel to the ground.
16 Copy and complete Table 33.1.

Table 33.1

Task of work	Force applied in newtons	Distance moved in metres	Work done in joules
Opening a door	9 N	0.8 m	
Pulling a wheeled suitcase		125 m	2500 J
Lifting a box of shopping	150 N		225 J
Pushing a toy	5 N	7 cm	
Driving a car along the road	750 N	20 km	

17 Joel is on the Moon in his spacesuit. His mass, including his spacesuit, is 150 kg. The gravitational field strength on the Moon is 1.6 N/kg.
  a) Calculate Joel's weight.
  b) Joel now climbs a ladder 8 m high into his spacecraft. How much work has he done?

Show you can...

Show you understand what work is by completing this task.

Write down an equation which links work, force and distance. Explain why when a force is applied to an object, work is not always done.

▲ **Figure 33.16** An unbalanced force will change the speed of a car.

○ **Forces and elasticity**

If one force only is applied to an object, for example a car, then the object will change speed or direction. If we want to change the shape of an object, we have to apply more than one force to it.

Figure 33.17 shows some examples of how balanced forces can change the shape of some objects. Because the forces balance, the objects remain stationary.

- Two balanced forces can stretch a spring.
- Two balanced forces can compress a beam.
- Three balanced forces cause a beam to bend.

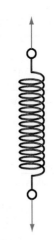

(a)

(b)

(c)

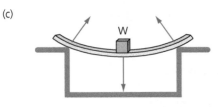

W

▲ **Figure 33.17** Balanced forces can change the shape of some objects.

Sometimes when an object has been stretched, it returns to its original length after the forces are removed. If this happens, the object experiences elastic deformation.

Sometimes an object that has been stretched does not return to its original length when the forces are removed. If the object remains permanently stretched, the object experiences inelastic deformation.

Elastic and inelastic deformations can be shown easily by stretching a spring in the laboratory. When small forces are applied and then removed, the spring returns to its original length and shape. When large forces are applied and then removed, the spring does not return to its original length.

You can also explore elastic and inelastic behaviour with an empty drinks can. When you squeeze the can gently, it springs back to its original shape when you remove your fingers. However, by applying larger forces you can change the can's shape permanently.

Stretching a spring

For a spring that is elastically deformed, the force exerted on a spring and the extension of the spring are linked by the equation:

$$F = k\,e$$

force = spring constant × extension

where force is in newtons, N

spring constant is in newtons per metre, N/m

extension is in metres, m.

The spring constant is a measure of how stiff a spring is. If k is large, the spring is stiff and difficult to stretch. When a spring has a spring constant of 180 N/m, this means that a force of 180 N must be applied to stretch the spring 1 m. The equation $F = ke$ can also be applied to the compression of a stiff spring by two forces. In this case, e is the distance by which the spring has been compressed (squashed).

The work done in stretching or compressing a spring (up to the limit of proportionality, see page 218) can be calculated using the equation:

$$E_e = \frac{1}{2}\,ke^2$$

elastic potential energy $= \dfrac{1}{2} \times$ **spring constant × (extension)²**

More details and an example are on page 218.

Example

George uses a spring to weigh a fish he has just caught. The spring stretches 8 cm. George knows that the spring constant is 300 N/m. Calculate the weight of the fish.

Answer

$F = k \times e$

$= 300 \times 0.08$

$= 24\,\text{N}$

Remember: you must convert the 8 cm to 0.08 m, because the spring constant is measured in N/m.

Investigate the relationship between force and extension for a spring

How much a spring stretches depends on the force applied to the spring. A 100 g mass hung from the end of a spring exerts a force of 1 newton on the spring.

You can investigate the extension of a spring using the apparatus shown in Figure 33.18. During the investigation you should wear a pair of safety glasses.

Method

1. Set up the retort stand, metre ruler and steel spring as shown in Figure 33.18. Make sure that the metre ruler is vertical.
2. Clamp the retort stand to the bench.
3. Measure the position of the bottom of the spring on the metre ruler, l_1.
4. Draw a suitable table to record the force acting on the spring, the metre ruler readings and the calculated values for the extension of the spring.
5. The first result to record is when no mass is attached to the spring. The force applied to the spring is then 0 and the extension of the spring is 0.
6. Hang a 100 g slotted mass hanger from the bottom of the spring. The force exerted on the spring by the mass hanger is 1 N.
7. Measure the new position of the bottom of the spring, l_2.
8. Calculate the extension of the spring: $e = l_2 - l_1$
9. Add a 100 g mass to the hanger. The total mass is now 200 g so the force exerted on the spring is 2 N.
10. Measure the new position of the bottom of the spring, l_3, and calculate the total extension $(l_3 - l_1)$.
11. Add a third, fourth, fifth and sixth 100 g mass to the hanger. Each time you add a mass, measure the new position of the bottom of the spring and calculate the total extension.

Analysing the results

1. Plot a graph of the extension of the spring against the force applied to stretch it.
2. Draw a straight line of best fit through the plotted points.

   Your graph should show the same pattern as the graph drawn in Figure 33.19.
3. What can you conclude from this investigation?

Taking it further

Using the same method, you could find out whether the extension of a rubber band follows the same pattern as that of a steel spring.

Questions

1. What precautions should be taken when carrying out this investigation to reduce the risk of injury?
2. Was the method given for this investigation a *valid* method? Give a reason for your answer.
3. A second set of results could have been taken without having to start again and repeat the investigation. Suggest how.
4. Suggest one way that the accuracy of the measurements taken during the investigation could have been improved.

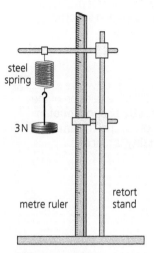

▲ **Figure 33.18** Apparatus for the investigation.

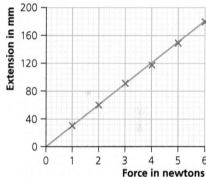

▲ **Figure 33.19** The relationship between the force applied to a spring and the extension of the spring.

> **KEY TERM**
>
> **Valid** A method of investigation which produces data that answers the question being asked.

573

Limit of proportionality

The graph in Figure 33.20 shows what happens if you put enough
weights on the spring to cause inelastic deformation. There comes a
point when the spring is permanently deformed, and it has gone beyond
the limit of proportionality. Now when the weights are removed, the
spring does not return to its original length.

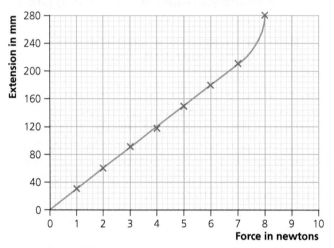

▶ **Figure 33.20** Identifying the limit of proportionality.

The spring constant for the spring can be calculated from the linear
section of the graph by dividing the force by the extension:

$$k = \frac{F}{e}$$

Elastic potential energy

When a spring is deformed elastically, energy is stored in the spring
as **elastic potential energy**. The further the spring is stretched, the
greater the amount of elastic potential energy stored in the spring.

However, if an object is deformed inelastically, there is no elastic
potential energy stored. For example, when you crush a drinks can,
work has been done moving atoms past each other to make a new
shape. The work done causes a small temperature rise in the can.

Test yourself

18 Each of the objects in Figure 33.21 has had its shape changed. Which of the objects have stored elastic potential energy? Give a reason for each of your answers.

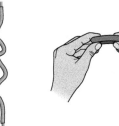

| A | B | C | D |
| twisted wire | a bent pencil rubber | a squashed piece of foam rubber | an overstretched spring |

▲ Figure 33.21

19 Explain what is meant by elastic deformation.

20 a) An object with an unknown weight is attached to the spring used in Figure 33.18. It causes the spring to extend by 136 mm. Use the graph in Figure 33.19 to calculate the weight of the object.

   b) Use the graph in Figure 33.19 to calculate the spring constant for the spring.

21 A force of 600 N causes the suspension spring of a car (Figure 33.22) to compress by 3 cm. Calculate the spring constant of the spring.

▲ Figure 33.22

22 Describe an experiment which you would use to determine the spring constant of a spring.

Show you can...

Show you understand what is meant by inelastic deformation by completing this task.

Design an experiment to show that a sheet of aluminium can be stretched both elastically and inelastically. Be careful to use exact language to describe your plan.

Chapter review questions

1 Which of these are scalar quantities and which are vector quantities?

force   speed   mass   weight   temperature

2 a) i) Gravity is an example of a non-contact force. What does the term 'non-contact force' mean?

ii) Give an example of another non-contact force.

b) A man has a mass of 85 kg. Calculate his weight. ($g = 9.8 \, \text{N/kg}$)

3 A boy with a weight of 600 N climbs up the Eiffel Tower. The Eiffel Tower has a height of 300 m. Calculate the work the boy does against gravity.

4 a) Explain what is meant by elastic deformation.

b) A spring is stretched elastically. How could you show that energy is stored in the spring?

5 A force of 10 N extends a spring by 2.5 cm. Calculate the spring constant of the spring in N/m.

6 In a tug of war contest, the relevant forces are as shown in Figure 33.23.

400 N   350 N

▲ Figure 33.23

a) What is the resultant force?

b) Explain why forces are vector quantities.

Practice questions

1 Which one of these quantities is a vector quantity?

| energy | time | velocity | [1 mark]

Give a reason for your choice. [1 mark]

2 A suitcase has a mass of 18 kg.

a) Calculate the weight of the case. [2 marks]

b) The case is lifted through a height of 2.1 m onto a luggage rack.

Calculate the work done in lifting it onto the luggage rack. [3 marks]

3 The graph in Figure 32.24 shows the relationship between the force applied to a spring and the extension of the spring.

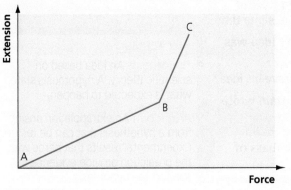

▲ Figure 33.24

a) State the unit for force. [1 mark]

b) What name is given to position B? [1 mark]

c) Between which letters is the spring undergoing elastic deformation? [1 mark]

d) What does a spring constant of 200 N/m mean? [2 marks]

4 Figure 33.25 shows a mass of 24 kg which is placed on top of an electronic newton balance. A light string is attached to the mass and pulled upwards with a force of 100 N.

a i) Calculate the weight of the mass. [2 marks]

ii) Explain why the reading on the newton balance is 135 N. [2 marks]

The end of the string is now moved so that it pulls the mass a the angle shown in Figure 33.25(b).

b i) Give a reason why the reading on the balance increases. [1 mark]

ii) Show that the vertical component of the force exerted by the string on the mass is 87 N. [2 marks]

iii) Use a scale drawing to calculate the horizontal component of the force exerted by the string on the mass. [3 marks]

iv) The mass remains at rest on the newton balance. Explain why. [2 marks]

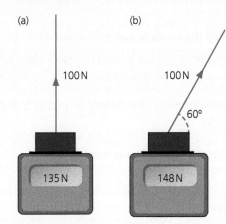

▲ Figure 33.25

Working scientifically: Hypotheses and predictions

Hypothesis, prediction and testing

About 400 years ago, the scientist Galileo came up with the idea that all objects falling through the air would accelerate at the same rate. However, most people did not believe Galileo. They had seen with their own eyes that different-sized objects took different times to fall the same distance.

To show that his idea was right, Galileo did an experiment. He dropped a small iron ball and a large iron ball from the top of the Leaning Tower of Pisa. The two objects hit the ground at almost, but not quite, the same time.

1 What was the observation that made people say that Galileo's idea was wrong?

2 Why would the result of Galileo's experiment, on its own, not prove his idea?

Liam did an investigation to test Galileo's idea. Before starting, Liam wrote this hypothesis:

> 'The acceleration of a falling object is independent of the mass of the object.'

Using his hypothesis Liam then made a prediction:

> 'Objects of different mass will take the same time to fall the same distance.'

Liam dropped a small sheet of lead. As soon as he let go, the sheet passed through the top light gate (see Figure 33.26) and the electronic timer started. The timer stopped when the sheet passed through the bottom light gate. Liam repeated this three times.

KEY TERMS

Hypothesis An idea based on scientific theory. A hypothesis states what is expected to happen.

Prediction An extrapolation arising from a hypothesis that can be tested. Experimental results that agree with the prediction provide evidence to support the hypothesis.

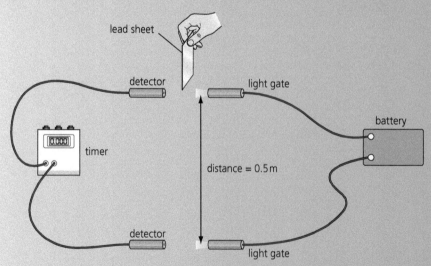

▲ **Figure 33.26** The apparatus used by Liam to test his hypothesis.

The three timer readings are shown in Table 33.2.

Table 33.2

Result	1	2	3
Time in s	0.326	0.330	0.319

3 Why did Liam repeat the time measurements?

4 Liam used the three time values to calculate a mean. What was the mean time calculated by Liam?

5 Were the time values recorded by Liam repeatable? Give a reason for your answer.

Liam then timed lead sheets of the same area but increasing mass falling the same distance.

The mean time for each sheet is given in Table 33.3.

Table 33.3

Sheet	Mass	Mean time in seconds
A	Smallest	0.320
B	Medium	0.325
C	Largest	0.322

6 Explain why the results from Liam's investigation support Galileo's idea about falling objects.

> **KEY TERM**
>
> **Repeatable** Measurements are repeatable when the same person repeating the investigation under the same conditions obtains similar results.

33B Observing and recording motion

A catamaran such as this is capable of speeds of up to 50 mph (22 m/s). The catamaran's design is a feat of excellent engineering, which is based on many of the principles that you will meet in this chapter. The boat must be streamlined and made of low density, high strength materials. It must be capable of high acceleration and rapid changes in direction.

Specification coverage

This chapter covers specification points 6.5.4.1 to 6.5.5.2, and is called Observing and recording motion.

It covers forces and motion, and momentum.

Previously you could have learnt:

> A force can cause an object to speed up, slow down or change the direction of a moving object.
> When the resultant force acting on an object is zero, the object remains stationary or it moves at a constant speed in a straight line.
> Speed = $\dfrac{\text{distance travelled}}{\text{time}}$

Test yourself on prior knowledge

1 A parachutist falls from the sky at a steady speed. Are the forces on him balanced? Explain your answer.
2 a) A car accelerates along a road. In which direction is the resultant force on the car?
  b) Eventually the car reaches a constant speed. What is the resultant force now?
  c) The driver takes her foot off the accelerator. Explain why the car slows down.
3 a) A boy goes for a walk and covers a distance of 10 km in 3 hours 20 minutes. Calculate his average speed.
  b) A woman plans a car journey. She will cover a distance of 560 km at an average speed of 80 km per hour. She wants to include two breaks in the journey of 30 minutes. How long do you expect the journey to take?

Forces and motion

○ Describing motion along a line

Table 33.4 shows some typical speeds of certain objects and people in motion. Some of these objects move at different speeds according to the circumstances. A jet plane cruises at speeds of about 200 m/s, but takes off and lands at about 60 m/s.

Table 33.4 Some typical speeds.

Moving object	Speed in m/s
Human walking	1.5
Human running	3.0
Human cycling	6.0
Car on the motorway	30.0
Express train	60.0
Jet plane	200.0

Human running speeds

A speed of 3 m/s represents the running speed of an average jogger we might see taking exercise in the park, but running speeds vary widely.

- We can run faster over a short distance. A 100 m sprinter runs much faster than a marathon runner.
- Young people run fast; we slow down as we get older.
- You cannot run so fast over rough terrain or when you are going uphill.

▲ **Figure 33.27** Hannah, Jenny and Natalia finish the 1500 m in 4 minutes and 5 seconds. What was their average speed?

Some people are extremely fit and strong. Calculate the speed of the athletes in Figure 33.27.

Average speed

When you travel in a fast car you finish a journey in a shorter time than when you travel in a slower car. If the speed of a car is 100 kilometres per hour (100 km/h), it will travel a distance of 100 kilometres in one hour.

Often, however, we use the equation to calculate an average speed because the speed of the car changes during the journey. When you travel along a motorway, your speed does not remain exactly the same. You slow down when you get stuck behind a lorry and speed up when you pull out to overtake a car.

The distance travelled by an object moving at a constant speed can be calculated using the equation:

$$s = vt$$

distance travelled = speed × time

where distance travelled is in metres, m

speed is in metres per second, m/s

time is in seconds, s.

TIP ✓

Typical values for speed are:
- walking 1.5 m/s
- running 3 m/s
- cycling 6 m/s

Example

A train travels 440 km in 3 hours. What is its average speed in m/s?

Answer

To solve this problem, you need to remember that 1 km = 1000 m and that 1 hour = 3600 s.

$$\text{average speed} = \frac{s}{t}$$

$$= \frac{440 \times 1000}{3 \times 3600}$$

$$= 41 \text{ m/s}$$

Vectors and scalars

The map in Figure 33.28 shows that Brussels, Paris and Liverpool are all a distance of about 300 km away from London. However, each city is in a different direction.

The displacement of Brussels from London is 300 km on a compass bearing of 110°. A helicopter can fly at about 300 km/h. So a helicopter can take us from London to one of the cities in an hour.

When a helicopter flies from London to Liverpool in an hour, its velocity is 300 km/h on a compass bearing of 330°.

Speed and **distance** are **scalar quantities**. **Displacement** and **velocity** are **vector quantities** because they are described by a magnitude and a direction.

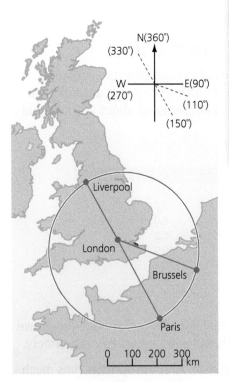

▲ **Figure 33.28** Brussels, Paris and Liverpool are all a distance of about 300 km away from London.

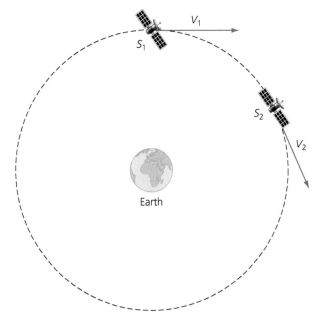

▲ **Figure 33.29** This satellite travels at a constant speed, but its velocity is always changing.

Changing velocity

If you slow down when you are running, both your speed and velocity change. However, you can also change your velocity even though your speed remains constant. You will have done this while playing a game. When you play a game of football or netball, you may suddenly change your direction when running, but keep running at the same speed. Because a velocity is described by direction as well as speed, your velocity has changed.

When an object moves around a circular path, the velocity of the object changes all the time, even though the speed is constant. The direction of the motion changes, so the velocity changes too.

Here are some more examples of changing velocity while travelling in a circular path.

- When a car travels round a bend at constant speed its velocity changes.
- When an aeroplane banks to change direction its velocity changes.
- A satellite travelling round the Earth in a circular orbit travels at a constant speed. However, its direction is always changing. This is illustrated in Figure 33.29.

> **TIP**
> An object that is travelling at a constant speed but has a changing velocity must be changing direction.

Distance–time graphs

When an object moves along a straight line, we can represent how far it has moved using a distance–time graph. Figure 33.30 is a distance–time graph for a runner. He sets off slowly and travels 20 m in the first 10 seconds. He then speeds up and travels the next 20 m in 5 seconds.

We can calculate the speed of the runner using the gradient of the graph.

> **Example**
>
> Calculate the speed of the runner using the distance–time graph (Figure 33.30) a) over the first 10 seconds, b) over the time interval 10 s to 15 s.
>
> **Answer**
>
> a) $\text{speed} = \dfrac{\text{distance}}{\text{time}}$
>
> $= \dfrac{20\,\text{m}}{10\,\text{s}} = 2\,\text{m/s}$

▲ **Figure 33.30** A distance–time graph.

b) $\text{speed} = \dfrac{40 - 20}{5}$

$= 4\,\text{m/s}$

You can see from the graph that these speeds are the gradients of each part of the graph.

When you set off on a bicycle ride, it takes time for you to reach your top speed. You accelerate gradually.

Figure 33.31 shows a distance–time graph for a cyclist at the start of a ride. The gradient gets steeper as time increases. This tells us that his speed is increasing. So the cyclist is accelerating.

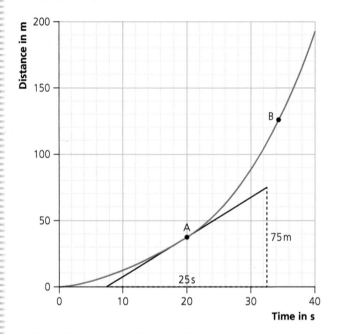

▲ **Figure 33.31** Calculating speed using a tangent to a curve.

We can calculate the speed of the cyclist at any point by drawing a tangent to the curve and then measuring the gradient.

Example

On the graph in Figure 33.31, a gradient has been drawn at point A, 20 seconds after the start of the ride. What is the speed at this time?

Answer

$\text{speed} = \text{gradient} = \dfrac{\text{vertical height (distance)}}{\text{horizontal length (time)}}$

$\text{speed} = \dfrac{75}{25}$

$= 3\,\text{m/s}$

Test yourself

1 A helicopter flies from London to Paris in 2 hours. Use the information in Figure 33.28 to calculate:
   a) the helicopter's speed
   b) the helicopter's velocity

2 A car travels 100m at a speed of 20m/s. Sketch a distance–time graph to show the motion of the car.

3 Curtis cycles to school. Figure 33.32 shows the distance–time graph for his journey.
   a) How long did Curtis stop at the traffic lights?
   b) During which part of the journey was Curtis travelling fastest?

4 Table 33.5 below shows average speeds and times recorded by top male athletes in several track events. Copy and complete Table 33.5.

   Table 33.5

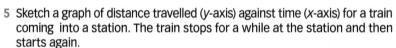

Event	Average speed in m/s	Time
100m		9.6 s
200m	10.3	
400m	8.9	
	7.1	3m 30s
10000m		29m 10s
	5.5	2h 7m 52s

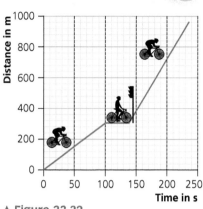

▲ Figure 33.32

5 Sketch a graph of distance travelled (*y*-axis) against time (*x*-axis) for a train coming into a station. The train stops for a while at the station and then starts again.

6 Which of the answers below is the closest to the speed at point B of Figure 33.31?

   4 m/s   10 m/s   15 m/s

7 Explain why it is possible to travel at a constant speed, but have a changing velocity.

8 Ravi, Paul and Tina enter a 30km road race. Figure 33.33 shows Ravi's and Paul's progress through the race.
   a) Which runner ran at a constant speed? Explain your answer.
   b) What was Paul's average speed for the 30km run?
   c) What happened to Paul's speed after 2 hours?

   Tina was one hour late starting the race. During the race she ran at a constant speed of 15km/h.
   d) Copy the graph and add to it a line to show how Tina ran.
   e) How far had Tina run when she overtook Paul?

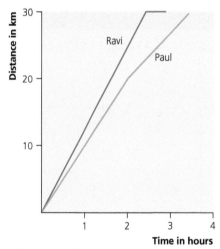

▲ Figure 33.33

Show you can...

Show that you understand distance–time graphs by completing this task.

Draw a distance–time graph to show this motion. A man walks for 30 seconds at 2 m/s; he stops for 20 s; he then runs at 6 m/s for 10 s.

Include as much information as possible in your graph. Explain what the gradient of the graph shows at each point.

▲ **Figure 33.34** A flea.

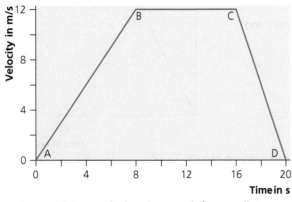

▲ **Figure 33.35** A Formula 1 racing car.

○ Acceleration

Starting from the grid, a Formula 1 car reaches a speed of 30 m/s after 2 s. A flea can reach a speed of 1 m/s after 0.001 s. Which accelerates faster?

Speeding up and slowing down

When a car is speeding up, we say it is **accelerating**. When it is slowing down we say it is **decelerating**.

A car that accelerates rapidly reaches a high speed in a short time. For example, a car might speed up to 12.5 m/s in 5 seconds. A van could take twice as long, 10 s, to reach the same speed. So the acceleration of the car is twice as big as the van's acceleration.

You can calculate the acceleration of an object using the equation:

$$a = \frac{\Delta v}{t}$$

$$\text{acceleration} = \frac{\text{change in velocity}}{\text{time taken}}$$

where acceleration is in metres per second squared, m/s^2

change in velocity (Δv) is in metres per second, m/s

time taken is in seconds, s.

$$\text{acceleration of car} = \frac{12.5}{5} = 2.5\,\text{m/s}^2$$

$$\text{acceleration of van} = \frac{12.5}{10} = 1.25\,\text{m/s}^2$$

Velocity–time graphs

It can be helpful to plot graphs of velocity against time.

Figure 33.36 shows the velocity–time graph for a cyclist as she goes on a short journey along a straight road.

The area under a velocity–time graph represents the distance travelled (or displacement).

- In the first 8 seconds, the cyclist accelerates up to a speed of 12 m/s (section AB of the graph).
- For the next 8 seconds, she cycles at a constant speed (section BC of the graph).
- Then for the last 4 seconds of the journey, she decelerates to a stop (section CD of the graph).

The gradient of the graph gives us the acceleration. In section AB she increases her speed by 12 m/s in 8 seconds.

$$\text{acceleration} = \frac{\text{change in velocity}}{\text{time}}$$

$$= \frac{12}{8} = 1.5\,\text{m/s}^2$$

▲ **Figure 33.36** A velocity–time graph for a cyclist on a straight road.

Velocity in m/s / Time in s

TIP
The gradient of a velocity–time graph represents the acceleration.

The area under section AB gives the distance travelled because:

$$distance = average\ speed \times time$$
$$= 6\,m/s \times 8\,s$$
$$= 48\,m$$

The average speed is 6 m/s as this is half of the final speed 12 m/s.

The area can also be calculated using the formula for the area of the triangle:

$$area\ of\ triangle = \frac{1}{2} \times base \times height$$
$$= \frac{1}{2} \times 8\,s \times 12\,m/s$$
$$= 48\,m$$

The following equation applies to an object that has a constant acceleration:

$$v^2 - u^2 = 2\,a\,s$$

where v is the final velocity in metres per second, m/s
  u is the initial (starting) velocity in metres per second, m/s
  a is the acceleration in metres per second squared, m/s^2
  s is the distance in metres, m.

Provided the value of three of the quantities is known, the equation can be rearranged to calculate the unknown quantity.

Example

The second stage of a rocket accelerates at 3 m/s^2. This causes the velocity of the rocket to increase from 450 m/s to 750 m/s. Calculate the distance the rocket travels while it is accelerating.

Answer

$v = 750\,m/s$, $u = 450\,m/s$, $a = 3\,m/s^2$
$v^2 - u^2 = 2\,a\,s$
$750^2 - 450^2 = 2 \times 3 \times s$
$562\,500 - 202\,500 = 6 \times s$
$s = \dfrac{360\,000}{6} = 60\,000\,m$

When the acceleration is not constant, you can work out the distance travelled by working out the area under a velocity–time graph. We do this by counting the squares, but we need to work out what distance each square represents.

Example

A hot air balloon rises from the ground. Figure 33.37 is a velocity–time graph which shows how the balloon's velocity changes with time. Use the graph to calculate the distance the balloon rises before stopping.

Answer

We count the squares under the graph. There are 20 whole squares. Then we have to estimate the area of the incomplete squares – this is about 7 (check that you agree.)

So the total area is 27 squares.

Each square represents a distance of 2 m/s × 4 s = 8 m.

Therefore the distance travelled is 27 × 8 m = 216 m.

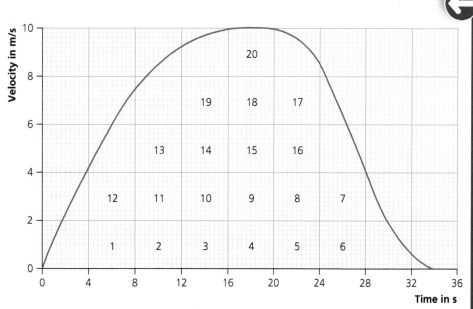

▲ Figure 33.37 A velocity–time graph, used to calculate distance travelled.

Show you can...

Show that you understand that velocity–time graphs are useful by completing this task.

A car accelerates from rest to 10 m/s over 20 s. It then travels at a constant speed for 15 s. Sketch a velocity–time graph to show this motion. Explain how the graph may be used to calculate the car's acceleration and the distance the car travels.

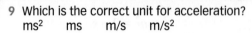

Test yourself

9 Which is the correct unit for acceleration?
ms² ms m/s m/s²

10 a) Write down the equation which links acceleration to a change of velocity and time.
   b) Use the information at the top of page 230 to calculate the acceleration of:
   i) the Formula 1 car
   ii) the flea

11 This question refers to the journey shown in Figure 33.36.
   a) Calculate the cyclist's deceleration over region CD of the graph.
   b) Use the area under the graph to calculate the distance covered during the whole journey.
   c) Calculate the average speed over the whole journey.

12 Table 33.6 shows how the speed, in m/s, of a Formula 1 racing car changes as it accelerates away from the starting grid at the beginning of a Grand Prix.

Table 33.6

Speed in m/s	0	10	20	36	49	57	64	69	72	72
Time in s	0	1	2	4	6	8	10	12	14	16

   a) Plot a graph of speed (y-axis) against time (x-axis).
   b) Use your graph to calculate the acceleration of the car at:
   i) 16 s
   ii) 1 s.
   c) Calculate how far the car has travelled after 16 s.

13 Copy the table and fill in the missing values.

Table 33.7

	Starting speed in m/s	Final speed in m/s	Time taken in s	Acceleration in m/s²
Cheetah	0		5	6
Train	13	25	120	
Aircraft taking off	0		30	2
Car crash	30	0		−150

14 Drag cars are designed to cover distances of 400 m in about 6 seconds. During this time the cars accelerate very rapidly from a standing start. At the end of 6 seconds, a drag car reaches a speed of 150 m/s.
   a) Calculate the drag car's average speed.
   b) Calculate its average acceleration.

15 An aeroplane that is about to take off accelerates along the runway at 2.5 m/s². It takes off at a speed of 60 m/s. Calculate the minimum length of runway needed at the airport.

Observing motion

A trainer who wants to know how well one of his athletes is running stands at the side of the track with a stopwatch in his hand. By careful timing, the athlete's speed at each part of the race can be analysed. However, if the trainer wants to know more detail, the measurements need to be taken with smaller intervals of time. One way to do this is by filming the athlete's movements. The following practicals show two further ways of analysing motion.

▲ Figure 33.38 This high-speed photograph allows the athlete to improve his technique by analysing his motion.

Light gates

Practical

The speed of a moving object can also be measured using light gates. Figure 33.39 shows an experiment to determine the acceleration of a rolling ball as it passes between two light gates. When the ball passes through a light gate, it cuts a beam of light. This allows the computer to measure the time taken by the ball to pass through the gate. By knowing the diameter of the ball, the speed of the ball at each gate can be calculated. You can tell if the ball is accelerating if it speeds up between light gates A and B. Follow the questions below to show how the gates can be programmed to do the work for you.

The measurements taken in an experiment are shown here.

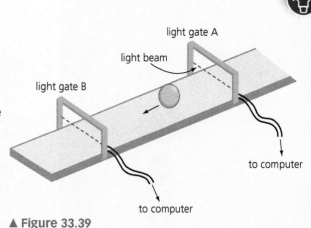

▲ Figure 33.39

Diameter of the ball	6.2 cm
Time for the ball to go through gate A	0.12 s
Time for the ball to go through gate B	0.09 s
Time taken for the ball to travel from gate A to gate B	0.23 s

1 Explain why it is important to adjust the light gates to the correct height.

2 Calculate the speed of the ball as it goes through:

   a) gate A

   b) gate B

3 Calculate the ball's acceleration as it moves from gate A to gate B.

Ticker timer

Practical

Changes to the speed and acceleration of objects in the laboratory can be measured directly using light gates, data loggers and computers. However, motion is still studied using the ticker timer (Figure 33.40), because it collects data in a clear way, which can be usefully analysed. A ticker timer has a small hammer that vibrates up and down 50 times per second. The hammer hits a piece of carbon paper, which leaves a mark on a length of tape.

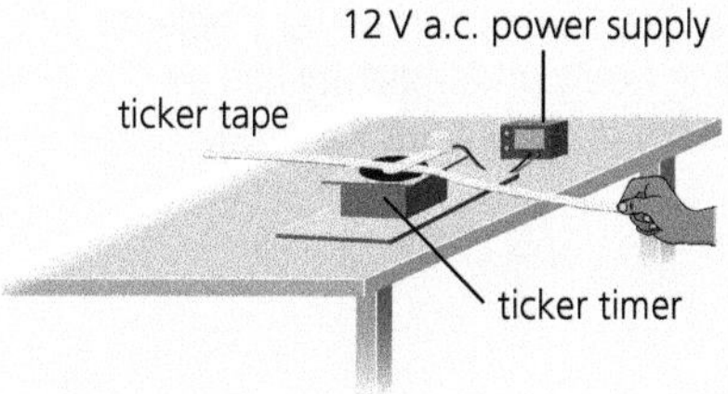

Figure 33.41 shows you a tape that has been pulled through the timer. You can see that the dots are close together over the region PQ. Then the dots get further apart, so the object moved faster over QR. The movement slowed down again over the last part of the tape, RS. Since the timer produces 50 dots per second, the time between dots is 1/50 s or 0.02 s. So we can work out the speed:

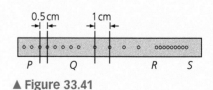

▲ Figure 33.40

$$\text{speed} = \frac{\text{distance between dots}}{\text{time between dots}}$$

Between P and Q, speed = $\dfrac{0.5\,\text{cm}}{0.02\,\text{s}}$

$= 25$ cm/s or 0.25 m/s

▲ Figure 33.41

Questions

1 Work out the speed of the tape in the region QR.

2 Which is the closest to the speed in the region RS?

   0.1 m/s   0.3 m/s   0.8 m/s

▲ **Figure 33.42** The skydiver has reached terminal velocity in a streamlined position. How could the skydiver slow down before opening the parachute?

○ Falling objects and parachuting

A falling object accelerates due to the pull of gravity. If there were no air resistance, an object dropped from any height would accelerate at 9.8 m/s² until it hit the ground. You can demonstrate that two objects of different mass accelerate at the same rate over a short distance. A marble and a 100 g mass dropped from a height of 2 m reach the ground at the same time. However, if you drop a feather it flutters towards the ground because the air resistance on it is about the same size as its weight.

Parachuting

The size of the air resistance on an object depends on the area of the object and its speed:

- The **larger the area**, the larger the air resistance.
- The **higher the speed**, the larger the air resistance.

Figure 33.43 shows how the velocity of a skydiver changes as she falls directly to the ground. The graph has five distinct parts.

TIP

Near the Earth's surface, any object falling freely under gravity has an acceleration of about 9.8 m/s².

TIP

It does not matter what the object is, if it has reached its terminal velocity then the resultant force must be zero.

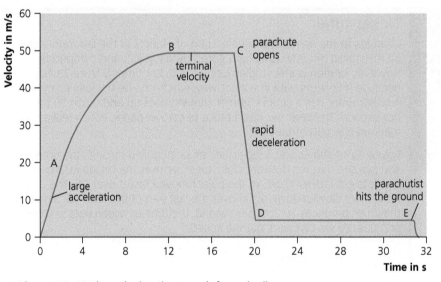

▲ **Figure 33.43** The velocity–time graph for a skydiver.

1 Origin to A. She accelerates at about 9.8 m/s² just after leaving the aeroplane.

2 AB. The effects of air resistance make her acceleration less and the gradient of the graph gets less. But she continues to accelerate as her weight is greater than the air resistance. There is a resultant force downwards.

3 BC. The air resistance has now increased so that it is the same size as her weight. The resultant force is zero, and she moves at a constant velocity. This is her terminal velocity.

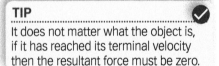

KEY TERM

Terminal velocity When the weight of a falling object is balanced by resistive forces.

4 CD. She opens her parachute. The increased surface area causes an increased air resistance. So she decelerates and slows down as there is a resultant force upwards.

5 DE. The air resistance on the parachute is now the same as her weight. She continues at a slower terminal velocity until she hits the ground.

▲ **Figure 33.44** A ball bearing falling through a thick liquid (like glycerine) reaches its terminal velocity before an identical ball bearing falling through the air reaches a faster velocity. Which is greater, the drag force in the glycerine or air resistance, when the two balls fall at their terminal velocities?

▲ **Figure 33.45** The top speed of a car is affected by the power of the engine and air resistance. By making the car streamlined, air resistance reduces and the top speed increases.

Practical

Terminal velocity and surface area

Figure 33.46 shows a model parachute made from a sheet of plastic.

Plan an experiment to find out whether the terminal velocity of the Plasticine figure depends on the area of the parachute.

Questions

1 Write a risk assessment in your plan.

2 What measurements will you make?

3 How will you know that the Plasticine figure has reached its terminal velocity?

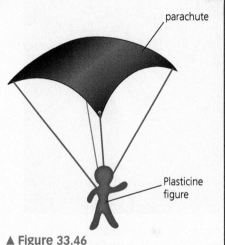

▲ **Figure 33.46**

4 Which one of the following is the independent variable in this experiment?
   - distance the parachute falls
   - area of the parachute
   - weight of the Plasticine figure
   - time taken to fall

5 What are the control variables in this experiment?

Test yourself

16 A parachutist falls at a constant speed. Which of the following statements is correct? Copy out the correct statement.
   - Her weight is much more than the air resistance.
   - Her weight is just a little more than the air resistance.
   - Her weight is the same size as the air resistance.

17 What is meant by terminal velocity?

18 Explain this observation: 'When a sheet of paper is dropped it flutters down to the ground, but when the same sheet of paper is screwed up into a ball, it accelerates rapidly downwards when dropped.'

19 This question refers to the velocity–time graph in Figure 33.43.
   a) What was the velocity of the skydiver when she hit the ground?
   b) Why is her acceleration over the part AB less than it was at the beginning of her fall?
   c) Use the graph to estimate roughly how far she fell during her dive. Was it nearer 100 m, 1 000 m, or 10 000 m?
   d) Draw diagrams to show the size of the skydiver's weight (W), and the air resistance (D) acting on her between these points:
      i) OA   ii) BC   iii) CD

Forces, accelerations and Newton's Laws of Motion

○ Newton's First Law: balanced forces

When the resultant force acting on an object is zero, the forces are balanced and the object does not accelerate. It remains **stationary** or continues to move in a **straight line** at a **constant speed**.

Figure 33.48 and 33.49 show some examples where the resultant force is zero.

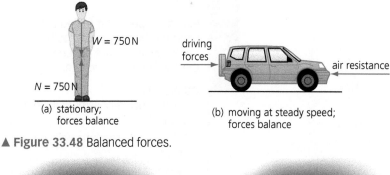

$W = 750\,N$

$N = 750\,N$

(a) stationary; forces balance

driving forces

air resistance

(b) moving at steady speed; forces balance

▲ **Figure 33.48** Balanced forces.

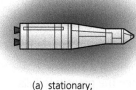

(a) stationary; no forces act

(b) moving at steady speed; no forces act

▲ **Figure 33.49** A resultant force of zero means there is no acceleration.

- In Figure 33.48(a) a person is standing still. Two forces act on him: his weight downwards and the normal contact force from the floor upwards. The forces balance; he remains stationary.
- In Figure 33.48(b) a car moves along the road. The forwards push from the road on the car is balanced by the air resistance on the car. The forces are balanced and so the car moves with a constant speed in a straight line.
- In Figure 33.49 a spacecraft is in outer space, so far away from any star that the gravitational force is zero. There are no frictional forces. So the resultant force is zero. The spacecraft is either at rest or moving in a straight line at a constant speed.

The speed and/or direction of an object will only change if a resultant force acts on the object. Newton's First Law does not apply to these objects.

○ Newton's Second Law: unbalanced forces

When an **unbalanced force** acts on an object it **accelerates**. The object could speed up, slow down or change direction.

Figure 33.50 shows two examples of unbalanced forces acting on a body

- In Figure 33.50 (a) spacecraft has turned its rockets on. There is a force pushing the craft forwards, so it accelerates.

- In Figure 33.50(b) driver has taken his foot off the accelerator while the car is moving forwards. There is an air resistance force that acts to decelerate the car.

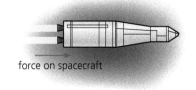

force on spacecraft

(a) acceleration

air resistance

(b) deceleration

▲ **Figure 33.50** Unbalanced forces cause (a) acceleration or (b) deceleration.

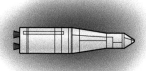

▲ **Figure 33.47** The Saturn V rocket is the largest rocket ever to take off. It took men to the Moon in 1969. Planning its flight required an understanding of Newton's laws.

Force, mass and acceleration

You may have seen people pushing a car with a flat battery. When one person tries to push a car, the acceleration is very slow. When three people give the car a push, it accelerates more quickly.

You will know from experience that large objects are difficult to get moving. When you throw a ball you can accelerate your arm more quickly if the ball has a small mass. You can throw a tennis ball much faster than you put a shot. A shot has a mass of about 7 kg so your arm cannot apply a force large enough to accelerate it as rapidly as a tennis ball.

Newton's Second Law states that:

- Acceleration is proportional to the resultant force acting on an object.

$$a \propto F$$

- Acceleration is inversely proportional to the mass of the object.

$$a \propto \frac{1}{m}$$

This can be written as an equation:

$$F = ma$$

resultant force = mass × acceleration

where force is in newtons, N

mass is in kilograms, kg

acceleration is in metres per second squared, m/s².

> **TIP**
> The symbol ∝ means proportional to.

> **TIP**
> In the equation $F = ma$, F means resultant force.

> **TIP**
> When you use the equation $F = ma$, the force must be in N, the mass in kg and the acceleration in m/s².

Example

▲ Figure 33.51

The mass of the car in Figure 33.51 is 1200 kg. Calculate the acceleration.

Answer

$$F = ma$$

$$1000 - 640 = 1200 \times a$$

$$a = \frac{360}{1200}$$

$$= 0.3 \, \text{m/s}^2$$

Note

- First we had to work out the resultant force.

- The acceleration is in the same direction as the resultant force. So the car speeds up or accelerates.

Investigate the effect of varying the force on the acceleration of an object of constant mass

Figure 33.52 shows one way that you can investigate the acceleration of an object.

The accelerating force is caused by the falling mass. A falling mass of 100g tied to the string exerts a force of 1 N.

A light gate is being used in this investigation but you could use a ticker timer and tape.

Use the method given below to investigate the following hypothesis:

> The acceleration of an object is directly proportional to the force applied to the object.

Using this hypothesis we can predict that:

> doubling the force applied to the object will double the acceleration of the object.

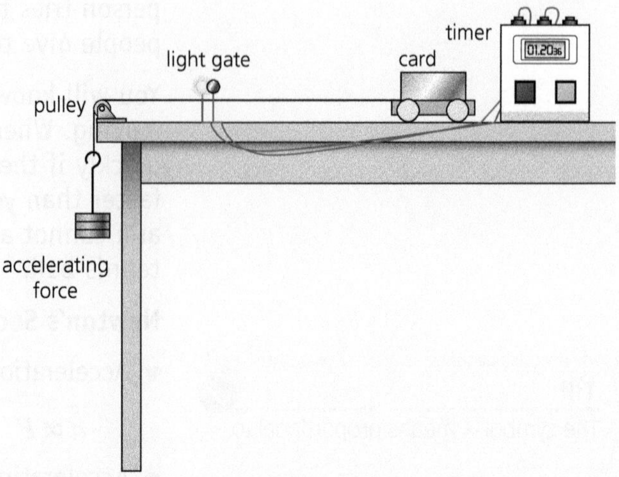

▲ **Figure 33.52** Investigating acceleration.

Method

1 Set up the equipment as shown in Figure 33.52.
2 The timer should start and then stop as the card passes through the light gate.
3 Draw a table to record all of the data collected. Include a column for the acceleration of the trolley.
4 Tie a mass of 100g to the string.
5 Add two extra 100g masses to the trolley.
6 Place the trolley so that the front of the trolley is 0.5 m away from the light gate. The string should be taut and the hanging mass at least 0.5 m above the floor. Hold the trolley stationary so its initial velocity is zero.
7 Let go of the trolley and use a stopwatch to time how long the trolley takes to travel to the light gate.

    Make sure that you are able to stop the trolley from falling to the floor and keep your feet away from the falling masses.
8 Write down the time taken for the card to pass through the light gate.
9 By taking a 100g mass from the top of the trolley and adding it to the hanging masses you increase the accelerating force. You can now gather data for a range of accelerating forces.

Calculating the acceleration

1 $$\text{final velocity of the trolley} = \frac{\text{length of card}}{\text{time taken to pass through the light gate}}$$

2 $$\text{acceleration} = \frac{\text{change of velocity}}{\text{time taken}} = \frac{\text{final velocity of the trolley}}{\text{time taken to reach the light gate}}$$

If you use a ticker timer, attach a 1 m length of ticker tape to the back of the trolley. Pass the tape through the timer. When you let go of the trolley a pattern of dots will be printed on the tape. Use this pattern to calculate the acceleration of the trolley.

Analysing the results

1 Plot a graph of acceleration against force.
2 Do the results from your investigation support the hypothesis? Explain the reason for your answer.

Taking it further

1 Plan and carry out an investigation to find out how the mass of an object affects the acceleration of the object.

2   Before starting your investigation, write your own hypothesis. Use your hypothesis to predict what will happen.

3   Write a risk assessment for your investigation.

Explain why it is necessary to store the extra 100g masses on top of the trolley which are later used to increase the accelerating force.

Questions

1   Why are the velocity values you calculate average values?

2   The trolley could have been run along a slightly raised runway rather than along a level bench top. Explain the effect this would have on the values calculated for acceleration.

3   In the first investigation, mass is kept constant while the accelerating force is changed. If you did the second investigation, the accelerating force will have been kept constant while the mass is changed. Why is it important that either the mass or the accelerating force is kept constant?

Show you can...

Show you understand the connection between force, mass and acceleration by completing this task.

Design a practical that enables you to show that an object accelerates at a greater rate when a greater force is applied to it. State clearly how you will measure the acceleration of the object.

Test yourself

20   Explain why Formula 1 racing cars have low masses.

21   Explain why you can throw a tennis ball faster than you can throw a 7 kg shot.

22   a) Calculate the resultant force acting on a mass of 8 kg which is accelerating at 2.5 m/s^2.

   b) Calculate the acceleration of a 3 kg mass that experiences a 15 N resultant force.

   c) Calculate the mass of an object which is accelerating at 4 m/s^2 when a resultant force of 10 N acts on it.

▲ **Figure 33.53** How do you know whether there are biscuits left without taking the lid off?

Inertia

Imagine that you are in a spacecraft on a long voyage to Mars. You are in the kitchen and there are two very large boxes of biscuits, marked with their mass: 14 kg. One box is empty, one box is full. How can you tell which is full without taking the lid off? Remember both boxes have no weight, because there is no gravitational pull on them.

The answer to the question above is that you give each of the boxes a small push. The empty one is easy to set in motion, but the one full of biscuits is much harder to start moving.

We say that the box full of biscuits has more inertia than the empty box.

The **inertial mass** of an object is a measure of how difficult it is to change its velocity.

We can define the inertial mass through the equation:

$$m = \frac{F}{a}$$

In the case of the biscuit tins, the one with the smaller acceleration (for the same applied force) has the larger inertial mass.

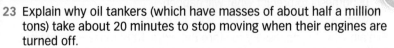

23 Explain why oil tankers (which have masses of about half a million tons) take about 20 minutes to stop moving when their engines are turned off.

24 A Formula 1 car has a mass of 730 kg. The car brakes from a speed of 84 m/s to 32 m/s in 1.3 s.

   a) Calculate the deceleration during this time.

   b) Calculate the size of the braking force on the car.

25 A spacecraft on the Moon has a weight of 48 000 N. The mass of the spacecraft is 30 000 kg.

   a) Calculate the Moon's gravitational field strength.

   b) When the spacecraft takes off, the force on the spacecraft due to the rockets is 63 000 N.

      i) Calculate the resultant force on the spacecraft.

      ii) Calculate the acceleration as the spacecraft takes off.

26 Explain why a passenger who is standing might lose his balance if a train accelerates quickly out of a station.

27 A parcel is on the seat of a car. When the car brakes suddenly, the parcel moves forwards and falls off the seat and onto the floor. Explain why.

○ Newton's Third Law of motion

Every force has a paired **equal and opposite** force. This law sounds easy to apply, but it requires some clear thinking. It is important to appreciate that the pairs of forces must act on *two* different objects, and the forces must be the same type of force.

An easy way to demonstrate Newton's Third Law is to connect two dynamics trolleys together with a stretched rubber band, as shown in Figure 33.54.

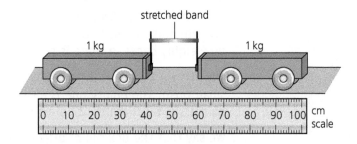

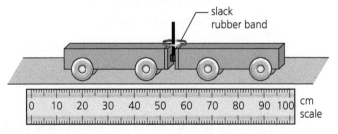

▲ **Figure 33.54** Demonstrating Newton's Third Law.

When the trolleys are released, they travel the same distance and meet in the middle. This is because trolley A exerts a force on trolley B, and trolley B exerts a force of the same size, in the opposite direction, on trolley A. Some examples of paired forces are given in Figure 33.55.

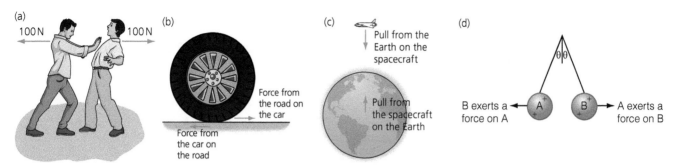

▲ **Figure 33.55** (a) If I push you with a force of 100N, you push me back with a force of 100N. (b) When the wheel of a car turns, it pushes the road backwards. The road pushes the wheel forwards with an equal and opposite force. (c) A spacecraft orbiting the Earth is pulled downwards by the Earth's gravity. The spacecraft exerts an equal and opposite gravitational force on the Earth. So if the spacecraft moves towards the Earth, the Earth moves too, but because the Earth is so massive its movement is very small. (d) Two balloons have been charged positively. They each experience a repulsive force from the other. These forces are of the same size, so each balloon (if of the same mass) is lifted through the same angle.

All of these are features of Newton's Third Law pairs:

- They act on **two separate bodies**.
- They are always of the **same type**, for example two electrostatic forces or two contact forces.
- They are of the **same magnitude**.
- They act along the **same line**.
- They act in **opposite directions**.

Test yourself

28 Use Newton's Third Law to explain the following:
   a) When you lean against a wall you do not fall over.
   b) When you are swimming you have to push water backwards so that you can move forwards.
   c) You cannot walk easily on icy ground.
29 A man with a weight of 850N jumps off a wall of a height 1.2 m. While he is falling, what force does he exert on the Earth?

Show you can...

Show you understand Newton's Third Law of motion by explaining three demonstrations of the law to a friend.

Forces and braking

When you begin to drive the most important thing you must learn is how to stop safely.

Frictional forces decelerate a car and bring it to rest. However, the driver also needs to use the brakes to slow the car in traffic or when traffic lights turn red. In an emergency the driver must apply the brakes as soon as possible. However, the average driver has a reaction time of about 0.7 s. This is the time taken by the driver to react to an emergency and move their feet from the accelerator pedal to the brake pedal.

TIP

The thinking distance is the **distance** covered in the time it takes to react, not the time it takes to react.

During this time the car carries on moving at a constant speed. The distance the car moves during the driver's reaction time is called the thinking distance.

Once the brakes have been applied, the car slows down and stops. The distance the car moves once brakes are applied is called the braking distance.

The stopping distance of a vehicle (Figure 33.56) is made up of the two parts – the thinking distance and the braking distance.

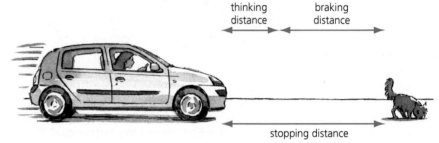

▲ **Figure 33.56** Stopping distance is thinking distance plus braking distance.

Speed affects both the thinking distance and the braking distance. Figure 33.57 shows the stopping distances for a typical car at different speeds, in good conditions, on a dry road. The force applied by the brakes is the same for each speed.

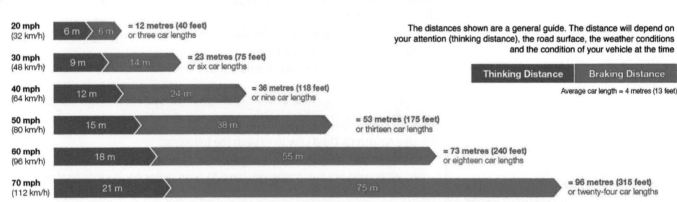

▲ **Figure 33.57** These stopping distances appear in the Highway Code. You need to know these to drive safely. The reaction time here is about 0.7 s.

Factors affecting reaction times

Reaction times vary from person to person. Typical values range from about 0.2 s to 0.9 s.

TIP

You should recall that a typical reaction time for drivers ranges from 0.2 s to 0.9 s.

Drivers' reaction times are much slower if they:

- have been drinking **alcohol**
- have been taking certain types of **drugs**
- are **tired**.

It is not just illegal drugs that slow reactions. Anybody taking a medicine and intending to drive should check the label.

People can also react slowly if they are **distracted**, either by thinking about something else or talking on their phone.

Factors affecting braking distances

Braking distances can be affected by several factors as well as speed:

- the **size of the braking force** – if you brake harder, a larger force acts to decelerate the car, so you stop in a shorter distance; but you have to be careful if the force is too large because a car can skid and you may lose control
- **weather conditions** – in wet or icy conditions there is less friction between the tyre and the road; this increases the braking distance
- the **vehicle** is poorly maintained – worn brakes or worn tyres increase the braking distance
- the **road surface** – a rough surface increases the friction between the road and tyres and reduces the braking distance.

Braking and kinetic energy

The work done by a braking force reduces a vehicle's kinetic energy, causing the vehicle to slow down and stop. As the kinetic energy of the vehicle goes down, the temperature of the brakes goes up. This is why friction brakes always get hot when they are applied. Large decelerations caused by large braking forces may lead to the brakes overheating.

See Chapter 33A for more information about work done.

Practical

Reaction and distraction

Method

Hold a ruler between the fingers of a partner's hand. Without giving a warning, let go of the ruler. Your partner should close their fingers as soon as they see the ruler fall.

Try it again, but this time with your partner listening to music or talking to a friend. What do you predict will happen?

If there is a difference in reaction time, how could you check that it was due to the music or talking and not some other factor?

Test yourself

30 a) What is meant by: **i)** thinking distance, **ii)** braking distance, **iii)** stopping distance?
   b) What is the connection between thinking distance, braking distance and stopping distance?

31 Which one of the following would affect the braking distance of a car?
   a tired driver   an icy road   a drunk driver

32 Copy and complete the following sentences:
   a) For a particular braking _____ the greater the _____ of a car, the greater the braking distance.
   b) A driver's reaction time may be increased if the driver is using a _____.

33 How could a road surface near a busy junction be changed to help reduce the braking distance of a car approaching the junction?

34 Worn tyres are dangerous. The graph in Figure 33.58 shows how the braking distance for a car depends on the depth of the tread on its tyres. Two road surfaces have been tested.

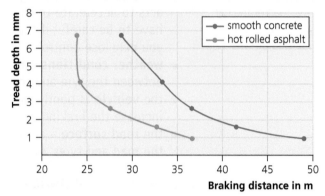

▲ **Figure 33.58** Braking distance against tread depth at 40mph

a) How do the curves show that worn tyres are dangerous?
b) Explain the importance of the road surface on braking distances.

35 a) Use Figure 33.57 to explain why it is so dangerous to drive at 40mph in a 30mph speed limit.
b) Explain why some councils impose 20mph speed limits in some areas.
c) Use the data in Figure 33.57 to show that the driver's reaction time is about 0.7 s.

36 Use the data in Figure 33.57 to calculate the deceleration of a car coming to a stop when travelling initially at:
a) 20mph
b) 60mph
[Hint: you might need to refer back to the equation for acceleration, earlier in this chapter.]

Show you can...

Show you understand about stopping distances by completing this task.

Write a paragraph to explain the factors which affect the stopping distance for a car.

Momentum

▲ **Figure 33.59** Why does a shotgun recoil when it is fired?

○ Momentum is a property of moving objects

Momentum is defined as the product of mass and velocity:

$$p = m \times v$$
momentum = mass × velocity
where momentum is in kilogram metres per second, kg m/s
  mass is in kilograms, kg
  velocity is in metres per second, m/s.

momentum
= 3 kg × –3 m/s
= –9 kg m/s

momentum
= 2 kg × +4 m/s
= +8 kg m/s

▲ **Figure 33.60** Momentum is mass times velocity.

Velocity is a vector quantity and therefore so is momentum. This means you must give a size and a direction when you talk about momentum. Look at Figure 33.60, where you can see two objects moving in opposite directions. One has positive momentum and the other negative momentum. Here we have chosen to define the positive direction to the right.

○ Conservation of momentum

Momentum is a very useful quantity when it comes to calculating what happens in collisions.

When two objects collide, the total momentum they have is the same after the collision as it was before the collision. This is **conservation of momentum**.

TIP ✓

Momentum is always conserved in a closed system of colliding particles.

▲ **Figure 33.61** The total momentum before a collision is the same as the total momentum after it.

KEY TERM ★

Closed system A system with no external forces acting on it.

This is always true, provided that the colliding objects are in a closed system. This means that when the objects collide, no forces outside of the system (external forces) act on the objects.

If an object explodes, the momentum of the object before the explosion is the same as the total momentum of the parts after the explosion. Again, this is always true in a closed system.

Figure 33.62 shows two ice hockey players colliding on the ice. When they meet they push each other. The blue player's momentum decreases and the red player's momentum increases by the same amount.

Before the collision the total momentum was:

> momentum of blue player = 100 × 5 = 500 kg m/s
> momentum of red player = 80 × 3 = 240 kg m/s

This makes a total of 740 kg m/s.

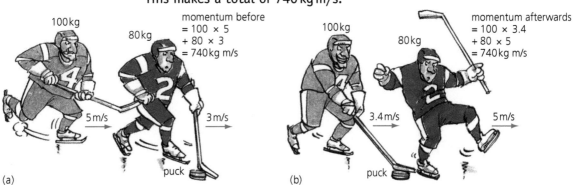

momentum before
= 100 × 5
+ 80 × 3
= 740 kg m/s

momentum afterwards
= 100 × 3.4
+ 80 × 5
= 740 kg m/s

(a)   (b)

▲ **Figure 33.62** Calculating momentum.

After the collision the total momentum was:

momentum of blue player = 100 × 3.4

= 340 kg m/s

momentum of red player = 80 × 5

= 400 kg m/s

This makes a total of 740 kg m/s which is the same as before.

Momentum as a vector

Figure 33.63 shows a large estate car colliding head-on with a smaller car. They both come to a halt. How is momentum conserved here?

Momentum is a vector quantity. In Figure 33.63(a) you can see that the red car has a momentum of +15 000 kg m/s and the blue car has momentum of –15 000 kg m/s. So the total momentum is zero before the collision and there is as much positive momentum as negative momentum. After the collision both cars have stopped moving and the total momentum remains zero.

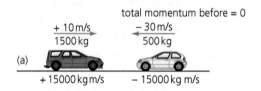

total momentum before = 0

+ 10 m/s
1500 kg

– 30 m/s
500 kg

(a)

+ 15000 kg m/s   – 15000 kg m/s

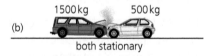

total momentum after = 0

1500 kg   500 kg

(b)

both stationary

▲ **Figure 33.63** Conserving momentum.

Chapter review questions

1   Figure 33.64 shows a distance–time graph for a car on a motorway.

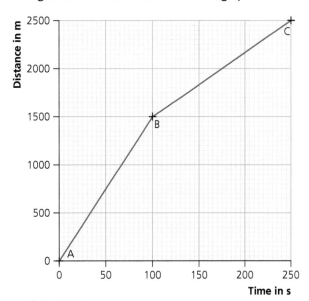

▲ Figure 33.64

a) Where is the car travelling faster, A to B or B to C?

b) i) How long did it take to travel from A to B?

ii) How far did the car travel from A to B?

iii) How far did the car travel from B to C?

c) Calculate the speed of the car between A and B.

2   A train travels at a constant speed of 45 m/s. Calculate:

a) how far it travels in 30 s

b) the time it takes to travel 9000 m.

3   A car has a mass of 1500 kg. It takes 6 s to increase its velocity from 5 m/s to 23 m/s.

a) i) Calculate the change of velocity.

ii) Calculate the car's acceleration.

b) When travelling at 23 m/s the driver takes his foot off the accelerator. The car takes 20 s to slow down to 15 m/s.

i) Calculate the deceleration of the car.

ii) Calculate the air resistance acting on the car to slow it down.

4   Explain why a skydiver wearing loose clothing and spreading out his arms and legs has a lower terminal velocity than a skydiver curled up into a ball.

5 Figure 33.65 shows a velocity–time graph for an aeroplane just before it takes off.

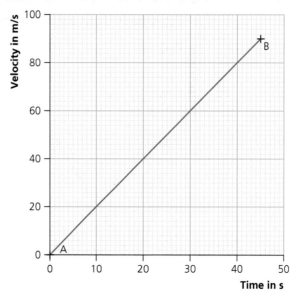

▲ Figure 33.65

a) Calculate the acceleration of the aeroplane.

b) Calculate the distance covered by the plane before it takes off at time B.

6 Calculate the work done in each of the following cases.

a) A car is accelerated by a force of 400N over a distance of 80m.

b) A girl with a weight of 470N climbs a staircase with a height of 3.6m.

c) You hold a weight of 200N stationary over your head for 5 seconds.

d) A boat moves a distance of 3000m when a drag force of 60000N acts on it.

7 A skydiver with a weight of 700N falls at a constant speed.

a) What is the drag force that acts on her?

b) When she opens the parachute, a drag force of 1500N acts on her. Explain what happens to her speed.

8 A tank moving at 6.5m/s has a momentum of 195000kgm/s. Calculate the mass of the tank.

9 A student investigated how the weight of a parachute affects how fast the parachute falls. The student used a cupcake case as the parachute. He kept the area of the case constant, but varied the weight by adding extra cases inside each other.

Table 33.8 shows the time taken by each set of cases to fall a height of 4m. The student measured the time of fall for each set of cases three times.

Table 33.8

Number of cake cases	Time of fall in s
1	2.7, 2.6, 2.6
1.5	2.2, 2.3, 2.2
2	2.0, 2.0, 1.9
3	1.5, 1.6, 1.7
4	1.4, 1.4, 1.4
6	1.3, 1.3, 1.2
8	1.1, 1.1, 1.2
10	1.1, 1.1, 1.0

extra case added to increase the weight

▲ Figure 33.66

a) Why did the student measure the time taken to fall three times for each set of cake cases?

b) Copy Table 33.8 and add two further columns to show:

   i) the average time of fall

   ii) the average speed of fall.

c) Plot a graph of the number of cases on the *y*-axis against the average speed of fall on the *x*-axis. Draw a line of best fit through the points and comment on any anomalous results.

d) Use the graph to predict the speed of fall for a weight of seven cases.

e) What can you conclude from the data displayed in the graph?

f) Assuming that the cases reach their terminal velocity quickly after release, comment on how the resistive force on the falling cupcake cases depends on their speed.

10 This is a question for some research; you can look on the web, or you might find some useful information from your parents or grandparents.

Driving is much safer now than it was 80 years ago. In the 1930s there were about 2.5 million cars on our roads and over 7000 people died each year in road accidents. Now there are over 32 million cars on the road and on average 2000 people die each year. While this is still a sad statistic, we are much safer now.

Write a paragraph to explain what measures have been put in place over the last 80 years to improve road safety.

Practice questions

1 A triathlon race has three parts: swimming, cycling and running.

Figure 33.67 shows the force responsible for the forward movement of the athlete in the swimming race.

▲ Figure 33.67

a) Copy Figure 33.67 and show three other forces acting on the athlete. [3 marks]

b) Table 33.9 shows the distance of each part of a triathlon race and the time an athlete takes for each part.

Table 33.9

Part of race	Distance in m	Time in s
Swimming	1500	1200
Cycling	40000	3600
Running	10000	2000

i) Calculate the athlete's average speed during the swim. [1 mark]

ii) Calculate the athlete's average speed for the whole race. [2 marks]

iii) The graph shows how the distance varied with time for the running part of the race.

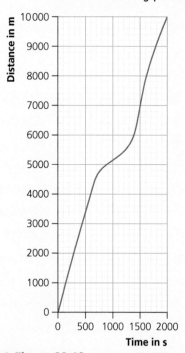

▲ Figure 33.68

c) Describe how the athlete's speed changed during this part of the race. [3 marks]

2 Table 33.10 gives information about a journey made by a cyclist.

Table 33.10

Time in hours	Distance in km
0	0
1	15
2	30
3	45
4	60
5	75
6	90

a) Plot a graph using the data in Table 33.10. [3 marks]

b) i) Use your graph to find the distance in kilometres that the cyclist travelled in 4.5 hours. [1 mark]

ii) Use the graph to find the time in hours taken by the cyclist to travel 35 kilometres. [1 mark]

c) Write down the equation which links average speed, distance moved and time taken. [1 mark]

3 A train travels between two stations. The velocity–time graph in Figure 33.69 shows the train's motion.

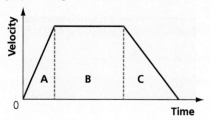

▲ Figure 33.69

a) How do you know that the train is decelerating in part C? [1 mark]

b) State the features of the graph that represent the distance travelled between the two stations. [1 mark]

c) A second train travels between the two stations at a constant velocity and does not stop. It takes the same time as the first train. Draw a copy of the graph and then draw a line showing the motion of the second train. [2 marks]

4 a) A lorry is travelling in a straight line and it is accelerating. The total forward force on the lorry is F and the total backward force is B.

i) Which is larger, force F or force B? Give a reason for your answer. [1 mark]

ii) Write down the equation that links acceleration, mass and resultant force. [1 mark]

iii) A resultant force of 15 000 N acts on a lorry. The mass of the lorry is 12 500 kg. Calculate the lorry's acceleration and give the unit. [3 marks]

b) The thinking distance is the distance a vehicle travels in the driver's reaction time. The braking distance is the distance a vehicle travels when the brakes are applied.

i) State one factor that increases the thinking distance. [1 mark]

ii) State one factor that increases the braking distance. [1 mark]

c) A council decides to impose a 20 mph speed limit in a town centre. A councillor says than drivers can react more quickly at 20 mph than they can at 30 mph.

What is wrong with this statement? [2 marks]

5 Figure 33.70 shows the minimum stopping distances, in metres, for a car travelling at different speeds on a dry road.

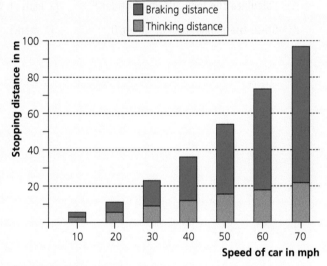

▲ Figure 33.70

a) Write an equation which links stopping distance, thinking distance and braking distance. [1 mark]

b) Describe the patterns shown in the graph. [2 marks]

c) Use the graph to estimate the stopping distance for a car travelling at 35 miles per hour. [1 mark]

d) To find the minimum stopping distance, several different cars were tested. Suggest how the data from the different cars should be used to give the values in the graph. [1 mark]

e) The tests were carried out on a dry road. If the road was icy, describe and explain what change if any there would be to:

i) the thinking distance [2 marks]

ii) the braking distance. [2 marks]

6 Figure 33.71 shows how the velocity of an aircraft changes as it accelerates along a runway. The aircraft takes off after 60 seconds.

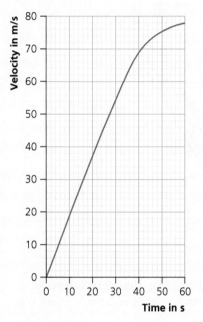

▲ Figure 33.71

a) Use the graph to find the average acceleration of the aircraft over the 60 seconds. [3 marks]

b) Explain why the acceleration is not constant, even though the engines produce a constant force. [3 marks]

c) Use the graph to estimate the length of the runway. [3 marks] **H**

7 A cyclist is travelling at 3 m/s and in 8 seconds increases his speed to 7 m/s.

a) Calculate his acceleration. [3 marks]

b) The mass of the cyclist and his bicycle is 80 kg. The force forwards on the bicycle is 60 N. Calculate the size of air resistance, *R*, acting on him. Assume he has the acceleration calculated in part (a). [3 marks]

▲ Figure 33.72

8 A group of students use a special track. The track is about 2 metres long and is horizontal. Two gliders, P and Q, can move along the track.

The surface of the track and the inside surface of the gliders are almost frictionless. Figure 33.73 shows that the gliders can move through two light gates, A and B.

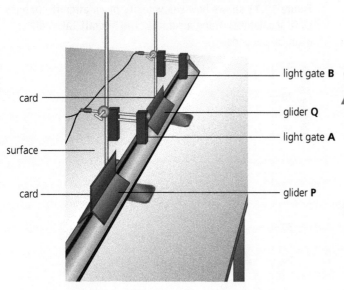

▲ Figure 33.73

The mass of glider P is 2.4 kg. This glider is moving towards Q at a constant velocity of 0.6 m/s. Glider Q is stationary.

Figure 33.74 shows a side view. Each glider has a card and a magnet attached. Light gate A records the time for which the card is in front of the light gate.

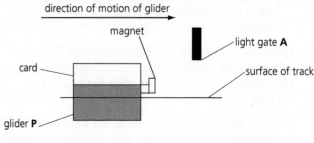

▲ Figure 33.74

a) i) Apart from the time recorded by the light gate A, state the other measurement that would be needed to calculate the velocity of glider P.   [1 mark]

   ii) Why does the surface of the track need to be frictionless and horizontal?   [1 mark]

b) Momentum is a vector quantity.

   i) State what is meant by a vector quantity. [1 mark]

   ii) Calculate the momentum of glider P.   [2 marks]

   iii) State the momentum of glider Q.   [1 mark]

Working scientifically:
Understanding variables

Independent, dependent and control variables

To investigate the idea of a crumple zone, Tracy used the apparatus shown in Figure 33.75 (She put some crumpled waste paper down to catch the weight too.) Tracy expected that in a collision the test material would slow down the trolley in the same way as the crumple zone slows down a car.

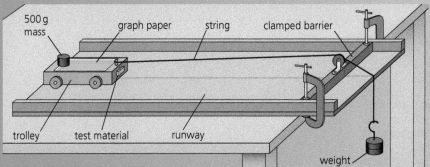

▲ Figure 33.75

A constant force was applied to the trolley by a falling weight. When the trolley hit the barrier, it decelerated and stopped. The 500 g mass on top of the trolley slid forwards until it also stopped.

Tracy marked the graph paper to show how far the 500 g mass moved. Tracy tested five different types of material for the crumple zone, each of the same area and thickness.

1 Why does the 500 g mass slide forwards when the trolley hits the barrier?

2 In this investigation, what is:

   a) the independent variable

   b) the dependent variable?

   Remember that the independent variable is the variable that affects the dependent variable.

By applying the same force to the trolley and starting it from the same position on the runway, the trolley always hit the barrier at the same speed. The speed at which the trolley hits the barrier is a control variable.

3 What else was a control variable in this investigation?

> **KEY TERMS** ⭐
>
> Independent variable The variable that you change. An investigation should only have one independent variable.
>
> Dependent variable The variable that changes because of a change made to the independent variable.
>
> Control variable A variable which is kept the same.

34 Waves

Electromagnetic waves, which travel at the speed of light, allow us to communicate with friends all around the world. These waves transfer energy for receivers to detect them, and information for us to understand the message.

Specification coverage

This chapter covers specification points 6.6.1.1 to 6.6.2.4 and is called Waves.

It covers waves in air, fluids and solids, and electromagnetic waves.

Previously you could have learnt:

› Sound waves are mechanical waves which travel through a medium such as air or water.

› Light waves are examples of electromagnetic waves which travel very quickly. These waves are able to travel through a vacuum as well as media such as air or glass.

When you drop a stone into a pond you see water ripples spreading outwards from the place where the stone landed (with a splash). As the ripples spread, the water surface moves up and down. These ripples are examples of **waves**.

The water ripples transfer energy and information. The energy moves outwards from the centre but the water itself does not move outwards. The shape of the waves provides us with the information about where the stone landed (if we did not see it land).

Test yourself on prior knowledge

1 Give three examples of waves which have not already been mentioned in the section above.

2 For each of your examples in Question 1, explain how the wave transfers energy and information.

3 How do thunderstorms help us to realise that light and sound transfer at different speeds?

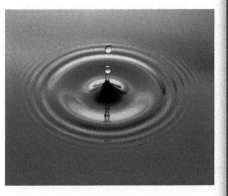

▲ Figure 34.1 Transverse waves in water.

Waves in air, fluids and solids

○ Transverse and longitudinal waves

A good way for us to visualise waves is to use a stretched 'slinky' spring. When two students stretch a slinky across the floor in a laboratory, they can transmit waves that travel slowly enough for us to see.

KEY TERM

Transverse wave A wave in which the vibration causing the wave is at right angles to the direction of energy transfer.

Transverse waves

Figure 34.2 shows a transverse wave. One student holds the end of the slinky stationary. The other end of the slinky is moved from side to side. A series of pulses move down the slinky, sending energy from one end to the other. The student holding the slinky still feels the energy as it arrives. None of the material in the slinky has moved permanently.

Water waves and **light waves** are two examples of transverse waves. The water shown in Figure 34.1 moves up and down. Energy is carried outwards by the wave, but water does not pile up at the edge of the pond.

direction of energy transfer

movements of hand from side to side

the tape moves from side to side

this end is held still

▲ Figure 34.2 The transverse waves transfer energy along the slinky from one end to the other. The coloured tape shows that the pulses cause the slinky to vibrate from side to side, just the same as the student's hand.

Longitudinal waves

Figure 34.4 shows a longitudinal wave. Energy is transmitted along the slinky by pulling and pushing the slinky backwards and forwards. This makes the slinky vibrate backwards and forwards. The vibration of the slinky is parallel to the direction of energy transfer. The coils are pushed together in some places (areas of **compression**). In other places the coils are pulled apart (areas of **rarefaction**). As energy is transferred, none of the material of the slinky moves permanently.

Sound is an example of a longitudinal wave. When a guitar string is plucked, energy is transferred through the air as the string vibrates backwards and forwards. However, the air itself does not move away from the string with the wave – there is not a vacuum left near the guitar.

direction of the vibration

this end is held still

R   C   R   C   R   C

direction of energy transfer

the coloured tape moves backwards and forwards

▲ **Figure 34.4** The coloured tape on the slinky shows that the pulses on the slinky move the coils backwards and forwards in the same direction as the student's hand moves.

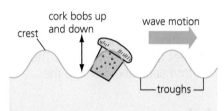

crest

cork bobs up and down

wave motion

troughs

▲ **Figure 34.3** Water waves cause a cork to bob up and down, as the peaks and troughs of the wave pass.

▶ **Figure 34.5** Earthquakes produce shock waves, which may be either longitudinal or transverse waves. Buildings are damaged by the transfer of energy, not by the movement of material.

Show you can...

A friend of yours has been away and missed the first lesson on waves. Your friend thinks that water waves are carried by the sea moving forwards.

Explain how we know that waves carry energy without the sea (or any other medium) moving along with the waves.

▲ **Figure 34.6** Waves in water.

Test yourself

1 Use diagrams to illustrate the nature of:
   a) transverse waves
   b) longitudinal waves
2 What do the terms 'area of compression' and 'area of rarefaction' mean?
3 Lucy is playing in her paddling pool. As she pushes a ball up and down in the water it makes a small wave (Figure 34.6).
   a) Which way does the water move as the waves move across the pool?
   b) Describe the motion of the other balls in the pool.
4 Describe how you would use a slinky to show that waves transfer both energy and information.

○ **Properties of waves**

We can learn more about the properties of waves by looking at water waves in a ripple tank as shown in Figure 34.7.

We can produce waves by lowering a dipper into the tank. A wooden bar is used to produce straight waves (plane waves) and a spherical dipper produces circular waves. By shining a light from above, we see the pattern of waves produced by the peaks and troughs of the waves (Figure 34.7).

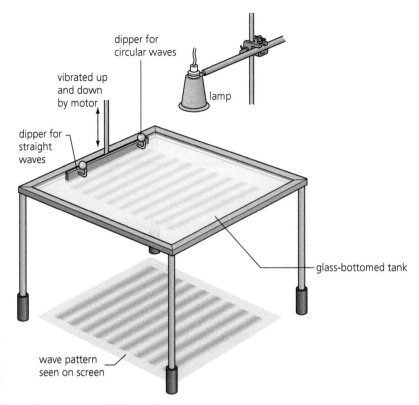

▲ **Figure 34.7** A ripple tank.

Describing waves

Waves are described by the terms amplitude (*A*), wavelength (*λ*), frequency (*f*) and period (T). *λ*, the symbol for wavelength, is a Greek letter pronounced *lambda*.

Figure 34.8 shows a transverse wave. This diagram helps us to understand the meaning of some terms.

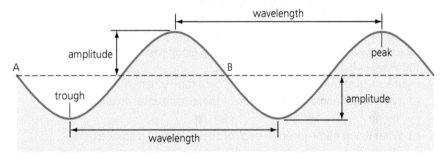

▲ **Figure 34.8** Features of a transverse wave.

- In Figure 34.8 the amplitude is the distance from a peak to the middle or from a trough to the middle.
- In Figure 34.8 a wavelength, *λ*, is the distance between two peaks or two troughs.
- Frequency is measured in **hertz** (Hz). A frequency of 1 Hz means a source is producing one wave per second. If a student, using a slinky, moves his hand from side to side and back twice each second, he produces two complete waves each second. The frequency of the waves is 2 Hz.
- The period, *T*, of a wave is the time taken to produce one wave.
- The frequency and time period of a wave are linked by this equation:

$$T = \frac{1}{f}$$

$$\text{period} = \frac{1}{\text{frequency}}$$

where period is in seconds, s

frequency is in hertz, Hz.

KEY TERMS

Amplitude The height of the wave measured from the middle (the undisturbed position of the water).

Wavelength The distance from a point on one wave to the equivalent point on the next wave.

Frequency The number of waves produced each second. It is also the number of waves passing a point each second.

Period The time taken to produce one wave.

Example

Calculate the period of a wave with a frequency of 10 Hz.

$$T = \frac{1}{10}$$
$$= 0.1\,s$$

You can remember this as follows: if a source produces 10 waves each second, the source takes 0.1 s to produce one wave.

Example

Sound travels at a speed of 330 m/s in air. Calculate the wavelength of a sound wave with a frequency of 660 Hz.

$$v = f\lambda$$
$$330 = 660 \times \lambda$$
$$\lambda = \frac{330}{660}$$
$$= 0.5\,m$$

Wave speed and the wave equation

The **wave speed** is the speed at which energy is transferred (or the wave moves) through the medium.

All waves obey this equation:

$$v = f\lambda$$

wave speed = frequency × wavelength

where wave speed is in metres per second, m/s

frequency is in hertz, Hz

wavelength is in metres, m.

Test yourself

5 Figure 34.9 shows wave pulses travelling at the same speed along two ropes. How are the wave pulses travelling along rope A different from the wave pulses travelling along rope B?

6 Figure 34.10 shows transverse waves on a rope of length 8 m.
  a) What is the name given to each of these horizontal distances?
     i) ae   ii) bf   iii) dg
  b) What is the name given to each of these vertical distances?
     i) ed   ii) ef
  c) Calculate:
     i) the wavelength of the wave ii) the amplitude of the wave
     iii) the horizontal distance ag

7 The periods of two waves are:
  a) 0.25 s   b) 0.01 s
  Calculate the frequency of each wave.

8 A stone is thrown into a pond and waves spread outwards. The waves travel with a speed of 0.4 m/s and their wavelength is 8 cm. Calculate the frequency of the waves.

9 Make a sketch to show the wavelength of a longitudinal wave.

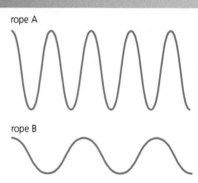

▲ Figure 34.9

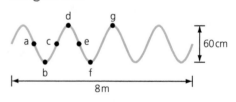

▲ Figure 34.10

Show you can...

Show you understand the key words which describe waves, by making a sketch which shows the wavelength and amplitude of a transverse wave.

(a)

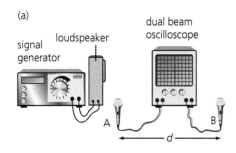

(b)

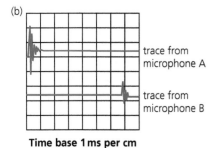

trace from microphone A

trace from microphone B

Time base 1 ms per cm

▲ **Figure 34.11** Measuring the speed of sound in a laboratory.

○ **Measuring wave speeds**

Three different methods of measuring wave speeds are described below.

Measuring the speed of sound in a laboratory

It is possible to measure wave speed using the equation:

$$\text{speed} = \frac{\text{distance travelled}}{\text{time taken}}$$

The problem with measuring the speed of sound in air is that sound travels quickly. So we must find a way to measure short times accurately in the laboratory. Figure 34.11 shows how we can do it.

● A loudspeaker is connected to a signal generator which produces short pulses of sound.
● Two microphones are placed near the loudspeaker but separated by a short distance, *d*. Each microphone is connected to an input of a dual beam oscilloscope.
● The oscilloscope can measure the time difference, *t*, between the sound reaching microphone A and microphone B.

Example

A student sets the microphones a distance of 220 cm apart. The time base on the oscilloscope is set to 1 ms per cm; this means that a distance of 1 cm on the horizontal scale corresponds to a time of 1 ms (0.001 s).

Using Figure 34.11(b) you can measure that the sound reaches microphone B 6.6 ms after the sound reaches microphone A.

So speed of sound $= \dfrac{d}{t}$

$$= \frac{2.2}{0.0066}$$

$$= 330 \, \text{m/s}$$

Remember: you must convert distance into metres and time into seconds.

▲ **Figure 34.12** Using echoes to measure the speed of sound in air.

Measuring the speed of sound outside

When outside, the echoes from a tall building can be used to measure the speed of sound in air.

1 Stand 40 m in front of a tall building and bang two blocks of wood together.
2 Each time you hear an echo, bang the blocks together again.
3 Have another student use a stopwatch to time how long it took to hear 10 echoes.

The results obtained by two students are given in Table 34.1.

● Calculate the mean value of the time for 10 echoes.
● How long does it take the sound to travel 80 m?
● Calculate the speed of sound given by these results.

Table 34.1

Trial number	Time in seconds
1	2.4
2	2.8
3	2.3

Make observations to identify the suitability of apparatus to measure the frequency, wavelength and speed of waves in a ripple tank and waves in a solid and take appropriate measurements

Waves in a ripple tank

Method

1 Set up the ripple tank as shown in Figure 34.7.
2 Pour enough water to fill the tank to a depth of about 5 or 6mm.
3 Adjust the wooden bar up or down so that it just touches the surface of the water.
4 Switch on the lamp and the electric motor.
5 Adjust the speed of the motor so that low frequency waves that can be counted are produced.
6 Move the lamp up or down so that a clear pattern can be seen on the floor.
7 If the pattern is difficult to see, placing a sheet of white paper or card on the floor under the tank and switching the laboratory lights off may help.
8 Use a metre ruler to measure across as many waves in the pattern as possible. Divide that length by the number of waves. This gives the wavelength of the waves.
9 Count the number of waves passing a point in the pattern over a given time (say 10 seconds).

Divide the number of waves counted by 10. This gives the frequency of the waves.

If you have connected the motor to a variable frequency power supply you can take the frequency directly from the power supply.

Analysing the results
Use the equation:

$$\text{wave speed} = \text{frequency} \times \text{wavelength}$$

and your values for wavelength and frequency to calculate the wave speed.

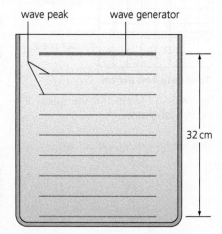

▲ **Figure 34.13** Eight wavelengths cover a distance of 32 cm, so one wavelength is equal to 32 ÷ 8 = 4 cm.

Waves in a (stretched string)

Figure 34.14 shows how the waves in a stretched string can be investigated. You can use either a string or an elastic cord for this investigation.

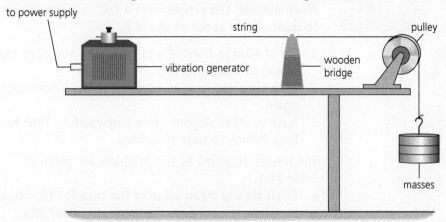

▲ **Figure 34.14** Investigating waves in a stretched spring.

Method

1 Switch on the vibration generator. The string (or elastic cord) will start to vibrate but not in any pattern.

2 Changing the mass attached to the string changes the tension in the string. Moving the wooden bridge changes the length of the string that vibrates.

3 Change the mass or move the wooden bridge until you see a clear wave pattern. The pattern will look like a series of loops. The length of each loop is half a wavelength.

4 Use a metre ruler to measure across as many loops as possible. Divide this length by the number of loops then multiply by two. Your final answer is equal to the wavelength of the wave.

5 The frequency of the wave is the same as the frequency of the power runin supply.

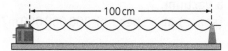

▲ **Figure 34.15** There are 10 loops in a distance of 100 cm. So the length of each loop is 10 cm making the wavelength of the wave equal to 20 cm.

Analysing the results

Use the equation:

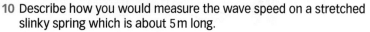

$$\text{wave speed} = \text{frequency} \times \text{wavelength}$$

and your values for wavelength and frequency to calculate the wave speed.

Take it further

Use the same apparatus to investigate how the velocity of a wave on a stretched string depends on the tension in the string. Identify suitable apparatus for making each of the measurements that will need to be taken.

Test yourself

10 Describe how you would measure the wave speed on a stretched slinky spring which is about 5 m long.

11 A student is using the apparatus in Figure 34.11. She moves the microphones closer together so that, on the oscilloscope, the horizontal distance between the traces from the two microphones is 4.2 cm. Calculate how far apart the microphones are. (The speed of sound in air is 330 m/s.)

Electromagnetic waves

Electromagnetic waves are transverse waves that transfer energy from the source of the waves to an absorb. There are different types of electromagnetic wave which produce different effects and have different uses. The wavelength of electromagnetic waves varies from about 10^{-12} m to over 1 km.

Although electromagnetic waves have a wide range of frequencies and wavelengths, they all have important properties in common:

- They are **transverse waves.**
- They transfer **energy** from one place to another.
- They obey the **wave equation:** $v = f\lambda$
- They can travel through a **vacuum.**
- They travel in a vacuum (space) at a speed of 300 000 000 m/s **(3×10^8 m/s).**

Electromagnetic waves form a continuous spectrum of wavelengths. Starting with the longest wavelength and going to the shortest, the groups of electromagnetic waves are: **radio, microwave, infrared, visible light** (red to violet), **ultraviolet, X-rays** and **gamma rays.**

Our eyes only detect visible light, so they detect a limited range of electromagnetic waves.

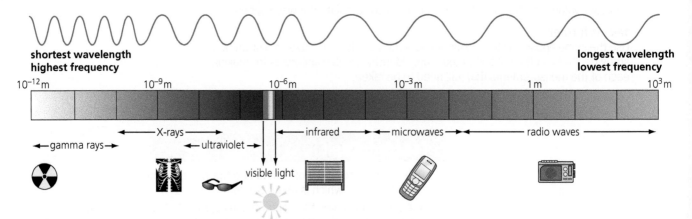

▲ **Figure 34.16** The waves in the electromagnetic spectrum have a range of different uses but also present a number of hazards.

TIP

It is easier to use standard form to solve problems when you are dealing with very large or very small numbers.

Example

A mobile phone uses electromagnetic waves with a frequency of 1.8×10^9 Hz. Calculate the wavelength of the waves.

$$v = f\lambda$$
$$3 \times 10^8 = 1.8 \times 10^9 \times \lambda$$
$$\lambda = \frac{3 \times 10^8}{1.8 \times 10^9}$$
$$\lambda = 0.17\,\text{m}$$

○ Properties of electromagnetic waves

Different wavelengths of electromagnetic waves are reflected, refracted, absorbed or transmitted differently by different substances and types of surface.

Refraction

When waves travel from one medium to another they usually change direction. This effect is called **refraction** (see Figure 34.17). Different wavelengths refract at different angles. For example, we see rainbows because the different colours of light (which have different wavelengths) are refracted through different angles (Figure 34.18).

Reflection, transmission and absorption

Different wavelengths of visible light behave differently when they are incident on various surfaces. A red shirt appears red because red light is reflected. All other colours (wavelengths) of light are absorbed.

A polished metal surface reflects most electromagnetic waves: wavelengths from radio waves to ultraviolet are reflected by metal. However, X-rays and gamma rays are able to pass through thin metal.

When a potato is cooked in a microwave oven, it absorbs microwaves. The wavelength of the microwaves is carefully chosen so that water molecules in the food absorb them.

▲ **Figure 34.17** This photograph shows an effect caused as light is refracted (changed in direction) by water in a glass.

▲ **Figure 34.18** Different colours of light refract through different angles.

Test yourself

12 Give five properties common to all electromagnetic waves.
13 Give the meanings of each of the following words when applied to electromagnetic waves: refraction, reflection, absorption and transmission.
14 a) A radio station transmits waves with a frequency of 100 MHz. Calculate the wavelength of the waves.
   b) A radio station transmits waves with a wavelength of 1500 m. Calculate the frequency of the waves.

TIP

Some electromagnetic waves have high frequencies. These can be given in kilohertz, kHz (1000 Hz, 10^3 Hz), megahertz, MHz (1 000 000 Hz, 10^6 Hz) and gigahertz, GHz (1 000 000 000, 10^9 Hz).

○ Refraction

Figure 34.19 shows how light is refracted as it passes through a rectangular block of glass.

● The angle between the incident ray and the normal is called the angle of incidence.
● The angle between the refracted ray and the normal is called the angle of refraction.

The light is refracted when it enters and leaves the glass because the **speed** of light in glass is different to the speed of light in air.

Light changes direction towards the normal when it enters the glass because light travels more slowly in glass than it does in air.

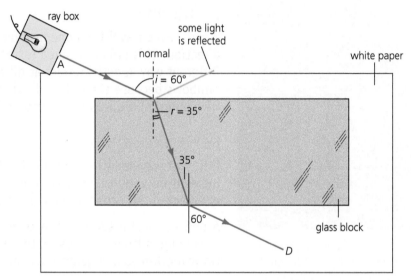

▲ **Figure 34.19** Refraction of light through a glass block.

Light changes direction away from the normal when it leaves the glass and enters the air because light travels faster in air than in glass.

Refraction is caused because waves change speed as they pass from one medium to another. All types of waves show refraction.

Water waves can be refracted. Figure 34.20 shows what happens when water waves cross from deep water to shallower water at an angle to the edge of the shallower water. In this case the waves:

● slow down
● become shorter in wavelength
● change direction.

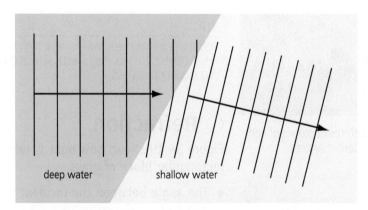

▲ **Figure 34.20** Refraction of water waves.

Test yourself

15 Draw diagrams to show how the water waves are refracted in each of the following cases.

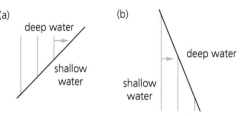

(a)

deep water

shallow water

(b)

deep water

shallow water

▲ **Figure 34.21**

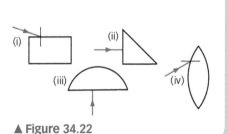

(i)

(ii)

(iii)

(iv)

▲ **Figure 34.22**

16 Draw diagrams to show the path of the light rays as they pass through the glass blocks shown in Figure 34.22.

○ **Producing electromagnetic waves**

Radio waves

Figure 34.23 shows how radio waves are transmitted and received.

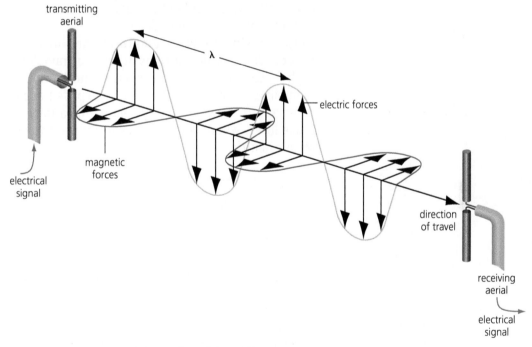

transmitting aerial

λ

electric forces

electrical signal

magnetic forces

direction of travel

receiving aerial

electrical signal

▲ **Figure 34.23** Transmitting and receiving radio waves.

1 A high frequency alternating current is supplied to the transmitting aerial. This makes electrons oscillate up and down the aerial.

2 An electromagnetic wave is emitted. This wave transfers energy in oscillating electric and magnetic fields. This is a transverse wave because the oscillations are at right angles to the direction of energy transfer.

3 When the electromagnetic wave reaches the receiving aerial, the electric and magnetic fields cause electrons to oscillate up and down the receiving aerial. This induces a current in an electrical circuit. **The alternating current has the same frequency as the radio wave itself.**

Radio waves also carry the information that we hear and see on radios and televisions.

Changes in atoms

Electromagnetic waves, ranging in wavelength from infrared to X-rays, can be generated or absorbed by changes in atoms.

Gamma rays

Gamma rays are generated when there are changes to the nucleus of an atom. When a gamma ray is emitted, the nuclear energy store of the atom decreases.

○ Hazards of electromagnetic waves

Too much exposure to some types of electromagnetic waves can be dangerous. The **higher the frequency** of the radiation, the **more damage** it is likely to do to the body. Ultraviolet, X-rays and gamma rays have the highest frequencies.

Ultraviolet, **X-rays** and **gamma rays** can all cause **mutations** to body cells. This can lead to **cancer**. We are most likely to be exposed to ultraviolet rays when we are in the sun. **Ultraviolet** waves can cause the **skin to age prematurely** and increase the risk of **skin cancer**. We take precautions against ultraviolet radiation by wearing high factor sun cream, sun glasses and by not exposing our skin for a long time. We are less likely to be exposed accidentally to X-rays and gamma rays. However, X-rays and gamma rays are used in medical procedures, where suitable precautions are taken to reduce the risk to our health.

Radiation dose

X-rays and gamma rays are ionising radiations. The damage that these radiations do to human tissue depends on the dose received. The effect of a radiation dose is measured in **sieverts** (Sv).

○ Uses and applications of electromagnetic waves

Electromagnetic waves have many practical applications. Some examples are given on the following pages.

● **Radio waves** are used to send **radio** and **television** signals from transmitters to our homes. These are called terrestrial signals as they go across the ground. Radio waves can be transmitted over long distances by reflecting them off the ionosphere (a layer in the Earth's upper atmosphere).

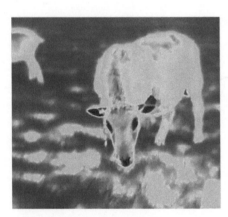

▲ **Figure 34.24** A photograph at night allows us to see the infrared radiation given out by warm bodies.

- **Microwaves** are used for **satellite communications** because they pass easily through the Earth's atmosphere. Many people now use satellite television. Microwaves are also used in mobile phone networks. These microwaves are transmitted using tall aerial masts.
- **Microwaves** are used in the home to **cook food** in microwave ovens. Some wavelengths of microwaves are absorbed by water molecules, and therefore cause heating.
- **Infrared waves** are used to provide **heating** from electrical heaters. We **cook** using infrared radiation in our ovens.
- We can also use **infrared cameras** to take photographs at night. Infrared has another popular use in remote controls.

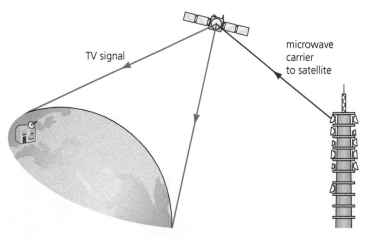

TV signal

microwave carrier to satellite

▲ Figure 34.25 A satellite dish can transmit a microwave signal that covers a large part of the Earth's surface.

▲ Figure 34.26 This person is using Bluetooth to link a mobile phone to a hands-free headset.

- We use **light waves** all the time to **see**. Light (and infrared) is also used in **fibre optic communication systems**. Information is coded into signals consisting of light (or infrared) pulses which are then transmitted along fibre optics. Many telephone links now use fibre optics rather than copper cables.
- Lasers that emit high-energy visible light are used for surgery.
- **Ultraviolet waves** are emitted by hot objects. Some substances can absorb the energy from ultraviolet radiation and then emit the energy as visible light. This is called fluorescence. Some types of **energy-efficient lamps** work by producing ultraviolet radiation. When electricity passes through a gas in the lamp, reactions occur and ultraviolet radiation is emitted. The ultraviolet radiation is absorbed by a chemical which covers the inside of the glass. The chemical fluoresces and visible light is emitted. Materials which fluoresce can also have applications in solving crimes (see Figure 34.28).

▲ Figure 34.27 Ultraviolet radiation can be used to aid the detection of counterfeit money.

▲ Figure 34.28 This man has been sprayed with 'smartwater'. Each batch of smartwater has a unique chemical signature. If a burglar gets covered with smartwater he can be linked to the scene of a particular crime when ultraviolet radiation is shone on him.

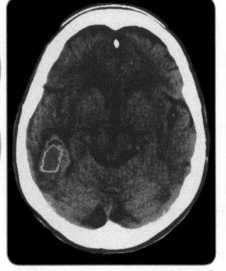

▲ Figure 34.29 A CT scan of the brain of a patient who has suffered a stroke. The area of the brain affected by the bleeding is shown in red. This is an example of an X-ray image.

TIP

X-rays are important in medical imaging, but both X-rays and gamma radiation are used in radiotherapy.

- **X-rays** can penetrate our bodies and therefore provide a valuable way to help doctors diagnose illness or damage to our bodies. In an X-ray photograph, bones, teeth and diseased tissues stand out because they absorb the X-rays. Many modern radiotherapy machines use high energy X-rays to kill cancer cells.

- **Gamma rays** are a penetrating radiation which can cause damage to body tissue. Sometimes this is useful. For example, gamma rays are used in radiotherapy to kill cancer cells.

Show you can...

Show you understand the nature of radio waves by completing this task.

Describe how an aerial transmits radio waves, and how an aerial receives radio waves.

Test yourself

17 Which parts of the electromagnetic spectrum have frequencies higher than ultraviolet waves?
18 Which part of the electromagnetic spectrum:
   a) is used in transmitting terrestrial signals
   b) could cause skin cancer
   c) is used in medical imaging?
19 Choose one part of the electromagnetic spectrum other than radio waves, and state one hazard of this wave and one application.

Investigate how the amount of infrared radiation absorbed or radiated by a surface depends on the nature of that surface.

Figure 34.30 shows a piece of apparatus called a 'Leslie cube'. The vertical sides of the cube have different colours or texture. The cube is used to compare the infrared radiation emitted from each surface.

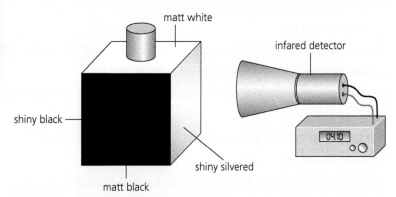

matt white

infared detector

shiny black

shiny silvered

matt black

▲ **Figure 34.30** Using a Leslie cube to compare the emission of infrared radiation from different surfaces.

Method

1 You will be using very hot or boiling water. Write a risk assessment for the investigation.

2 Place the Leslie cube onto a heat proof mat.

3 Fill the cube with very hot water and replace the lid of the cube.

4 Hold an infrared detector close to one of the sides, wait for the reading to settle and then record the reading in a suitable table.

5 Move the detector and take a reading from each side of the cube. Make sure that the detector is always the same distance from each side of the cube.

If a Leslie cube is not available, beakers covered in different materials, for example aluminium foil and black sugar paper, could be used. Fill the beakers with hot water and place a cardboard lid on each beaker. Take the temperature of the water in each beaker after 15 minutes. The cooler the water the faster the surface emits radiation.

Analysing the results

1 Draw a bar chart to show the amount of infrared radiation emitted for each type of surface.

2 What can you conclude from the results of your investigation?

Questions

1 Why is it important that the detector is held the same distance from each side of the cube?

2 Why is it appropriate to use the readings to draw a bar chart and not a line graph?

Taking it further

Figure 34.31 shows one way of comparing how well two different surfaces absorb infrared radiation. Before doing this investigation predict which drawing pin will drop off first. Give a reason for your prediction.

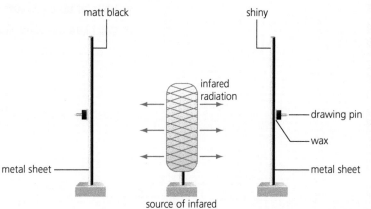

matt black

shiny

infared radiation

drawing pin

wax

metal sheet

metal sheet

source of infared

▲ **Figure 34.31** The drawing pin that drops off first is the one attached to the sheet with the surface that absorbs infrared radiation the fastest.

Chapter review questions

1 a) Explain the difference between a longitudinal and a transverse wave.

   b) Give two examples of each type of wave.

2 Name a type of electromagnetic wave that:

   a) is emitted by hot objects

   b) is used to communicate with satellites

   c) causes a sun tan.

3 A tuning fork vibrates at a frequency of 512 Hz. The speed of sound is 330 m/s.

   a) Calculate the period of the vibration of the tuning fork.

   b) Calculate the wavelength of the sound waves produced by the tuning fork.

4 A ship is close to a cliff and sounds its fog horn. An echo is heard.

   a) What causes the echo?

   Sound travels at 330 m/s. The echo is heard after 4 seconds.

   b) Calculate the distance from the ship to the cliff.

5 End A of the rope in Figure 34.32 is shaken up and down at a rate of five oscillations every 2 seconds.

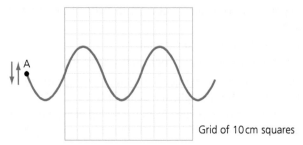

Grid of 10 cm squares

▲ Figure 34.32

   a) Determine the amplitude of the wave motion (maximum displacement to one side).

   b) Determine the wavelength.

   c) Calculate the time period.

   d) Calculate the frequency.

   e) How far do the waves travel during one time period?

   f) Calculate the wave speed.

Practice questions

1 a) Use the correct words from the box to complete the sentence.

absorbs	ionises	reflects	transmits

When X-rays enter the human body, soft tissue
_____ X-rays and bone _____
X-rays. [2 marks]

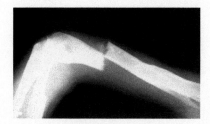

▲ Figure 34.33

b) Which one of the following is another use of X-rays?
- scanning unborn babies
- providing a sun tan
- killing cancer cells [1 mark]

c) Table 34.2 shows the dose received by the human body when different parts are X-rayed.

Table 34.2

Part of body X-rayed	Dose received by the human body in millisieverts
Chest	3
Skull	15
Back	120

How many chest X-rays would give a patient the same dose as one back X-ray? [2 marks]

2 A teacher and two students are measuring the speed of sound.

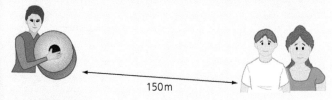

150m

▲ Figure 34.34

The teacher makes a sound by hitting two cymbals together. Each student starts a stopwatch when they see the teacher hit the cymbals. They each stop their stopwatch when they hear the sound.

a) Describe how a sound wave moves through the air. [3 marks]

b) The teacher repeats the experiment and the students record the readings on their stopwatch in Table 34.3.

Table 34.3

Student	Time in seconds
1	0.44, 0.46, 0.44, 0.48, 0.43
2	0.5, 0.6, 0.4, 0.4, 0.6

i) State the resolution of the first student's readings. [1 mark]

ii) State the equation linking speed, distance travelled and time taken. [1 mark]

c) The teacher was standing 150m from the students. Use the experimental data recorded by each student to calculate:

i) the average time recorded by each student [2 marks]

ii) the speed of sound calculated by each student. [2 marks]

Write each answer to an appropriate number of significant figures.

3 Figure 34.35 represents the electromagnetic spectrum.

radio waves	microwaves	infrared	A	ultraviolet	B	gamma rays

▲ Figure 34.35

a) Name the two parts of the spectrum labelled A and B. [2 marks]

b) Which electromagnetic radiation is used for night vision equipment? [1 mark]

c) Which electromagnetic radiation is used for transmitting signals to a satellite? [1 mark]

d) Which type of electromagnetic wave has the greatest frequency? [1 mark]

e) Exposure to excessive electromagnetic radiation can be harmful to the human body. For two named types of radiation, describe:

i) a harmful effect for each type [2 marks]

ii) how the risks of exposure can be reduced [2 marks]

4 Explain why a red flower looks red when seen in white light. [2 marks]

5 Figure 34.36 shows a wave on the sea.

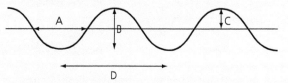

▲ Figure 34.36

a) i) Which letter shows the wavelength of the wave? [1 mark]

ii) Which letter shows the amplitude of the wave? [1 mark]

b) What type of wave is shown in the diagram? Describe how energy is transferred by the water. [2 marks]

c) A man watches some waves pass his boat. One wave passes him every 4 s.
Calculate the frequency of the waves. [2 marks]

Working scientifically:
Communication in science

Communicating scientific results and developments

Reports of events or scientific developments in the newspapers, on television and on the internet may often be oversimplified, inaccurate, biased or simply confusing. Following the nuclear accident at Fukushima in Japan, newspaper reports about the dangers of radioactivity caused a lot of uncertainty and concern for many people living in Japan. Recent reports in the media suggest a link between the increase in the numbers of children having cancer of the thyroid and the nuclear accident. Health scientists say this link has not been substantiated.

Peer review

One of the problems with media reports is that they are not peer-reviewed.

Peer review is a process scientists use to try and make sure that scientific results, developments and reports are correct. The work of one group of scientists will be studied and checked by another group of scientific experts. They will look at the evidence and the way in which the evidence was obtained. Peer review is a way of having someone who understands the subject checking over another person's work. Peer review usually, but not always, takes place before scientific work is published and made available to a wider audience.

Particles measured travelling faster than the speed of light

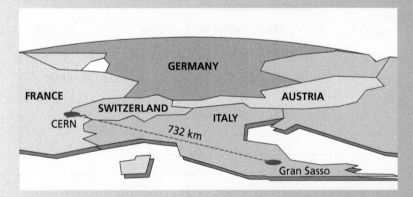

▲ Figure 34.37 The beam of neutrinos travelled from Switzerland to Italy.

In 2011, a group of physicists published the results of an experiment called OPERA. The experiment had been carried out over three years. It involved timing a beam of particles, called neutrinos, travelling 730 km from the CERN laboratory in Switzerland to the Gran Sasso laboratory in Italy. The results were amazing; they suggested that the particles could travel faster than the speed of light!

Having measured over 16 000 events, the physicists decided to publish their results for checking and review by other groups of physicists. In 2012 a mistake was found in the experiment which had caused an error in the original results. The time taken by the neutrinos to travel the 730 km was in fact always 60 nanoseconds (60×10^{-9} s) more than the time recorded in the experiment.

1 Calculate how long it would take a neutrino travelling at the speed of light to go from CERN to Gran Sasso.

2 What type of error did the physicists in the OPERA experiment make?

3 Why was it important that the physicists published the results of the OPERA experiment?

Fact or opinion?

Headlines in newspapers are designed to grab your attention, to make you want to read the rest of the article. Mobile phones and phone masts are often in the headlines.

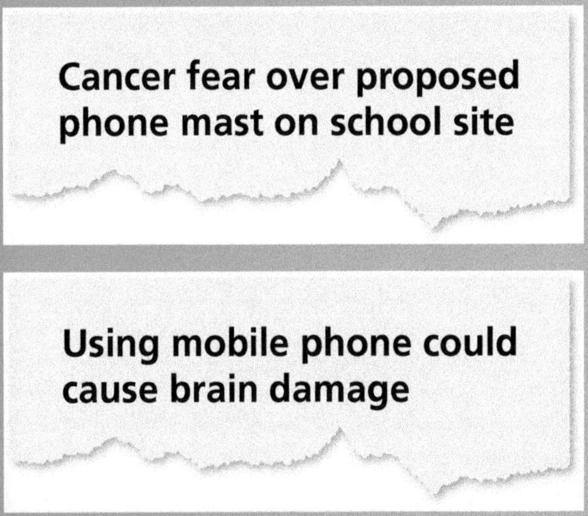

Cancer fear over proposed phone mast on school site

Using mobile phone could cause brain damage

▲ Figure 34.38

4 Do the newspaper headlines give facts or opinions?

There has been a lot of scientific research into the potential health effects of using a mobile phone. Some studies have suggested a link between mobile phone use and health problems such as tiredness, headaches and brain tumours. A study of 750 people in Sweden led to the suggestion that using a mobile phone over a period of at least 10 years increases the risk of a tumour on the nerve between the ear and the brain by four times. However, the evidence from these and other studies is not conclusive. What the evidence does suggest is a need for more research and for continued peer review of new evidence.

5 What do scientists mean by conclusive evidence?

6 Do you think that the number of people studied in Sweden was enough to give firm evidence of a link between mobile phone use and developing a tumour? Give reasons for your answer.

35 Magnetism and electromagnetism

At home you might use a magnet for a purpose as simple as attaching your shopping list to the fridge door. The photograph opposite shows a completely different application of magnetism. Here you can see the central view of a particle detector at the Large Hadron Collider in the European Centre for Nuclear Research (CERN). The particle detector is an arrangement of eight magnetic coils (each weighing 100 tonnes), which are used to deflect high-energy particles into the central detector. The coils are cooled to a temperature of $-269\,°C$, at which temperature the coils are superconducting – this means that the coils have no electrical resistance. A current of about 21 000 A is used to produce very strong magnetic fields.

Specification coverage

This chapter covers specification points 6.7.1.1 to 6.7.2.3 and is called Magnetism and electromagnetism.

It covers permanent and induced magnetism, magnetic forces and fields, and the motor effect.

Prior knowledge

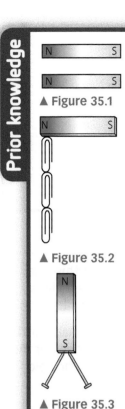

▲ Figure 35.1

▲ Figure 35.2

▲ Figure 35.3

Previously you could have learnt:

> Some materials are magnetic. The most common magnetic materials are iron and steel.

> Magnets have two poles, north and south.

> Magnets attract magnetic materials at a distance. Magnetism is a non-contact force.

> Two like poles repel each other: a north pole repels a north pole; a south pole repels a south pole.

> Two unlike poles attract: a north pole attracts a south pole.

Test yourself on prior knowledge

1 Copy Figure 35.1 and add arrows to show the direction of the magnetic forces acting on each magnet.

2 In Figure 35.2, three steel paper clips are attracted to a magnet. Copy the sentences below and fill in the gaps to explain why this happens.
  • The magnetic field of the magnet ____ the paper clips.
  • The top of each clip becomes a ____ magnetic pole and the bottom of each paper clip becomes a ____ magnetic pole. Each clip attracts the one below. The size of the magnetic attraction is greater than the ____ of each clip.

3 Explain why the steel pins repel each other in Figure 35.3.

Permanent and induced magnetism, magnetic forces and fields

◯ Poles

KEY TERMS

Magnetic Materials that are attracted by a magnet.

North-seeking pole The end of the magnet that points north.

South-seeking pole The end of the magnet that points south.

Some metals, for example **iron**, **steel**, **cobalt** and **nickel**, are magnetic. A magnet will attract them. If you drop some steel pins on the floor, you can pick them up using a magnet. A magnetic force is an example of a **non-contact force**, which acts over a distance.

In Figure 35.4, you can see a bar magnet which is hanging from a fine thread. When it is left for a while, one end always points north. This end of the magnet is called the north-seeking pole. The other end of the magnet is the south-seeking pole. We usually refer to these poles as the **north** and **south poles** of the magnet.

The magnetic forces on steel pins, iron filings and other magnetic objects are always greatest when they are near the poles of a magnet. Every magnet has two poles which are equally strong.

When you hold two magnets close together you find that two north poles (or two south poles) repel each other, but a south pole attracts a north pole (Figure 35.5).

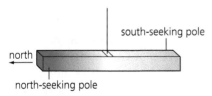

▲ **Figure 35.4** This suspended bar magnet has a north pole and a south pole.

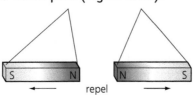

(a) like poles repel

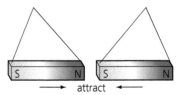

(b) unlike poles attract

▲ **Figure 35.5** Two like poles will repel, while unlike poles attract.

Magnetic fields

> **TIP**
> A magnetic field is the area around a magnet in which a magnetic force acts on magnetic objects or other magnets.

There is a **magnetic field** in the area around a magnet. In this area, a force acts on a magnetic object or another magnet. If the field is strong, the force is big. If the field is weak, the force is small.

The direction of a magnetic field can be found by using a small plotting compass. The compass needle always points along the direction of the field. Figure 35.6 shows how you can investigate the magnetic field near to a bar magnet using a compass.

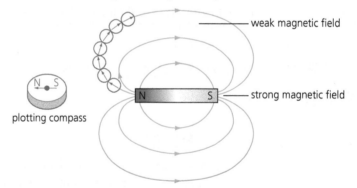

plotting compass

weak magnetic field

strong magnetic field

▲ **Figure 35.6** Investigating the magnetic field near to a bar magnet using a compass.

We use magnetic field lines to represent a magnetic field. Magnetic field lines always start at a north pole and finish at a south pole. When the field lines are close together, the field is strong. The further apart the lines are, the weaker the field is. A magnetic field is strongest at the poles of a magnet and gets weaker as the distance from the magnet increases.

Magnetic field lines are not real, but are a useful model to help us to understand magnetic fields.

Compass

▲ **Figure 35.7** A compass, which you can use to find north when you are walking.

Figure 35.7 shows a compass that you can use to find north when you are walking. A compass contains a small **bar magnet** that can rotate. When a compass is held at rest in your hand, the needle always settles along a north-south direction. This behaviour provides evidence that the Earth has a magnetic field. The pole that points towards north is the a north-seeking pole. Since unlike poles attract, this tells us that at the magnetic north pole, there is a south-seeking pole. Figure 35.8 shows the shape of the Earth's magnetic field.

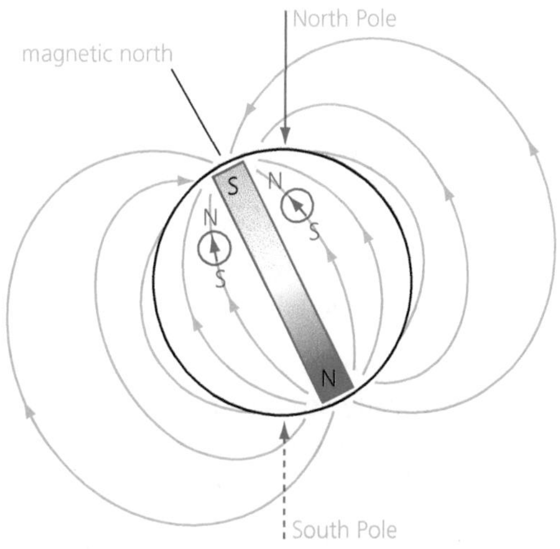

North Pole

magnetic north

South Pole

Permanent and induced magnets

Some magnets are permanent magnets. Permanent magnets produce their own magnetic field. They always have a north and south pole. You can check to see if a magnet is a permanent magnet by placing it near to another magnet that you know is permanent. If both magnets are permanent, then you will be able to see that they can repel each other, as well as attract.

An induced magnet is a material that becomes magnetic when it is placed in a magnetic field. Induced magnets are **temporary magnets**. An induced magnet is always attracted towards a permanent magnet. This is because the induced magnet is magnetised in the direction of the permanent magnet's field. When the induced magnet is taken away from the permanent magnetic field, it will lose all (or most) of its magnetism quickly.

KEY TERMS

Permanent magnet A magnet which produces its own magnetic field. It always has a north pole and a south pole.

Induced magnet A magnet which becomes magnetic when it is placed in a magnetic field.

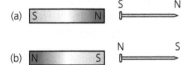

▲ **Figure 35.9** The nail becomes magnetised by the magnet's field. The nail is always attracted to the magnet.

Show you can...

Complete this task to show you understand the nature of magnets.

Plan an experiment to show the difference between permanent and induced magnets.

Test yourself

1 Which of the following items will a bar magnet pick up?
brass screw   steel pin
iron nail   aluminium can

2 a) Figure 35.10 shows a bar magnet surrounded by four plotting compasses. Copy the diagram. Mark in the direction of the compass needle for the positions B, C and D.

  b) Which is the north pole of the magnet, X or Y? Give the reason for your answer.

3 Two bar magnets have been hidden in a box. Use the information in Figure 35.11 to suggest how they are arranged inside the box.

4 a) Explain what is meant by a permanent magnet.

  b) Explain what is meant by an induced magnet.

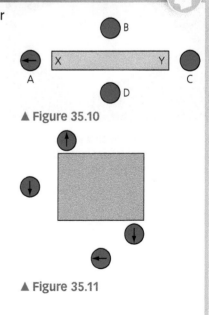

▲ Figure 35.10

▲ Figure 35.11

Electromagnetism and the motor effect

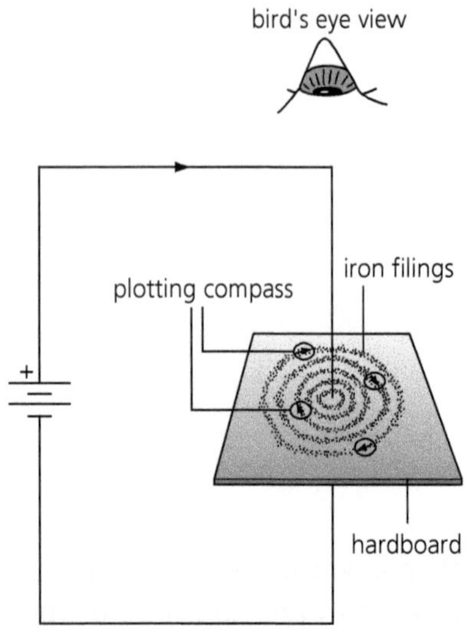

bird's eye view

plotting compass

iron filings

hardboard

▲ **Figure 35.12** This experiment shows there is a magnetic field around a current-carrying wire.

◯ The magnetic field near a straight wire conducting an electrical current

An electric current in a conducting wire produces a magnetic field around the wire.

In Figure 35.12, a long straight wire carrying an electric current is placed vertically so that it passes through a horizontal piece of hardboard. Iron filings have been sprinkled onto the board to show the shape of the field. Here is a summary of the important points of the experiment:

- When the current is small, the magnetic field is too weak to notice. However, when a large current is used, the iron filings show a **circular magnetic field pattern** (see Figure 35.13).
- The magnetic field gets weaker further away from the wire.
- The direction of the magnetic field can be found using a compass. If the **current** direction **is reversed**, the direction of the **magnetic field is reversed**.

Figure 35.14 shows the pattern of magnetic field lines surrounding a wire. When the current flows into the paper (shown ⊗) the field lines point in a clockwise direction around the wire. When the current flows out of the paper (shown ⊙) the field lines point anti-clockwise.

The right-hand grip rule will help you to remember this (Figure 35.15). Put the thumb of your right hand along a wire in the direction of the current. Your fingers will point in the direction of the magnetic field.

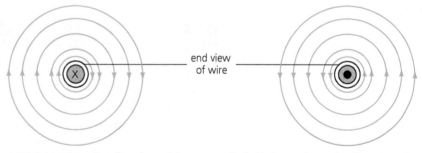

end view of wire

▲ **Figure 35.14** The direction of the magnetic field depends on the direction of the current.

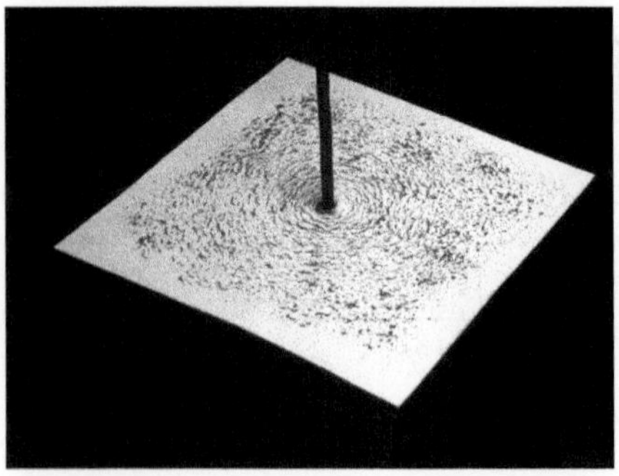

▲ **Figure 35.13** The circular shape of the magnetic field is shown by the pattern of iron filings.

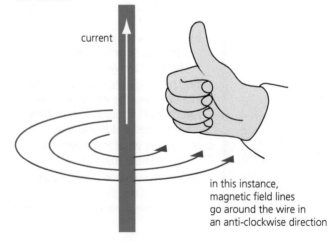

current

in this instance, magnetic field lines go around the wire in an anti-clockwise direction

▶ **Figure 35.15** The right-hand grip rule.

The magnetic field of a solenoid

Figure 35.16 shows the magnetic field that is produced by a current flowing through a long coil of wire or solenoid. The magnetic field from each loop of wire adds on to the next. The result is a magnetic field that is like that of a long bar magnet.

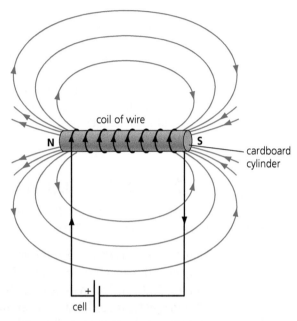

coil of wire

N   S

cardboard cylinder

+
cell

▲ **Figure 35.16** The magnetic field around a solenoid is similar in shape to that of a bar magnet.

The strength of the magnetic field produced by a solenoid can be **increased** by:

- using a **larger current**
- using **more turns** of wire
- putting the turns **closer** together
- putting an **iron core** into the middle of the solenoid to make an **electromagnet**.

▲ **Figure 35.17** An electromagnet in action.

Electromagnets

Figure 35.17 shows a practical laboratory electromagnet. It is made into a strong magnet by two coils with many turns of wire which increase the magnetising effect of the current. When the current is switched off, the magnet loses its magnetism and so the iron filings fall off.

Test yourself

5 List four ways to increase the strength of an electromagnet.
6 Why must you use insulated wire to make an electromagnet?
7 Figure 35.18 represents a wire placed vertically, with the current flowing out of the paper towards you. Copy the diagram and draw the magnetic field lines round the wire, showing how the field strength decreases with distance away from the wire.

▲ **Figure 35.18**

Test yourself

8 Figure 35.19 shows a long Perspex tube with wire wrapped round it to make a solenoid.

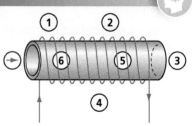

a) Copy the diagram and mark in the direction of the compass needles 1–6 when the current flows through the wire.

b) Which end of the solenoid acts as the south pole?

▲ Figure 35.19

c) What happens when the current is reversed?

d) Copy the diagram again, leaving out the compasses. Sketch the shape of the magnetic field.

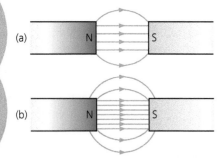

(a)

(b)

▲ Figure 35.20 Lines of magnetic flux between two pairs of magnets. The flux density is greater in (b) than in (a).

KEY TERM

Flux density The number of lines of magnetic flux in a given area.

○ Magnetic flux density

We represent magnetic fields by drawing lines that show the direction of a force on a north pole. These lines are also known as lines of **magnetic flux**.

Figure 35.20 shows the lines of magnetic flux between two pairs of magnets. The magnets in Figure 35.20(b) are stronger than the magnets in Figure 35.20(a). This means that they will exert a stronger attractive force on a magnetic material such as an iron nail. We show that the magnets are stronger by drawing more lines of magnetic flux for the area of the magnets.

The strength of the magnetic force is determined by the flux density, B.

The flux density is measured in tesla, T. A laboratory bar magnet produces a flux density of about 0.1 T near to its poles.

Calculating the force on a wire

The force on a wire of length L at right angles to a magnetic field and carrying a current, I, is given by the equation $F = BIL$:

$$F = BIL$$

force = magnetic flux density × current × length

where force is in newtons, N
    magnetic flux density is in tesla, T
    current is in amperes, A
    length is in metres, m.

Example

In Figure 35.21 the wire carries a current of 3.0A. Calculate the force acting on the wire.

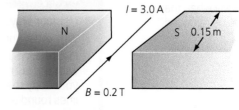

Answer

$F = BIL$
  $= 0.2 \times 3.0 \times 0.15$
  $= 0.09\,\text{N}$

▲ Figure 35.21

▲ **Figure 35.22** The aluminium foil carrying a current is pushed out of the magnetic field.

The motor effect

In Figure 35.22 you can see a piece of aluminium foil between the poles of a strong magnet. A current through the foil has caused it to be pushed down, away from the poles of the magnet. Reversing the current would make the foil move upwards, again away from the poles of the magnet. This is called the motor effect. It happens because of an interaction between the two magnetic fields: one from the permanent magnet and one produced by the current in the foil.

Combining two magnetic fields

In Figure 35.23 you can see the way in which the two fields combine. By itself the field between the poles of the magnet would be of constant strength and direction. The current around the foil produces a circular magnetic field. In one direction the magnetic field from the current squashes the field between the poles of the magnet. It is the squashing of the field that produces a force on the foil, upwards in this case.

The size of the force acting on the foil depends on:

- the **magnetic flux density** between the poles
- the size of the **current**
- the **length of the foil** between the poles.

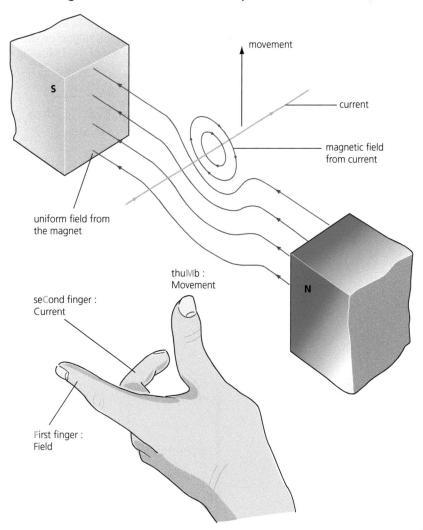

▶ **Figure 35.23** You can use the left-hand rule to predict the direction of movement of the wire.

The left-hand rule

To predict the direction in which a straight conductor moves in a magnetic field you can use **Fleming's left-hand rule** (Figure 35.23). Spread out the first two fingers and the thumb of your left hand so that they are at right angles to each other. Let your first finger point along the direction of the magnet's field (north to south), and your second finger point in the direction of the current (positive to negative). Your thumb then points in the direction in which the wire moves.

This rule works when the field and the current are at right angles to each other. When the field and the current are parallel to each other, there is no force on the wire and it stays where it is.

Try using the left-hand rule to show that the direction of the force on the conductor will reverse if:

- the direction of the magnetic field is reversed
- the direction of the current is reversed.

TIP ✓

The key thing to remember is how to use the left-hand rule to help you recall that the magnetic field current and the movement of the wire are all at right angles to each other.

Test yourself

9 State two ways of increasing the force on a conductor carrying a current in a magnetic field.

10 How is it possible to position a wire carrying a current in a magnetic field so that there is no force acting on the wire?

11 Use the left-hand rule to predict the direction of the force on the wire in each of the following cases.

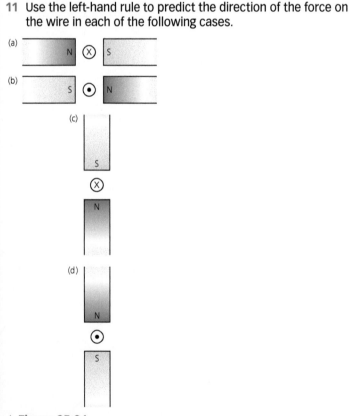

▲ Figure 35.24

12 A wire of length 0.2 m is placed at right angles to a region with a magnetic flux density 2.0 T. A current of 4.5 A is passed through the wire. Calculate the force acting on the wire.

Electric motor

Figure 35.25 shows a coil of wire that is able to rotate about an axle. When a current flows, as shown in the diagram, there is an downward force on the side CD and a upward force on the side AB. So the coil begins to rotate anti-clockwise, but the coil will rotate no further than a vertical position.

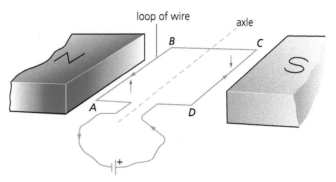

▲ **Figure 35.25** When the current flows, the coil begins to rotate clockwise.

Figure 35.26 shows how we can design a motor so that direct current can keep a coil rotating all the time.

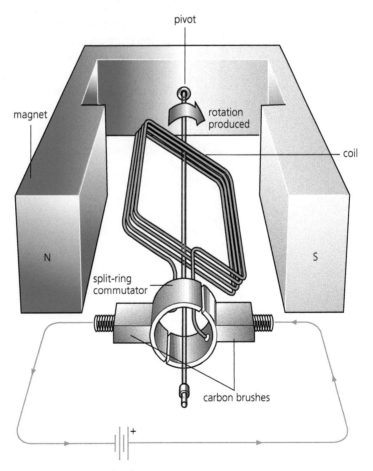

▲ **Figure 35.26** A simple motor.

The coil is kept rotating by using a **split-ring commutator**, which rotates with the coil between the carbon brush contacts. As the coil passes the vertical position, the two halves of the commutator change contact from one carbon brush to the other. This causes the direction of the current in the coil to reverse, so that the forces continue to act on the coil in the same direction. The coil will continue to rotate clockwise as long as there is a current.

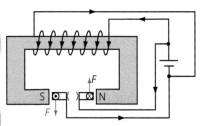

▲ **Figure 35.27** A model electric motor.

Test yourself

13 Look at Figure 35.25. Explain why there is no force acting on the side BC.

14 Give three ways in which you can increase the speed of rotation of the coil of an electric motor.

15 Figure 35.27 shows a model electric motor. In the diagram the current flows round the coil in the direction A to B to C to D.
   a) What is the direction of the force on:
      i)   side AB   ii) side CD?
   b) In which direction will the coil rotate?
   c) In which position is the coil most likely to stop?

16 Figure 35.28 shows an electric motor that is made using an electromagnet. The arrows show the direction of the forces on the coil.
   a) The battery is now reversed. What effect does this have on:
      i)   the polarity of the magnet
      ii) the direction of the current in the coil
      iii) the forces acting on the coil?
   b) Explain why this motor works using an a.c. supply as well as a d.c. supply.
   c) Why are the electromagnet and coil run in parallel from the supply rather than in series?

▲ **Figure 35.28** The current and direction of forces in an electric motor.

Show you can...

Show you understand the purpose of the parts of a motor by explaining how the split-ring commutator keeps an electric motor spinning in the same direction all the time.

Chapter review questions

1 a) Draw accurately the magnetic field pattern around a bar magnet.

  b) Explain how you would use a compass to plot this magnetic field pattern.

2 Draw carefully the shape of a magnetic field close to:

  a) a long wire   b) a long solenoid

  when each is carrying a current.

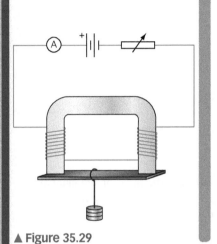

▲ Figure 35.29

3 Figure 35.29 shows an experimental arrangement to investigate the action of an electromagnet. Coils of wire have been wound round a C-shaped soft iron core.

At the bottom of the C-core, a soft iron bar stays in place because of the attraction of the electromagnet.

The strength of the electromagnet is measured by increasing the weight hanging on the bar until it falls off.

Table 35.4 below shows some results for this experiment.

Table 35.4

Maximum weight on the iron bar in N	0	0.6	1.2	1.8	2.3	2.7	3.0	3.2	3.3	3.4	3.4
Current in the magnet coils in A	0	0.2	0.4	0.6	0.8	1.0	1.2	1.4	1.6	1.8	2.0

  a) Plot a graph of the maximum weight supported by the electromagnet (y-axis) against the current in the magnet coils (x-axis).

A second student does the same experiment, but uses more turns of wire on her electromagnet.

  b) Sketch a second graph, using the same axes, to show how the maximum weight supported changes with the current now.

Practice questions

1 The diagram shows a magnetic screwdriver that has picked up and is holding a small metal screw.

a) What type of material is the screw made from? [1 mark]

aluminium   copper   steel

b) The magnetic force between the screw driver and screw is an example of a non-contact force.

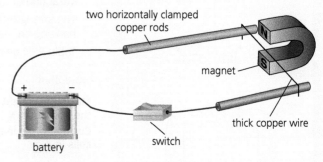

▲ Figure 35.30

Which **one** of the following is also a non-contact force?

air resistance   friction   gravity   [1 mark]

c) Explain how you can use a bar magnet to show that an unmarked bar of material is either a bar magnet or an unmagnetised bar of iron. [3 marks]

2 Describe what an induced magnet is. [1 mark]

3 a) Describe what a magnetic field is. [1 mark]

b) State the features of the magnetic field around a bar magnet. [2 marks]

c) Describe the magnetic field around a straight wire that is conducting an electrical current. [2 marks]

4 State three ways in which the magnetic field produced by a solenoid can be increased. [3 marks]

H 5 Figure 35.31 shows a diagram of an electric motor. As the current flows from the battery, the coil and split-ring commutator spin.

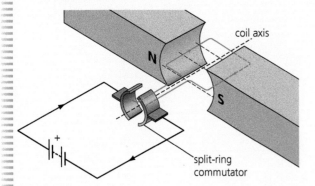

▲ Figure 35.31

a) Explain why the coil spins when a current flows through it. [3 marks]

b) Without changing the coil, give two ways in which it could be made to spin faster. [2 marks]

c) Give two ways in which the coil could be made to spin in the opposite direction. [2 marks]

6 A student investigates how a thick copper wire can be made to move in a magnetic field. Figure 35.32 shows the apparatus.

two horizontally clamped copper rods

magnet

thick copper wire

switch

battery

▲ Figure 35.32

The wire is placed between the poles of the magnet.

a) Use the information in the diagram to predict the direction of motion of the wire. [1 mark]

b) Explain what happens to the motion of the wire when the magnet is turned so the N pole is below the wire. [2 marks]

Appendix: the periodic table

Key

relative atomic mass
atomic symbol
name
atomic (proton) number

| 1 | H hydrogen 1 |

1	2											3	4	5	6	7	0
																	4 **He** helium 2
7 **Li** lithium 3	9 **Be** beryllium 4											11 **B** boron 5	12 **C** carbon 6	14 **N** nitrogen 7	16 **O** oxygen 8	19 **F** fluorine 9	20 **Ne** neon 10
23 **Na** sodium 11	24 **Mg** magnesium 12											27 **Al** aluminium 13	28 **Si** silicon 14	31 **P** phosphorus 15	32 **S** sulfur 16	35.5 **Cl** chlorine 17	40 **Ar** argon 18
39 **K** potassium 19	40 **Ca** calcium 20	45 **Sc** scandium 21	48 **Ti** titanium 22	51 **V** vanadium 23	52 **Cr** chromium 24	55 **Mn** manganese 25	56 **Fe** iron 26	59 **Co** cobalt 27	59 **Ni** nickel 28	63.5 **Cu** copper 29	65 **Zn** zinc 30	70 **Ga** gallium 31	73 **Ge** germanium 32	75 **As** arsenic 33	79 **Se** selenium 34	80 **Br** bromine 35	84 **Kr** krypton 36
85 **Rb** rubidium 37	88 **Sr** strontium 38	89 **Y** yttrium 39	91 **Zr** zirconium 40	93 **Nb** niobium 41	96 **Mo** molybdenum 42	[98] **Tc** technetium 43	101 **Ru** ruthenium 44	103 **Rh** rhodium 45	106 **Pd** palladium 46	108 **Ag** silver 47	112 **Cd** cadmium 48	115 **In** indium 49	119 **Sn** tin 50	122 **Sb** antimony 51	128 **Te** tellurium 52	127 **I** iodine 53	131 **Xe** xenon 54
133 **Cs** caesium 55	137 **Ba** barium 56	139 **La*** lanthanum 57	178 **Hf** hafnium 72	181 **Ta** tantalum 73	184 **W** tungsten 74	186 **Re** rhenium 75	190 **Os** osmium 76	192 **Ir** iridium 77	195 **Pt** platinum 78	197 **Au** gold 79	201 **Hg** mercury 80	204 **Tl** thallium 81	207 **Pb** lead 82	209 **Bi** bismuth 83	[209] **Po** polonium 84	[210] **At** astatine 85	[222] **Rn** radon 86
[223] **Fr** francium 87	[226] **Ra** radium 88	[227] **Ac*** actinium 89	[261] **Rf** rutherfordium 104	[262] **Db** dubnium 105	[266] **Sg** seaborgium 106	[264] **Bh** bohrium 107	[277] **Hs** hassium 108	[268] **Mt** meitnerium 109	[271] **Ds** darmstadtium 110	[272] **Rg** roentgenium 111	[285] **Cn** copernicium 112	[286] **Uut** ununtrium 113	[289] **Fl** flerovium 114	[289] **Uup** ununpentium 115	[293] **Lv** livermorium 116	[294] **Uus** ununseptium 117	[294] **Uuo** ununoctium 118

* The Lanthanides (atomic numbers 58 – 71) and the Actinides (atomic numbers 90 – 103) have been omitted.

Relative atomic masses for **Cu** and **Cl** have not been rounded to the nearest whole number.

Glossary

Abiotic factors The non-living parts of the environment.

Accurate A measurement or calculated value that is close to the true value.

Acid Solution with a pH less than 7; produces H+ ions in water.

Acid rain Precipitation that is acidic as a result of air pollution.

Activation energy The minimum energy particles must have to react.

Active site The region of an enzyme that binds to its substrate.

Active transport The net movement of particles from an area of low concentration to an area of higher concentration using energy.

Adrenaline A hormone produced by your adrenal glands that causes an increase in heart rate ready for a 'fight or flight' response.

Adrenal glands Glands that produce adrenaline.

Aerobic In the presence of oxygen.

Aerobic respiration Respiration using oxygen.

Alkali Solution with a pH more than 7; produces OH– ions in water.

Alkali metals The elements in Group 1 of the periodic table (including lithium, sodium and potassium).

Alkanes A homologous series of saturated hydrocarbons with the general formula C_nH_{2n+2}.

Alleles Two versions of the same gene, one from your mother and the other from your father.

Alloy A mixture of a metal with small amounts of other elements, usually other metals.

Alpha particle A particle formed from two protons and two neutrons.

Alternative energy resources Resources other than fossil fuels. The resources may or may not be renewable. Nuclear power is not a renewable energy resource, but tidal power is. Alternative energy resources do not contribute to global warming.

Alveoli Tiny air sacs found in the lungs through which gases exchange between blood and air.

Amino acids The sub-units of protein.

Amplitude The height of the wave measured from the middle (the undisturbed position of the water).

Anaerobic In the absence of oxygen.

Angle of reflection The angle between the reflected ray and the normal.

Anode An electrode where oxidation takes place (oxidation is the loss of electrons) – in electrolysis, it is the positive electrode.

Anomalous A result that doesn't fit the expected pattern.

Antibiotic A group of medicines, first discovered by Sir Alexander Fleming, that kill bacteria and fungi but not viruses.

Antibiotic-resistant bacterium A bacterium that cannot be killed by antibiotics.

Antibody A protein produced by lymphocytes that recognises pathogens and helps to clump them together.

Antigen A protein on the surface of a pathogen that antibodies can recognise as foreign.

Antimicrobial secretions Chemicals that destroy pathogens.

Antitoxin A protein produced by your body to neutralise harmful toxins produced by pathogens.

Aqueous Dissolved in water.

Artery A large blood vessel that takes blood from the heart.

Asexual reproduction Reproduction involving one parent, giving genetically identical offspring.

Atom The smallest part of an element that can exist. A particle with no electric charge made up of a nucleus containing protons and neutrons surrounded by electrons in energy levels.

Atomic number Number of protons in an atom.

Atrium (plural atria) An upper chamber of the heart surrounded by a thin wall of muscle.

Avogadro constant The number of atoms, molecules or ions in one mole of a given substance (the value of the Avogadro constant is 6.02×10^{23}).

Axon The extension of a nerve cell along which electrical impulses travel.

Barrier A contraceptive method that prevents sperm reaching an egg.

Battery Two or more chemical cells connected together.

Behavioural adaptation An advantage to an organism as a result of behaviour, such as a courtship display.

Benign A non-cancerous tumour that does not spread.

Beta particle A fast moving electron.

Biconcave Describes a shape with a dip that curves inwards on both sides.

Bile A green-coloured liquid produced by your liver, stored by your gall bladder and released into your small intestine to help break down fats.

Binary fission The asexual reproduction of bacteria.

Biodiversity A measure of the different species present in a community.

Biofuel Fuel produced from biological material. Biofuels are provided by trees such as willow that can be grown specifically as energy resources.

Bioleaching The use of bacteria to produce soluble metal compounds from insoluble metal compounds.

Biomass A resource made from living or recently living organisms.

Biotic factors The living parts of the environment.

Blood plasma The straw-coloured liquid that carries our blood cells and dissolved molecules.

Braking distance The distance a car travels while the car is stopped by the brakes.

Breeding programme Activity of zoos to breed captive animals together to increase their gene pool.

Burette A glass tube with a tap and scale for measuring liquids to the nearest $0.1\,cm^3$.

Calibrate Mark a scale onto a measuring instrument so that you can give a value to a measured quantity.

Capillaries Tiny blood vessels found between arteries and veins that carry blood into tissues and organs.

Carbohydrates Biological molecules containing carbon, hydrogen and oxygen.

Carbon footprint The amount of carbon dioxide and other greenhouse gases given out over the full life cycle of a product, service or event.

Carbon neutral Fuels and processes whose use results in zero net release of greenhouse gases to the atmosphere.

Carcinogen A cancer-causing substance.

Catalyst A substance that speeds up a chemical reaction but is not used up.

Categoric variable A variable with values that are given a name or label.

Cathode An electrode where reduction takes place (reduction is the gain of electrons) – in electrolysis it is the negative electrode.

Causation The act of making something happen.

Cell Two electrodes in an electrolyte used to generate electricity.

Cell cycle A series of steps in which the genetic material doubles and the cell divides into two identical cells.

Central nervous system (CNS) The brain and spinal cord.

Centre of mass The point through which the weight of an object can be taken to act.

Chemical bond The attraction between atoms that holds them together to form molecules.

Cholesterol An important biological molecule for cell membranes but leads to atherosclerosis if found in high levels in the blood.

Chromosome Structure containing DNA, found in the nucleus of eukaryotic cells.

Cilia Tiny hair-like projections from ciliated cells that waft mucus out of the gas exchange system.

Clone An organism produced asexually that has identical genes to its parent.

Cloning The asexual reproduction of an organism to produce genetically identical offspring.

Closed system (in Chemistry) A system where no substances can get in or out.

Closed system (in Physics) A system with no external forces acting on it.

Combustion Burning.

Common ancestor An organism from which others have evolved.

Communicable A disease that can be transmitted from one organism to another.

Community A group of two or more populations of different species that live at the same time in the same geographical area.

Competition The contest between organisms for resources such as food and shelter.

Complete combustion When a substance burns with a good supply of oxygen.

Compound Substance made from different elements chemically bonded together.

Computer modelling Using computer software to theoretically examine or test.

Concentrated A solution in which there is a lot of solute dissolved.

Concentration gradient A measurement of how a concentration of a substance changes from one place to another.

Condenser The part of the apparatus that causes the solvent to condense to a liquid.

Conservation Protecting an ecosystem or species of organism from reduced numbers and often extinction.

Conservation of mass In a reaction, the total mass of the reactants must equal the total mass of the products.

Consumer Any organism in a feeding relationship that eats other organisms for food.

Contact force A force that can be exerted between two objects when they touch.

Continuous (data) Data that come in a range and not in groups.

Continuous variable A variable with numerical values obtained by either measuring or counting.

Control variable This is the variable that can affect the outcome of an investigation and therefore must be kept constant or monitored. If all control variables are kept constant then the experiment is a fair test.

Coronary arteries Arteries that supply the heart muscle with oxygenated blood.

Corpus luteum After ovulation the empty follicle turns into the corpus luteum and releases progesterone.

Correlation When an action and outcome are linked but the action does not necessarily cause the outcome.

Covalent bond Two shared electrons joining atoms together.

Cracking The thermal decomposition of long alkanes into shorter alkanes and alkenes.

Cross-pollination When pollen from one plant fertilises ova from a different plant.

Cystic fibrosis (CF) A genetic disorder in which those affected inherit a recessive allele from both parents.

Daughter cells The cells produced during mitosis.

Defecation Removing solid waste from the body.

Deficiency A lack or shortage.

Deforestation Clearing trees from an area that will then be used for other purposes.

Delocalised Free to move around.

Denatured A permanent change to an enzyme as a result of extremes of pH and temperature, which stop it working.

Dendrites The branched beginnings of neurones, which can detect neurotransmitters and start another electrical impulse.

Deoxygenated Without oxygen.

Dependent variable The variable that is measured or recorded for each change of the independent variable.

Desalination Process to remove dissolved substances from sea water.

Diabetes A non-communicable disease that reduces control of blood glucose concentrations.

Diatomic molecule A molecule containing two atoms.

Differentiate To specialise, or adapt for a particular function.

Diffusion The net movement of particles from an area of high concentration to an area of lower concentration.

Dilute A solution in which there is a small amount of solute dissolved.

Diploid Describes a cell or nucleus of a cell that has a paired set of chromosomes, e.g. 46 in humans.

Directly proportional When two quantities are directly proportional, doubling one quantity will cause the other quantity to double; when a graph is plotted, the graph line will be straight and pass through the origin (0, 0).

Discharge Gain or lose electrons to become electrically neutral.

Discontinuous (data) Data that come in groups and not a range.

Displacement A distance travelled in a defined direction.

Displacement reaction Reaction where a more reactive element takes the place of a less reactive element in a compound.

Displayed formula Drawing of a molecule showing all atoms and bonds.

Dissipate To scatter in all directions or to use wastefully; the energy has spread out and heats up the surroundings.

DNA (deoxyribonucleic acid) The genetic information found in all living organisms.

Dominant An allele that will show a characteristic if inherited from one or both parents.

Double blind trial A medical experiment in which the patients and doctors do not know who has been given the drug and who has been given the placebo.

Double helix The characteristic spiral structure of DNA.

(Dynamic equilibrium) System where both the forward and reverse reactions are taking place simultaneously and at the same rate.

Ecosystem The interaction of a community of living (biotic) organisms with the non-living (abiotic) parts of their environment.

Effector A muscle or a gland.

Efficacy How effective a drug is.

Elastic deformation When an object returns to its original length after it has been stretched.

Electric current A flow of electrical charge; the size of the electric current is the rate at which electrical charge flows round the circuit.

Electrolysis Decomposition of ionic compounds using electricity.

Electrolyte A liquid that conducts electricity.

Electron A negatively charged particle that orbits the nucleus of an atom.

Electronic structure The arrangement of electrons in the shells (energy levels) of an atom.

Electron microscope A microscope that uses electron beams in place of light to give higher magnification.

Element A substance containing only one type of atom; a substance that cannot be broken down into simpler substances by chemical methods.

Endothermic reaction Reaction where thermal energy is transferred from the surroundings to the chemicals and so the temperature decreases.

Energy level (shell) The region an electron occupies surrounding the nucleus inside an atom.

Environmental variation Differences in organisms as a result of the environment in which they live.

Enzymes Molecules that act as catalysts in biological systems.

Epidermis The outermost layer of cells of a plant.

Ethene A plant hormone that ripens fruit.

Ethical issue An idea some people disagree with for religious or moral reasons.

Eukaryote An organism that is made of eukaryotic cells (those that contain a nucleus).

Eukaryotic cells Cells that contain a nucleus.

Evaporation When water changes state to form water vapour (a gas). (This is the opposite of condensation.)

Evolution The change in a species over a long time.

Excess When the amount of a reactant is greater than the amount that can react.

Excretion The removal of substances produced by chemical reactions inside cells from cells or organisms.

Exothermic reaction Reaction where thermal energy is transferred from the chemicals to the surroundings and so the temperature increases.

Extension The difference between the stretched and unstretched lengths of a spring.

Extinct When no members of a species remain alive.

Extrapolation (or prediction) Assuming that an existing trend or pattern continues to apply in an unknown situation.

Extreme environment A location in which it is challenging for most organisms to live.

Extremophile An organism that lives in an extreme environment.

Fair test Only the independent variable affects the dependent variable.

Fermentation The chemical breakdown of glucose into ethanol and carbon dioxide by respiring microorganisms such as yeast.

Filtrate Liquid that comes through the filter paper during filtration

Finite resource A resource that cannot be replaced once it has been used.

Flammability How easily a substance catches fire; the more flammable, the more easily it catches fire.

Fluid A liquid or a gas; a fluid flows and can change shape to fill any container.

Flux density The number of lines of magnetic flux in a given area.

Follicle A structure in an ovary in which an ovum matures.

Follicle-stimulating hormone (FSH) A hormone produced by the pituitary gland that causes an ovum to mature in an ovary and the production of oestrogen.

Force A push or a pull.

Formulation A mixture that has been designed as a useful product.

Fossil record All of the fossils that have been discovered so far.

Fraction A mixture of molecules with similar boiling points.

Fractional distillation A method used to separate miscible liquids with different boiling points.

Frequency (Hz) The number of waves produced each second or the number of waves passing a point each second. The unit of frequency is hertz; 1 hertz means there is 1 cycle per second.

Fullerenes Family of carbon molecules each with carbon atoms linked in rings to form a hollow sphere or tube.

Functional adaptation An advantage to an organism as a result of a process, such as the production of poisonous venom.

Gametes Sex cells, e.g. sperm, ova and pollen.

Gamma ray An electromagnetic wave.

Geiger-Müller (GM) tube A device which detects ionising radiation; an electronic counter can record the number of particles entering the tube.

Gene A section of a chromosome made from DNA that carries the code to make a protein.

Genetically modified (GM) crops Crops that contain genes from other species.

Genetic engineering A scientific technique in which a gene is moved from one species to another.

Genetic modification As genetic engineering.

Genetic variation Inherited differences in organisms.

Genome One copy of all the DNA found in an organism's diploid body cells.

Genotype The genetic make-up of an organism represented by letters.

Genus The second smallest group of classifying organisms.

Giant lattice A regular structure containing a massive number of particles that continues in all directions throughout the structure.

Gland A structure in the body that produces hormones.

Global warming An increase in the temperature at the Earth's surface.

Glucagon A hormone produced in the pancreas that raises blood glucose by breaking down glycogen stored in the liver.

Glycogen An insoluble store of glucose in the liver.

Greenhouse gas A gas that absorbs long wavelength infrared radiation given off by the Earth but does not absorb the Sun's radiation.

Group The name given to each column in the periodic table.

Haemoglobin The molecule in red blood cells that can temporarily bind with oxygen to carry it around your body.

Half-life The average time taken for the number of nuclei in a radioactive isotope to halve; in one half-life the activity or count rate of a radioactive sample also halves.

Halides Compounds made from Group 7 elements.

Halogens The elements in Group 7 of the periodic table (including fluorine, chlorine, bromine and iodine).

Haploid Cells that have half the 'normal' number of chromosomes; for example, sperm and egg cells in humans have 23 chromosomes. Gametes are haploid.

Haploid Describes a cell or nucleus of a gamete that has an unpaired set of chromosomes (i.e. only half the normal number).

Hazard Something that could cause harm.

Heart bypass A medical procedure in which a section of less important artery is moved to allow blood to flow around a blockage in a more important artery.

Heterozygous A genotype with one dominant and one recessive allele.

Homeostasis The maintenance of a constant internal environment.

Homologous series A family of compounds with the same general formula and similar chemical properties.

Homozygous dominant A genotype with two dominant alleles.

Homozygous recessive A genotype with two recessive alleles.

Hormone A chemical (produced in a gland in mammals) that moves around an organism to change the function of target cells, tissues or organs.

Humid Describes an atmosphere with high levels of water vapour.

Hydrocarbon A compound containing hydrogen and carbon only.

Hypothesis An idea based on scientific theory that states what is expected to happen or explains how or why something happens.

Immiscible Liquids that do not mix together and separate into layers.

Inbreeding Artificial selection using parents from a small, closely related group, which reduces variation.

Incomplete combustion The burning of fuel without sufficient oxygen, which produces poisonous carbon monoxide.

Independent variable The variable that you change. An investigation should only have one independent variable.

Induced magnet A magnet which becomes magnetic when it is placed in a magnetic field.

Inelastic deformation When an object does not return to its original length after it has been stretched.

Inert electrodes Electrodes that allow electrolysis to take place but do not react themselves.

Inertia Inactivity. Objects remain in their existing state of motion – at rest or moving with a constant speed in a straight line – unless acted on by an unbalanced force.

Infectious Describes a pathogen that can easily be transmitted, or an infected person who can pass on the disease.

Inorganic Non-living.

Insecticide A chemical that kills insects.

Insoluble Cannot dissolve.

Insulin A hormone produced in your pancreas that lowers blood glucose by converting it to glycogen and storing it in the liver.

Interdependence All the organisms in a community depend upon each other and because of this changes to them or their environment can cause unforeseen damage.

Intermolecular forces Weak forces between molecules.

Interval The difference between one value in a set of data and the next.

Intrauterine In the uterus. An intrauterine contraceptive device is implanted in the uterus.

Invasive species An organism that is not native and causes negative effects.

Inversely proportional When two quantities are inversely proportional, doubling one quantity will cause the other quantity to halve.

In vitro fertilisation (IVF) A medical procedure in which ova are fertilised outside of a woman, then placed into her uterus to develop into a baby.

Ion An electrically charged particle containing different numbers of protons and electrons.

Ionic bonding The electrostatic attraction between positive and negative ions.

Ionic equation Balanced equation for reaction that omits any spectator ions.

Ionising radiation UV rays, X-rays and gamma rays that can cause mutations to DNA.

Isotopes Different forms of a particular element; isotopes have the same number of protons but different numbers of neutrons.

Joule The unit of work.

Kingdom The largest group of classifying organisms, e.g. the animal kingdom.

Leachate A solution produced by leaching or bioleaching.

Leaching The use of dilute acid to produce soluble metal compounds from insoluble metal compounds.

Life cycle assessment An examination of the impact of a product on the environment throughout its life.

Limiting factor Anything that reduces or stops the rate of a reaction.

Limiting reactant The reactant in a reaction that determines the amount of products formed. Any other reagents are in excess and will not all react.

Limit of proportionality The point beyond which a spring will be permanently deformed. Elastic deformation stops and inelastic deformation starts.

Lipids Fats or oils, which are insoluble in water.

Lock and key hypothesis A model that explains the action of enzymes.

Longitudinal wave A wave in which the vibration causing the wave is parallel to the direction of energy transfer.

Luteinising hormone (LH) A hormone produced by the pituitary gland that stimulates ovulation.

Lymphocyte A type of white blood cell that produces antibodies to help clump pathogens together to make them easier to destroy.

Magnetic Materials that are attracted by a magnet.

Malaria A communicable disease, caused by a protist transmitted in mosquitos, which attacks red blood cells.

Malignant A cancerous tumour that can spread to other parts of the body.

Malleable Can be hammered into shape.

Mass extinction A large number of extinctions occurring at the same time (humans are the latest cause of a mass extinction).

Mass number Number of protons plus the number of neutrons in an atom.

Meiosis Cell replication that produces four non-identical haploid cells from one diploid cell.

Meniscus The curve at the surface of a liquid in a container.

Menopause The point in a woman's life, usually between 45 and 55, when she stops menstruating and therefore cannot become pregnant.

Meristem An area of a plant in which rapid cell division occurs, normally in the tip of a root or shoot.

Metabolism The sum of all the chemical reactions that happen in an organism.

Metallic bonding The attraction between the nucleus of metal atoms and delocalised electrons.

Methicillin-resistant *Staphylococcus aureus* (MRSA) A bacterium that has evolved resistance to antibiotics.

Mineral ions Substances that are essential for healthy plant growth, e.g. nitrates and magnesium.

Miscible Liquids that mix together.

Mitochondrion A small cell organelle, in which respiration occurs, found in the cytoplasm of eukaryotic cells.

Mitosis Cell replication that produces two identical copies of a diploid cell.

Mixture More than one substance that are not chemically joined together.

Mole Measurement of the amount of a substance.

Molecule Particle made from atoms joined together by covalent bonds.

Monomer The building block (molecule) of a polymer.

Motor effect The force produced between a conductor carrying a current within a magnetic field and the magnet producing the field.

Motor neurone A neurone that carries an electrical impulse away from the central nervous system to an effector (muscle or gland).

Mucus A sticky substance that traps pathogens.

Mutation A permanent change to DNA, which may be advantageous, disadvantageous or have no effect. Any genetic differences between individuals of one species, or even genetic differences between species, were originally caused by mutations.

Myelin sheath The insulating cover along an axon, which speeds up the electrical impulse.

Nanoscience The study of nanoparticles.

Negative feedback control A homeostatic mechanism by which the body detects a change and makes an adjustment to return itself to normal.

Net Overall.

Neutralisation A reaction that uses up some or all of the H^+ ions from an acid.

Neutron A neutral particle found in the nucleus of an atom.

Noble gases The elements in Group 0 of the periodic table (including helium, neon and argon).

Non-communicable disease A disease that is not passed from person to person.

Non-contact force A force that can sometimes be exerted between two objects that are physically separated.

Non-ohmic The current flowing through a non-ohmic resistor is not proportional to the potential difference across it; the resistance changes as the current flowing through it changes.

Non-renewable energy resources Energy resources which will run out and cannot be replenished.

Normal distribution Data that are more common around a mean and form a bell-shaped graph.

North-seeking pole The end of the magnet that points north.

Nucleus Central part of an atom containing protons and neutrons.

Oestrogen A female sex hormone produced in the ovaries that controls puberty and prepares the uterus for pregnancy.

Ohmic The current flowing through an ohmic conductor is proportional to the potential difference across it; if the p.d. doubles, the current doubles but he resistance stays the same.

Ore A rock from which a metal can be extracted for profit.

Organelle A small structure with a specific function in a cell.

Osmosis The net diffusion of water from an area of high concentration of water to an area of lower concentration of water across a partially permeable membrane.

Ova (singular ovum) Eggs.

Ovulation The release of an egg from an ovary.

Oxidation A reaction that uses oxygen.

Oxygenated Rich in oxygen.

Oxygen debt The amount of extra oxygen the body needs after exercise to break down the lactic acid.

Oxyhaemoglobin The name given to the substance formed when haemoglobin in your red blood cells temporarily binds with oxygen.

Painkiller A drug that treats the symptoms of disease, such as a headache, but does not kill any pathogens that may be causing the disease.

Palisade mesophyll Tissue found towards the upper surface of leaves with lots of chloroplasts for photosynthesis.

Partially permeable Allowing only substances of a certain size through.

Pathogen A disease-causing microorganism (e.g. a bacterium or fungus).

Peer review A process by which scientists check each other's work.

Peer reviewed When scientific research is studied and commented on by experts in the same area of science to check that it is scientifically valid.

Period (in Chemistry) The name given to a row in the periodic table.

Period (in Physics) The time taken to produce one wave.

Permanent magnet A magnet which produces its own magnetic field. It always has a north pole and a south pole.

Phagocyte A type of white blood cell that engulfs pathogens.

Phenotype The physical characteristics of an organism as described by words.

Phloem Living cells that carry sugars made in photosynthesis to all cells of a plant.

Photosynthesis A chemical reaction that occurs in the chloroplasts of plants and algae and stores energy in glucose.

Phytomining The use of plants to absorb metal compounds from soil as part of metal extraction.

Pipette A glass tube used to measure volumes of liquids with a very small margin of error.

Pituitary gland The master gland in your brain that produces a number of hormones, including ADH, TSH, FSH (in women) and LH (again in women.)

Placebo A medicine that has only psychological effects.

Planet A large body which orbits the Sun.

Platelets Small structures (not cells) in your blood that fuse together to form a scab.

Polymer A long chain molecule in which lots of small molecules (monomers) are joined together.

Population The total number of all the organism of the same species or the same group of species that live in a particular geographical area.

Potable water Water that is safe to drink.

Potential difference (p.d.) A measure of the electrical work done by a cell (or other power supply) as charge flows round the circuit; potential difference is measured in volts (V).

Power The rate at which energy is transferred.

$$\text{Power} = \frac{energy\ transferred}{time}$$

Precipitate A solid formed when solutions are mixed together.

Precipitation Rain, snow, hail, sleet.

Precise A set of measurements of the same quantity that closely agree with each other.

Prediction An extrapolation arising from a hypothesis that can be tested. Experimental results that agree with the prediction provide evidence to support the hypothesis.

Primary consumer An animal that feeds on producers.

Producer Any organism that photosynthesises (a plant or alga.)

Product The substance or substances produced by an enzyme reaction.

Progesterone A female sex hormone produced in the ovaries that prepares the uterus for pregnancy.

Prokaryotes Prokaryotic organisms (bacteria).

Prokaryotic cells Describes single-celled organisms that do not contain a nucleus.

Proteins Polymer molecules made from lots of different amino acids joined together.

Proton A positively charged particle found in the nucleus of an atom.

Punnett square A grid that makes determining the chance of inheriting a characteristic easier to understand.

Pure substance A single element or compound that is not mixed with any other substance.

Quadrat A square frame used in biological sampling.

Random errors Errors that vary unpredictably. You can reduce the effect of random errors by taking more measurements and calculating a new mean.

Range The maximum and minimum values used or recorded.

Reactivity series An arrangement of metals in order of their reactivity.

Receptor A cell or group of cells at the beginning of a pathway of neurones that detects a change and generates an electrical impulse.

Recessive An allele that will show a characteristic only if inherited from both parents.

Recycle Changing a waste product into new raw materials to make another.

Redox reaction A reaction where both reduction and oxidation take place.

Reduction A reaction where a substance loses oxygen and/or gains electrons.

Reflex arc The pathway of neurones in a reflex action.

Reflex response An automatic response that you do not think about.

Relative atomic mass The average mass of atoms of an element taking into account the mass and amount of each isotope it contains on a scale where the mass of ^{12}C (on a scale where the mass of a ^{12}C atom is 12).

Relative formula mass The sum of the relative atomic masses of all the atoms shown in the formula (often referred to as *formula mass*).

Relay neurone A neurone that carries an electrical impulse around the central nervous system (brain and spinal cord.)

Renewable energy resources Energy resources which will never run out and are (or can be) replenished as they are used.

Renewable resource A resource that we can replace once we have used it.

Repeatable A measurement is repeatable if the same person uses the same method and equipment and gets the same results.

Reproducible Measurements are reproducible when similar results are achieved when an experiment is repeated by another person or a different method is used.

Residue Solid left on the filter paper during filtration.

Resistor A component that acts to limit the current in a circuit; when a resistor has a high resistance, the current is low.

Resolution (in Biology) When using a microscope, resolution is the shortest distance between two points that can be seen as two distinct separate points.

Resolution (in Chemistry) The smallest change a piece of apparatus can measure.

Respiration The release of energy from glucose.

Resultant force A number of forces acting on an object may be replaced by a single force that has the same effect as all the forces acting together. This single force is called the resultant force.

Ribosome A small cell organelle in the cytoplasm in which proteins are made.

Right-hand grip rule A way to work out the direction of the magnetic field in a current-carrying wire if you know the direction of the current.

Risk The probability that something unpleasant will happen as the result of doing something.

Risk assessment An analysis of an investigation to identify any hazards, possible risks associated with each hazard, and any control measures you can take to reduce the risks.

Risk factor Any aspect of your lifestyle or substance in your body that increases the risk of a disease developing.

Sampling The process of recording a smaller amount of information to make wider conclusions.

Saturated (in the context of organic chemistry) A molecule that only contains single covalent bonds. It contains no double covalent bonds.

Saturated (in the context of solutions) A solution in which no more solute can dissolve at that temperature.

Scalar quantity A quantity with only magnitude (size) and no direction.

Secondary consumer An animal that feeds on primary consumers.

Secondary tumour A tumour that develops when malignant cancer spreads.

Selective breeding When breeders choose two parents with particular characteristics that they want in the offspring.

Self-pollination When pollen from one plant fertilises ova from the same plant.

Sensory neurone A neurone that carries an electrical impulse from a receptor towards the central nervous system.

Separating funnel Glass container with a tap used to separate immiscible liquids.

Sink A long-term store of a substance, often carbon.

Solenoid A long coil of wire.

Soluble Can dissolve.

South-seeking pole The end of the magnet that points south.

Species Organisms that can breed with each other to create fertile young.

Species The smallest group of classifying organisms, all of which are able to interbreed to produce fertile offspring.

Specific heat capacity The energy needed to raise the temperature of 1 kg of substance by 1 °C.

Spectator ions Ions that do not take part in a reaction and do not appear in the ionic equation for the reaction.

Spongy mesophyll Tissue found towards the bottom surface of leaves with spaces between the cells to allow gases to diffuse.

States of matter These are solid, liquid and gas.

Statin Drug that reduces blood cholesterol.

Stem cell An undifferentiated cell that can develop into one or more types of specialised cell.

Stent A small medical device made from mesh that keeps arteries open.

Step-down transformer A transformer that decreases potential difference and increases current.

Step-up transformer A transformer that increases potential difference and decreases current.

Stopping distance The sum of the thinking distance and braking distance.

Strong acid Acid in which all the molecules break into ions in water.

Structural adaptation An advantage to an organism as a result of the way it is formed, like the streamlining seen in fish.

Substrate The molecule or molecules on which an enzyme acts.

Sustainable Describes an activity that can continue without damaging the environment.

Sustainable development Using resources to meet the needs of people today without preventing people in the future from meeting theirs.

Synapse A gap between the axon of one nerve and the dendrites of another where neurotransmitters transmit the impulse.

Systematic error A consistent error, usually caused by the measuring instruments, when all of the data is higher or lower than the true value; data with a systematic error will give a graph line that is higher or lower than it should be.

Terminal velocity When the weight of a falling object is balanced by resistive forces.

Tertiary consumer An animal that feeds on secondary consumers.

Testosterone A male sex hormone produced in the testes that controls puberty.

Thermal decomposition Reaction where high temperature causes a substance to break down into simpler substances.

Thinking distance The distance a car travels while the driver reacts.

Thyroid gland A gland in your neck that produces thyroxine to regulate how quickly your body uses energy and controls your metabolic rate.

Thyroid-stimulating hormone (TSH) A hormone produced by your pituitary gland that stimulates the production of thyroxine by the thyroid gland.

Tissue fluid The liquid that surrounds ('bathes') cells in the body tissues. It is formed from plasma that diffuses through the capillary walls.

Toxin A poison that damages tissues and makes us feel ill.

Transect A line along which systematic sampling occurs.

Transgenic Describes a genetically engineered organism.

Translocation The movements of sugars made in photosynthesis from the leaves of plants.

Transparent An object which allows us to see clearly through it; glass is transparent.

Transpiration The gradual release of water vapour from leaves to continue the 'pull' of water up to them from the soil.

Transverse wave A wave in which the vibration causing the wave is at right angles to the direct of energy transfer.

True value The value that would be obtained in an ideal measurement.

Turbidity The cloudiness of a solution.

Turgid Describes cells that have a lot of water in their vacuole. The pressure created on the cell wall keeps the cell rigid.

Type 1 diabetes A medical condition that usually develops in younger people, preventing the production of insulin.

Type 2 diabetes A medical condition that usually develops in later life, preventing the person producing enough insulin or preventing cells from responding to insulin.

Uncertainty The range of measurements within which the true value can be expected to lie.

Unsaturated (in the context of organic chemistry) A molecule that contains one or more double covalent bonds.

Vaccine A medicine containing an antigen from a pathogen that triggers a low level immune response so that subsequent infection is dealt with more effectively by the body's own immune system.

Valid A method of investigation which produces data that answers the question being asked.

Valid results Results from a fair test in which only one variable is changed.

Value judgement An assessment of a situation that may be subjective, based on a person's opinions and/or values.

Variation The differences that exist within a species or between different species.

Vasectomy A contraceptive medical procedure during which a man's sperm ducts are blocked or cut.

Vector An organism that spreads a communicable disease.

Vector quantity A quantity with both magnitude (size) and direction.

Vein A large blood vessel that returns blood to the heart.

Velocity A speed in a defined direction.

Ventilation Breathing in (inhaling) and out (exhaling).

Ventricle A lower chamber of the heart surrounded by a thicker wall of muscle.

Villi (singular villus) Tiny finger-like projections that increase the surface area of the small intestine.

Virtual image An image formed by light rays which appear to diverge from a point.

Viscosity How easily a liquid flows; the higher the viscosity, the less easily it flows.

Water stress A shortage of fresh water.

Wavelength The distance from a point on one wave to the equivalent point on the next wave

Weak acid Acid in which only a small fraction of the molecules break into ions in water.

Work When a force causes an object to move. Work = force × distance

Xylem Dead plant cells joined together into long tubes through which water flows during transpiration.

Yield The amount of an agricultural product.

Zero error When a measuring instrument gives a reading when the true value is zero.

Index

Acknowledgements

The Publisher would like to thank the following for permission to reproduce copyright material:

AQA material is reproduced by permission of AQA.

Photo credits:

p.1 © Fuse via Thinkstock/Getty Images; **p.2** *tl* © Ablestock.com via Thinkstock/Getty Images, *tm* © Fuse via Thinkstock/Getty Images, *tr* © winbio – iStock via Thinkstock/Getty Images, *bl* © Jimmyhuynh – iStock via Thinkstock/Getty Images, *bm* © Tanchic – iStock via Thinkstock/Getty Images, *br* © magicflute002 – iStock via Thinkstock/Getty Images; **p.3** © BSIP SA/Alamy; **p.4** © Dr Tony Brain & David Parker/Science Photo Library; **p.5** © Dr Gopal Murti/Science Photo Library; **p.6** © Biophoto Associates/Science Photo Library; **p.9** © Nick Dixon; **p.10** ©Mike Watson – Moodboard via Thinkstock/Getty Images; **p.12** *l* © imageBROKER / Alamy, *m* © Deco Images II/Alamy, *r* © BSIP SA/Alamy; **p.13** *tl* © Don Fawcett/Science Photo Library, *bl* © Dr Keith Wheeler/Science Photo Library; **p.16** © Chrispo – HRF Fotolia.com; **p.17** *t* © PHOTOTAKE Inc./Alamy, *b* © Cre8tive Studios/Alamy, **p.18** © Francis Leroy, Biocosmos/Science Photo Library; **p.19** © CNRI/Science Photo Library; **p.21** all © Pr. G Gimenez-Martin/Science Photo Library; **p.22** © Anatomical Travelogue/Science Photo Library; **p.23** *tr* © Alexis Rosenfeld/Science Photo Library, *ml* © Nick Dixon; **p.24** *tl* © Sebastian Kaulitzki/Getty Images, *mr* © David McNew/Getty Images; **p.26** © Science Vu, Visuals Unlimited/Science Photo Library; **p.28** © buyit – iStock –Thinkstock/Getty Images; **p.29** © aimy27feb – iStock – Thinkstock/Getty Images; **p.34** © Power and Syred/Science Photo Library; **p.41** all © Nick Dixon; **p.42** © Eduard Lysenko – iStock – Thinkstock/Getty Images; **p.44** © David M. Martin, MD/Science Photo Library; **p.45** © Gastrolab/Science Photo Library; **p.46** © David M. Martin, MD/Science Photo Library; **p.52** © Ralph Hutchings, Visuals Unlimited; **p.55** Susumu Nishinaga/Science Photo Library; **p.56** *tl* © Power and Syred/Science Photo Library, *ml* © Paul Rapson/Alamy Stock Photo; **p.58** *tl* © Antonia Reeve/Science Photo Library, *tr* © Dieter Meyrl – iStock – Thinkstock/Getty Images, *bl* © egal – iStock –Thinkstock/Getty Images; **p.59** © Andres Rodriguez – Fotolia; **p.60** © stockdevil – iStock – Thinkstock/Getty Images; **p.62** © CNRI/Science Photo Library; **p.67** © Stéphane Bidouze – Fotolia; **p.70** © Power and Syred/Science Photo Library; **p.78** © saknakorn – iStock – Thinkstock/Getty Images; **p.79** © Peter D Noyce/Alamy Stock Photo; **p.80** *tl* © RioPatuca Images – Fotolia, *bl* © Lowell Georgia/Science Photo Library; **p.81** © NORM THOMAS/SCIENCE PHOTO LIBRARY; **p.82** © Piotr Marcinski – iStock –- Thinkstock/Getty Images; **p.84** © Henrik Larsson – Fotolia; **p.85** © sale123 – iStock – Thinkstock/Getty Images; **p.87** © Biology Media/Science Photo Library; **p.88** © Creative – Fotolia; **p.89** *tl* © National Library of Medicine/Science Photo Library, *ml* © St Mary's Hospital Medical School/Science Photo Library; **p.90** *ml* © sybanto – iStock – Thinkstock/Getty Images, *bl* © Colin Cuthbert/Science Photo Library; **p.93** © Nixxphotography – iStock – Thinkstock/Getty Images; **p.94** © dina2001 – iStock – Thinkstock/Getty Images; **p.95** © Nigel Cattlin/Science Photo Library; **p.96** © Biophoto Associates/Science Photo Library; **p.98** © Ruslan Kurbanov – Fotolia; **p.104** © PaulPaladin – Fotolia; **p.106** © Dinadesign – Fotolia; **p.107** © Vladimir Wrangel – Fotolia; **p.109** *tl* © John Fryer/Alamy Stock Photo, *bl* © Stéphane Bidouze – Fotolia; **p.116** © Richard Grime; **p.118** *ml* © johny007pan – Fotolia, *bl* © Dionisvera – Fotolia; **p.126** © SCIENCE PHOTO LIBRARY; **p.130** © sciencephotos / Alamy; **p.131** *l* © sciencephotos / Alamy, *m* © TREVOR CLIFFORD PHOTOGRAPHY/SCIENCE PHOTO LIBRARY, *r* © The Open University; **p.132** © MARTYN F. CHILLMAID/SCIENCE PHOTO LIBRARY; **p.133** © 2005 Richard Megna - Fundamental Photographs; **p.135** © SSPL/Science Museum/Getty Images; **p.137** *l* © SCIENCE PHOTO LIBRARY, *m* © ANDREW LAMBERT PHOTOGRAPHY/SCIENCE PHOTO LIBRARY, *r* © hriana - Fotolia; **p.138** © ANDREW LAMBERTPHOTOGRAPHY/SCIENCE PHOTO LIBRARY; **p.139** © ANDREW LAMBERT PHOTOGRAPHY/SCIENCE PHOTO LIBRARY; **p.140** all © Richard Grime; **p.145** © Richard Grime; **p.147** © Rich Legg - iStock via Thinkstock/Getty Images; **p.148** © Vera Kuttelvaserova - Fotolia; **p.152** © CHARLES D. WINTERS/SCIENCE PHOTO LIBRARY; **p.158** © Dzarek - iStock via Thinkstock/Getty Images; **p.159** © Martinan - Fotolia; **p.161** *tr* © Alexandru Dobrea - Hemara via Thinkstock/Getty Images, *bl* © ulkan - iStock via Thinkstock/Getty Images; **p.166** © Igor_M - iStock via Thinkstock/Getty Images; **p.167** © koosen - iStock via Thinkstock/Getty Images; **p.169** © karaboux - Fotolia; **p.172** *l* © koosen - iStock via Thinkstock/Getty Images, *r* © CHARLES D. WINTERS/SCIENCE PHOTO LIBRARY; **p.174** © Chad Baker - DigitalVision via Thinkstock/Getty Images; **p.175** © Richard Grime; **p.176** © psphotograph - iStock via Thinkstock/Getty Images; **p.177** © Blend Images / Alamy Stock Photo; **p.189** © studiomode / Alamy Stock Photo; **p.191** © W. Oelen via Wikipedia Commons (https://creativecommons.org/licenses/by-sa/3.0/deed.en); **p.193** © Richard Grime; **p.201** © ANDREW LAMBERT PHOTOGRAPHY/SCIENCE PHOTO LIBRARY; **p.202** *t* © TREVOR CLIFFORD PHOTOGRAPHY/SCIENCE PHOTO LIBRARY, *b* © MARTYN F. CHILLMAID/SCIENCE PHOTO LIBRARY; **p.203** *t* © Alvey & Towers Picture Library / Alamy, *b* © CHARLES D. WINTERS/SCIENCE PHOTO LIBRARY; **p.208** *l* © Isidre blanc via Wikipedia (http://creativecommons.org/licenses/by-sa/3.0/), *r* © Alexandru Dobrea - Hemara via Thinkstock/Getty Images; **p.210** © Martin Shields / Alamy Stock Photo; **p.211** © Getty Images/iStockphoto/Thinkstock; **p.216** © GEOFF KIDD/SCIENCE PHOTO LIBRARY; **p.218** © Hodder Education; **p.228** *tr* © Prill Mediendesign & Fotografie - iStock via Thinkstock/Getty Images, *bl* © PRHaney via Wikipedia (http://creativecommons.org/licenses/by-sa/3.0/), *br* © Richard Grime; **p.230** © David J. Green / Alamy Stock Photo; **p.231** *t* Courtesy of Nephron via Wikipedia Commons (https://creativecommons.org/licenses/by-sa/3.0/deed.en), *b* © studiomode / Alamy Stock Photo; **p.232** *t* © Solent News & Photo Agency/REX Shutterstock, *m* © ANDREW LAMBERT PHOTOGRAPHY/SCIENCE PHOTO LIBRARY, *b* © MBI / Alamy Stock Photo; **p.233** © Creatas Images - Creatas via Thinkstock/Getty Images ; **p.241** © ANDREW LAMBERT PHOTOGRAPHY/SCIENCE PHOTO LIBRARY; **p.243** © Zoonar RF – Thinkstock; **p.258** © zentilia - Fotolia; **p.259** © U. Gernhoefer – Fotolia; **p.267** © Mikael Damkier – Fotolia; **p.272** © Nick England; **p.273** *m* © PHB.cz – Fotolia, *b* © WILLIAM WEST/AFP/Getty Images; **p.277** © Alex White – Fotolia; **p.279** *t* Vladislav Gajic – Fotolia, *b* © Nigel Hicks / Alamy; **p.280** *tl* © ADRIANO PECCHIO – iStock – Thinkstock, *mr* © Laurence Gough – Fotolia, *ml* © Saskia Massink – Fotolia; **p.291** © SeanPavonePhoto – Fotolia; **p.305** © SeanPavonePhoto – Fotolia; **p.306** © Fatbob – Fotolia; **p.309** © TebNad – Fotolia; **p.318** © Andrew Dunn / Alamy Stock Photo; **p.335** © Ljupco Smokovski – 123RF; **p.337** © Photography by Courtesy Decoding Science, Bond Life Sciences Center ; **p.347** © Nailia Schwarz – 123RF; **p.356** *t* © Rick Hyman – iStock – Thinkstock, *m* © sumnersgraphicsinc – Fotolia, *b* © Monkey Business – Fotolia; **p.357** © Valua Vitaly – Fotolia; **p.367** © ron sumners – iStock –Thinkstock/Getty Images; **p.371** *tl* © Dmitry Lobanov – Fotolia, *bl* © Digital Vision – iStock – Thinkstock/Getty Images; **p.372** © Jeff Crow/REX Shutterstock; **p.375** © Fuse – Thinkstock/Getty Images; **p.376** © Tomasz Trojanowski – Hemara – Thinkstock/Getty Images; **p.377** © Startraks Photo/REX Shutterstock; **p.378** © Michael Blann – iStock – Thinkstock/Getty Images; **p.381** *tr* Keystone/Getty Images, *bl* © SEBASTIAN KAULITZKI/Getty Images; **p.382** © Cathy Yeulet; **p.383** © Alexis Rosenfeld/Science Photo Library; **p.384** © Suzz – Fotolia; **p.385** *tl* © Karsten Koehler – Fotolia, *tr* © kicsilcsi – iStock – Thinkstock/Getty Images, *ml* © Sergey Lavrentev – Fotolia; **p.391** © Luk Cox – Fotolia; **p.392** © caia image/Alamy Stock Photo; **p.393** *bl* © RusN – iStock – Thinkstock/Getty Images, *br* © Ozgur Coskun/Alamy Stock Photo; **p.395** *t* © CNRI/SCIENCE PHOTO LIBRARY; © Science Photo Library; **p.397** © Luk Cox – Fotolia; **p.400** © Robert Adrian Hillman/Alamy Stock Photo; **p.401** © MagicBones – iStock – Thinkstock/Getty Images; **p.402** © Christine Woodward; **p.403** *tl* © Purestock – Thinkstock/Getty Images, *bl* © Hogan Imaging – Fotolia; **p.405** *tl* © WilleeCole – iStock – Thinkstock/Getty Images, *bl* © wolfavni – Fotolia; **p.406** *ml* © Makoto Iwafuki/Eurelios/Science Photo Library, *bl* © Beboy – Fotolia; **p.413** © M.Rosenwirth – Fotolia; **p.414** *ml* © Ersier Dmitry – Hemara – Thinkstock/Getty Images, *bl* © Georgios Kollidas – Fotolia; **p.415** © Robin Weaver/Alamy Stock Photo; **p.416** *l* © Scott_Walton – iStockphoto – Thinkstock/Getty Images, *r* © Herve Conge, ISM/Science Photo Library;

Acknowledgements

p.417 *tr* © Tom McHugh/Science Photo Library, *ml* © Reuters/Corbis; **p.418** © CDC/Science Photo Library; **p.419** © Photos.com – Thinkstock/Getty Images; **p.422** © Bob Ainsworth – iStock – Thinkstock/Getty Images; **p.423** © ReformBoehm – iStock – Thinkstock/Getty Images; **p.424** *ml* © British Library/Science Photo Library, *bl* © EcoView – Fotolia, *blm* © lightpoet – Fotolia, *bm* © Matt_Gibson – iStock – Thinkstock/Getty Images, *br* © Fotolia, *br* © abzerit – iStock – Thinkstock/Getty Images; **p.426** *tl* © iosifbudau – iStock – Thinkstock/Getty Images, *br* © Robert Harding World Imagery/Alamy Stock Photo; **p.428** © Ch'ien Lee/Minden Pictures/Corbis; **p.429** *tr* © randimal – iStock – Thinkstock/Getty Images, *mr* © sirichai_raksue – iStock – Thinkstock/Getty Images, *br* © Fuse – Thinkstock/Getty Images; **p.430** © leonmaraisphoto – Fotolia; **p.431** © antpkr – iStock –Thinkstock/Getty images; **p.432** *tl* © National Geographic Image Collection/Alamy Stock Photo, *ml* © veneration – Fotolia, *mr* © o01shorty10o – iStock – Thinkstock/Getty Images, *bl* © jez_ bennett – iStock –Thinkstock/Getty images; **p.433** © chris jewiss – Fotolia; **p.434** *t* © MaxMichelMannt – iStock –Thinkstock/Getty images, *m* © Elen11 – iStock – Thinkstock/Getty Images, *b* © ian600f – iStock –Thinkstock/Getty images; **p.435** *t* © Alex Waters/Alamy Stock Photo, *b* © Juulijs – Fotolia; **p.436** *tl* © schaef – Fotolia, *ml* © MattiaPovolo – Fotolia; **p.437** *tr* © Gregory Dimijian/Science Photo Library, *ml* © errni – Fotolia, *bl* © manfredxy – Fotolia; **p.438** *t* © Norma Cornes – Hemara – Thinkstock/Getty Images, *b* © andreanita – Fotolia; **p.439** *t* © Frederick R. McConnaughey/Science Photo Library, *m* © Dr Ken MacDonald/Science Photo Library; **p.441** © Achim Prill – iStock – Thinkstock/Getty Images; **p.442** © Fuse via Thinkstock/Getty Images; **p.444** © Georgios Kollidas – iStock – Thinkstock/Getty Images; **p.445** © Science Photo Library; **p.447** © Martyn F. Chillmaid/Science Photo Library; **p.450** © micro_photo – iStock – Thinkstock/Getty Images; **p.458** © Fotolia; **p.460** *t* © Victorburnside – iStock – Thinkstock/Getty Images, *m* © paulprescott72 – iStock Editorial – Thinkstock/Getty Images, *b* © John Moore – iStock Editorial – Thinkstock/Getty Images; **p.461** *t* © steve young/Alamy Stock Photo, *b* © wusuowei – Fotolia; **p.462** *t* © monticello – iStock – Thinkstock/Getty Images, *m* © Global_Pics – iStock – Thinkstock/Getty Images; **p.463** *t* © Purestock – Thinkstock/Getty Images, *m* © BANNER – Foto lia; **p.464** *t* © Horizons WWP/Alamy Stock Photo, *m* © Nancy Sefton/Science Photo Library; **p.465** *t* © NASA/Science Photo Library, *m* © NASA/Science Photo Library; **p.466** *t* © Danicek – Fotolia, *b* © Yiming Chen – Getty Images; **p.467** *t* © Fotolia, *b* © sherez – Fotolia; **p.470** © fotoVoyager/Getty Images; **p.473** © Linda Macpherson - Hemara via Thinkstock/Getty Images; **p.478** © bulentozber - iStock via Thinkstock/Getty Images; **p.481** © Mediablitzimages /Alamy Stock Photo; **p.486** *t* © MARTYN F. CHILLMAID/SCIENCE PHOTO LIBRARY, *m* © GIPhotoStock/SCIENCE PHOTO LIBRARY, *b* © MARTYN F. CHILLMAID/SCIENCE PHOTO LIBRARY; **p.497** © TomasSereda - iStock via Thinkstock/Getty Images; **p.504** *l* © ANDREW LAMBERT PHOTOGRAPHY/SCIENCE PHOTO LIBRARY, *r* © ANDREW LAMBERT PHOTOGRAPHY/SCIENCE PHOTO LIBRARY; **p.509** © Keith Brofsky - Photodisc via Thinkstock/Getty Images; **p.511** *t* © Alistair Heap / Alamy Stock Photo, *b* © Steve Mann - Hemara via Thinkstock/Getty Images; **p.513** *tl* © Kevin Britland / Alamy Stock Photo, *tr* © Nigel Cattlin / Alamy Stock Photo, *ml* © FOOD DRINK AND DIET/MARK SYKES / Alamy Stock Photo, *mr* © René van den Berg / Alamy Stock Photo; **p.516** *from top to bottom* © MARTYN F. CHILLMAID/SCIENCE PHOTO LIBRARY, © MARTYN F. CHILLMAID/SCIENCE PHOTO LIBRARY, © SCIENCE PHOTO LIBRARY, © ANDREW LAMBERT PHOTOGRAPHY/SCIENCE PHOTO LIBRARY; **p.520** *from top to bottom* © ANDREW LAMBERT PHOTOGRAPHY/SCIENCE PHOTO LIBRARY, © MARTYN F. CHILLMAID/SCIENCE PHOTO LIBRARY, © TREVOR CLIFFORD PHOTOGRAPHY/SCIENCE PHOTO LIBRARY, © GIPhotoStock/SCIENCE PHOTO LIBRARY, © Richard Grime; **p.522** © dell640 - iStock via Thinkstock/Getty Images; **p.524** © Byelikova_Oksana - iStock via Thinkstock/Getty Images; **p.525** © Alexey Stiop – Fotolia; **p.528** *ml* © tfoxfoto - iStock via Thinkstock/Getty Images, *br* © Ingram Publishing via Thinkstock/Getty Images; **p.529** *t* © Ingram Publishing via Thinkstock/Getty Images, *b* The cover of volume 81 of Environment International was published in AQA GCSE Chemistry, Copyright Elsevier; **p.530** *l* © Science & Society Picture Library via Getty Images, *r* © Richard Grime; **p.531** *ml* © Louis Levy, 1906, *m* © Steve Speller / Alamy Stock Photo, *mr* © Joseph Nickischer - iStock via Thinkstock/Getty Images, *bl* © Mr.Lukchai Chaimongkon - iStock via Thinkstock/Getty Images; **p.532** © erectus - Fotolia; **p.533** © GIPhotoStock X / Alamy Stock Photo; **p.537** *l* © INTERFOTO / Alamy Stock Photo, *r* © Peter Barritt / Alamy Stock Photo; **p.543** © NASA; **p.545** © Harvepino - iStock via Thinkstock/Getty Images; **p.547** © jimiknightley –iStock via Thinkstock/Getty Images; **p.548** © Ros Drinkwater / Alamy Stock Photo; **p.549** © Chalabala - iStock via Thinkstock/Getty Images; **p.551** © Michael Blann - DigitalVision via Thinkstock/Getty Images; **p.556** *l* © Richard Grime, *r* © sciencephotos / Alamy Stock Photo; **p.560** © pilip76 - iStock via Thinkstock/Getty Images; **p.562** © mikhail mandrygin – 123RF; **p.565** *t* © Daniel Vorley/LatinContent/Getty Images, *m* © Toutenphoton – Fotolia; **p.570** © kris Mercer / Alamy Stock Photo; **p.575** © vladru – Thinkstock; **p.580** © Christopher Ison / Alamy Stock Photo; **p.582** © OLIVIER MORIN/AFP/Getty Images; **p.586** *t* © Copyright 2006 Carolina K. Smith - Fotolia, *m* © jpmatz – Fotolia; **p.588** © Gustoimages/Getty Images; **p.591** © Mariusz Blach – Fotolia; **p.592** © NASA; **p.600** © bytesurfer – Fotolia; **p.601** © Volvo Car UK Ltd; **p.610** © Martin Dohrn / Science Photo Library; **p.611** © Martin Dohrn / Science Photo Library; **p.612** © art_zzz – Fotolia; **p.619** *t* © alexsalcedo – Fotolia, *b* © Jose Manuel Gelpi – Fotolia; **p.622** © Cultura RM Exclusive/Joseph Giacomin – Getty Images; **p.623** © PAUL RAPSON/SCIENCE PHOTO LIBRARY; **p.624** *br* © FRANCK FIFE/AFP/Getty Images, *bl* © gloszilla - Fotolia, *t* © nevodka – Thinkstock; **p.627** © dule964 – Fotolia; **p.630** © CERN; **p.632** © AlexStar – iStock via Thinkstock – Getty Images; **p.634** © sciencephotos / Alamy; **p.635** © SSPL/Getty Images; **p.637** © Andrew Lambert Photography / Science Photo Library.

t = top, *b* = bottom, *l* = left, *r* = right, *m* = middle

Every effort has been made to trace all copyright holders, but if any have been inadvertently overlooked, the Publisher will be pleased to make the necessary arrangements at the first opportunity.